Y0-BTA-971

POLYMER STABILIZATION

POLYMER STABILIZATION

Edited by

W. LINCOLN HAWKINS

Bell Telephone Laboratories, Incorporated
Murray Hill, New Jersey

WILEY-INTERSCIENCE, a Division of John Wiley & Sons, Inc.
New York · London · Sydney · Toronto

Library of Congress Catalogue Card Number: 70-154324

ISBN 0-471-36300-6

Printed in the United States of America.

10 9 8 7 6 5 4 3 2 1

PREFACE

The primary purpose of this book is to present a fundamental approach to the stabilization of polymers. In the chapters that follow, stabilization is discussed as it relates to each of the principal factors responsible for polymer deterioration. The basic mechanisms by which stabilizers inhibit or retard deterioration are developed wherever these reactions have been established by reliable experimental data. In those instances where mechanisms have not been established, the discussion of stabilization is based on scientific investigation of the protection of polymers against the exposure environment. It is anticipated that established theories of polymer stabilization can be applied to the scientific choice of stabilizers for specific applications. Thus the ultimate purpose of this volume is to present to the student of polymer chemistry and to the practical polymer chemist the basic principles of stabilization that are essential in a scientific approach to the protection of polymers.

W. LINCOLN HAWKINS

November 1971
Murray Hill, New Jersey

CONTENTS

POLYMER STABILIZATION

1

ENVIRONMENTAL DETERIORATION OF POLYMERS

W. LINCOLN HAWKINS

Bell Telephone Laboratories, Incorporated, Murray Hill, New Jersey

A. INTRODUCTION

A fundamental consideration of polymer stabilization must be based on a sound understanding of the mechanisms by which polymers deteriorate. In the following discussion deterioration is interpreted as an irreversible process in which useful polymer properties degenerate when exposed to the environment. Although in most instances deterioration involves the rupture of primary chemical bonds, cleavage of secondary bonds can also be responsible for polymer deterioration. Since mechanisms through which polymers fail in service vary with exposure conditions as well as with polymer structure, detailed mechanisms of polymer deterioration are included throughout the book when these reactions are essential to the understanding of stabilization. Certain aspects of polymer deterioration are generally applicable, however, and these are discussed in the following sections of this chapter.

Both natural and synthetic polymers deteriorate as a result of conditions encountered during exposure. The changes in composition and structure lead to the degeneration of useful properties and, ultimately, to failure of the polymer. The cracking of rubber, the yellowing of cotton fabrics, and the embrittlement of plastics are typical examples of polymer deterioration. These and similar changes occur under a wide variety of conditions, generally described as the exposure environment.

B. GENERAL ASPECTS OF POLYMER DETERIORATION

Polymers vary widely in their vulnerability to deterioration when exposed to the same environment. These variations in stability are caused primarily by differences in chemical structure, but they also result from trace impurities in the polymer. Impurities that promote deterioration can be foreign materials, such as residues from polymerization catalysts, or they can be an integral part of polymer molecules. Partially degraded molecules are a common impurity in polymers and these modified molecules often initiate deterioration. In many instances, only traces of these impurities have a pronounced effect on polymer stability. Stability is also dependent on the physical structure or morphology of the polymer.

1. The Exposure Environment

Both chemical reactants and energy sources contribute to the deterioration of polymeric materials. Although a wide variety of chemical reactants in the environment is responsible for polymer deterioration, oxygen is by far the most important. All polymers react with oxygen under extreme conditions, that is, at combustion temperatures, and many oxidize extensively at much lower temperatures. Reaction with oxygen accounts for the vast majority of polymer failures occurring under service conditions. This generalization includes the attack of ozone, as a form of oxygen, on unsaturated polymers under stress.

Water ranks next to oxygen in importance as a chemical reactant in the exposure environment. Although its effect is most evident in the hydrolysis of condensation polymers, certain addition polymers are also vulnerable, and hydrolysis of side groups, for example, ester groups, can adversely affect important properties. Hydrolysis of polymers, as with their low-molecular-weight analogs, is often catalyzed by traces of acids or bases. Industrial smog contains a wide variety of chemical reactants, many of which contribute to polymer deterioration. In addition, many chemicals to which polymers may be exposed in unusual applications should be considered as components of the environment. These include dyes, adhesives, solvents, detergents, metals, and other foreign materials.

The absorption of several types of energy by polymers initiates or accelerates those chemical reactions responsible for deterioration. Various forms of radiation and mechanical stress contribute to polymer failure. Acceleration of deterioration by heat is a general phenomenon responsible for the degradation of polymers both in the presence and absence of chemical reactants. Ultraviolet radiation initiates the deterioration of many polymers, and mechanical stress also contributes to failure, particularly in combination with thermal energy during processing or fabrication.

Deterioration takes place during two general periods in the lifetime of a polymer. The first occurs as the polymer is fabricated into the form in which it is to be used. This period is characterized by exposure to relatively high temperatures over short intervals of time. Deterioration of some polymers can, and probably does, occur even during synthesis, but protection at this stage is generally impractical, since stabilizers that inhibit deterioration usually retard polymerization. Conditions encountered during fabrication can be extreme, both in temperature and in mechanical stress. Sensitizing groups are often introduced into a small fraction of the total number of molecules, and these trace impurities can accelerate deterioration of the polymer during its service life.

Long-term aging is the second important period of exposure. During this period the polymer is exposed to conditions of actual use, and deterioration is more gradual. Exposure conditions usually vary through cycles in contrast to the more constant conditions of fabrication. This phase also includes storage. In many applications polymers may be stored at temperatures above those encountered during the service life. Also, on a time scale, storage can be a significant portion of the overall life of a polymer. Although temperature and mechanical stress are usually lower during aging than during fabrication, the time of exposure is much longer; and in practical applications, failure usually occurs during this period. Deterioration is usually slow and can escape detection for long periods of time. When visible effects become apparent, many polymers may already have failed for the application intended. Polymer stabilization should take into account both types of exposure, that is, fabrication conditions and those of long-term aging.

2. Polymer Failure

The degree to which a stabilizer extends the useful life of a polymer is the ultimate measure of its effectiveness. It is therefore important that the concept of failure be clearly understood, particularly with reference to test procedures devised to evaluate the effectiveness of stabilizers (Chapter 10). Failure in service must be related to the end use for which the polymer is intended. It therefore follows that the time to failure can vary, dependent on the application. Polymers are used in a wide variety of applications in which the material is chosen for its mechanical, dielectric, or cohesive strength; its physical appearance, including transparency and color stability; its chemical purity; and combinations of these properties. Failure occurs when any important property is altered beyond the maximum limit specified in the design. Mechanical strength normally is associated with bulk properties of the polymer. Appearance of the polymer product, except for transparent materials, is generally a surface property, and electrical failure may be determined by either a bulk or a surface dielectric property.

Since polymer deterioration resulting from reaction with chemical reactants usually starts first at the polymer surface and then penetrates into the bulk, extensive degradation, as evidenced by cracking, crazing, discoloration, or change in texture, can occur before bulk properties are altered significantly. Hence failure can occur in applications based on appearance of the product even though bulk properties such as elongation, tensile strength, modulus, brittleness, and so on, are essentially unchanged. Conversely, where mechanical strength is the critical property, surface appearance could be grossly altered before the material fails in service. Certain bulk properties, however, are surface sensitive; for example, impact strength is lowered by cracks at the surface. It is therefore most important that tests for polymer stability or stabilizer effectiveness take into account the critical property that must be preserved for successful application of the polymer. Various types of failure can be related in a general way to the several basic mechanisms of polymer deterioration.

C. GENERAL FACTORS IN POLYMER DETERIORATION

The chemical composition of polymers deteriorates through a complex sequence of reactions. The molecular weight is changed considerably in most of these reactions, but deterioration can occur with no significant change in the size of polymer molecules. Changes in molecular weight result from chain scission and crosslinking, both reactions occurring simultaneously in many polymers. The relative rates of these two overall reactions are dependent on polymer structure and on reaction conditions.

Chemical bonds in polymers are broken under a variety of conditions, for example, mechanical stress, heat, ionizing radiation, and chemical reaction, to form free radicals as the first (nonmolecular) products. These reactive fragments of polymer molecules are relatively short lived and react rapidly with other polymer molecules or with other available reactants. Chain scission occurs when bonds of the backbone chain are broken irreversibly. Because of the restricted motion in a polymer matrix, recombinations of radicals can occur to reverse the process of chain scission. However, when a

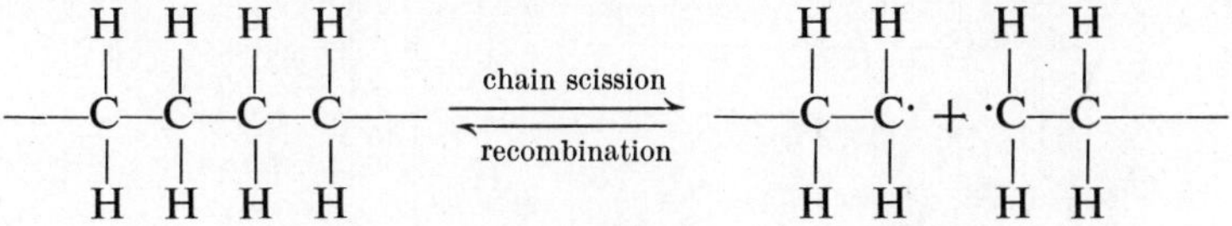

$$—\underset{\underset{H}{|}}{\overset{\overset{H}{|}}{C}}—\underset{\underset{H}{|}}{\overset{\overset{H}{|}}{C}}—\underset{\underset{H}{|}}{\overset{\overset{H}{|}}{C}}—\underset{\underset{H}{|}}{\overset{\overset{H}{|}}{C}}— \underset{\text{recombination}}{\overset{\text{chain scission}}{\rightleftharpoons}} \underbrace{—\underset{\underset{H}{|}}{\overset{\overset{H}{|}}{C}}—\underset{\underset{H}{|}}{\overset{\overset{H}{|}}{C}}\cdot + \cdot\underset{\underset{H}{|}}{\overset{\overset{H}{|}}{C}}—\underset{\underset{H}{|}}{\overset{\overset{H}{|}}{C}}—}_{\text{Alkyl radicals}}$$

chemical reactant such as oxygen is accessible, rapid addition to alkyl radicals occurs and the original molecules cannot then reform. Additions of

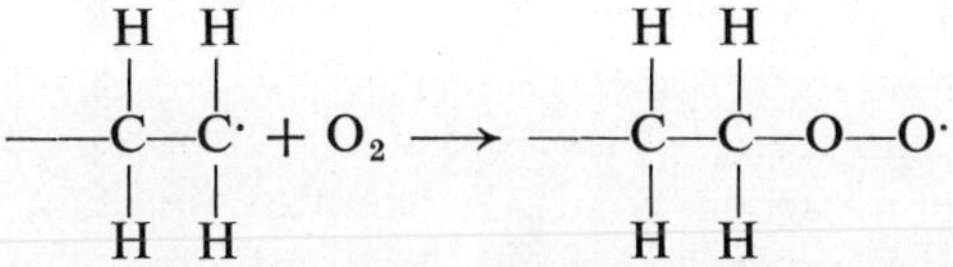

$$—\underset{\underset{H}{|}}{\overset{\overset{H}{|}}{C}}—\underset{\underset{H}{|}}{\overset{\overset{H}{|}}{C}}\cdot + O_2 \longrightarrow \underbrace{—\underset{\underset{H}{|}}{\overset{\overset{H}{|}}{C}}—\underset{\underset{H}{|}}{\overset{\overset{H}{|}}{C}}—O—O\cdot}_{\text{Peroxy radical}}$$

peroxy radicals with each other or with alkyl radicals can occur, however, by reactions such as,

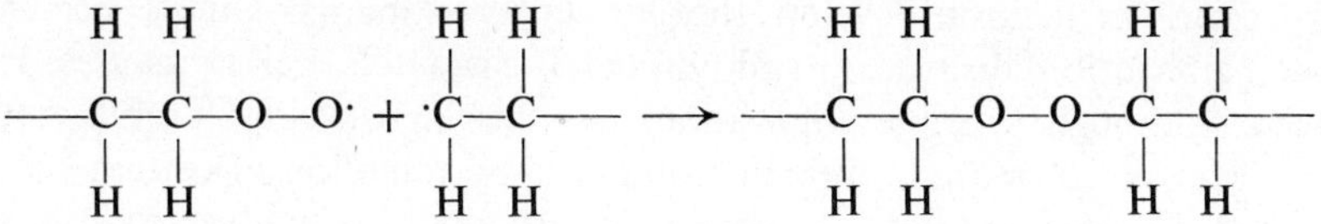

$$\underbrace{—\underset{\underset{H}{|}}{\overset{\overset{H}{|}}{C}}—\underset{\underset{H}{|}}{\overset{\overset{H}{|}}{C}}—O—O\cdot}_{\text{Peroxy radical}} + \underbrace{\cdot\underset{\underset{H}{|}}{\overset{\overset{H}{|}}{C}}—\underset{\underset{H}{|}}{\overset{\overset{H}{|}}{C}}—}_{\text{Alkyl radical}} \longrightarrow \underbrace{—\underset{\underset{H}{|}}{\overset{\overset{H}{|}}{C}}—\underset{\underset{H}{|}}{\overset{\overset{H}{|}}{C}}—O—O—\underset{\underset{H}{|}}{\overset{\overset{H}{|}}{C}}—\underset{\underset{H}{|}}{\overset{\overset{H}{|}}{C}}—}_{\text{Peroxide}}$$

Except where recombination of radicals takes place, chain scission has as its end effect reduction in molecular weight of the polymer.

Chain scission can be a random reaction, occurring at any position along the backbone chain and resulting in a broad spectrum of molecular weights, or it may be a sequential reaction of depolymerization, proceeding stepwise along the chain to give a high yield of monomer. Depolymerization is the predominant reaction in polytetrafluoroethylene, poly(α-methylstyrene), and the polyacetals. Random scission predominates in polyethylene, polypropylene, and many condensation polymers. Hydrolysis of condensation polymers is a typical chain scission reaction.

Radicals also form in polymers by cleavage of bonds which are not in the backbone chain. A common example is the cleavage of a carbon–hydrogen bond to form a polymer radical and a proton. The latter may combine with another proton from a neighboring molecule to form molecular hydrogen and a second radical. Combination of these radicals would then lead to a crosslinked structure. In the presence of oxygen, peroxy crosslinks can form

```
   H  H  H  H
   |  |  |  |                  H  H  H  H
 --C--C--C--C--                |  |  |  |
   |  .  |  |                --C--C--C--C--
   H     H  H                  |  |  |  |
                               H  |  H  H
        +              →          |
                               H  |  H  H
   H     H  H                  |  |  |  |
   |     |  |                --C--C--C--C--
 --C--Ċ--C--C--                |  |  |  |
   |  |  |  |                  H  H  H  H
   H  H  H  H
```

between polymer molecules. These crosslinks are less stable than those composed of carbon–carbon bonds. In contrast to chain scission, crosslinking increases the molecular weight. Vulcanization of rubber, although not a deterioration process, is a classic example of polymer crosslinking.

Both chain scission and crosslinking have a gross effect on the mechanical properties of polymers. Reduction in molecular weight through chain scission leads to loss in modulus, tensile strength, and many other related properties. As network structures develop through crosslinking, polymers become brittle, elongation decreases, and ultimately insoluble gel structures are formed. Although in some applications crosslinking develops useful properties (e.g., the increase in modulus obtained by radiation of certain polymers and the drying of paints), crosslinking usually has an adverse effect on polymer properties.

Many other reactions responsible for deterioration occur along the main chain and at side groups of polymer molecules with little effect on molecular weight. When these reactions involve only the elimination of low-molecular-weight fragments as in the dehydrochlorination of poly(vinyl chloride) molecular weight of the polymer is essentially unchanged. Reactions of side groups or polymer branches usually do not involve cleavage of the backbone chain. Polar groups and color-producing, unsaturated sequences can also form in a polymer without significantly affecting the molecular weight. Typical examples of these reactions are discussed in Chapter 3.

1. The Effects of Chemical Structure

The rate at which a polymer deteriorates is dependent on the strength of chemical bonds in the structure. However, the energy required to dissociate individual bonds can vary considerably, reflecting the complexity and inhomogeneity of macromolecules. Individual bonds may be more susceptible to dissociation because of molecular irregularities such as random branching or the presence of sensitizing groups, and these weak links are favorable sites for the initiation of deterioration reactions. The classic maxim—*a chain is no stronger than its weakest link*—is indeed applicable.

Chemical bonds in polymer molecules often have thermal dissociation energies significantly lower than those of their low-molecular-weight analogs. The lower activation energy for bond cleavage in polymers has been attributed by Rice[1] to a low-energy, free-radical reaction, typical of chain mechanisms discussed later in this chapter. The presence of branching and certain side groups in polymer molecules also contributes to lowering of bond strengths. In the following polymer series, thermal stability decreases as branch groups

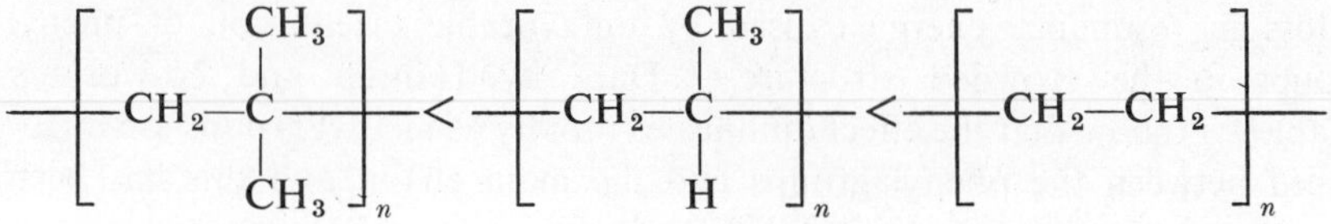

$$\left[-CH_2-\underset{CH_3}{\overset{CH_3}{\underset{|}{\overset{|}{C}}}}-\right]_n < \left[-CH_2-\underset{H}{\overset{CH_3}{\underset{|}{\overset{|}{C}}}}-\right]_n < \left[-CH_2-CH_2-\right]_n$$

Polyisobutylene   Polypropylene   Polymethylene

are added to the basic polymethylene chain. Additional relations between chemical structure and thermal stability are developed in the discussion of high-temperature reactions (Chapter 3).

The ease of hydrogen abstraction from polymer molecules (carbon–hydrogen bond cleavage) is usually the rate-controlling step in polymer oxidation. The rate at which these reactions occur therefore depends on the type of carbon–hydrogen bonds present. The strength of carbon–hydrogen bonds in polymers, as in low-molecular-weight compounds, increases in the following order:

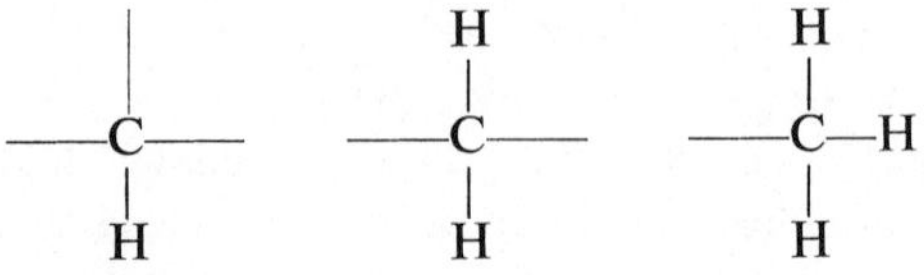

$$\begin{array}{ccc} \begin{array}{c} | \\ -\!\!-C-\!\!- \\ | \\ H \end{array} & \begin{array}{c} H \\ | \\ -\!\!-C-\!\!- \\ | \\ H \end{array} & \begin{array}{c} H \\ | \\ -\!\!-C-H \\ | \\ H \end{array} \\ \mathbf{1} & \mathbf{2} & \mathbf{3} \end{array}$$

Thus hydrogens at branch points in polymer chains (**1**) are more readily abstracted than those of methylene groups (**2**) and the hydrogens of methyl groups (**3**) occurring at chain or branch ends are the least labile. This accounts for the relative instability to oxidation of polypropylene in contrast to linear polyethylene.

Other more subtle factors influence the ease of hydrogen abstraction. For example, both polystyrene and polypropylene have only one labile

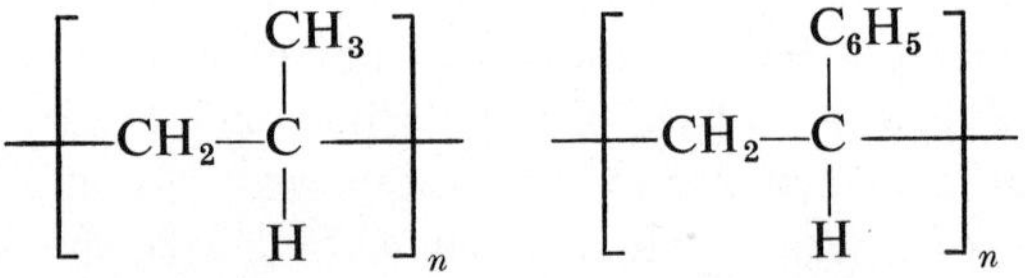

$$\left[-CH_2-\underset{H}{\overset{CH_3}{\underset{|}{\overset{|}{C}}}}- \right]_n \qquad \left[-CH_2-\underset{H}{\overset{C_6H_5}{\underset{|}{\overset{|}{C}}}}- \right]_n$$

Polypropylene Polystyrene

hydrogen in each repeating unit. In polystyrene this is an allylic hydrogen, which should be more reactive than the labile hydrogen of polypropylene. Therefore it might be expected that the oxidation rate of polystyrene would be much more rapid with abstraction of the labile hydrogen being the rate-controlling step. However, the oxidative stability of polystyrene is much greater than that of polypropylene. The unusual stability of polystyrene may result from shielding of labile hydrogens by the bulky phenyl groups or from a loss in resonance energy caused by unfavorable orientation of phenyl groups in the crowded structure.[2,3] Data by Hansen and co-workers (Table 1.1) show that the effect diminishes rapidly as methyl groups are introduced between the phenyl groups and the main chain, and this has been interpreted as support for the shielding theory.

When elements other than carbon and hydrogen are present in a polymer, the dissociation energy of the additional bonds becomes a contributing factor in determining stability. For example, polytetrafluoroethylene, which is similar to polyethylene except that all hydrogens are replaced by fluorine, is significantly more stable to pyrolysis and, whereas polyethylene burns when ignited, polytetrafluoroethylene does not support its own combustion. The strength of carbon–hydrogen bonds is also influenced by the proximity of sensitizing groups in the polymer molecule. Hydrogens on carbon atoms

Table 1.1 Oxidative Stability of Polystyrene of Related Polymers[3]

Polymer	Structure	Induction Period (hr)	
		At 80°C	At 110°C
Polystyrene	$—CH_2—CH(C_6H_5)—$	—	>10,000
Poly(3-phenyl-1-propene)	$—CH_2—CH(CH_2C_6H_5)—$	>10,000	1,900
Poly(4-phenyl-1-butene)	$—CH_2—CH((CH_2)_2C_6H_5)—$	500	30
Poly(6-phenyl-1-hexene)[a]	$—CH_2—CH((CH_2)_4C_6H_5)—$	200	13

[a] Tacky at room temperature.

adjacent to carbonyl, carboxyl, and other electron-withdrawing groups are more easily abstracted by radicals.

2. The Effects of Physical Structure

Deterioration depends not only on the chemical effects described in the preceding section; physical structure, or polymer morphology, is also an important variable. The effects of physical structure are related to the arrangement of molecules in ordered (crystalline) and disordered (amorphous) regions of the polymer matrix. Many polymers are semicrystalline and thus have both ordered and disordered regions. Density of individual polymers is directly related to their degree of crystallinity. In many instances, polymer density can be modified by annealing, crosslinking, and deterioration.

The maximum rate of oxidation, hydrolysis, and reaction with other chemicals develops only when permeation of the reactant into the polymer matrix is adequate to sustain this rate. When it is not, deterioration becomes diffusion controlled. It follows, therefore, that factors influencing permeability have an important effect on polymer deterioration. As reaction proceeds, the diffusion of volatile products out of the polymer mass and progressive changes in morphology can reduce the rate at which the reactant diffuses into the polymer. Thus deterioration of a polymer, although not diffusion controlled initially, may become so in later stages of the reaction.

Permeability of reactants into a polymer matrix is dependent on the density of the material which varies with the degree of crystallinity and the compactness of amorphous and crystalline regions. Comprehensive reviews have been published on permeation[4–6] and crystallinity[7–9] in polymers, and the effect of these physical factors on deterioration has been studied extensively in the oxidation of polyolefins.[10–13] The oxidation rates of linear and branched polyethylenes, as shown in Figure 1.1, are quite similar at 140°C. Since this is above the melting temperature of these polymers, both are completely disordered, and differences in crystallinity evident in the solid state have no effect. At 100°C, however, both polymers are in the semicrystalline state and the amount of oxygen absorbed before reaction subsides is quite different. The amount of oxygen reacting with each polymer is roughly proportional to the degree of crystallinity, and, as a result, the branched polyethylene reacts with more oxygen than the linear modification.

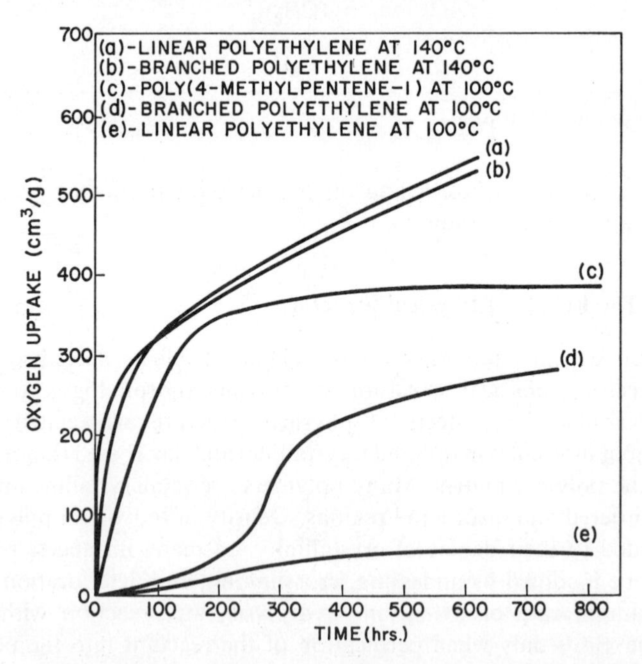

Fig. 1.1 Effect of morphology on the oxygen uptake of polyolefins.[11]

It is apparent that oxidation of semicrystalline polyethylene is restricted to the amorphous regions. Studies[10,12] on the oxidation of single crystals prepared by solution crystallization of linear polyethylene have confirmed this hypothesis, the nearly completely crystalline modification reacting with less oxygen than the semicrystalline form prepared by compression molding. Apparently the oxidation rate of the crystalline region of polyethylene is a diffusion-controlled process. Similar data[14] with polypropylene indicate that, since the crystalline regions of this polymer also are impermeable to oxygen, its oxidation too occurs primarily in amorphous regions. Density of the crystalline phase of poly(4-methylpentene-1), however, is approximately the same as that of the disordered region,[15] and this polymer reacts with approximately the same amount of oxygen as does molten (amorphous) polyethylene before oxidation subsides (Fig. 1.1). Thus compactness of crystalline structure as well as the degree of crystallinity have an effect on polymer stability.

The surface-to-volume ratio of a polymer sample is important in determining the rate of deterioration. In samples too thick to permit diffusion throughout the polymer mass, the reaction rate is limited by the concentration gradient of oxygen in the sample. The concentration of oxygen and the reaction rate are highest at the surface, and at oxygen pressures up to 1 atm the rate is proportional to the square root of oxygen pressure.[10] When reactants cannot diffuse completely into the polymer matrix, the processes of oxidation, hydrolysis, and so on, are limited to surface layers with little change in bulk properties.

In many polymers the degree of crystallinity can be increased by annealing in an inert atmosphere. The density increases with annealing and permeation of reactants into the polymer is reduced. The effect of annealing[12] on oxygen consumption is shown in the comparison of the cumulative oxygen uptake of compression-molded and annealed linear polyethylene curves *b* and *c*, respectively of Figure 1.2. The same polymer, in the form of a fine powder, prepared by crystallization from 0.04% xylene solution at 85°C, as shown in curve *d*, reacts with even less oxygen before the reaction subsides. Cross-linking of linear polyethylene at 100°C with a 100-Mrad radiation dose results in the modification seen in curve *a*, which is much more permeable to oxygen than the other preparations,[12] and the cumulative oxygen uptake is considerably greater.

Crystallization of ultrahigh-molecular-weight polyethylenes is constrained in the latter stages of the process.[16] Severance of a few chain segments in the disordered region during the early stages of oxidation results in a process referred to as chemicrystallization.[12,17] If the oxidized samples are melted and remolded, mechanical strength is partially recovered. Kafavian[18] accounts for this recovery by suggesting that broken chain ends are redistributed more uniformly throughout the polymer mass.

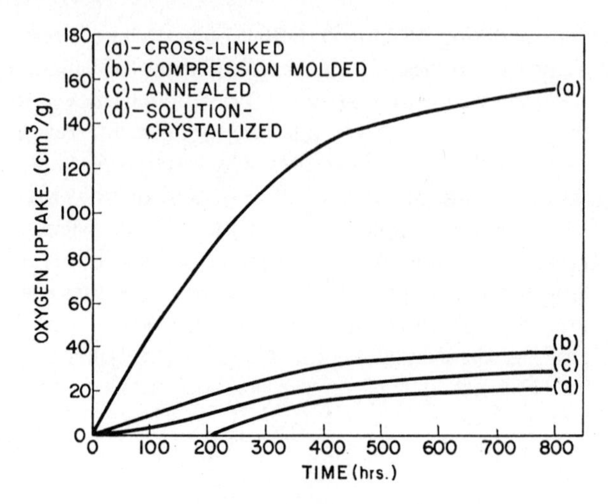

Fig. 1.2 Dependence of oxygen uptake at 100°C on the morphology of linear polyethylene.[10]

3. Chain Mechanisms in Polymer Deterioration

The ease with which many polymers deteriorate results from a chain reaction, usually free radical in nature. Chain reactions are known to occur in the thermal oxidation of polyolefins and probably contribute to other deterioration reactions including pyrolysis[1] and photooxidation (Chapter 4). Although chain reactions greatly increase the rate and extent of deterioration, they also afford a convenient point at which effective protection can be accomplished with catalytic amounts of stabilizers. There are three primary stages in a chain reaction: initiation, propagation, and termination. Chain branching is an additional initiation step, occurring when intermediate products decompose into reactive products to start new oxidation chains. Each of these stages has been established in the oxidation of low-molecular-weight hydrocarbons and evidence has been obtained to support a similar mechanism in the oxidation of polyolefins. These examples of chain reactions are described in Section D.1 of this chapter in the discussion of autoxidation.

Free-radical chain reactions are initiated when primary bonds in a polymer are ruptured to form radicals. These radicals are reactive intermediates which continue the chain mechanism by further reaction with other polymer molecules. Whereas initiation is normally a slow reaction, the propagation stage is usually rapid. Hundreds or even thousands of cycles can occur in the propagation stage before termination takes place. Thus initiation of the reaction in one molecule can result in deterioration of many neighboring molecules. The longer the reaction chain, the more opportunity there will be for chain termination by antioxidants to be an effective stabilization reaction.

Normal termination of chain reactions takes place slowly, usually through a complex series of reactions. Ultimately propagating radicals combine or disproportionate to form inert products. Proposed mechanisms for chain termination, reported elsewhere[19,20] are considered beyond the scope of this book, since for a successful polymer application, stabilization—an induced termination reaction—must inhibit deterioration before normal termination takes place.

D. THERMAL EFFECTS IN POLYMER DETERIORATION

Thermal effects are a major factor in polymer deterioration. Exposure to relatively high temperatures during processing and to more moderate temperatures during long-term aging accounts for the failure of many polymers. Absorption of thermal energy is the primary environmental factor responsible for pyrolysis in vacuum or in an inert atmosphere. However, thermal effects are most generally evident in the temperature dependence of other reactions that result in polymer deterioration. Reaction rates increase with temperature as defined by the activation energy of the particular process. Deterioration, however, usually occurs through a complexity of reactions so that the observed activation energy is that of the composite mechanism and not of individual reactions. Temperature dependence is observed in all types of polymer deterioration, but the effects have been studied most extensively in the autoxidation of polyolefins.

1. Autoxidation

Autoxidation, defined as the thermal oxidation that takes place between room temperature and about 150°C, proceeds by a typical free-radical chain mechanism. Most polymers undergo autoxidation, although there is considerable variation in resistance to this type of deterioration. The reaction is usually autocatalytic, following a rate curve similar to that shown in Figure 1.3. Autoxidation is initiated slowly, accelerates through an autocatalytic stage, and eventually subsides. The following kinetic scheme has been proposed[21–23] to account for the autoxidation of low-molecular-weight hydrocarbons.

$$RH \longrightarrow R^{\cdot} \qquad \text{Initiation}$$

$$\left.\begin{aligned} R^{\cdot} + O_2 &\longrightarrow ROO^{\cdot} \\ ROO^{\cdot} + RH &\longrightarrow ROOH + R^{\cdot} \end{aligned}\right\} \qquad \text{Propagation}$$

$$\left.\begin{aligned} ROOH &\longrightarrow RO^{\cdot} + HO^{\cdot} \\ RO^{\cdot} + RH &\longrightarrow ROH + R^{\cdot} \\ HO^{\cdot} + RH &\longrightarrow HOH + R^{\cdot} \end{aligned}\right\} \qquad \text{Chain branching}$$

$$2\ ROO^{\cdot} \longrightarrow \text{inert products} \qquad \text{Termination}$$

$$ROO^{\cdot}(RO^{\cdot}, HO^{\cdot}, \text{etc.}) + HA \longrightarrow ROOH = A^{\cdot} \qquad \text{Inhibition}$$

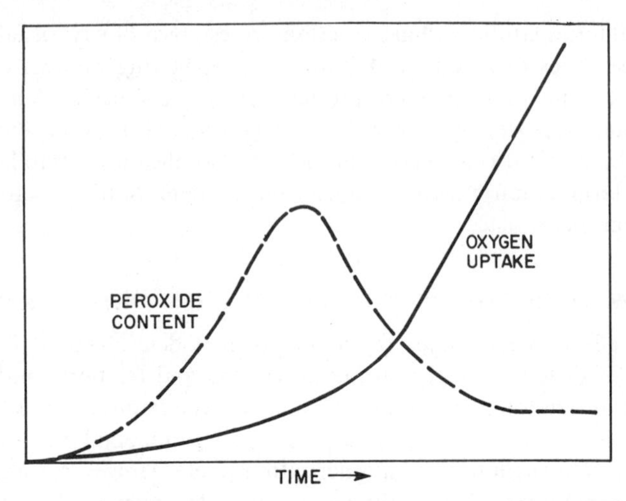

Fig. 1.3 Relation between oxidation rate and hydroperoxide accumulation.

Two important observations contributed to development of this mechanism. In the first it was established that oxidation is initiated by additives that decompose into free radicals capable of extracting hydrogen from the hydrocarbon substrate. The second showed that the reaction is inhibited by additives that can compete with the substrate in donating labile hydrogens to propagating radicals. To be most effective, hydrogen abstraction from the chain terminator (HA) should yield as the byproduct a radical (A˙), which cannot continue the chain reaction. This does not exclude the possibility, however, that some terminators may react by chain transfer to decrease the length of oxidative chains sufficiently to give effective stabilization. A detailed discussion of stabilization by chain termination appears in Chapter 2.

Radicals that initiate autoxidation in polymers are formed by thermal energy, ultraviolet radiation, bombardment by high-energy particles, mechanical stress, metal catalysis, and the addition of initiators. When polymer bonds are ruptured under any of these conditions and oxygen is present in the environment, autoxidation can occur. Many polymers, however, oxidize under relatively mild conditions in which the absorbed energy does not reach the level of the dissociation energies of carbon–carbon or carbon–hydrogen bonds as determined for simple paraffins. Under these conditions, it seems unlikely that simple bond cleavage is the initiation step. It has been suggested[24,25] that traces of hydroperoxides introduced into the polymer during processing are the source of initiating radicals. Such secondary degradation products as carbonyl or carboxyl groups could also contribute to initiation by sensitizing adjacent bonds and thus lowering the dissociation energy. Hydroperoxide decomposition into radicals is accelerated by various catalysts, notably the transition metals and derivatives of these.[26] Polymers are normally in contact with metal surfaces during processing and traces of iron, manganese, copper, and so on, may be incorporated into the material, acting eventually as oxidation catalysts. Metallic impurities derived from polymerization catalysts are often catalysts for autoxidation, and removal of these residues is essential to obtain maximum stability of the polymer.

A typical rate curve for hydrocarbon oxidation is shown schematically in Figure 1.3. Reaction starts slowly, accelerates through a period characterized as the autocatalytic stage, and then slowly subsides. The transition from acceleration through deceleration can result in a period of essentially constant reaction rate—the steady-state rate. Hydroperoxide concentration (Fig. 1.3) starts at a very low level (essentially 0 within the sensitivity of measurement), increases as autocatalysis begins, and reaches a maximum as reaction passes into the steady-state rate. Competitive reactions of hydroperoxide formation and decomposition occur throughout the reaction, and the maximum concentration is reached when the rate of decomposition equals that of formation. It has been suggested[25] that a bimolecular decomposition occurs as the concentration of hydroperoxide increases. It is evident that the homo-

$$\begin{array}{ccc} 2\,R\text{—}O\text{—}O\text{—}H & \underset{\text{high temperature or dilution}}{\overset{\text{low temperature or concentration}}{\rightleftharpoons}} & R\text{—}O\text{—}\overset{\displaystyle H}{\overset{|}{O}}\cdots\cdots H\text{—}O\text{—}O\text{—}R \\ \downarrow & & \downarrow \\ 2\,RO^{\cdot} + 2\,HO^{\cdot} & & RO^{\cdot} + H_2O + ROO^{\cdot} \\ \text{Monomolecular} & & \text{Bimolecular} \end{array}$$

lytic decomposition of hydroperoxides can yield a variety of free radicals, all potential initiators of new oxidative chains. Chain branching through initiation of new chains by these radicals accounts for the autocatalytic phase of autoxidation. The same reaction could be the primary initiation step if traces of hydroperoxides had been introduced into the polymer during processing. This mechanism was developed in studies on the oxidation of low-molecular-weight hydrocarbons, but evidence obtained through infrared analysis[27] of oxidizing hydrocarbon polymers supports the assumption that these polymers oxidize by a similar mechanism. Termination of chain reactions takes place slowly, usually through a complex series of reactions.[19,20] Ultimately propagating radicals combine to form unreactive molecular products. Disproportionation reactions in which two radicals react to yield a saturated and an unsaturated molecular product constitute another important termination mechanism.

Autoxidation subsides before conversion of the polymer to carbon dioxide and water has been completed. Although polymer properties are changed considerably at this stage, only a fraction of the oxygen required for complete oxidation has reacted and the gross physical structure may appear unchanged. Initiation occurs first at the most reactive sites and the number of these reacting at any given temperature is fixed. In the absence of additional initiation reactions, which would take place if the temperature were raised, autoxidation subsides. Additional energy must be absorbed to continue oxidation at a significant rate. There is evidence[27,29] that some oxidation products function as stabilizers, and it is quite probable that autotermination also contributes to the deceleration of autoxidation.

2. Nonoxidative Thermal Deterioration

Although reaction with oxygen is the more general phenomenon, thermal deterioration also occurs in the absence of oxygen or other chemical reactants. There are two basic mechanisms for thermal deterioration under nonoxidative conditions. Dependent on polymer structure, and to a lesser extent on reaction conditions, deterioration can occur through elimination of low-molecular-weight, volatile fragments leaving the backbone chain essentially intact. Alternatively, rupture of the polymer chain can occur, either by a random process or by stepwise depolymerization. Elimination of low-molecular-weight fragments usually occurs at lower temperatures than are required for the cleavage of polymer chains. As the temperature is increased, all polymers eventually dissociate. Volatile products are formed and, dependent on polymer structure, there may be varying proportions of a carbonaceous residue.

Elimination of low-molecular-weight fragments without cleavage of the backbone chain is restricted to a relatively small group of polymers. Structure of the polymer must be such that reactions are possible with activation energies lower than those required to break bonds along the main chain. Splitting of hydrogen chloride from poly(vinyl chloride) is an important example of this type of reaction. As in autoxidation, a chain reaction may be responsible for the relative ease of hydrogen chloride elimination. These reactions are discussed in Chapter 3 as they relate to stabilization against thermal degradation in the absence of oxygen.

Rupture of the backbone chain occurs by depolymerization or by a random process, perhaps initiated at weak bonds in the chain. Both reactions can occur in a single polymer with the relative rates dependent on temperature. When depolymerization, often referred to as "unzipping," is the primary reaction, high yields of the monomer result. Stabilization against this reaction has been accomplished in the polyacetals by modification of the polymer structure.[30] Some polymers undergo what is essentially a depolymerization, although the product is not identical with the true monomer. Formation of levoglucosan in the vacuum distillation of wood (cellulose pyrolysis) is a classic example.[31]

As the temperature increases, random cleavage of bonds in the main chain occurs in the polymers that, because of their structure, do not depolymerize readily. Fragments of molecular weight greater than the monomer result, as in the formation of wax-like products from polyethylene on heating in vacuum. Emphasis on pyrolytic reactions of polymers and methods for inhibiting deterioration at high temperatures in an essentially oxygen-free environment has assumed greater importance as the temperatures used in fabrication have increased. Stabilization by additives is difficult, however, under these conditions. Extraterrestrial applications of polymers to reduce the heat generated in space vehicles during reentry into the earth's atmosphere and other severe service conditions, have led to development of a class of heat-resistant polymers (Chapter 3).

E. OTHER RADIATION EFFECTS IN POLYMER DETERIORATION

Radiation effects in polymer deterioration are conventionally separated into those resulting from photooxidation or from ionizing radiation. From the viewpoint of deterioration, this subdivision is perhaps one of convenience more than of basic principle. The approach to stabilization, however, is sufficiently different to warrant consideration of these radiation effects in separate chapters.

The ultraviolet region of the electromagnetic spectrum (solar radiation) provides the energy source for photo reactions, whereas other regions (x-ray and γ-ray) are responsible for the high-energy reactions induced by ionizing radiation. In addition, a variety of particles of finite mass, as described in Chapter 6, may induce deterioration. The ultimate effects of photo and ionizing radiation on polymers are often quite similar, varying in degree more than in basic character. For example, chain scission and crosslinking occur under both reaction conditions. Although reaction mechanisms have not been developed completely, free-radical reactions involving cleavage of carbon–carbon and carbon–hydrogen bonds are known to occur during photooxidation and under conditions of high-energy radiation. Other similarities in reaction mechanism will be apparent in the chapters dealing with stabilization under specific radiation conditions. There is perhaps an analogy between the thermal reactions of deterioration, ranging from autoxidation to combustion or pyrolysis, and radiation effects from the more moderate conditions of ultraviolet radiation to the more intense reactions initiated by ionizing radiation.

1. Deterioration by Ultraviolet Radiation

Photooxidation is the principal reaction in the degradation of polymers during outdoor weathering. Although polymers vary widely in their resistance to photooxidation, almost every polymer eventually deteriorates on continued exposure to solar radiation. Adequate resistance to or protection against photooxidation is a most important factor limiting the application of polymers in products that must be exposed out of doors.

Photooxidation is initiated by energy absorbed from light sources, either sunlight or artificial light. Although the ultraviolet region of the solar spectrum extends from 40 to 4000 Å, radiation below about 2900 Å is almost completely filtered out by the atmosphere, and the total amount that does reach the earth's surface represents less than 7% of the total solar radiation.[32] Yet this range of the electromagnetic spectrum is the energy source on which life processes depend—compensation, certainly, for the adverse effect on polymer weathering.

The intensity of ultraviolet radiation to which polymers are exposed in natural weathering is never constant. In addition to the obvious variations that occur with location and season, such uncertain factors as cloud cover, smog density, and temperature greatly influence the intensity of incident radiation and hence the rate of photooxidation. These variables account for much of the lack of agreement among outdoor weathering data reported in the literature. Careful control of samples during preparation for exposure and analysis of the voluminous data on weather variables with appropriate compensation could do much toward obtaining more consistent, reliable, and useful results. Artificial light sources have an effect on polymers similar to solar radiation to the extent that they also radiate at frequencies where the polymer absorbs damaging radiation. However, artificial light sources never accurately reproduce the solar spectrum. Their use as a means of intensifying ultraviolet radiation and thus accelerating tests for photooxidation are discussed in detail in Chapter 10.

The mechanism of photooxidation has not been developed to an extent comparable to that of autoxidation. It has been generally assumed that similar mechanisms apply to both reactions. Initiation of photooxidation apparently occurs at a more rapid rate, and there is evidence for a chain reaction although the chain length is apparently not nearly as long as in autoxidation. A typical rate curve for photooxidation of a polymer does not show the characteristic autocatalysis evident in autoxidation (Figure 1.4). Reaction starts rapidly and continues at an essentially constant rate until deceleration occurs. In other words, deterioration by ultraviolet radiation appears, by simple analysis of rate data, not to include the autocatalytic phase, characteristic of autoxidation. This raises a question about the potential role of

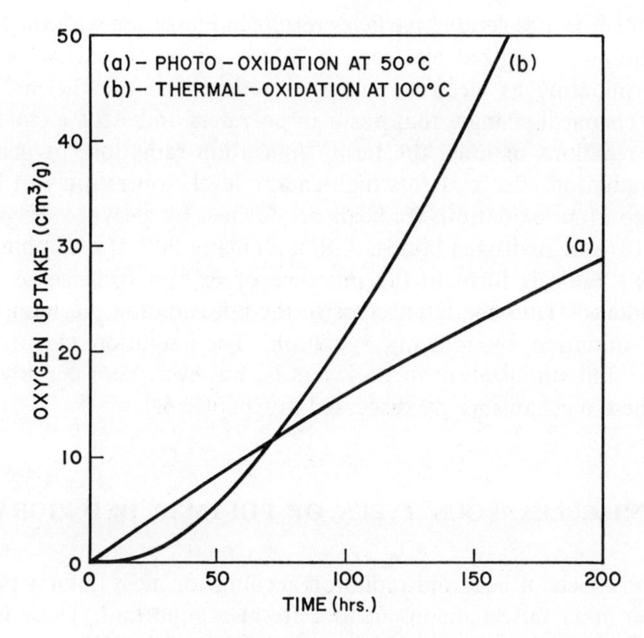

Fig. 1.4 Comparison between the rates of photooxidation and autoxidation.

hydroperoxides in the photooxidation of polymers. The presence of hydroperoxides has been demonstrated in the photooxidation of simple organic compounds, but recent studies[33] have failed to detect the presence of these intermediates in the photooxidation of polyethylene. However, hydroperoxides absorb ultraviolet radiation and are rapidly decomposed by the absorbed energy. If this decomposition occurs by a homolytic cleavage, then new radicals would be formed which, in turn, would initiate new oxidative chains. Oxidation would then rapidly attain the maximum rate, accounting for the more rapid initial rate observed in photooxidation. Recent developments in the mechanism for the oxidative photodegradation of polymers, with particular emphasis on the initiation step, are discussed in Chapter 4.

2. Deterioration by High-Energy Radiation

The unique combination of strength with lightness makes polymers of interest in extraterrestrial applications, both as structural and as dielectric materials. The high-energy radiation encountered in space, however, imposes a serious limitation on these applications. Deterioration of polymers by high-energy radiation also limits the use of these materials in proximity to sources of x-rays and γ-rays or to various types of accelerators.

Bombardment by accelerated particles results in ionization with the formation of additional energized electrons in what can be considered a chain reaction, terminating as electron energy is reduced to the thermal level. Because the chemical changes that occur in polymers under these conditions result from reactions of ions, the term "ionization radiation" is generally applied to radiation effects at this high-energy level. Ionization can be the initiation step in autoxidation. Radicals are formed by cleavage of carbon–carbon and carbon–hydrogen bonds. Although many radical recombinations occur, peroxy radicals form in the presence of oxygen from some of the radicals produced. Thus the latter steps in the autoxidation mechanism can occur after initiation by ionizing radiation. The radiation chemistry of deterioration and stabilization in polymers is, however, considerably more complex. These mechanisms are discussed in Chapter 6.

F. MISCELLANEOUS TYPES OF POLYMER DETERIORATION

Although the effects of heat and radiation account for most failures in polymers, several other miscellaneous factors are also important. These include reaction with a variety of chemical reagents and attack by living organisms. Mechanical stress also contributes to polymer deterioration, even in the absence of chemical reactants.[34] To attain the maximum service life, polymers usually require stabilization against these factors when they are a part of the environment.

1. Chemical Agents

Many chemical reactants attack polymers and most reactions result in deterioration of useful properties. Oxygen, water, and ozone are the principal reactants present in the normal environment, and these reagents are responsible for the failure of most polymers in service. In certain specialized applications, however, polymers are exposed to a variety of reagents that are not components of the normal environment or to these reagents at concentrations considerably higher than normal. Considering the many types of polymers that are available commercially and the multitude of reactants to which they may be exposed, it is apparent that most chemical effects are quite specific and too numerous for individual consideration. However, certain general principles apply to deterioration resulting from exposure to chemical agents. Exposure to water is a fairly general cause of polymer failure. Less common reagents that attack polymers include acids, bases, dyes of various types, adhesives, solvents, and nonsolvents.

Chemical agents deteriorate polymers either by cleavage of primary chemical bonds or by the irreversible rupture of secondary valence forces. For attack to proceed into the polymer bulk, the reactant must permeate the structure. Since the permeation process is dependent on polymer structure, nature of the reactant, and temperature, it follows that these variables are also important in determining the rate of deterioration. Because deterioration resulting from exposure to reagents other than oxygen is more specific, these reactions are included in Chapter 8 as they apply to stabilization against these effects.

2. Biological Agents

Although most synthetic polymers are organic, they exhibit a high degree of resistance to biological attack. In contrast, many natural fibers decay rapidly when exposed to aerobic cultures.[35,36] Organisms that cause polymer deterioration include bacteria, fungi, marine animals, and borers of various types. The organism may consume the polymer if it has food value, but many borers attack polymers for no evident purpose. Protection against biological attack has been approached both by modification of polymer structure and by the use of additives, as described in Chapter 9.

3. Effects of Mechanical Stress

Mechanical stress accelerates polymer deterioration in many ways. Application of stress renders some unsaturated polymers vulnerable to attack by ozone (Chapter 5). Autoxidation of polyolefins is also accelerated by placing the samples under stress.[37] These reactions and bond rupture under high stress involve cleavage of primary bonds. Autoxidation, initiated in part by bond cleavage under stress, contributes to the deterioration that occurs during fabrication.

Deterioration under stress also takes place under conditions that apparently do not involve rupture of primary bonds. A typical example is the failure that occurs in polyolefins when stressed samples are exposed to nonsolvents, a process referred to as environmental stress cracking. The mechanism is not completely understood, but it is generally recognized that this type of cracking occurs only when a critical stress is applied.[38] Resistance to environmental stress cracking is improved by relatively minor changes in polymer structure. For example, a change in the molecular weight distribution of polyethylene to reduce the proportion of low-molecular-weight polymer results in increased resistance to environmental stress cracking. Copolymers of ethylene with acrylic acid are also more resistant to this type of cracking than the homopolymer of ethylene.

Table 1.2 Relative Stability of Some Unstabilized Polymers to Various Environmental Factors

KEY: E = excellent
G = good
F = fair
P = poor

		Environmental Factors				
PLASTICS		Photo-oxidation	Thermal Oxidn	Ozone Attack	Combustion	Moisture
Polyethylene (branched)	$-[-CH_2-CH_2-]-_n$	P	F	E	P	E
Polypropylene	$-[-CH(CH_3)-CH_2-]-_n$	P	P	E	P	E
Polyisobutylene	$-[-C(CH_3)_2-CH_2-]-_n$	P	G	E	P	E
Polystyrene	$-[-CH(C_6H_5)-CH_2-]-_n$	P	G	E	P	E
Poly(vinyl chloride)	$-[-CHCl-CH_2-]-_n$	F*	F*	E	G	G
Polytetrafluoroethylene	$-[-CF_2-CF_2-]-_n$	E	E	E	E	E
Polyoxymethylene	$-[-CH_2-O-]-_n$	P	P	G	P	F
Poly(phenylene oxide)	$-[-C_6H_4-O-]-_n$	P	F	E	F	G
Polyacrylonitrile	$-[-CH_2-CH(CN)-]-_n$	F	G	E	G	G
Poly(vinyl acetate)	$-[-CH_2-CH(OAc)-]-_n$	F	F	E	P	G
Poly(methyl methacrylate)	$-[-CH_2-C(CH_3)(COOCH_3)-]-_n$	E	G	E	P	G
Poly(hexamethylene adipamide)	$-[-NH-(CH_2)_6-NHCO-(CH_2)_4-CO-]-_n$	F	F	E	P**	F
Poly(ethylene terephthalate)	$-[-CH_2-CH_2-OOC-C_6H_4-COO-]-_n$	F	G	E	F	G
Poly(m-phenylene isophthalamide)	$-[-NH-C_6H_4NHCO-C_6H_4-CO-]-_n$	P	E	E		G

Cellulose propionate	$-[-O-C_6H_7(OCOCH_2-CH_3)_3-]-_n$	P	F	E	P	P
Polycarbonate	$-[-O-C_6H_4-C(CH_3)_2-C_6H_4-OCO-]-_n$	F	G	E	F	G
Polysulfone	$-[-O-C_6H_4-C(CH_3)_2-C_6H_4-O-C_6H_4-SO_2-C_6H_4-]-_n$	P	E	E	F	G
ABS resin		P	P	E	P	G
ELASTOMERS						
Polyisoprene	$-[-CH_2C(CH_3)=CH-CH_2-]-_n$	P	P	P	P	G
Polybutadiene	$-[-CH_2-CH=CH-CH_2-]-_n$	P	P	P	P	G
Polychloroprene	$-[-CH_2C(Cl)=CH-CH_2-]-_n$	P	G	G	G	G
SBR rubber		P	P	P	P	G
Ethylene-propylene rubber		P	F	E	P	E
THERMOSETS						
Phenol-formaldehyde (wood filled)		F	G	E	F	P
Phenol-formaldehyde (mica filled)		G	G	E	G	G
Urea-formaldehyde resin		E	E	E	G	E
Polyurethane (ester type)		F	F	E	P	F
Poly(dimethyl siloxane)	$-[-Si(CH_3)_2-O-]-_n$	E	E	E	G	E
Diallyl phthalate resin		G	G	E	F	C

*Based in part on HCl evolution

**Rating modified to account for dripping

G. SUSCEPTIBILITY OF SPECIFIC POLYMERS TO VARIOUS ENVIRONMENTAL FACTORS

In the preceding sections it has been shown that the rate at which a polymer deteriorates is dependent on its chemical and physical structure and on the conditions of exposure. Stabilization is intended to increase the useful life of polymers; hence the choice of an appropriate stabilizer is related to the inherent vulnerability of the basic polymer structure. In the following chapters, stabilization is discussed with reference to specific factors responsible for deterioration. Although several relationships between polymer structure and stability will be developed, it is appropriate at this point to present a general summary of the effect of chemical structure of unprotected polymers on their rates of deterioration.

In Table 1.2, the relative stability to important environmental factors is given for several selected polymers. It should be emphasized that these ratings do not indicate the stability of commercial materials, which usually contain stabilizers or which may be modified by addition of plasticizers, fillers, pigments, and so on. The exception of phenol–formaldehyde resins containing wood or mica as the filler is included to show the range of stability that can be obtained by varying the filler in this polymer.

Since it is the purpose of this summary to relate polymer structure to stability, the ratings are based on significant changes in structure and not on failure of the material in a specific design. Thus stability to photooxidation, based on outdoor exposure in a temperate climate, is rated on changes in visual appearance or physical strength, known to reflect changes in chemical structure. Resistance to thermal oxidation is based on the relative rates of autoxidation at or near 100°C, a temperature range at which variations in thermal stability are quite evident. The discoloration of poly(vinyl chloride) resulting from the loss of hydrogen chloride occurs under conditions of thermal and photooxidation, and this change in chemical structure has been taken into account in rating this polymer. Comparable test data have been used to rate the resistance to ozone attack, and the relative ease of combustion is based on oxygen-ratio data (Chapter 10). Stability to moisture reflects the combined effects of absorption and hydrolysis with emphasis on the rupture of bonds in the backbone chain.

Only a limited number of polymers are included in Table 1.2, and these have been selected as representative of their respective types. Since, however, important variations do occur in any class of polymers, these relations cannot be extrapolated to all members of the class. For example, significant differences are observed in the relative stability of linear and branched polyethylenes, the latter being used as the example in the table. The stability of polyamides to moisture is exemplified by the 6,6-polymer, but other polyamides differ to a considerable degree in resistance to moisture, as described in Chapter 8. However, despite the sometimes important variations among individual members in a class of polymers, the relations developed in Table 1.2 show a basic pattern of vulnerability of typical polymer structures to specific environmental factors and establish the relative need for adequate protection of individual polymers or classes of polymers under each exposure condition.

The final section of this chapter is a transition from the general discussion of polymer deterioration into mechanisms of stabilization. The historical development of stabilization concepts, as exemplified by the development of antioxidant theory, serves as an excellent background for this. The basic theory of stabilization in polymers has developed extensively since its inception over a century ago, and reaction mechanisms, described in the chapters that follow, have been proposed to explain several deterioration processes in addition to autoxidation.

H. THE DEVELOPMENT OF ANTIOXIDANT THEORY

The deterioration of rubber was studied as early as the mid-nineteenth century, when it was shown that the aging phenomenon coincided with oxygen absorption[39] and was retarded by phenolic substances.[40] Later it was found that the degradation involved peroxide intermediates,[41] was an autocatalytic process,[42] and was inhibited by aromatic amines.[43] Finally the theory of antioxidant action evolved from the discovery by Moureu and Dufraisse[44] that the oxidation of acrolein was suppressed by pyrogallol. They referred to protectants as anti-oxygens and regarded the protective action as "negative catalysis." Their papers[45] on the subject were plentifully peppered with lyrical lines and philosophical asides. For example, in quoting Dumas, they suggested that, "For such a delicate subject [negative catalysis] one always risks saying too much, however little one says." The remark proved to be prophetic, because their studies inspired Taylor[46] to liken negative catalysis to "a warden caring for a hundred lunatics. The warden would be powerless were all the lunatics simultaneously violent. Only at intervals does an occasional lunatic become a candidate for a padded cell. The warden cares for him, the gentler ninety-nine do not require attention." Christiansen[47] deplored such attribution of intelligence to molecules and proposed instead that the so-called catalytic effect could be described far better as a chain reaction of the type devised by Bodenstein[48] to account for the photochemical synthesis of hydrogen chloride. This view was supported by Backström[49] who showed that photooxidation of benzaldehyde produced as much as 10,000 moles of benzoic acid per quantum of energy absorbed. But the concept of negative catalysis lingered on since, in a 1930 review of rubber chemistry, Fisher[50] concluded that, "The action of anti-agers is probably one of negative catalysis rather than of preferential absorption." However, it soon became clear that free radicals[51] and hydroperoxides[52–54] were key intermediates in hydrocarbon oxidations, and eventually a general theory of antioxidant action emerged largely from work at the Natural Rubber Producers' Research Association in England.[55,56]

According to this theory, inhibitors substituted for the substrate function in sacrificial reactions with peroxy radicals, forming hydroperoxides and antioxidant derivatives incapable of promoting further reaction. The oxidation rate was suppressed by interruption of the reaction chain at an early stage. The hydroperoxide product was regarded merely as a potential but minor source of chain initiation. Still later studies have demonstrated the extraordinary protection obtainable from synergistic systems based on sulfur compounds that decompose hydroperoxides by nonradical mechanisms[57,58] as well as on metal deactivators[59] that suppress the catalytic decomposition of hydroperoxides to radical products.

REFERENCES

1. F. O. Rice and K. K. Rice, Eds., *The Aliphatic Free Radicals*, Johns Hopkins Press, Baltimore, 1935.
2. L. A. Wall, M. R. Harvey, and M. Tryon, *J. Phys. Chem.*, **60,** 1306 (1956).
3. R. H. Hansen, W. H. Martin, and T. DeBenedictis, *Trans. Inst. Rubber Ind.* (*London*), **39** (6), T301, (1963).
4. C. E. Rogers, "The Solution, Diffusion and Permeation of Gases and Vapors in High Polymers," in D. Fox, M. M. Labes, and A. Weissberger, Eds., *Physics and Chemistry of the Organic Solid State*, Pt. II, Interscience, New York, 1965, p. 510.
5. V. Stannett and H. Yasuda, "Permeability," in R. A. V. Raff and K. W. Doak, Eds., *Crystalline Olefin Polymers*, Pt. II, Interscience, New York, 1964, p. 131.
6. R. M. Barrer, *Diffusion in and through Solids*, Cambridge Press, London, 1941.
7. H. D. Keith, "The Crystallization of Long-Chain Polymers," in D. Fox, M. M. Labes, and A. Weissberger, Eds., *Physics and Chemistry of the Organic Solid State*, Pt. I, Interscience, New York, 1963, p. 461.
8. H. D. Keith and F. J. Padden, Jr., *J. Appl. Phys.*, **34,** 2409 (1963); **35,** 1270, 1286 (1964).
9. R. L. Miller, "Crystalline and Spherulitic Properties," in R. A. V. Raff and K. W. Doak, Eds., *Crystalline Olefin Polymers*, Pt. I, Interscience, New York, 1965, p. 577.
10. F. H. Winslow and W. L. Hawkins, "Stability and Structure Relationships," in R. A. V. Raff and K. W. Doak, Eds., *Crystalline Olefin Polymers*, Pt. I, Interscience, New York, 1965, p. 819.
11. W. L. Hawkins, W. Matreyek, and F. H. Winslow, *J. Polym. Sci.*, **41,** 1 (1959).
12. F. H. Winslow, M. Y. Hellman, W. Matreyek, and S. M. Stills, *Polym. Eng. Sci.*, **6** (3), 1 (1966).

13. F. H. Winslow, C. J. Aloisio, W. L. Hawkins, W. Matreyek, and S. Matsuoka, *Chem. Ind.*, **1963,** 533.
14. F. H. Winslow and W. Matreyek, *ACS Polym. Preprints*, **3** (1), 229 (1962).
15. J. H. Griffith and B. G. Ranby, *J. Polym. Sci.*, **44,** 369 (1960).
16. S. Matsuoka, *Amer. Phys. Soc. Bull.*, **7,** 207 (1962).
17. F. H. Winslow, C. J. Aloisio, W. L. Hawkins, and W. Matreyek, *Chem. Ind.*, **1963,** 1465.
18. G. Kafavian, *J. Polym. Sci.*, **24,** 499 (1967).
19. G. A. Russell, *Chem. Ind.*, **1956,** 1483; *J. Amer. Chem. Soc.*, **77,** 4583 (1955); *J. Amer. Chem. Soc.*, **78,** 1047 (1956); *J. Amer. Chem. Soc.*, **79,** 3871 (1957); *J. Chem. Ed.*, **36,** 111 (1957).
20. H. S. Blanchard, *J. Amer. Chem. Soc.*, **81,** 4548 (1959).
21. J. L. Bolland, *Quart. Rev.* (*London*), **3,** 1 (1949).
22. C. E. Frank, *Chem. Rev.*, **46,** 155 (1950).
23. L. Bateman, *Quart. Rev.* (*London*), **8,** 147 (1954).
24. L. Bateman, M. E. Cain, T. Colcough, and J. I. Cunneen, *J. Chem. Soc.*, **1962,** 3570.
25. J. R. Shelton and D. N. Vincent, *J. Amer. Chem. Soc.*, **85,** 2433 (1963).
26. W. L. Hawkins and F. H. Winslow, *Trans. Plastics Inst.* (*London*), **29**(81), 82 (1961).
27. M. G. Chan and W. L. Hawkins, *Polym. Eng. Sci.*, **1967,** 3.
28. A. Robertson and W. A. Waters, *J. Chem. Soc.*, **1948,** 1574.
29. M. F. R. Mulcahy and I. C. Watt, *Proc. Roy. Soc.* (*London*), **216A,** 10, 30 (1953).
30. P. G. Kelleher and B. D. Gesner, *Polym. Eng. Sci.*, **10**(1), 38 (1970).
31. R. J. Dimler, H. A. Davis, and G. E. Hilbert, *J. Amer. Chem. Soc.*, **68,** 1377 (1946).
32. N. Z. Searle and R. C. Hirt, *J. Opt. Soc. Amer.*, **55,** 1413 (1965).
33. F. H. Winslow, W. Matreyek, and A. M. Trozzolo, *ACS Polym. Preprints*, **10,** 1271 (1969).
34. N. Grassie, Ed., *Chemistry of High Polymer Degradation Processes*, Interscience, New York, 1956, p. 101.
35. L. R. Snoke, *Bell Syst. Tech. J.*, **36,** 1095 (1957).
36. P. L. Steinberg, *Bell Syst. Tech. J.*, **40,** 1369 (1961).
37. F. S. Kaufman, Jr., "A New Technique for Evaluating Outdoor Weathering Properties of High Density Polyethylene," in M. R. Kamal, Ed., *Appl. Polym. Symp. No. 4*, Interscience, New York, 1967, p. 131.
38. J. B. Howard and H. M. Gilroy, *SPE* (*Soc. Plastics Engrs.*), *J.*, **24** (1), 68 (1968).
39. A. W. Hoffman, *J. Chem. Soc.*, **13,** 87 (1861).
40. W. L. Semon, "History and Use of Materials which Improve Aging," in C. C. Davis and J. T. Blake, Eds., *The Chemistry and Technology of Rubber*, Reinhold, New York, 1937, p. 414.
41. S. J. Peachy, *J. Soc. Chem. Ind.* (*London*), **31,** 1103 (1912).
42. W. Ostwald, *J. Soc. Chem. Ind.* (*London*), **32,** 179 (1913).
43. W. Ostwald and W. Ostwald, German Patent, 221,310 (Nov. 1908); through *Chem. Abstr.*, **4,** 2746 (1910).
44. C. Moureu and C. Dufraisse, *Bull. Soc. Chim. Fr.*, **31,** 1152 (1922).
45. C. Moureu and C. Dufraisse, *Chem. Rev.*, **3,** 113 (1926).
46. H. S. Taylor, *J. Phys. Chem.*, **27,** 322 (1923).
47. J. A. Christiansen, *J. Phys. Chem.*, **28,** 145 (1924).
48. M. Bodenstein, *Z. Phys. Chem.*, **84,** 329 (1913).
49. H. L. J. Bäckström, *J. Amer. Chem. Soc.*, **49,** 1460 (1927).
50. H. L. Fisher, *Chem. Rev.*, **7,** 130 (1930).
51. H. L. J. Bäckström, *Z. Phys. Chem. B.*, **25,** 122 (1934).

52. H. Hock and W. Susemihl, *Ber.*, **66,** 61 (1933).
53. R. Criegee, H. Pilz, and H. Flygare, *Ber.*, **72,** 1799 (1939).
54. E. H. Farmer, G. F. Bloomfield, A. Sundralingham, and D. A. Sutton, *Trans. Faraday Soc.*, **38,** 348 (1942).
55. J. L. Bolland, *Quart. Rev.* (*London*), **3,** 1 (1949).
56. L. Bateman, *Quart. Rev.* (*London*), **8,** 147 (1954).
57. G. H. Denison, Jr., *Ind. Eng. Chem.*, **36,** 477 (1944).
58. W. L. Hawkins, V. L. Lanza, B. B. Loeffler, W. Matreyek, and F. H. Winslow, *J. Appl. Polym. Sci.*, **1,** 43 (1959).
59. R. H. Hansen, T. DeBenedictis, and W. M. Martin, *Polym. Eng. Sci.*, **5** (4), 223 (1965).

2

STABILIZATION AGAINST THERMAL OXIDATION

J. REID SHELTON

Case Western Reserve University, Cleveland, Ohio

A. INTRODUCTION

All organic materials are susceptible to oxidative degradation in varying degrees and under appropriate conditions. Natural and synthetic polymers are particularly sensitive to degradation by elemental oxygen as a result of the drastic changes in the tensile strength, hardness, elongation, and other properties of these high-molecular-weight materials that can be produced by a small amount of oxygen. Reactions leading to scission of molecular chains as well as crosslinking reactions occur in varying proportions, along with the formation of peroxides and other oxy functional groups at various points along the polymer chain. Since a single scission can reduce the size of a polymer molecule by one-half, and the joining together of molecular chains leads to aggregation and gelation, it is evident that the resulting effects upon viscosity and strength properties will become significant even though the bulk of the material remains unchanged.

Stabilization against thermal oxidation is thus essential for all organic polymers, since they are inevitably exposed to oxygen during synthesis, fabrication, storage, and use. The amount and type of stabilizer required will depend upon the chemical structure of the particular polymer, the temperature, the oxygen concentration, and the time of exposure likely to be encountered in the various stages of manufacture and in the intended application. Even relatively stable polymers undergo significant deterioration on long-term aging at ordinary temperatures in air. If the material contains certain metal-ion impurities or is likely to be exposed to ultraviolet light or to ozone, as in outdoor exposure, additional special stabilizers are required to provide adequate protection, as will be described in subsequent chapters.

This chapter is concerned with the thermal autoxidation reaction, the mechanisms by which various types of stabilizers inhibit or retard the reaction, and the various factors that influence the susceptibility of polymers to oxidative degradation. Thermal degradation in the absence of oxygen is discussed in the following chapter, and high-temperature oxidation (combustion) is covered in Chapter 7; this review of stabilization against thermal oxidation is therefore restricted to reactions occurring at temperatures below those at which pyrolysis and combustion occur and to the effect of temperature upon rate of oxidation, the resulting degradation, and the ability of various antioxidant types to provide stabilization under these conditions.

Many reviews[1–9] of thermal oxidation and antioxidant action provide historical coverage of the subject which will not be duplicated here. Stabilization of elastomers has been particularly well covered because of the greater susceptibility of the unsaturated diene polymers to oxidative degradation.[10–19]

B. THERMAL AUTOXIDATION

The oxidation of hydrocarbons and related organic materials with elemental oxygen is an autocatalytic process in which the major primary product of the reaction is a hydroperoxide that decomposes under appropriate conditions to give free radicals capable of initiating the free-radical chain reaction. The decomposition of the hydroperoxide is accelerated by heat, light, and the presence of certain metal ions. Since stabilization against oxidative photodegradation is covered in a subsequent chapter, the present discussion is confined to the problem of stabilization against thermal oxidation, including the effect of metal-ion catalysts.

In recognition of the autocatalysis of the oxidation by hydroperoxides formed in the reaction, the process is commonly referred to as autoxidation. The characteristic S-shaped curve for oxygen absorption as a function of time is illustrated by curve *d* in Figure 1.1 for the oxidation of branched polyethylene at 100°C. The reaction begins slowly and increases in rate as the hydroperoxide product contributes to an ever-increasing rate of initiation until a maximum rate is attained, and then the reaction subsides and levels off as the polymer becomes modified by extensive oxidation.

1. Free-Radical Chain Mechanism of Autoxidation

Our present knowledge of the reactions and mechanisms of autoxidation is largely based on extensive studies of the oxidation of relatively simple hydrocarbons and their derivatives. The detailed studies of the autoxidation of ethyl linoleate and other model compounds at the laboratories of the Natural Rubber Producers' Research Association (NRPRA)[1,2,17,19] are particularly noteworthy. Bateman[2] once described the reaction of molecular oxygen with olefins as one of the most thoroughly understood chemical processes. However uncertainties about the extent to which the behavior of model compounds represents the reactions of high-molecular-weight materials of similar structure make us less confident of our understanding of the chemistry of the oxidation of polymers. Furthermore the uninhibited oxidation of pure compounds is considerably modified in the presence of added stabilizers.[6,14,20]

Nevertheless, fundamental reactions appear to be common to all autoxidation mechanisms. We will begin with a consideration of the simple reaction scheme that is now generally accepted for the autoxidation of a pure hydrocarbon in the absence of added initiators or stabilizers. Under these conditions only peroxide initiation need be considered, although free radicals generated in a polymer by other processes such as mechanical shear, high-energy radiation, ultraviolet light, or thermal decomposition of weak bonds in the polymer or added radical precursor, will also initiate the same free-radical chain mechanism. Both first- and second-order decomposition of the hydroperoxide have been observed in the autoxidation of simple compounds in the liquid phase, with the bimolecular process becoming dominant as the concentration of peroxide increases.[1,2]

$$\mathrm{ROOH} \rightarrow \mathrm{RO\cdot} + \mathrm{\cdot OH}$$

$$\mathrm{2ROOH} \rightarrow \mathrm{RO\cdot} + \mathrm{RO_2\cdot} + \mathrm{H_2O}$$

Random introduction of hydroperoxide at isolated sites in a solid polymer would be expected to favor unimolecular decomposition, at least in the early stages of autoxidation. The following simplified reaction scheme is consistent with either or both of the above peroxide initiation reactions.

Initiation:

$$n\mathrm{ROOH} \xrightarrow{k_i} \mathrm{RO\cdot}, \mathrm{RO_2\cdot}, \ldots$$

Propagation:

$$\mathrm{RO_2\cdot} + \mathrm{RH} \xrightarrow{k_p} \mathrm{ROOH} + \mathrm{R\cdot}$$

$$\mathrm{R\cdot} + \mathrm{O_2} \xrightarrow{\text{fast}} \mathrm{RO_2\cdot}$$

Termination:

$$\mathrm{2RO_2\cdot} \xrightarrow{k_t} \text{nonradical products}$$

$$\mathrm{2R\cdot} \xrightarrow{k_t'} \mathrm{R{-}R}$$

$$\mathrm{RO_2\cdot} + \mathrm{R\cdot} \xrightarrow{k_t''} \mathrm{RO_2R}$$

At partial pressures of oxygen equivalent to air or higher concentrations, the rate of combination of R· with O_2 is so fast that $[\mathrm{R\cdot}] \ll [\mathrm{RO_2\cdot}]$, and the only termination reaction of consequence will be the bimolecular combination of two $RO_2\cdot$ radicals. Under such conditions an equation for the rate of oxidation derived from the above mechanism shows that the effect of oxygen concentration upon the rate of oxidation is negligible in the case of a pure hydrocarbon, except at low oxygen pressures (less than 100 mm) or with a very reactive substrate at higher temperatures.[1,2] Under these special conditions the steady-state concentration of R· becomes significant and all three termination reactions would have to be included in the derivation of a rate equation, which would then include O_2 concentration terms indicating a limitation of rate at lower oxygen concentrations.

If a steady-state concentration of $RO_2\cdot$ is attained, the rate of radical generation in the initiation process must be exactly balanced by the rate at which free radicals are removed in the termination reaction. (The two propagation reactions can be ignored, since $RO_2\cdot$ formed in one is consumed in the other.) Thus the rate of initiation, R_i, can be set equal to the rate of termination, R_t, and solved for the steady-state concentration $[RO_2\cdot]$:

$$R_i = R_t = 2k_t[RO_2\cdot]^2$$

$$[RO_2\cdot] = \left(\frac{R_i}{2k_t}\right)^{1/2}$$

The rate of oxidation, R_{ox}, will be controlled by the rate at which $RO_2\cdot$ abstracts hydrogen from the hydrocarbon undergoing autoxidation to form ROOH and R·, which consumes O_2 in a very fast reaction. The rate equation can thus be derived by substitution of the steady-state radical concentration $[RO_2\cdot]$ in the expression for the rate-limiting propagation reaction:

$$R_{ox} = k_p[RO_2\cdot][RH]$$

$$= k_p\left(\frac{R_i}{2kt}\right)^{1/2}[RH]$$

Thus the rate of uninhibited autoxidation can be expected to be independent of oxygen concentration in the case of pure materials, provided there is no significant limitation upon the availability of oxygen attributable to concentration or to rate of diffusion (in the case of polymers). The situation is quite different, however, if oxidizable impurities are present or if stabilizers against thermal oxidation are added.[6,14,20]

2. Retarded Autoxidation Mechanism

The problem of stabilization of polymers and other organic materials against thermal oxidation can be approached in several ways. Modifications of polymer composition and structure to obtain materials that exhibit inherent thermal oxidative stability are discussed in the following section of this chapter. However polymers with the desired properties for most applications usually require the addition of some type of antioxidant to minimize oxidative degradation during fabrication, storage, and use. These stabilizers against thermal oxidation may function in a number of different ways to inhibit, or at least to retard, the oxidation reaction. The various possible mechanisms of antioxidant action are discussed in some detail later in the chapter, but first we must consider how certain commonly used types of antioxidants, such as aryl amines and hindered phenols, participate in initiation, propagation, and termination reactions and thus alter the simple free-radical chain mechanism described in the preceding section.

There are two major classes of stabilizers against thermal oxidation: the preventive antioxidants, which in some way inhibit or retard the formation of free radicals in the initiation step, and the chain-breaking antioxidants, which interrupt the propagation cycle by reacting with the free radicals, $R\cdot$ or $RO_2\cdot$, and thus introduce new termination reactions. It is evident that the first type will slow oxidation without changing the mechanism, but the second type introduces competing reactions that bring about changes in the reaction sequence, which makes the mechanism of retarded autoxidation considerably more complex than the uninhibited reaction.

The desired function of the added stabilizer is to prevent, or at least to retard, the inevitable deterioration of properties caused by reactions involving either the propagating radicals, $R\cdot$ or $RO_2\cdot$, or the peroxides that are formed as the primary oxidation products. However the compounds used as antioxidants are frequently capable of reacting in more than one way. For example, they may also react directly with molecular oxygen (which is in fact a biradical, $\cdot O_2\cdot$) and contribute to the formation of free radicals that are active in the initiation mechanism. If the antioxidant functions by reaction with $RO_2\cdot$ to interrupt the propagation sequence, a new radical derived from the antioxidant is formed which may either trap another $RO_2\cdot$ to complete the termination process or react in some way to reinitiate the reaction by a chain-transfer process. The following reaction sequence includes the various ways in which antioxidants may participate in the mechanism of retarded thermal autoxidation:[14]

Initiation

(peroxide decomposition)	$n\mathrm{ROOH} \rightarrow \mathrm{RO}\cdot, \mathrm{RO_2}\cdot, \ldots$
(O_2 attack on hydrocarbon)	$\mathrm{RH} + \mathrm{O_2} \rightarrow \mathrm{R}\cdot + \mathrm{HO_2}\cdot$
(O_2 attack on antioxidant)	$\mathrm{AH} + \mathrm{O_2} \rightarrow \mathrm{A}\cdot + \mathrm{HO_2}\cdot$

Propagation

(same as uninhibited)	$\mathrm{RO_2}\cdot + \mathrm{RH} \rightarrow \mathrm{ROOH} + \mathrm{R}\cdot$
	$\mathrm{R}\cdot + \mathrm{O_2} \rightarrow \mathrm{RO_2}\cdot$

Chain Transfer

(with antioxidant)	$\mathrm{RO_2}\cdot + \mathrm{AH} \rightarrow \mathrm{ROOH} + \mathrm{A}\cdot$
(in part)	$\mathrm{A}\cdot + \mathrm{RH} \xrightarrow{\mathrm{O_2}} \mathrm{AO_2H} + \mathrm{RO_2}\cdot$

Termination

(by antioxidant)	$\mathrm{RO_2}\cdot + \mathrm{A}\cdot \rightarrow \mathrm{RO_2A}$
	$2\mathrm{A}\cdot \rightarrow \mathrm{A{-}A}$
(as in uninhibited autoxidation)	$2\mathrm{RO_2}\cdot \rightarrow$ nonradical products
	$\mathrm{RO_2}\cdot + \mathrm{R}\cdot \rightarrow \mathrm{ROOR}$
	$2\mathrm{R}\cdot \rightarrow \mathrm{R{-}R}$

Peroxide Destruction

(preventive antioxidant)	$\mathrm{ROOH} + \mathrm{AH} \rightarrow$ nonradical products

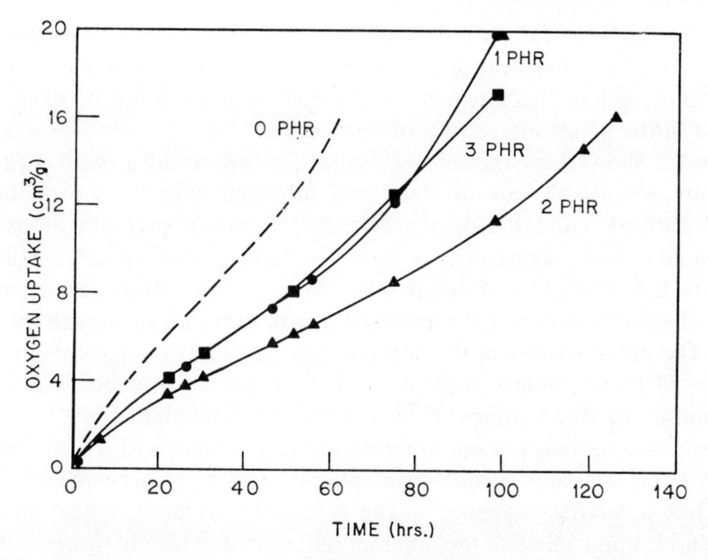

Fig. 2.1 Oxygen absorption of hevea black stocks containing phenyl-2-naphthylamine (90°C, 760 mm O_2).[21]

The reaction scheme presented above for retarded autoxidation includes four ways in which the antioxidant may participate in the mechanism:

1. Initiation by direct attack of oxygen on the antioxidant to produce chain-initiating free radicals.
2. Chain transfer with the antioxidant in which the radical derived from the antioxidant reacts in some way to reform a propagation free radical.
3. Termination by hydrogen donation to $RO_2\cdot$ as in the first step of chain transfer, followed by reaction of the antioxidant radical with a second $RO_2\cdot$, thus terminating two kinetic chains per molecule of antioxidant consumed.
4. Prevention of peroxide initiation by decomposing hydroperoxides to form stable products rather than free radicals.

Complete inhibition of oxidation in polymers by the addition of stabilizers against thermal autoxidation is not to be expected. Rather one expects to observe a marked reduction in rate of oxidation during the so-called induction period, which is in reality a period of retarded autoxidation extending over the useful life of the material. The rate of oxidation in this period is decreased by increasing the amount of stabilizer up to an optimum concentration. Excess stabilizer beyond the optimum usually acts as a prooxidant increasing the observed rate of oxygen absorption as illustrated in Figure 2.1.

The rate of the retarded autoxidation in the presence of antioxidants is sensitive to increased concentration of oxygen in contrast to the uninhibited reaction in which the rate is independent of oxygen concentration except at very low oxygen concentrations or with very reactive materials at higher temperatures, where the rate may be limited by availability of oxygen as discussed in the preceding section of this chapter.

Figure 2.2 shows how the rate of oxygen absorption of a natural rubber vulcanizate containing carbon black and inhibited with 2,2,4-trimethyl-6-phenyl-1,2-dihydroquinoline increases with the partial pressure of oxygen when heated in various mixtures of oxygen and nitrogen at a total pressure of 1 atm. The use of samples of different thicknesses in this study demonstrated that the observed rate was not limited by rate of diffusion of oxygen into the sample. The observed rate in the constant-rate stage in this case was directly proportional to the square root of the oxygen partial pressure, consistent with initiation by direct attack of oxygen on the antioxidant.

In other cases the relation was more complex, consistent with a combination of first power and square-root terms plus initiation by peroxide or other process not involving oxygen.[21,22] The variations evidently reflect changes in the relative importance of the different initiation and termination reactions included in the above reaction scheme. Indeed when a rate equation was derived from this sequence of reactions, in which steady-state and other simplifying conditions were assumed, it was found to include both first-power and square-root terms involving oxygen along with terms independent of oxygen, consistent with the various relations actually observed.[21] Let us therefore take a critical look at the possible initiation, propagation, and termination reactions listed and seek to evaluate their relative degrees of importance in the mechanism of retarded autoxidation.

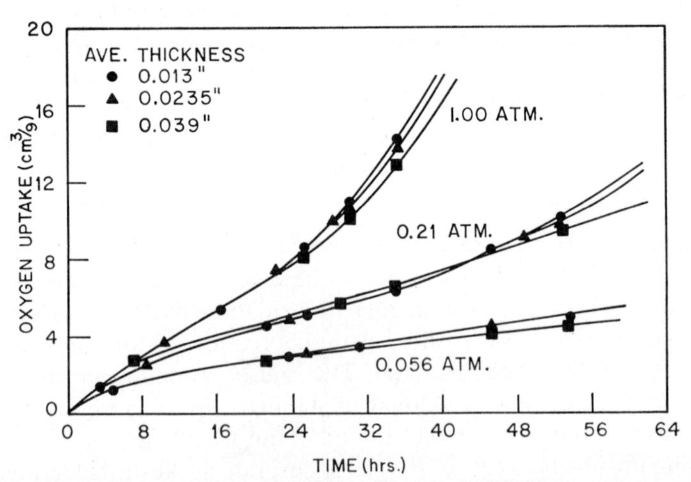

Fig. 2.2 Oxygen absorption of inhibited hevea black stock at 100°C and various partial pressures of oxygen with nitrogen at total pressure of 760 mm.[22]

a. Initiation. The dependence of the rate of retarded thermal autoxidation on oxygen concentration clearly suggests an initiation reaction involving direct oxygen attack either on the organic substrate or the added stabilizer. Direct attack of oxygen on the polymer, either by addition to a reactive double bond or by abstraction of a reactive hydrogen is possible, but it would not be expected to occur as readily as the oxidation of an amine or phenolic antioxidant. Furthermore the observed prooxidant effect of higher concentrations of such antioxidants above the optimum concentration clearly indicates the formation of chain-initiating radicals by reactions such as the following:

$$AH + O_2 \rightarrow A\cdot + HO_2\cdot$$

$$A\cdot + O_2 \rightarrow AO_2\cdot$$

The occurrence of such a hydrogen-abstraction reaction associated with the prooxidant effect at higher concentrations of antioxidant and at higher reaction temperatures has been demonstrated by kinetic deuterium isotope effects observed under these conditions. In studies involving deuterated amines and phenols in styrene–butadiene rubbers[23,24] and in purified polyisoprene[25] the normal isotope effects observed show a more rapid oxidation in the presence of the deuterated species, Ar_2ND and ArOD, since they are less efficient antioxidants than the corresponding >N—H and —O—H compounds. At higher antioxidant concentrations and at higher temperatures, however, the net isotope effect observed was in the reverse direction, and the samples containing the deuterated antioxidants, AD, showed a slower rate, since they are also less reactive toward elemental oxygen than the normal AH compounds in the above initiation reaction. This behavior is illustrated in Figure 2.3[25] in the case of *N*,*N*′-diphenyl-*p*-phenylenediamine at a concentration of 13.2×10^{-5} mole/g (3 pph) in purified polyisoprene, which shows an isotope effect $k_D/k_H = 0.64$, which corresponds to a reversed isotope effect $k_H/k_D = 1.56$.

Figure 2.3 reveals the presence of two stages of retarded autoxidation that occur prior to the start of the autocatalytic stage (not shown). The change in rate from a slower initial stage to a second stage of more rapid oxidation has been shown to be related to peroxide concentration by taking a sample oxidized at 90°C beyond the point at which the rate increased, and heating to a higher temperature (150°C) in a nitrogen atmosphere for several hours to decompose peroxides before returning it to the oxidation apparatus. Two stages of oxidation were again observed with the break between the two stages occurring at approximately the same oxygen uptake on reoxidation of the same sample. The data for both an amine and a phenolic antioxidant in purified *cis*-1,4-polyisoprene are presented in Table 2.1.[25] Samples heated

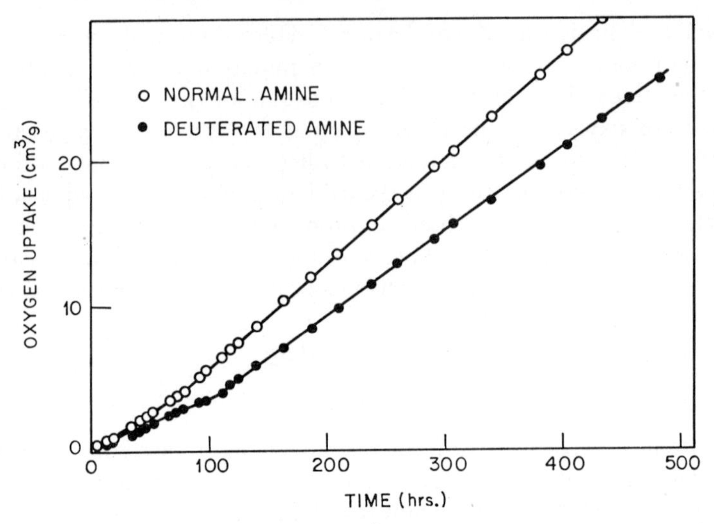

Fig. 2.3 Oxidation of polyisoprene (75°C, 760 mm O_2); inhibitor, *N,N'*-diphenyl-*p*-phenylene-diamine, 3 pph.[25]

in nitrogen at the oxidation temperature of 90°C attained the higher rate soon after they were returned to the oxidation apparatus, showing that little peroxide decomposition was occurring at that temperature.

Recent unpublished work[26] on quantitative determinations of hydroperoxide concentration as a function of time of oxidation showed that hydroperoxide accumulated during the first slow stage of retarded autoxidation reaching a maximum value near the break in the oxygen-absorption curve. Thus little contribution of peroxide to initiation is involved in the initial stage of retarded autoxidation of purified polyisoprene containing a phenolic antioxidant. During the second stage, however, peroxides decomposed faster than they were formed and the increased rate shows that this decomposition contributes significantly to initiation of new kinetic chains. Some product of oxidation of the polymer or of the antioxidant evidently accelerates the rate of hydroperoxide decomposition, but the identity of this material and the mechanism of the reaction has not yet been established.

It thus appears that the most important initiation reaction in the first stage of retarded autoxidation is the reaction of the antioxidant itself with molecular oxygen to form radicals capable of initiating the autoxidation of the polymer. Peroxide initiation begins to contribute significantly in the second stage of more rapid, but still retarded autoxidation. If the reaction is continued to the point at which the added stabilizer concentration has been reduced to an ineffective level, a further increase in rate is observed as the reaction enters the autocatalytic stage and the hydroperoxide concentration starts to rise again from a minimum value.[26] Only peroxide initiation will be significant in this final autocatalytic stage of thermal autoxidation.

Table 2.1 Oxygen Absorption of Preoxidized *cis*-1,4-Polyisoprene[25]

Sample	Treatment	Rate at 90°C, ml O_2/(g)(hr) (22°C, 760 mm) Initial Stage	Second Stage	Absorption at Break, ml O_2/g (22°C, 760 mm)
N-Phenyl-2-naphthylamine, 8.82 × 10^{-5} mole/g	Initial oxidation.	0.180	0.256	4.0
	Oxidized, then heated 26.9 hr at 150°C in N_2 atm.	0.200	0.296	2.8
		0.253	0.426	4.4
	Oxidized, then heated 83.5 hr at 90°C in N_2 atm	0.266	0.449	1.0
		0.372	0.540	1.1
2,6-di-*t*-butyl-4-methylphenol, 4.41 × 10^{-5} mole/g	Initial oxidation	0.060	0.092	2.5
	Oxidized, then heated 28.5 hr at 150°C in N_2 atm.	0.070	0.089	2.2
		0.080	0.110	1.7
	Oxidized, then heated 86.9 hr at 90°C in N_2 atm.	0.076	0.117	0.3
		—	0.124	—

b. Propagation and Chain Transfer. The propagation cycle is the same for retarded autoxidation as for uninhibited autoxidation except for the additional possibility of chain transfer involving the antioxidant. The rate-controlling step is the reaction of $RO_2\cdot$ with the polymer to form $R\cdot$, which rapidly reacts with O_2 to reform $RO_2\cdot$ and continue the process. The length of the kinetic chain, which can be defined as the number of oxygen molecules reacting per initiation step, depends upon molecular structure, temperature, oxygen pressure, and the presence or accumulation of inhibitors.[7] Very long kinetic chains are to be expected in uninhibited autoxidation, with the reaction of more than a hundred molecules of oxygen from a single initiation before termination occurs.[17] In the presence of an efficient stabilizer of the chain-breaking type, the kinetic chain length will be reduced to a very low value approaching unity. The dependence of rate of propagation in autoxidation on polymer structure is discussed in the next section of this chapter.

The mechanism of retarded autoxidation presented earlier in this discussion included a chain-transfer process involving the reaction of a chain-propagating peroxy free radical with an antioxidant containing a readily extractable hydrogen, and the subsequent reaction of the antioxidant-derived free radical with the polymer and oxygen to reform a peroxy radical. The effective length of the kinetic chain is thus extended as though no interruption had occurred. The extensive kinetic and product studies of retarded autoxidation which Bickel and Kooyman[27] carried out in the presence of amine and phenolic inhibitors showed a number of different kinetic relationships relative to inhibitor concentration. In certain cases the observed relation was interpreted as consistent with a chain transfer of the type:

$$RO_2\cdot + AH \rightarrow RO_2H + A\cdot$$

$$A\cdot + RH \rightarrow AH + R\cdot$$

The evidence cited above in the discussion of initiation by direct reaction of antioxidant with oxygen to form chain-initiating free radicals clearly supports the chain-transfer mechanism, since the same antioxidant derived free radical would be formed. The direct reaction of this radical with polymer to regenerate the antioxidant by abstraction of hydrogen with formation of a new radical, $R\cdot$, as in the above sequence is possible with a very reactive material. However, it seems more probable that the antioxidant radical would react with oxygen either by addition or by abstraction of a second hydrogen with the formation of a peroxy radical ($AO_2\cdot$ or $HO_2\cdot$) capable of reinitiating the propagation cycle as in the following scheme:

$$A\cdot + O_2 \rightarrow AO_2\cdot \text{ (or } HO_2\cdot + \text{oxidized A)}$$

$$AO_2\cdot \text{ (or } HO_2\cdot) + RH \rightarrow AO_2H \text{ (or } HO_2H) + R\cdot$$

$$R\cdot + O_2 \rightarrow RO_2\cdot$$

c. Termination. The relative importance of the possible termination reactions in thermal autoxidation varies depending upon the nature and concentration of the stabilizer, as well as such factors as temperature, oxygen concentration, and polymer structure. With an effective chain-breaking antioxidant at optimum or higher concentration, essentially all kinetic chains will be intercepted by the stabilizer. In the case of the hydrogen donor types, the following reaction sequence has been established[4,6,14,25] and will be discussed in greater detail in a Section E.2, which deals with mechanisms of stabilization against thermal oxidation by chain-breaking antioxidants.

$$RO_2\cdot + AH \rightarrow RO_2H + A\cdot$$

$$RO_2\cdot + A\cdot \rightarrow RO_2A$$

$$2A\cdot \rightarrow A\text{—}A$$

The hydrogen transfer reaction is the kinetically significant step as shown by kinetic deuterium isotope effects observed[23–25] and is the same as the first step in the chain transfer as described above. At or below the optimum antioxidant concentration, the new radical derived from the antioxidant will mainly combine with a second $RO_2\cdot$ to terminate a second kinetic chain, rather than to react with O_2 to complete a chain transfer. Termination by combination of two antioxidant radicals, as in the third reaction above, may also occur.

In uninhibited autoxidation, or with preventive type antioxidants that interfere with initiation rather than participating in termination, the only termination reaction of importance will be the bimolecular combination of two peroxy free radicals. Under some conditions, as previously described, terminations involving coupling of carbon radicals, R·, with each other and with $RO_2\cdot$ may also be involved. Even in the presence of chain-breaking types of antioxidants at lower concentrations, the bimolecular combination of $RO_2\cdot$ radicals characteristic of uninhibited autoxidation may contribute significantly to termination.

The nature of the bimolecular termination reaction of peroxy radicals is now well established.[28] Russell[29] first proposed the formation of a tetroxide as an unstable intermediate that decomposes via a cyclic transition state in the reaction of secondary and primary peroxy radicals to form the observed products. A kinetic isotope effect was observed in the termination step for the oxidation of ethylbenzene and various deuterated derivatives at 60°C with initiation by α,α′-azodiisobutyronitrile (AIBN). The α-phenylethylperoxy free radicals interacted 1.9 times faster than the α-deutero derivative, indicating that a C–H bond is broken in the transition state of the termination reaction:

$$2\,(CH_3)(C_6H_5)CHO_2\cdot \longrightarrow \left[(CH_3)(C_6H_5)C(H)\text{–}O\text{–}O\text{–}O\text{–}O\text{–}CH(CH_3)(C_6H_5)\right] \longrightarrow$$

$$(CH_3)(C_6H_5)C{=}O + O_2 + HO\text{–}CH(CH_3)(C_6H_5)$$

This mechanism would account for the observed lower activation energy and more rapid termination for *sec*-$RO_2\cdot$ as compared to *tert*-$RO_2\cdot$, as well as the observed deuterium isotope effect. Rate constants ($2k_t$) for bimolecular combinations of peroxy radicals increase[28] in the order:

$$tert\text{-}RO_2\cdot(10^4) < sec\text{-}RO_2\cdot(10^6) < prim\text{-}RO_2\cdot(10^8)$$

The slower reaction in the case of the *tert*-peroxy radical may be attributed to the formation of a more stable tetroxide that is unable to decompose by the cyclic mechanism illustrated above for a *sec*-peroxy radical. Formation of a tetroxide intermediate has been established by Bartlett and co-workers,[30] who have monitored the *t*-$BuO_2\cdot$ radical concentration present in equilibrium with the tetroxide at very low temperatures by the use of electron paramagnetic resonance spectroscopy. The decomposition of this tetroxide requires a significant activation energy for decomposition (10–15 kcal/mole), and below −85°C equilibrium is attained and can be shifted by temperature change.

$$2t\text{-}RO_2\cdot \leftrightarrows t\text{-}ROOOOt\text{-}R$$

$$\underbrace{\boxed{t\text{-}RO\cdot + O_2 + \cdot Ot\text{-}R}}_{\text{Cage}} \begin{array}{l} \xrightarrow{\text{combination}} t\text{-}ROOt\text{-}R + O_2 \\ \xrightarrow[\text{diffusion}]{} 2t\text{-}RO\cdot + O_2 \end{array}$$

At −70°C the more stable trioxide, which breaks down on subsequent warming to −30°C, is produced.

$$t\text{-}RO\cdot + t\text{-}RO_2\cdot \underset{-30°C}{\overset{-70°C}{\rightleftarrows}} t\text{-}ROOOt\text{-}R$$

In the autoxidation of model compounds or polymers that can form tertiary peroxy radicals, the formation of significant amounts of tertiary alkoxy radicals as indicated above will result in some β-scission of *t*-$RO\cdot$ to form other radicals which will react with oxygen to form new peroxy radicals. Thus a certain fraction of chain terminations involving primary or secondary peroxy radicals is to be expected along with termination involving the more numerous tertiary peroxy radicals.

Electron spin conservation rules suggest that the oxygen evolved in the self-reaction of secondary peroxy radicals should be formed in an excited singlet state rather than the normal triplet biradical state, $\cdot O_2\cdot$, if the tetroxide is an intermediate and decomposes via the Russell cyclic transition state described above. A small amount of oxygen evolved from the *s*-butyl peroxy radical coupling reaction has been trapped with 9,10-diphenylanthracene to form the transannular peroxide.[31] This reaction is considered to be a diagnostic test for singlet oxygen and thus supports the Russell mechanism for termination in the case of *s*-$RO_2\cdot$ radicals.

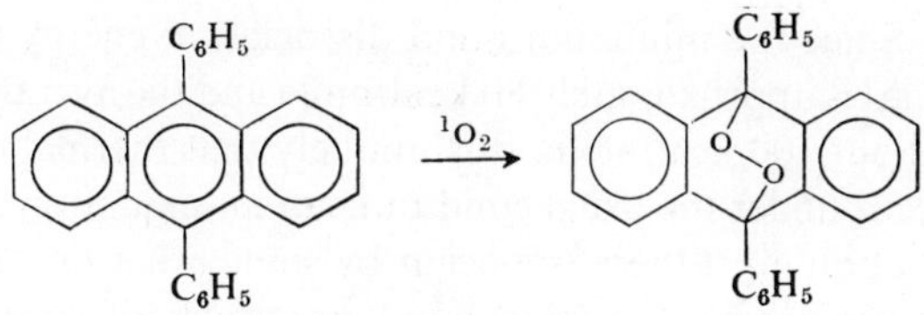

C_6H_5 C_6H_5 $\xrightarrow{^1O_2}$ C_6H_5 O O C_6H_5

Under similar conditions no transannular peroxide can be detected in the self-reaction of *t*-butyl peroxy radicals or in the case of an $HO_2\cdot$ free radical. In these cases the O_2 is presumably formed in the ordinary triplet ground state.

C. DEPENDENCE OF OXIDATIVE STABILITY ON POLYMER STRUCTURE

Polymeric materials exhibit a wide range of susceptibilities to thermal autoxidation that can be explained in part by differences in structure. Physical factors such as the degree of crystallinity and permeability to oxygen also introduce considerable variations among the polymers and even in a given polymer under different conditions. Stabilization against thermal oxidation can be enhanced by modification of the polymer structure to eliminate the more reactive sites and to introduce groups that contribute to oxidative stability. Addition of stabilizers can compensate for the inherent susceptibility of most organic materials to oxidation and also provide additional protection for polymers that are relatively resistant to oxidation under the usual conditions of fabrication and use.

1. Relation of Structure and Reactivity in Autoxidation

The extensive literature on autoxidation noted in the introduction to this chapter[1–19] provides a background of knowledge on relative reactivities of both model compounds and polymers. The relation of structure and reactivity in oxidation of elastomers has received special attention.[18] One can readily formulate qualitative generalizations that are useful guides in selecting polymers that would be expected to exhibit good oxidation resistance. For example, saturated polymers are more resistant to oxidation than polymers of dienes that retain unsaturation in the polymer chain. Branched polymers are more reactive than linear, and crystalline polymers are usually resistant to oxidation at temperatures lower than the melting point. The presence of substituent groups and the introduction of crosslinks also alter the relative oxidizability of the material.

Bond dissociation energies are helpful in considering the reactivity of various structures. Since the minimum bond dissociation energy for normal C–C bonds in alkanes is approximately 80 kcal/mole and the average value for the C–H bond is about 100 kcal/mole, it is unlikely that thermal cleavage of these bonds will occur under the usual conditions of autoxidation at ordinary temperatures.[7] Initiation must therefore occur by some other process, such as cleavage of the lower energy peroxide bond or attack of oxygen on the reactive N–H or O–H bond of aryl amines or phenols added as antioxidants in the case of retarded autoxidation as described above. Once free radicals are formed to initiate the reaction chain, abstraction of suitably reactive hydrogens does occur in the chain-propagating reaction:

$$RO_2\cdot + RH \rightarrow ROOH + R\cdot$$

Bond dissociation energies are thus particularly significant in this reaction which controls the rate of propagation. Table 2.2 gives the bond dissociation energies for the C–H bond in a number of model compounds. Reactivity increases as the bond dissociation energies decrease in the order:

$$\text{Primary, } RCH_2\text{—H (98 kcal)} > \text{secondary, } R_2CH\text{—H (94.5 kcal)} > \text{tertiary, } R_3C\text{—H (91 kcal)}$$

Table 2.2 Bond Dissociation Energies[32]

R—H	$D_{R—H}$ (kcal/mole)
CH_3—H	104
C_2H_5—H	98
n-C_3H_7—H	98
$(CH_3)_2CH$—H	94.5
$C_2H_5CH(CH_3)$—H	94.6
$(CH_3)_3C$—H	91
c-C_5H_9—H	93
c-C_6H_{11}—H	94
CH_2=CH—H	104
CH_2=$CHCH_2$—H	85
C_6H_5—H	104
$C_6H_5CH_2$—H	85
$CNCH_2$—H	86
$CH_3C(O)$—H	88
CH_3COCH_2—H	92
$CH_3CH(OH)$—H	90

Table 2.3 Relations Between Bond Dissociation and Activation Energies[2,5]

Alkene	E_p (kcal/mole)	ΔE_p (exptl)	$\Delta D_{C-H} = \Delta E_p/0.4$ (kcal/mole)
$CH_2{=}CHCH_2{-}H$	13.5	0	0
$CH_2{=}CHCHR{-}H$	11.5	2.0	5
$RCH{=}CHCHR{-}H$	10.5	3.0	8
$CH_2{=}CHCHC_6H_5{-}H$	10	3.5	9
$RCH{=}CHCH{-}H$ (with $CH{=}CHR$ on the CH)	6	7.5	19

The effect of unsaturation on reactivity of adjacent C–H bonds is evident in the higher bond dissociation energies for vinylic and aromatic hydrogens (104 kcal), and the marked decrease in bond dissociation energy for the highly reactive allylic and benzylic hydrogens (85 kcal/mole). The large decrease in bond dissociation energy when the radical is formed on a carbon α to a double bond or an aromatic ring can be attributed to the lower energy state resulting from the dissociation to form a radical stabilized by resonance delocalization with the π-electron system. Similar activation is observed when hydrogen is α to other unsaturated systems such as carbonyl, $RCOCH_2{-}H$, and nitrile, $RCH(CN){-}H$, or to an atom bearing unshared electrons as in an alcohol or ether, $RCH(OR){-}H$.

The relation between activation energies for propagation and changes in bond dissociation energies has been investigated in the autoxidation of olefins.[1,2] The compounds listed in Table 2.3 are arranged in the order of increasing reactivity as the energy of activation for propagation, E_p, and bond dissociation energies decrease. The magnitude of the change, ΔE_p, in these experimentally observed activation energies and the corresponding decrease, ΔD_{C-H}, in bond dissociation energies are shown relative to the allylic C–H bond in propylene. The activating effect of alkyl, aryl, and additional unsaturation on the reactivity of the α hydrogen is apparent.

Consideration of the differences in bond dissociation energies and the related energies of activation enable us to explain the relative susceptibility of most polymers to thermal autoxidation. Thus, linear polyethylene (polymethylene) is the most stable among the aliphatic hydrocarbon polymers, and the presence of tertiary C–H groups along the chain in branched polyethylene and polypropylene results in increased susceptibility to oxidation. The presence of unsaturation in polyisoprene and polybutadiene elastomers activates the allylic hydrogens as potentially reactive sites. The presence of 1,2 and 3,4 monomer units in such polymers gives rise to even more active tertiary allylic hydrogen sites for reaction.

From the data of Tables 2.2 and 2.3, one would expect a tertiary benzylic hydrogen to be at least as reactive as allylic hydrogen and more reactive

Table 2.4 Bond Dissociation Energies in Radicals[32]

Bond in Radical	$D_{C-H \text{ or } C-C}$ (kcal/mole)
$CH_3CH_2CH(\dot{C}H_2)—H$	36
$CH_3CH_2\dot{C}HCH_2—H$	40
$(CH_3)_2C(\dot{C}H_2)—H$	35
$(CH_3)_2\dot{C}CH_2—H$	42
$CH_3CH(\dot{C}H_2)—CH_3$	24
$CH_3\dot{C}HCH_2—CH_3$	27
$\dot{C}H_2CH_2—C_2H_5$	21
$\dot{O}CH_2—H$	22
$\dot{O}CH_2—CH_3$	13
$\dot{O}C(CH_3)_2—CH_3$	6

than the tertiary hydrogen of polypropylene. In fact, as noted in Chapter 1, polystyrene is considerably more resistant to oxidation than polypropylene. Steric and inductive effects of the phenyl group have been suggested as explanations, although the anomaly has not yet been fully explained.

Once a free radical is formed on a polymer chain, C–H and C–C bonds on the carbon adjacent to the radical site are subject to cleavage. The much lower bond dissociation energies required to break these bonds are shown by the bond dissociation energies calculated from the heats of formation for a number of simple radicals listed in Table 2.4.[32] The reduction in the energy required to break these particular bonds results from the gain in energy when a new double bond is formed either to carbon as in the first example below, or to oxygen in the case of alkoxy radicals undergoing β-scission as in the second example.

$$RCH(R')—\dot{C}(R)—CH(H)R \xrightarrow{-H\cdot} RR'CH—C(R){=}CHR$$

$$RCH(R')—\dot{C}(R)—CH(H)R \xrightarrow{-R'\cdot} RCH{=}C(R)—CH_2R$$

$$R—C(H)(R)—O\cdot \xrightarrow{-R\cdot} R—C(H){=}O$$

$$R—C(H)(R)—O\cdot \xrightarrow{-H\cdot} R—C(R){=}O$$

Chain scission is thus an expected degradation reaction in polymers whenever alkyl or alkoxy free radicals are formed. The R· radicals formed in autoxidation rapidly combine with O_2 to form $RO_2\cdot$ although some scission probably occurs in the case of tertiary carbon radicals. Reaction of $RO_2\cdot$ with double bonds to form epoxides is one source of alkoxy free radicals,[33–36] and termination by coupling of two $RO_2\cdot$ gives O_2 and $2RO\cdot$ as previously described. Decomposition of hydroperoxide formed on the polymer is another source of alkoxy radicals. Thus scission reactions may be associated with all three processes in autoxidation–initiation, propagation, and termination.

The relation between olefin structure and autoxidizability for a series of model compounds related in structure to natural rubber has been compared quantitatively by evaluating the quantity $k_p k_t^{-1/2}$ based on the assumption that the rate of combination of two alkenyl peroxy radicals in the termination step might be expected to change little with olefin structure. Although this assumption is not in fact generally true,[17,28] the relative ease of hydrogen atom removal obtained on this basis is of value in calculating the approximate change in rate to be expected with various structural changes.

Bolland's rules[1,17] relative to the effect of substitution in propene $CH_3CH{=}CH_2$ upon the hydrogen abstraction reaction at 45°C follow:

1. Replacement of one or two hydrogen atoms in the CH_3 group and/or the CH_2 group by alkyl groups increases reactivity by $(3.3)^n$, where n is the total number of substituents.
2. Replacement of a hydrogen atom in the CH_3 group by a phenyl group increases reactivity 23-fold.
3. Replacement of a hydrogen atom in the CH_3 group by an alk-1-enyl group increases reactivity 107-fold.
4. The value of k_p for an α-methylenic group contained in a cyclic structure is 1.7 that for the same group contained in an analogous acyclic structure.

Application of the above rules to the predominant structures present in (2.1)—natural rubber (*cis*-1,4-polyisoprene), (2.2)—1,4-polybutadiene, (2.3)—styrene–butadiene copolymer (SBR), and (2.4)—polyisobutylene–isopropene copolymer (butyl rubber) predicts oxidizability ratios of 60:20:1 for structures (2.1):(2.2 or 2.3):(2.4).[37]

```
     C           C           C
     |           |           |
 —C—C=C—C—C—C=C—C—C—C=C—C—                 (2.1)
 —C—C=C—C—C—C=C—C—C—C=C—C—                 (2.2)

             C6H5
              |
 —C—C=C—C—C—C—C—C=C—C—C—C=C—C—             (2.3)

  C     C     C        C     C
  |     |     |        |     |
 —C—C—C—C—C—C—C—C=C—C—C—C—                 (2.4)
  |     |     |        |     |
  C     C     C        C     C
```

These ratios would apply only to the pure polymers free from impurities, catalysts, or inhibitors, and in the amorphous state. The numbers are only of semiquantitative significance in view of the assumptions noted above, but they correctly indicate the greater reactivity of natural rubber as compared to polybutadiene and SBR, as well as the much lower reactivity of butyl rubber in autoxidation. The effect of crystallinity in rubbers is apparent in the decreased reactivity of gutta percha (*trans*-1,4-polyisoprene) to oxidation at ordinary temperatures as compared to natural rubber, since the trans polymer is crystalline at room temperature and the cis polymer is not. Both rubbers oxidize at comparable rates above 50°C where both are essentially noncrystalline.[37]

2. Polymers that Exhibit Inherent Oxidative Stability

It is clear from the preceding discussion of the relation of structure and reactivity that, if one desires to synthesize a polymer that is resistant to oxidation, it is important to avoid both unsaturation and branching. Olefinic double bonds activate the allylic hydrogens on the α carbon atom toward abstraction by peroxy free radicals. Other unsaturated functional groups such as carbonyl, $>C=O$, and nitrile, $-C\equiv N$, also activate hydrogen on the adjacent carbon by their ability to stabilize the resulting free radical. Introduction of alkyl groups on a linear polymer chain forms tertiary $\rightarrow C-H$ groups, which are more reactive than secondary, $>CH_2$, or primary, $-CH_3$, groups. Introduction of an aryl group on the polymer chain not only forms a tertiary $\rightarrow C-H$ site but also activates the position by resonance stabilization of the resulting benzylic free radical. The presence of heteroatoms such as oxygen, sulfur, and nitrogen also activates an adjacent C–H group, since these atoms with unshared electrons can stabilize an odd electron on the carbon bearing the heteroatom.

Linear polyethylene (polymethylene) thus represents a relatively stable aliphatic hydrocarbon polymer structure. The methylene groups are still susceptible to hydrogen abstraction, however, by reaction with $RO\cdot$ and $RO_2\cdot$ radicals. Consequently complete replacement of hydrogen by fluorine as in polytetrafluoroethylene provides a means of obtaining a polymer that is quite inert to oxidation.

Introduction of chlorine also reduces the susceptibility of a polymer to oxidative attack. Thus poly(vinyl chloride) and poly(vinylidene chloride) are more resistant to oxidation than polyolefins, but dehydrohalogenation to form unsaturation is a major problem, discussed in the next chapter. Similarly, the chlorine in neoprene, a polymer of 2-chloro-1,3-butadiene, provides enhanced oxidative stability as compared to polybutadiene and natural rubber. Partial saturation of diene rubbers by addition of either HCl or Cl_2 also increases their resistance to oxidation. The presence of halogen on a carbon also bearing hydrogens does, however, activate that position to hydrogen abstraction, since the unshared electrons of the halogen stabilize an odd electron on the same carbon. The effect of halogen on flame retardance is discussed in Chapter 7.

Polystyrene exhibits surprising stability relative to other hydrocarbon polymers in view of the anticipated activating effect of the phenyl group upon the tertiary benzylic hydrogen. It has been suggested that the lack of reactivity arises partly from shielding effects of the bulky phenyl group and partly from loss in resonance stabilization of the radical that would be formed because of unfavorable orientation resulting from crowding of the phenyl groups on every second carbon of the chain.[38] It is also possible that small amounts of phenol could be formed by decomposition of benzylic hydroperoxide by a polar mechanism to provide some antioxidant protection. In contrast polyisopropylstyrene is highly vulnerable to oxidation, consistent with the expected activating effect of the phenyl group.[39]

Completely aromatic polymers are quite stable, since hydrogens on the aromatic ring, like vinylic hydrogens, are unreactive toward abstraction by alkoxy and peroxy free radicals under ordinary conditions. An interesting example is poly(*p*-phenylene) prepared by polymerization of benzene by a combination of a Lewis acid and an oxidizing agent, for example, $AlCl_3$ plus $CuCl_2$.[40,41] Ferric chloride alone gives polymer with benzene, since the catalyst is both a Lewis acid and an oxidant. The mechanism proposed is an electrophilic substitution plus oxidative dehydrogenation to give an aromatic polymer consisting mainly of *para*-phenylene units:

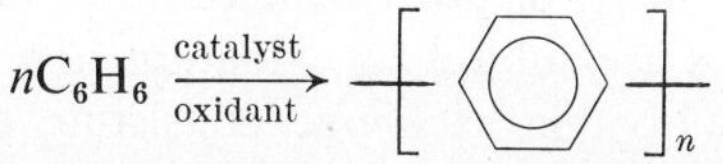

$$nC_6H_6 \xrightarrow[\text{oxidant}]{\text{catalyst}} \left[-C_6H_4- \right]_n$$

A number of fused ring polymers of the aromatic–heterocyclic type have been synthesized and found to exhibit unusually good thermal and radiation stability, along with inherent oxidative stability. These and other polymers with high thermal stability are discussed in Chapter 3, and the interrelation between thermal degradation and thermal oxidation is considered in more detail in the final section of that chapter. The polybenzimidazoles formed by condensation of polycarboxylic acids with aromatic polyamines are an important example of this type of polymer:[42]

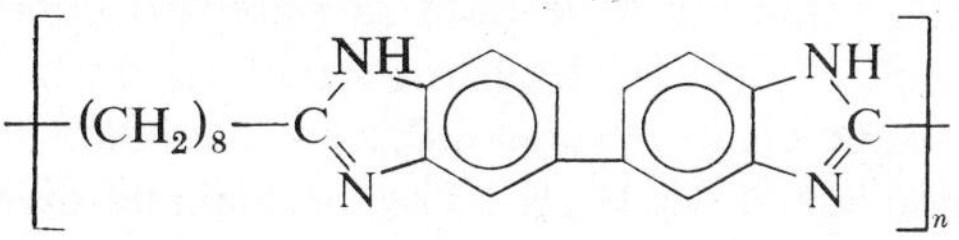

$$\left[-(CH_2)_8-C(=N-)(NH-)\text{—bibenzimidazole—}C(=N-)(NH-)- \right]_n$$

Ladder polymers are an interesting structural type in which two parallel polymer chains are connected at frequent and usually regular intervals by short chains, which serve as crosslinks corresponding to the rungs of a ladder. These polymers may show unusual resistance to degradation of properties even though the structures involved would be expected to exhibit normal reactivity toward oxygen. Chain scissions will occur as a result of oxidative attack just as in a comparable linear polymer, but the adjacent crosslinks will prevent cleavage into segments and the tensile properties of the material will not be significantly degraded. Polymers with extended ladder structures have been prepared by acid-catalyzed cyclization reactions involving 1,2 and 3,4 polymers of conjugated diene monomers and also by intramolecular aldol condensations in the case of poly(methyl vinyl ketone) as illustrated:[43]

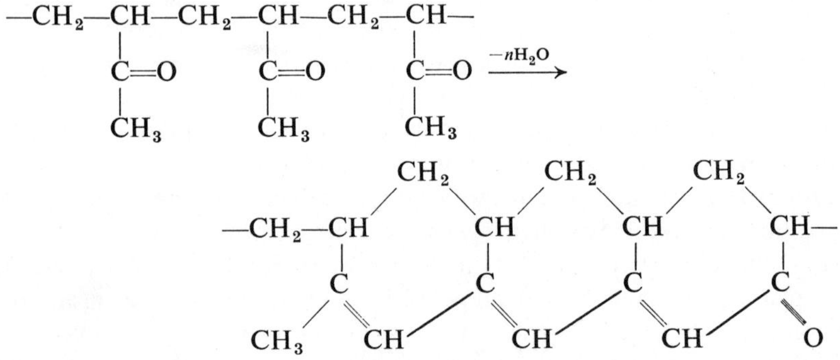

$$—CH_2—\underset{\underset{\underset{CH_3}{|}}{C=O}}{\underset{|}{CH}}—CH_2—\underset{\underset{\underset{CH_3}{|}}{C=O}}{\underset{|}{CH}}—CH_2—\underset{\underset{\underset{CH_3}{|}}{C=O}}{\underset{|}{CH}}— \xrightarrow{-nH_2O}$$

```
                 CH2        CH2        CH2
                /   \      /   \      /   \
   —CH2—CH           CH          CH          CH—
        |            |           |           |
        C            C           C           C
       / \\         / \\        / \\          \\
   CH3     CH ——     CH ——      CH ——          O
```

Ladder polymer

Organosilicon polymers are among the most stable toward oxidative degradation. This is especially true if the R groups in the polysiloxane structure illustrated below are phenyl or other oxidation-resistant groups. Even when the substituent groups are subject to oxidation, the reaction occurs in the side chain and the inorganic backbone consisting of (Si–O)_n chains will not be affected. Branching and crosslinking is also possible in silicone polymers by replacement of one or both of the R groups on any silicon atom by an —OSi— branch on the linear structure illustrated:

$$—\underset{R}{\overset{R}{Si}}—O—\underset{R}{\overset{R}{Si}}—O—\underset{R}{\overset{R}{Si}}—O—\underset{R}{\overset{R}{Si}}—O—$$

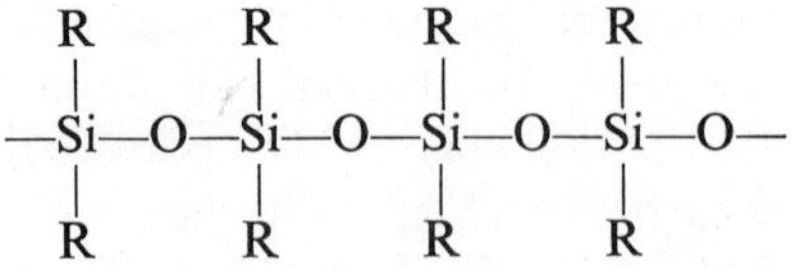

Polymers of unusual stability have been prepared by condensation of a phthalocyaninsilandiol, $PcSi(OH)_2$, to give a structure in which each silicon atom in the polysiloxane chain is surrounded by the bulky, but essentially planar phthalocyanine ligand, $C_{32}H_{16}N_8$. The thermal and oxidative stability

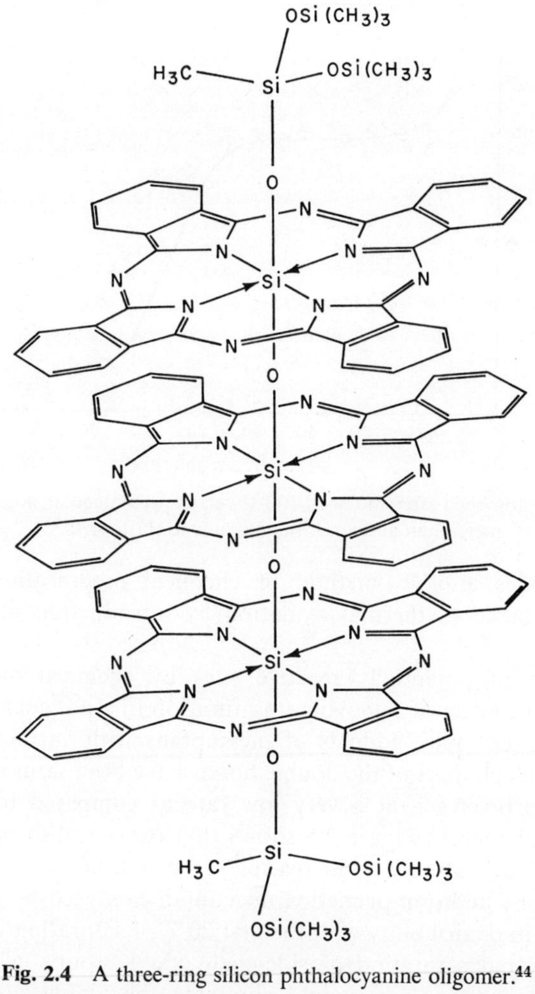

Fig. 2.4 A three-ring silicon phthalocyanine oligomer.[44]

of the ligand is thus combined with that of the inorganic backbone in the resulting structure which resembles a bulky string of large flat beads as illustrated for three units in Figure 2.4.

The resistance of polymers to oxidative degradation can thus be increased by a variety of chemical modifications. The introduction of crosslinks in polyethylene by radiation, or by use of peroxides, to form radicals that combine to form C–C links between chains will offset an equivalent amount of chain scission resulting from oxidation. The use of sulfur vulcanization to increase the strength and durability of natural and synthetic rubber is essential to produce a usable material. The grafting of side chains onto a

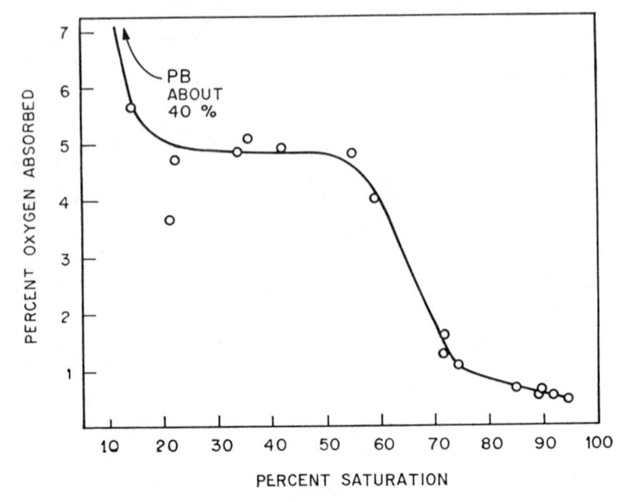

Fig. 2.5 Oxygen absorbed after 140 hr at 100°C versus percentage of saturation of methyl mercaptan adduct of polybutadiene elastomers.[45]

polymer provides another method of chemical modification to provide increased resistance to thermal oxidation by introduction of less reactive material.

Deactivation of potentially reactive sites by chemical modification is illustrated by the effect of removing unsaturation in a polymer such as polybutadiene. For example, adducts of mercaptans with various unsaturated elastomers in which most of the double bonds have been saturated have been shown to absorb oxygen at a very low rate as compared to the original unsaturated polymers.[45] Figure 2.5 shows that oxygen absorption after 140 hours at 100°C decreased with increasing saturation of the double bonds in polybutadiene by addition of methyl mercaptan catalyzed by peroxide. The large decrease in oxidizability with the first 20% of saturation was attributed to the more rapid saturation of side-chain vinyl groups as compared to internal double bonds. This would eliminate the highly reactive tertiary allylic hydrogen sites at the branch points by converting them to saturated tertiary hydrogens.[14] An isopropyl mercaptan adduct of polybutadiene showed even greater selectivity in saturation of the vinylic groups and a correspondingly greater resistance to oxidation.[45]

$$—CH_2—\overset{\displaystyle H}{\underset{\displaystyle CH=CH_2}{C}}—CH_2— \xrightarrow[\text{peroxide}]{\text{RSH}} —CH_2—\overset{\displaystyle H}{\underset{\displaystyle CH_2—CH_2—SR}{C}}—CH_2—$$

3. Limitations of Stabilization by Modification of Polymer Structure

Even though one can prepare polymers that exhibit inherent oxidative stability, as described in the preceding section, few of these polymers are used extensively. Higher costs limit them to specialty applications where their unique stability is required. Part of the increased cost is in the difficulty of fabrication attributable to decreased solubility and higher softening temperatures. Indeed such materials as poly(*p*-phenylene) and the phthalocyaninsilicone polymer cannot be molded by conventional methods.

Even more important than cost and processing difficulties are the effects of chemical modifications upon essential properties of the materials. Silicone rubbers are used for both high- and low-temperature applications, since their properties change little with temperature, but the strength properties at ordinary temperatures are quite inferior to natural and synthetic rubbers, which are used for most purposes in spite of their susceptibility to oxidative degradation. Saturation of the double bonds in elastomers such as polybutadiene or polyisoprene with halogen, hydrogen halide, or mercaptan does decrease oxidizability as expected; but these modifications inevitably change other essential properties, including the resilience that is essential for use as a rubber.

In view of these limitations upon stabilization by modification of polymer structure, it is clear that the use of added stabilizers to protect against thermal oxidation is essential for most polymeric materials. The choice of stabilizers must vary with the nature of the material and the conditions it will be subjected to in manufacture, storage and use. The various types of antioxidants and the mechanisms by which they function are discussed in detail later in this chapter.

4. Effect of Crystallinity on Oxidative Stability

The dependence of stability on polymer structure includes the effect of physical structure, or morphology, in addition to the effects of chemical structure and composition. Some polymers are essentially amorphous in that the molecular chains are highly disordered. Others show an ordered arrangement in which the polymer chains are oriented to form crystalline regions. The degree of orientation depends upon the linearity of the molecular chain, the regularity of the repeat units, and the nature of any functional groups or other substituents on the chain. Many polymers are semicrystalline with varying amounts of ordered (crystalline) and disordered (amorphous) regions of the solid polymer matrix as noted in Chapter 1. Since the density of individual polymers is directly related to their degree of crystallinity, the permeability to oxygen is altered and the oxidation reaction frequently becomes diffusion controlled in the crystalline regions.

For example, linear and branched polyethylene oxidize similarly at temperatures above the crystalline melting point as shown in Figure 1.1, but at 100°C the oxidation of the linear polymer is slower and the plot of oxygen absorption versus time levels off at a much lower total oxygen uptake. The cumulative oxygen uptake is approximately inversely proportional to the crystalline content of the polymers consistent with the inability of oxygen to permeate the crystalline phase so that oxidation of polyethylene is restricted to the amorphous regions. Crystalline polypropylene is similarly resistant to oxidation, but the more bulky chain of poly(4-methylpentene-1) gives a crystalline phase of about the same density as the amorphous material and both phases appear to oxidize at essentially the same rate.[7,46]

The crystalline phase of linear hydrocarbon polymers consists of crystallites in which the long chains are folded at the edges of the crystal to permit the chain to reenter the ordered region. Thus oxidation will occur not only in the disordered regions but also on the folds on the exposed surfaces of the crystalline regions. Degradation of the molecular chain may thus occur even in the case of molecules that are mainly in the crystalline phase, but oxidation will be limited in most cases to the portions of the chain that are in contact with the amorphous phase.

Whan an antioxidant is present in a polymer, crystallization from the melt would be expected to concentrate the antioxidant in the amorphous regions. This, of course, is a desirable situation, since the ordered region is usually protected by lack of permeation of oxygen and the antioxidant concentration is effectively increased in the amorphous phase where it is needed. It also follows, however, that relative efficiencies of antioxidants determined at elevated temperatures (above the crystalline melting point of the polymer) may fail to rate inhibitors correctly for applications at ordinary temperatures at which the polymer may be mainly crystalline.[7]

D. USE OF STABILIZERS AGAINST THERMAL OXIDATION

The selection of specific stabilizers for a particular polymeric material requires a knowledge of many factors. The environmental factors to which the material will be exposed in manufacture and in the intended use are the major consideration. Protection against photodegradation, ozone, ionizing radiation, and thermal degradation in the absence of oxygen are discussed in other chapters. In the case of stabilization against thermal autoxidation the major concern, of course, is, exposure to elevated temperatures, since the oxidation reaction is accelerated and many antioxidants that are effective at ordinary temperatures are sufficiently volatile to be lost under such conditions. Some applications require resistance to laundry detergents and dry cleaning fluids, which may leach out the stabilizers and leave the material unprotected. Solubility of the stabilizer in the polymer is important, and color contributed by the stabilizer either before or after oxidation limits the choices for many applications. Metal impurities, which often vary in concentration in different batches of a given polymer, may contribute to discolorations; and in the case of transition metals, their prooxidant effect may mask the normal antioxidant effectiveness of the stabilizer.

There is no simple way to approach this problem, since each polymer and each application must ultimately be considered individually. If the polymer is to be used in contact with food products, only stabilizers approved by the federal Food and Drug Administration can be used in the United States. Nevertheless, knowledge of the mechanisms involved and of the effect of changes in structure upon antioxidant activity enables one to select stabilizers in a systematic fashion, rather than by trial-and-error methods. These factors are considered in some detail, following a brief discussion of the use of stabilizers for protection of polymeric materials in the various stages of production, storage, and use.

1. Role in Polymer Synthesis and Fabrication

Polymers that are particularly susceptible to oxidation must be protected by stabilizers as soon as they are formed and before they first come in contact with oxygen. The unsaturated elastomers, including natural rubber and the synthetic polymers of isoprene and butadiene, are in this category. A good antioxidant for polymer stabilization may be of little value in a vulcanized rubber under many conditions of use.[16] Consequently, only a minimum amount (less than 1 pph rubber) is employed, with the expectation that additional stabilizers will be added during compounding prior to vulcanization to provide the desired aging resistance to the final product.

An example of a widely used antioxidant for polymer stabilization during synthesis and fabrication is 2,6-di-*t*-butyl-4-methylphenol. It is nonstaining and effective under mild conditions even though its volatility limits its effectiveness at higher temperatures and for longer times of storage or use.

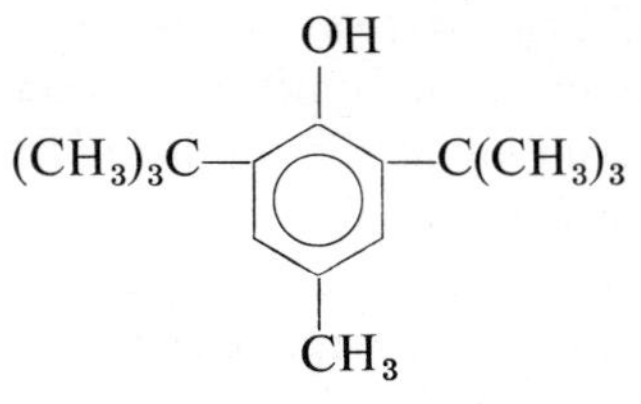

$$(CH_3)_3C-C_6H_2(OH)(CH_3)-C(CH_3)_3$$

A listing of typical examples of a number of different types of stabilizers which are used to protect polymers during synthesis and fabrication is available in *Rubber Reviews* for 1963.[16] The following abbreviated list includes representative structures of the various classes of stabilizers reported to be useful for this purpose.

Class	Representative Structures
I. Polyhydric phenols (*ortho* and *para* types)	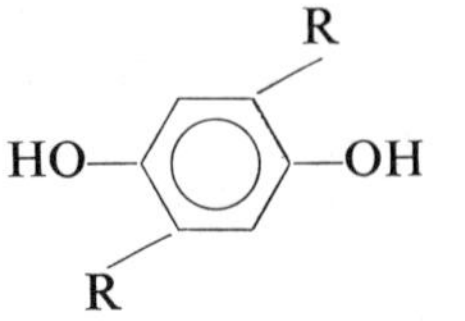 R / HO—⟨○⟩—OH / R
II. Hindered bisphenols (*ortho* and *para* types)	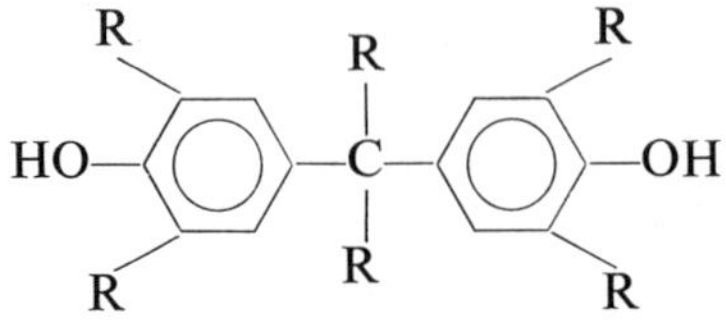 R, R, R / HO—⟨○⟩—C(R)(R)—⟨○⟩—OH / R, R
III. Metal complexes of phenols	
IV. Phosphite esters (and other group V elements)	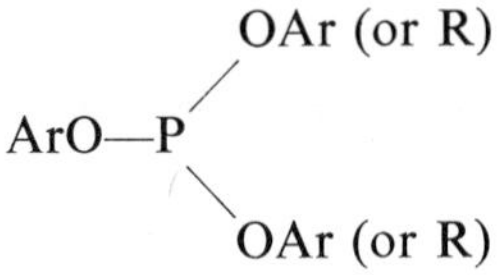 ArO—P(OAr (or R))(OAr (or R))
V. Hindered monohydric phenols (nondiscoloring)	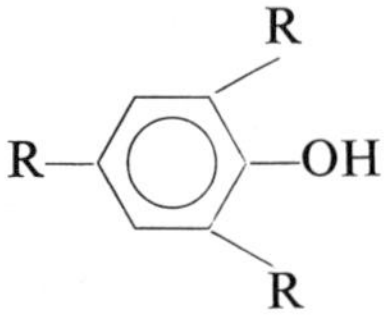 R / R—⟨○⟩—OH / R
VI. Secondary arylamines (discoloring)	Ar—N(H)—Ar
VII. Sulfur compounds	

It is important to have a general knowledge of the mechanisms by which various types of stabilizers function, plus a background of experience relating to the effects of structural variations on volatility, solubility, color formation, and efficiency in various types of materials; knowing these things, one can select possible compounds or combinations to use for stabilization of a given polymer with a reasonable probability of success. In the end, however, the value of a polymer stabilizer must be judged by its ability to prevent or at least to minimize undesirable changes in the physical properties of the polymer that otherwise occur during separation from latex or solution and drying, during storage prior to fabrication, and during compounding, extrusion, or molding.

Physical properties that may be used as a measure of polymer stability in the presence and absence of added stabilizers include plasticity of the polymer, viscosity of polymer solutions, and formation of insoluble gel. Changes in these properties may be studied by accelerated aging tests at elevated temperatures; however correlation with actual use is generally poor unless the temperature is reasonably close to the actual temperatures encountered in processing, storage, and fabrication.

Because of the higher temperature required to soften the more highly crystalline material, the processing of polypropylene requires higher temperatures than, for example, branched polyethylene. Consequently a stabilizer that adequately protects polyethylene may fail to provide the necessary protection for polypropylene during extrusion or injection molding.

The amount of stabilizer required to provide adequate protection for initial polymer stabilization prior to final fabrication is usually quite small. Even though exposure to high temperatures may be involved in processing and drying, the time of such exposure is short. Similarly, prolonged shelf life of the bulk polymer prior to fabrication is not required, since efficient warehouse management will provide for rapid turnover of oxidizable materials such as uncompounded synthetic rubber polymers. A shelf life of six to twelve months at ordinary temperatures without significant changes in plasticity or gel content is generally considered adequate.[16]

2. Role in Long-Term Aging and Use

Additional stabilizer is frequently required to provide protection against thermal autoxidation if the material is likely to be subjected to long-term aging either in storage or in the intended application. If the material is produced specifically for a known use, adequate stabilizers can be added during the synthesizing process. A general purpose polymer, on the other hand, may be incorporated into a variety of products that are subject to a wide range of environmental conditions and times of exposure to oxidative conditions. It is thus necessary to add other specific stabilizers appropriate to the intended use at the time of fabrication. For example, in the case of elastomers, antioxidants are added during compounding prior to vulcanization in sufficient concentration (typically 1–3 pph polymer) to protect the vulcanizate during long-term aging both under storage conditions and under the expected environmental conditions of use.

Accelerated tests are frequently employed to compare the effectiveness of various possible stabilizers in providing the desired protection.[14,16] Physical properties such as tensile strength, ultimate elongation at break, stress required to produce a given elongation (e.g., modulus at 100% elongation), hardness, resilience, resistance to flex cracking, and so on, are measured on fresh samples and on samples aged for various periods of time at a specified temperature and oxygen concentration. Such tests are useful in screening possible stabilizers for a given material, but correlation with performance in actual service is often poor. Consequently the test conditions should simulate the expected service conditions as much as possible. Oven aging in air at temperatures in the range of 70 to 100°C is frequently employed. It is preferable to isolate the samples in separate containers within the oven to minimize migration of volatile antioxidants from one sample to another.

Actual measurement of the rate of oxygen absorption by a polymer containing various stabilizers is one of the best techniques for studying antioxidants in polymers.[21,47–49] Correlations of changes in physical properties with the amount of oxygen absorbed at a given temperature are usually quite good. Figure 2.6[48] shows the changes in modulus at 200% elongation of a natural rubber vulcanizate as a function of oxygen absorbed at four different temperatures.

Changes in the ratio of chain scission to crosslinking at different temperatures and different oxygen concentrations make it difficult to project results obtained at elevated temperatures in oxygen to prediction of aging behavior at lower temperatures in air. Correlations have been obtained by making measurements at different temperatures and oxygen concentrations.[22,48]

Adequate stabilization of polymers against thermal oxidation during long-term aging and use under normal conditions can be provided by addition of

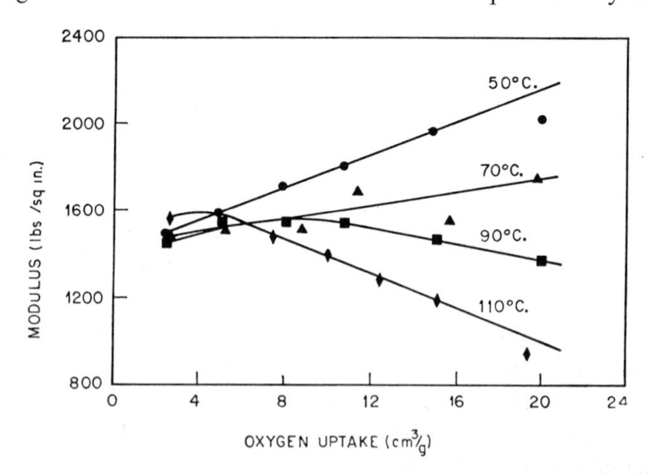

Fig. 2.6 Modulus at 200% elongation of hevea black stock versus oxygen absorption at several temperatures.[48]

appropriate types of antioxidants, but more exacting conditions create special problems. At high temperatures or under evacuation at reduced pressures, antioxidants may be lost by volatilization. Exposure to steam increases volatility by a steam-distillation effect. Constant leaching with water, as in radiator or other hoses, may remove significant amounts of stabilizer. Dry-cleaning solvents are particularly harmful, since they can extract organic compounds of the type commonly used for stabilization of polymers. Repeated exposure to either dry-cleaning or laundering processes is known to have a detrimental effect upon rubber carpet backings, latex threads and foamed rubber pads used in many garments, and upon polymers used in applications such as bonded fibers. The use of antioxidants of increased molecular weight to reduce both volatility and solubility helps to minimize the effect of losses of stabilizers by continued or repeated exposure to these special conditions of aging and use.

A promising approach to the problem of reducing losses of stabilizers by volatilization or by extraction with water or other solvents is the introduction of antioxidants chemically bound to the polymer chain. Workers at the NRPRA laboratory have reported the incorporation of network-bound antioxidants in natural rubber.[50] It was observed that the autoxidation of squalene (a low-molecular-weight polyisoprene) containing *N*,*N*-diethyl-*p*-nitrosoaniline (DENA) was strongly inhibited only after the oxidation had proceeded for some time at 100°C. Since this time could be decreased by pre-heating in the absence of oxygen, a reaction of DENA with squalene was suspected and subsequently confirmed by reaction of the nitroso compound with a model compound, 2-methyl-2-pentene:

$$+ (C_2H_5)_2NC_6H_4N{=}O \longrightarrow NHC_6H_4N(C_2H_5)_2$$

Although the mechanism of the above reaction is still not fully understood, it is apparent that an analogous reaction of DENA with natural rubber would produce a network-bound *p*-phenylenediamine. The effectiveness of such reaction products are compared in Table 2.5 with related conventional *p*-phenylenediamine derivatives, both before and after extraction with an azeotrope of methanol, acetone, and chloroform to remove any stabilizer not bound to the network. The other nitroso compounds besides DENA were *p*-nitrosodiphenylamine (NDPA) and *p*-nitrosophenol (NP). The conventional antioxidants were:

N-Isopropyl-*N*′-phenyl-*p*-phenylenediamine (IPPD)
Polymerized 2,2,4-trimethyl-1,2-dihydroquinoline (TMQ)
2,6-Di-*t*-butyl-*p*-cresol (BHT)

The times for the absorption of 1% by weight of oxygen by samples of the sulfenamide-accelerated sulfur vulcanizates containing the appropriate additives are listed in Table 2.5. The nitroso compounds showed antioxidant

Table 2.5 Oxygen Absorption of Sulfenamide-Accelerated Sulfur Vulcanizates[50]

Additives		Hours to 1% w/w Absorption	
Type	Identity	Unextracted	Extracted
Nitroso	DENA	39	30
	NDPA	60	53
	NP	31	30
Conventional	IPPD	47	4
	TMQ	53	5
	BHT	47	4

activity comparable to that of the conventional antioxidants prior to extraction. The high degree of retention of this activity after prolonged azeotrope extraction is in marked contrast to the severe losses of effectiveness observed with the conventional materials. It is apparent that the reduction in mobility resulting from incorporation of the reaction product of the nitroso compounds into the network did not significantly decrease their effectiveness as antioxidants.

A comprehensive summary of antioxidant types that have proved effective in vulcanized rubber is included in a discussion of antioxidants and antizonants for general purpose elastomers in *Rubber Reviews* for 1963.[16] The most important classes are the secondary diaryl amines and hindered phenols that function as free-radical chain stoppers. The best of the amine antioxidants are somewhat more effective than the phenols, but they also give more discoloration and staining. The many types of compounds that exhibit antioxidant activity are described and compared in the review with references to the original literature.[16] The various classifications are summarized in the following list with typical structures to illustrate some of the more important types of stabilizers against thermal autoxidation, which have proved to be effective in fabricated polymers and vulcanizates during long-term aging and use.

Familiarity with the chemical types of compounds that have been observed to be effective as stabilizers against thermal oxidation in rubber and other polymeric materials enables one to suggest possible types of compounds for use in a new material or a new application. A knowledge of the mechanisms

Class	Representative Structures
I. Secondary Diarylamines	Ar—NH—Ar
A. Phenyl naphthylamines *N*-Phenyl-2-naphthylamine	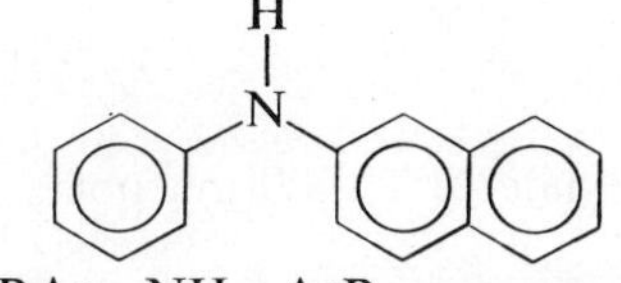
B. Substituted Diphenylamines	RAr—NH—ArR
C. *para*-Phenylenediamines	Ar—NH—C_6H_4—NH—Ar or R or R
N,*N*′-Diphenyl-*p*-phenylenediamine	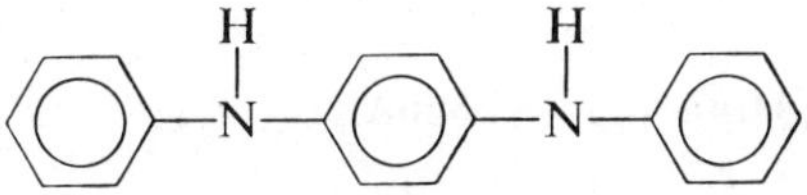
II. Ketone–Amine Condensates	
A. Dihydroquinolines (Acetone–primary arylamine reaction products) 6-Ethoxy-1,2-dihydro-2,2,4-trimethylquinoline	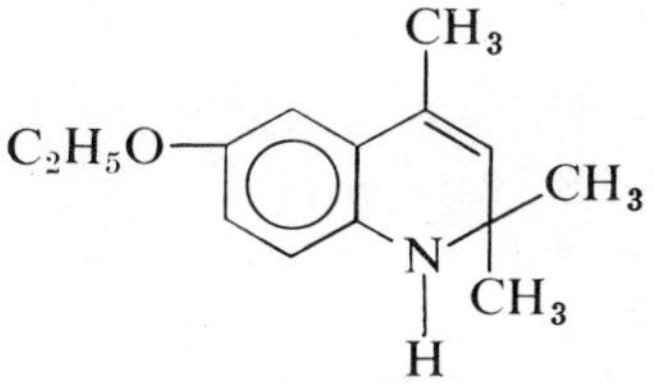
B. Secondary Diarylamine Reaction Products (diphenylamine with acetone)	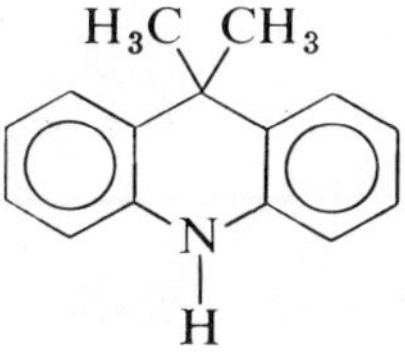
III. Aldehyde–amine Condensates (aldol with 1-naphthylamine)	
IV. Alkyl Aryl Secondary Amines	Ar—N(H)—R—N(H)—Ar
V. Primary Arylamines	H_2N—Ar—NH_2
VI. Hindered Phenols	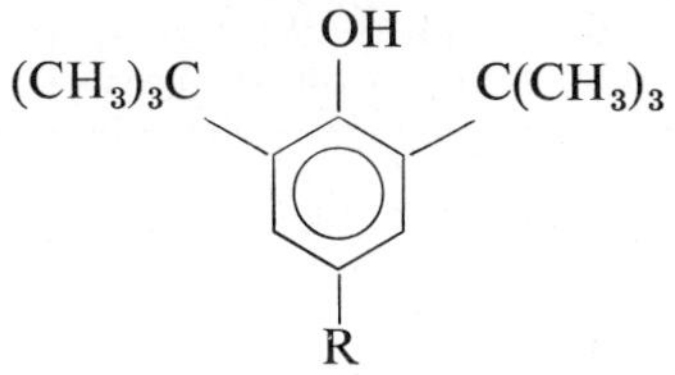

(continued)

Class	Representative Structures
VII. Hindered Thio bisphenols	HO—(2,6-R₂-phenylene)—S—(2,6-R₂-phenylene)—OH
VIII. Hindered bisphenols A. Ortho, ortho′	(2-OH, 3-R, 5-R-phenyl)—CH_2—(2-OH, 3-R, 5-R-phenyl)
B. Para, para′	HO—(R₂-phenylene)—CH_2—(R₂-phenylene)—OH
IX. Polyhydric Phenols	HO—(R₂-phenylene)—OH
X. Sulfur and Phosphorus Compounds A. 2-Mercaptobenzimidazole	benzimidazole—C—SH (N, N—H)
B. Zinc dimethyldithiocarbamate	$(CH_3)_2N—C(=S)—S—Zn—S—C(=S)—N(CH_3)_2$
C. Metal dialkyl dithiophosphates	$(RO)_2P(=S)—S—M—S—P(=S)(OR)_2$

involved with various types of antioxidants permits one to choose the appropriate type or combinations of more than one type of stabilizer with a reasonable probability of finding compounds that will give the desired protection. The remainder of this chapter is devoted to discussion of the mechanisms by which the various types of antioxidants function and the way in which combinations of stabilizers can be used to provide additional protection against oxidative degradation.

E. MECHANISMS OF STABILIZATION BY CHAIN-BREAKING ANTIOXIDANTS

The free-radical chain mechanism of autoxidation is described earlier in this chapter and compared with the retarded autoxidation mechanism that is found in the presence of antioxidants. Consideration of the various ways in which a stabilizer might function to inhibit or retard the process suggests two mechanistically distinct classes of antioxidants.[9] The first and most obvious way to retard the oxidation reaction would be to interrupt the chain process by intercepting either the R· or RO_2· free radicals that propagate the kinetic chain. Antioxidants that function in this way may be classified as chain-breaking antioxidants. The mechanisms by which this can he accomplished form the subject of this section of the chapter.

The second classification of antioxidants includes the compounds that act in some way to retard the initiation process. Thus they are appropriately designated as preventive antioxidants. The mechanisms by which this can be accomplished are discussed later; the use of combinations of stabilizers against thermal oxidation is the subject of the final section of the chapter.

Chain-breaking antioxidants that interrupt the normal propagation processes may react with either of the two radicals involved (R· or RO_2·); most antioxidants of this type react more readily with peroxy radicals, RO_2·, although in some cases the carbon radical, R·, may be involved. The antioxidant may react with the propagating radical by addition or coupling, by electron transfer, or more commonly by hydrogen transfer. Although it is possible to terminate the chain directly, in most cases a new radical is formed which either continues the chain at a reduced rate (chain transfer) or terminates a second kinetic chain.[51] Three types of chain-breaking antioxidants are considered; they are classified by their mode of action as free-radical traps, electron donors, and hydrogen donors.

1. Free-Radical Traps

Any substance capable of reacting with free radicals to form products that do not reinitiate the oxidation reaction could be considered to function as a free-radical trap. For example, quinones are known to scavenge alkyl free radicals as evidenced by their effectiveness as inhibitors of polymerization. Many polynuclear hydrocarbons show activity as inhibitors of oxidation and are thought to function by trapping free radicals.[51] Addition of a free radical, R·, to a quinone or to a polynuclear aromatic hydrocarbon would generate a more stable free radical, which would be correspondingly less capable of reinitiating the autoxidation process. Addition of R· to quinones on either

O or C produces adduct radicals that can undergo subsequent dimerization, disproportionation, or reaction with a second R· to form stable products.

Carbon black is an effective inhibitor of oxidation in many systems, and its structure is best described as agglomerates of complex condensed polycyclic aromatic rings with hydrogen and reactive functional groups around the edges of the essentially planar molecules. The hypothetical structures shown in Figure 2.7, illustrate the typical functional groups believed to be present in many carbon blacks.[52] The phenolic groups could account for part of the activity as an antioxidant as will be described later, but the presence of both quinone and polynuclear aromatic structures suggests the ability of carbon blacks to trap free radicals by addition to form stabilized radicals and thus to terminate the kinetic chain.

Stable free radicals introduced into the material undergoing autoxidation can terminate the kinetic chain by direct combination with one of the propagating free radicals. Stable dialkyl nitroxide radicals such as di-*t*-butyl nitroxide (**1**) and 2,2,6,6-tetramethyl-4-pyridone nitroxide (**2**) react rapidly with R· radicals to give stable molecular products.[51]

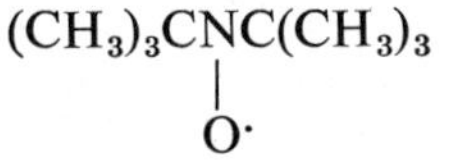

$(CH_3)_3CN(O\cdot)C(CH_3)_3$

1

O
‖
CH_3 — N — CH_3
CH_3 | CH_3
O·

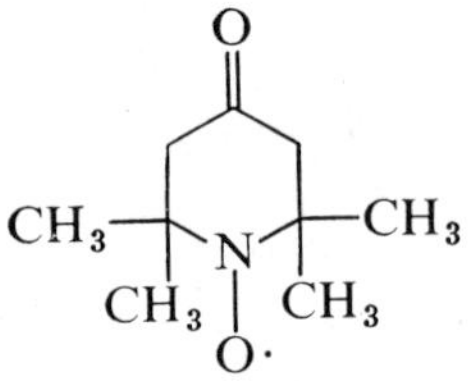

2

$$R_2'NO\cdot + R\cdot \longrightarrow R_2'NOR$$

The dialkyl nitroxide radicals are of little practical value as stabilizers against thermal oxidation compared to the more conventional chain-breaking antioxidants described later because the former must compete with molecular oxygen for the R· radicals. Consequently the rate of inhibited oxidation is proportional to the partial pressure of oxygen and inversely proportional to the nitroxide radical concentration, consistent with reaction exclusively with R· and not $RO_2\cdot$ radicals.

Diaryl nitroxides are even more stable than the dialkyl nitroxides and are somewhat better antioxidants. There is kinetic evidence that radicals such as 4,4′-dimethoxydiphenyl nitroxide (**3**) react with both R· and $RO_2\cdot$ but the latter reaction is much slower.[51] The antioxidant efficiency of these radicals is less than that of the corresponding diphenylamine (**4**) and hydroxylamine (**5**) which function as hydrogen donors and ultimately produce nitroxide radicals that contribute part of the chain-stopping capacity of these antioxidants. Similarly the ArO· radicals produced from phenols by hydrogen

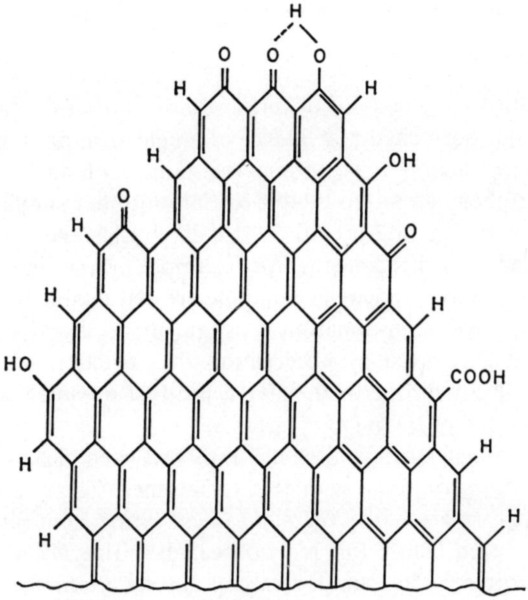

Fig. 2.7 Typical structures believed to be present in carbon blacks.[52]

transfer also function as free-radical traps and thus contribute half of the total antioxidant efficiency observed for the hindered phenols.

$$ArO\cdot + RO_2\cdot \longrightarrow RO_2ArO$$

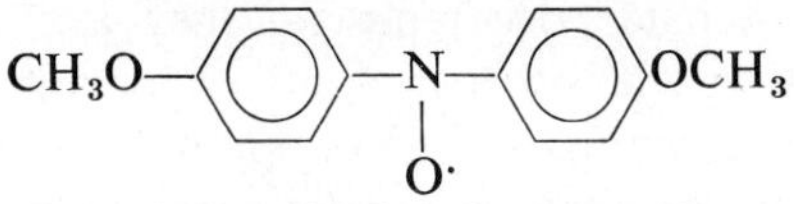

$CH_3O-C_6H_4-N(O\cdot)-C_6H_4-OCH_3$

3

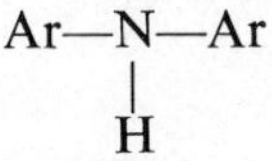

Ar—N(H)—Ar **4** Ar—N(OH)—Ar **5**

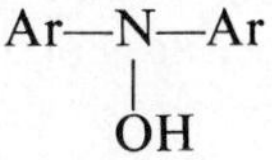

Another stable free radical frequently employed to trap other free radicals is 1,1-diphenylpicrylhydrazyl (**6**)

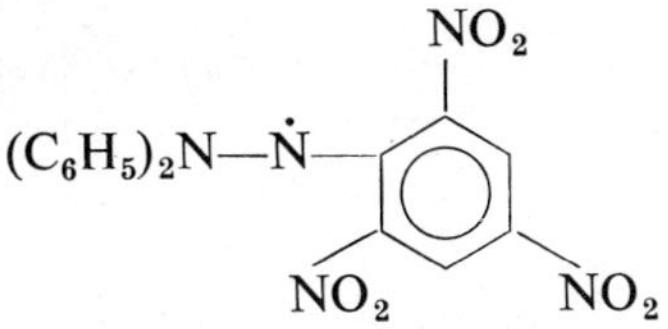

$(C_6H_5)_2N-\dot{N}-C_6H_2(NO_2)_3$

6

2. Electron Donors

The known antioxidant activity of tertiary amines that contain no reactive N–H function has been cited as evidence of an electron-donation mechanism of chain-stopping action.[53] However, the failure of such compounds as pyridine and triphenylamine to inhibit oxidation under similar conditions is not consistent with this interpretation. It is likely that the hydrogens on the carbon attached to nitrogen in, for example, *N*,*N*-dimethylaniline and *N*,*N*,*N'*,*N'*-tetramethyl-*p*-phenylenediamine are extractable by $RO_2\cdot$ and thus function by the hydrogen-donation mechanism as described below. This interpretation is supported by observation that reaction of *N*,*N*-dimethylaniline with *t*-butyl hydroperoxide gave a product in which a peroxy group was attached to one of the methyl carbons.[54]

The electron-transfer mechanism has thus not been clearly established as yet for the reaction of $RO_2\cdot$ with tertiary amines. There is some physical evidence for the formation of a Würster radical cation,[9] but further investigation will be required before this reaction can be either accepted or rejected as a chain-stopping mechanism in retarded autoxidation.

$$RO_2\cdot + Ar\ddot{N}R_2 \longrightarrow RO_2\!:^{-}Ar\dot{N}R_2^{+}$$

An electron-transfer mechanism has been proposed to account for the chain-stopping activity of zinc dialkyl dithiocarbamates and related metal salts of dialkyl dithiophosphates and xanthates. Since these stabilizers function mainly as peroxide decomposers, they are considered in more detail in Section F along with other preventive type antioxidants.

The most clear-cut examples of inhibition by one-electron transfer are found in studies of metal-catalyzed oxidations. Under certain conditions, transition metals may inhibit rather than catalyze oxidations. Cobalt, manganese, and copper are notable examples.[51]

$$RO_2\cdot + Co^{2+} \rightarrow RO_2\!:^{-}\ Co^{3+}$$

3. Hydrogen Donors

The antioxidants most commonly used as stabilizers against thermal oxidation in polymers are either secondary aryl amines or hindered phenols. The presence of a reactive N–H or O–H functional group in these materials suggests that they compete with the polymer substrate for the peroxy radicals, $RO_2\cdot$ and thus terminate the propagation reaction by transfer of hydrogen to form RO_2H and a stabilized $Ar_2N\cdot$ or $ArO\cdot$ radical. These radicals would be expected to function as free-radical traps and thus terminate a second kinetic chain.

$$Ar_2N{-}H + RO_2\cdot \rightarrow RO_2H + Ar_2N\cdot$$

$$Ar_2N\cdot + RO_2\cdot \rightarrow Ar_2NO_2R$$

$$ArOH + RO_2\cdot \rightarrow RO_2H + ArO\cdot$$

$$ArO\cdot + RO_2\cdot \rightarrow RO_2ArO$$

This hydrogen-donation mechanism has generally been considered the most probable explanation of the antioxidant activity of aryl amines and phenols in polymers and other organic materials. The mechanism is consistent with most kinetic evidence and with the observed reaction products derived from the antioxidant.[27] If this assumption is correct, it is to be expected that substitution of a labile hydrogen by deuterium should give rise to a kinetic isotope effect. The rate of chain termination would be slower in the case of N–D or O–D functional groups reacting with $RO_2\cdot$ as compared to abstraction of ordinary hydrogen from an aryl amine or phenol. Thus the observed rate of oxidation of the substrate would be increased in the presence of the deuterated antioxidant as compared to the rate observed with an equivalent concentration of the normal form of the same antioxidant. Isotopic evidence for the various reactions of antioxidants during inhibited autoxidations has been obtained by many workers and is summarized in a recent book on autoxidation of hydrocarbons and polyolefins.[55]

Initial attempts to observe kinetic deuterium isotope effects in retarded autoxidation were unsuccessful. Hammond and co-workers[56] looked for isotope effects in the oxidation of both cumene and tetralin in the presence of deuterated *N*-methylaniline and deuterated diphenylamine as compared to the normal undeuterated antioxidants. The observed rates of oxidation in chlorobenzene at 62.5°C initiated by azobisisobutyronitrile (AIBN) were not measurably changed by the substitution of deuterium for hydrogen in the antioxidant. Similarly, no change in the induction period for the oxidation of gasoline inhibited with *N*,*N′*-diphenyl-*p*-phenylenediamine was observed when the N–H hydrogen was replaced by deuterium.[53]

An alternative inhibition mechanism was therefore proposed by Hammond on the basis of the failure to observe a kinetic deuterium isotope effect, together with the observation that with certain weak antioxidants such as *N*-methylaniline the rate of oxidation varied inversely with the square root of the inhibitor concentration. Since this kinetic variation suggested some sort of reversible interaction between inhibitor and peroxy radicals, the reversible formation of a complex of peroxy radical with the inhibitor was proposed as the kinetically controlling step, followed by a rapid reaction of the complex with a second peroxy radical.

$$RO_2\cdot + AH \rightleftarrows (RO_2AH)\cdot$$

$$(RO_2AH)\cdot + RO_2\cdot \xrightarrow{\text{rapid}} \text{inactive product}$$

Alternative explanations of the observed kinetics have heen considered. For example, Hammond and Nandi[59] considered the possibility that hydrogen abstraction from the antioxidant, AH, by $RO_2\cdot$ radical to form RO_2H and $A\cdot$ might be reversible. Addition of hydroperoxide would then be expected to favor the reverse reaction, but no effect upon the rate consistent with such an expectation was observed. Bickel and Kooyman[27] had previously noted a kinetic relation similar to Hammond's observation in their extensive investigation of the effect of amine and phenolic antioxidants upon rates of oxidation. They pointed out that this result could be explained on the basis of a chain-transfer process involving the antioxidant as illustrated in the mechanism of retarded autoxidation presented earlier in this chapter. Several workers[57,58] have reported results that were considered to support the Hammond mechanism of reversible formation of a complex of $RO_2\cdot$ and antioxidant as the kinetically significant process that preceded a rapid hydrogen transfer to form the observed products. The mechanism was considered favorably for a number of years and was generally accepted in academic circles, even though contrary evidence was reported as early as 1958 in studies of the mechanism of antioxidant action in elastomers.[23]

The previously developed[47] volumetric oxygen-absorption method of studying oxidation of elastomers was employed to investigate the importance of hydrogen abstraction from the antioxidant by peroxy radicals in the mechanism of the chain-stopping action of amine and phenolic inhibitors. Deuterium was substituted for the active hydrogen of N-phenyl-2-naphthylamine and 2,6-di-*t*-butyl-4-methylphenol by reacting the compounds with methylmagnesium iodide and then with deuterium oxide. The oxidation rates of butadiene–styrene polymer inhibited by 3 pph of the deuterated materials were measured and compared with the rates observed using the untreated antioxidant. Figure 2.8 shows the first example of a significant kinetic deuterium isotope effect observed in autoxidation in the presence of a deuterated antioxidant. The observed rates in the constant-rate portion of each plot show a ratio of $k_D/k_H = 1.8 \pm 0.3$ with a deuterium substitution of not less than 60%, based on a mass spectrometer analysis of a mixture of CH_3D and CH_4 obtained by reaction of a sample of the deuterated antioxidant with methylmagnesium iodide.

The possibility that an inhibitor will contribute to initiation in addition to its normal role in termination (described above) is a complicating factor,

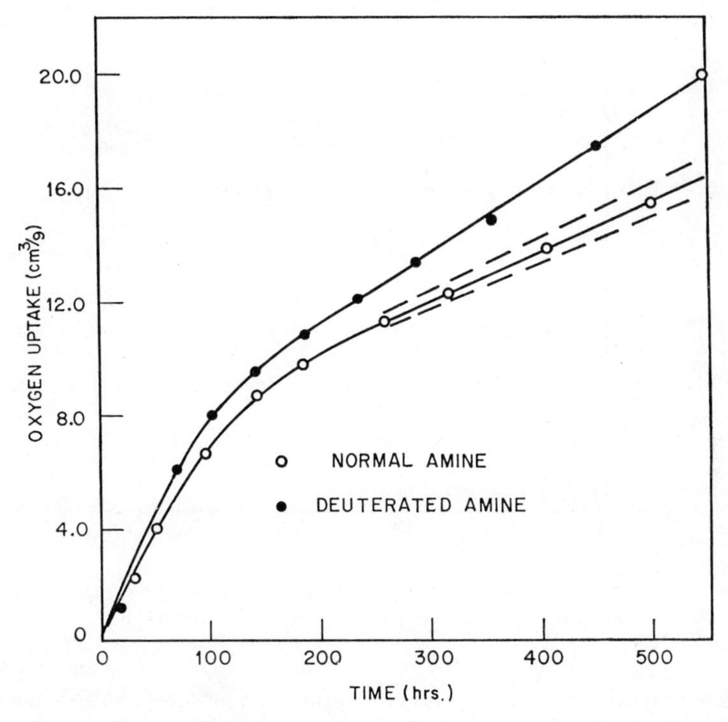

Fig. 2.8 Kinetic isotope effect in the oxidation of butadiene–styrene polymer inhibited by normal and deuterated *N*-phenyl-2-naphthylamine (90°C, 1 atm O_2).[23]

since both processes will exhibit an isotope effect. Initiation by abstraction of hydrogen from the N–H or O–H group of the antioxidant by O_2 will be faster than from the corresponding N–D or O–D group just as abstraction of hydrogen by $RO_2\cdot$ is faster than abstraction of deuterium from the corresponding amine or phenol. The deuterium isotope effect as measured by rate of oxygen absorption is in opposite directions for the two competing processes, since the AH form is both a better initiator (increased rate of oxidation) and a better terminator (decreased rate of oxidation) as compared to the AD form of the antioxidant. The observed ratio of rates, k_D/k_H, thus reflects the net effect of deuterium substitution upon the two processes.

Any change that increases the participation of the antioxidant in the initiation process without a corresponding increase in termination will therefore diminish the observed isotope effect as measured by the ratio of k_D/k_H observed. Thus when the concentration of *N*-phenyl-2-naphthylamine was increased from 3 pph in SBR polymer, as used to obtain the data of Figure

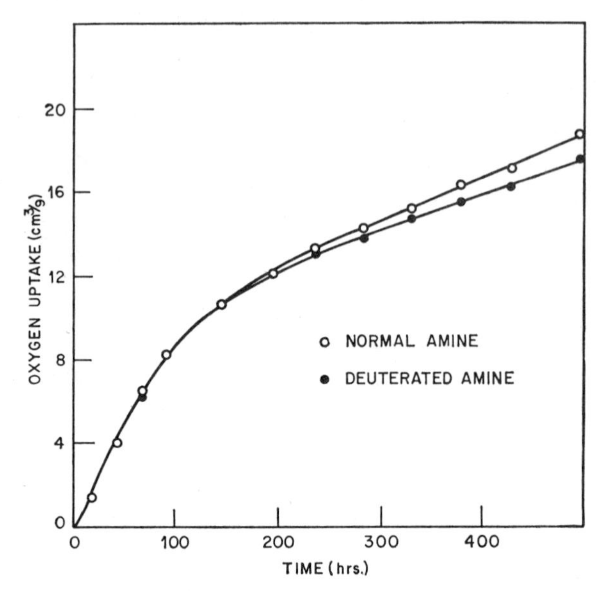

Fig. 2.9 Comparison of oxygen uptake of SBR polymers containing 5 pph normal and deuterated *N*-phenyl-2-naphthylamine (90°C, 1 atm O_2).[24]

2.8, to 5 pph (which is well above the optimum concentration for this polymer), the antioxidant effect was offset by the prooxidant effect and only a small net isotope effect was observed. The ratio of $k_D/k_H = 0.86 \pm 0.05$ calculated from the constant-rate portions of the curves of Figure 2.9 shows a reversal of the isotope effect; that is, the polymer inhibited by the normal undeuterated form of the antioxidant oxidized somewhat faster.

Higher temperatures of oxidation increase the participation of the antioxidant in initiation as compared to its role in termination. This is consistent with a higher activation energy for abstraction of hydrogen by O_2 as compared to the more reactive $RO_2\cdot$ radical. This temperature effect is clearly evident when the oxidation rates observed at 80 and 90°C for SBR polymer containing 3 pph normal and deuterated diphenylamine are compared in Figures 2.10 and 2.11. At the lower temperature the two competing effects canceled each other, so that no net effect of deuterium substitution in the antioxidant upon the rate of oxidation is observed in Figure 2.10. At the higher temperature a significant effect is observed but in the reverse direction, $k_D/k_H = 0.78 \pm 0.06$, consistent with a dominance of the prooxidant effect of diphenylamine under these conditions.

The diphenylamine was deuterated by refluxing with a mixture of heptane and D_2O for an hour. The amine dissolved in the heptane layer at reflux

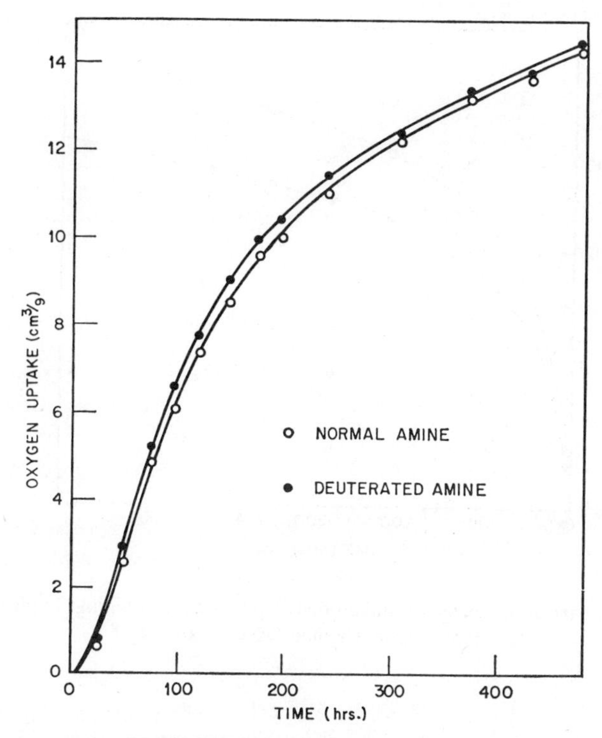

Fig. 2.10 Comparison of oxygen uptake of SBR polymers containing 3 pph normal and deuterated diphenylamine (80°C, 1 atm O_2).[24]

temperature, but crystallized out on cooling and was pressure filtered and dried with a stream of nitrogen. Infrared analysis for undeuterated N–H at 3.02 μ showed 90% deuterium substitution. The amazing ease of exchange of H and D under these conditions in the presence of an organic phase in which the antioxidant can be dissolved emphasizes the necessity of excluding all water from the system under investigation. The formation of water and ROOH in an oxidation mixture could rapidly replace the deuterium on a deuterated antioxidant and thus account for the early failures to observe isotope effects.[53,56] This explanation was rejected[59] when first suggested on the basis of the above observation, but subsequent studies by Ingold and Howard[60] clearly demonstrated that large isotope effects can be demonstrated if excess D_2O is incorporated into the oxidation mixture to keep the antioxidant in a fully deuterated condition.

The isotope effect shown in Figure 2.8 was the first verification of the hydrogen-donation mechanism of antioxidant action by this technique. The validity of this interpretation of the experimental observation was questioned

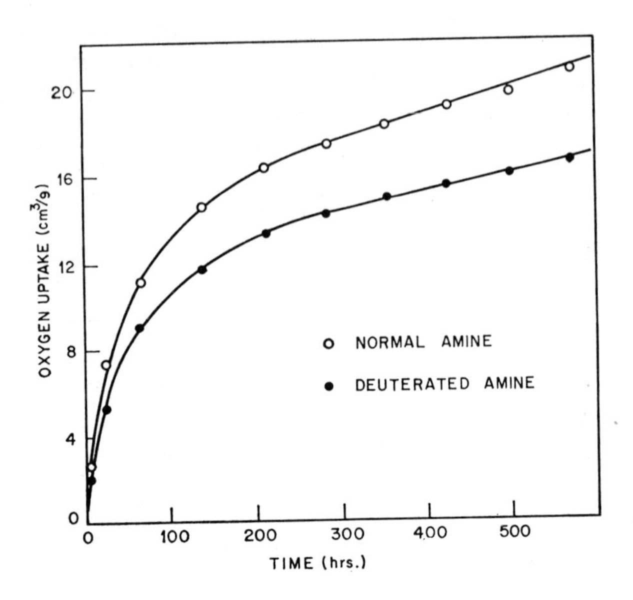

Fig. 2.11 Comparison of oxygen uptake of SBR polymers containing 3 pph normal and deuterated diphenylamine (90°C, 1 atm O_2).[24]

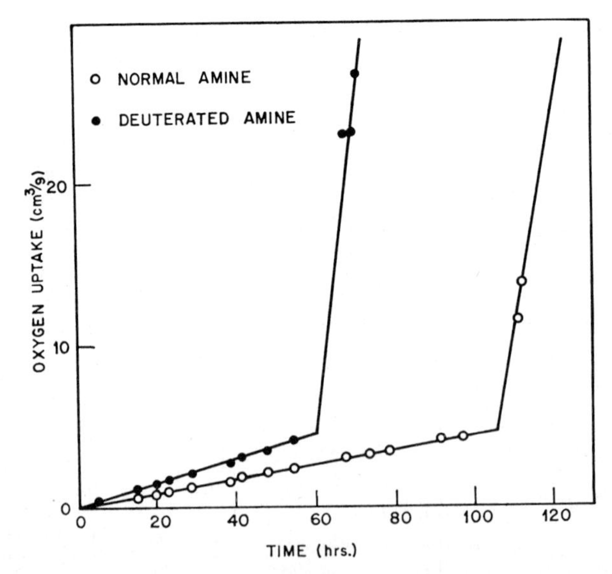

Fig. 2.12 Oxidation of purified polyisoprene containing 1 pph *N*-phenyl-2-naphthylamine (75°C, 1 atm O_2).[25]

by some[4] on the basis that the constant-rate portion of the oxygen-absorption curve was only attained after more than 100 hours at 90°C and after the absorption of more than 8 ml O_2/gram of polymer. The work of Lorenz and Parks,[61] in which vulcanized rubber samples were extracted with solvents prior to oxidation, indicated that the initial stage of more rapid oxidation was caused by the presence of extractable nonrubber components. It was thus evident that studies in a pure hydrocarbon system would be required to definitely establish the mechanism of the chain-stopping action of amine and phenol type stabilizers against thermal oxidation.

A synthetic *cis*-1,4-polyisoprene was chosen for this purpose and subjected to an elaborate purification procedure carried out in a glove-box with a nitrogen atmosphere.[25] Addition of 1 pph *N*-phenyl-2-naphthylamine and oxidation at 75°C gave the desired initial linear rate of oxidation as shown in Figure 2.12 with a deuterium isotope effect demonstrated during the first 60 hours of oxidation. Similar results were observed with 1 pph of 2,6-di-*t*-butyl-4-methylphenol at 90°C as shown in Figure 2.13. The significance of the break between the first and second linear stages of retarded autoxidation was discussed earlier with regard to changes in the initiation process that occur as a result of the accumulation of peroxides and other oxidation products. In this discussion we consider only the significance of the observed

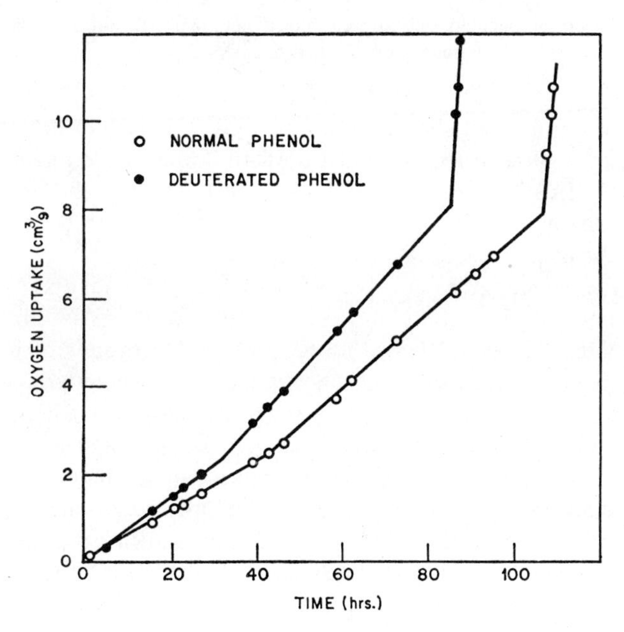

Fig. 2.13 Oxidation of purified polyisoprene containing 1 pph 2,6-di-*t*-butyl-4-methylphenol (90°C, 1 atm O_2).[25]

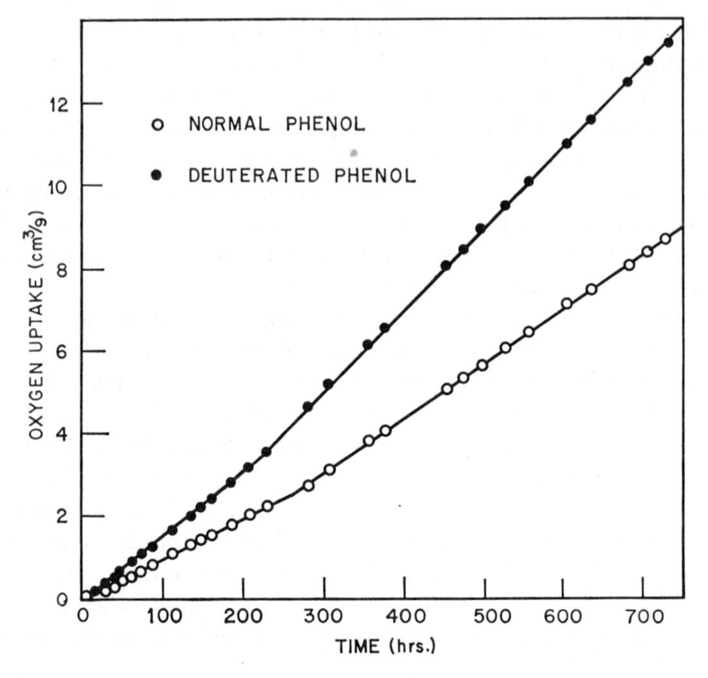

Fig. 2.14 Oxidation of purified polyisoprene containing 2 pph 2,6-di-*t*-butyl-4-methylphenol (75°C, 1 atm O_2).[25]

isotope effects. Figure 2.14 shows significant isotope effects in both the first and second stages of retarded autoxidation with 2 pph of the phenolic antioxidant at 75°C. In all these examples, the sample containing the less efficient deuterated antioxidant oxidized more rapidly, consistent with the hydrogen-donation mechanism.

Tables 2.6 to 2.8 record the data obtained with three antioxidants (*N*-phenyl-2-naphthylamine, 2,6-di-*t*-butyl-4-methylphenol, and *N*,*N*′-diphenyl-*p*-phenylenediamine) at several temperatures and concentrations of inhibitor. It should be noted that the isotope effects obtained in the second stage (which corresponds to the constant-rate stage of the prior study with SBR polymer) agree quite well with the values obtained in the first stage of oxidation. Values of less than unity for R_D/R_H obtained at higher temperatures and higher concentrations of antioxidant confirm the prooxidant effect involving direct attack of oxygen on the inhibitor to produce radicals that are able to initiate the free-radical oxidation mechanism. Examples of these reversed isotope effects are shown in Figure 2.15 and were seen in Figure 2.3.

The mechanism of chain-stopping action by amine and phenolic inhibitors in the retarded autoxidation of polymers is now clearly established on the

Table 2.6 Oxygen Absorption Rates for *cis*-1,4-Polyisoprene Inhibited with *N*-Phenyl-2-naphthylamine[25]

Temperature (°C)	Inhibitor	Inhibitor Concentration (moles/g × 10^5)	Rate, ml O_2/g/hr (22°C, 760 mm)		Isotope Effect, R_D/R_H	
			Initial Stage	Second Stage	Initial Stage	Second Stage
90	InH	8.82	0.174	0.253	0.86	0.89
	InD		0.150	0.226		
75	InH	4.41	0.0436	—	1.71	—
	InD		0.0744	—		
75	InH	8.82	0.0130	0.0172	1.03	0.98
	InD		0.0134	0.0169		
75	InH	13.2	0.0513	0.0709	0.72	0.84
	InD		0.0370	0.0596		
60	InH	8.82	0.00215	—	2.14	—
	InD		0.00460	—		
60	InH	13.2	0.0147	0.0183	0.86	0.84
	InD		0.0126	0.0154		

Table 2.7 Oxygen Absorption Rates for *cis*-1,4-Polyisoprene Inhibited with 2,6-Di-*t*-butyl-4-methylphenol[25]

Temperature (°C)	Inhibitor	Inhibitor Concentration (moles/g × 10^5)	Rate, ml O_2/g/hr (22°C, 760 mm)		Isotope Effect, R_D/R_H	
			Initial Stage	Second Stage	Initial Stage	Second Stage
90	InH	4.41	0.0584	0.0861	1.27	1.25
	InD		0.0750	0.1077		
90	InH	13.2	0.0273	0.0393	0.79	0.82
	InD		0.0216	0.0321		
75	InH	8.82	0.00968	0.0132	1.56	1.49
	InD		0.0151	0.0197		
75	InH	13.2	0.00710	0.00962	0.92	0.95
	InD		0.00653	0.00913		
60	InH	4.41	0.00251	—	1.76	—
	InD		0.00441	0.00549		
60	InH	13.2	0.00124	—	1.16	—
	InD		0.00144	—		

Table 2.8 Oxygen Absorption Rates for *cis*-1,4-Polyisoprene Inhibited with N,N′-Diphenyl-*p*-Phenylenediamine[25]

Temperature (°C)	Inhibitor	Inhibitor Concentration (moles/g × 10^5)	Rate, ml O_2/g/hr (22°C, 760 mm) Initial Stage	Second Stage	Isotope Effect, R_D/R_H Initial Stage	Second Stage
75	InH	4.41	0.0216	—	1.18	—
	InD		0.0255	—		
75	InH	13.2	0.0592	0.0776	0.64	0.68
	InD		0.0381	0.0526		

basis of the deuterium isotope effects observed in the initial stage of retarded autoxidation of purified *cis*-1,4-polyisoprene. The even larger effects obtained by Howard and Ingold[62] with pure compounds in solution by addition of D_2O to prevent loss of deuterated antioxidant by exchange with water and hydroperoxides confirms the general applicability of the rate-controlling hydrogen-donation mechanism. There is therefore no need to postulate a reversible complexing of $RO_2\cdot$ with the inhibitor. If such complexing occurs, it would be an additional but not essential feature, which may occur prior to the rate-controlling hydrogen transfer reaction.

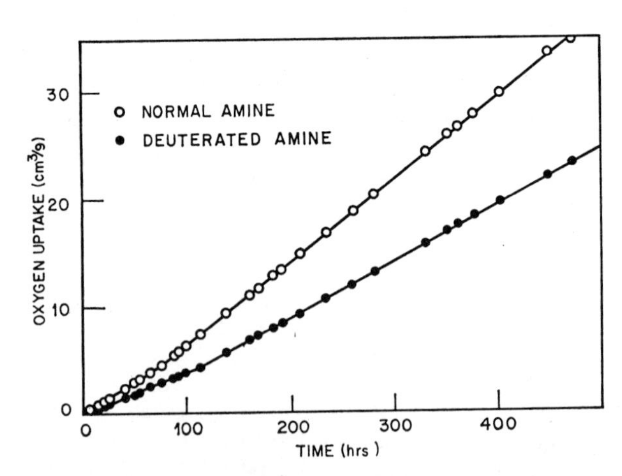

Fig. 2.15 Oxidation of purified polyisoprene containing 3 pph *N*-phenyl-2-naphthylamine (75°C, 1 atm O_2).[25]

4. Correlation of Structure and Reactivity of Stabilizers

The activities of amines as antioxidants and antiozonants for rubber have been investigated by Furukawa and co-workers[63] and discussed in relation to their chemical structures. The effectiveness of the various amine derivatives were evaluated by several methods: (*a*) the relative ability to protect a rubber solution against viscosity changes attributed to the action of oxygen or ozone, (*b*) the rate of reaction with peroxy radicals generated by reaction

Table 2.9 Amine Derivatives and Their Oxidation Potentials[63]

Number[a]	Amine Derivative	Oxidation Half-Wave Potential (SCE)
1	Aniline	0.65[b]
2	*p*-Methoxyaniline	0.47
3	*p*-Hydroxyaniline	0.45
4	*p*-Phenylaniline	0.65[b]
5	*N*-Methylaniline	0.75[b]
6	*N*,*N*-Dimethylaniline	—
7	Dicyclohexylamine	0.70
8	Diphenylamine	0.62[b]
9	Di-*p*-tolylamine	—
10	*p*-Hydroxydiphenylamine	0.58
11	*N*-Phenyl-2-naphthylamine	0.54
12	*p*-Phenylenediamine	0.18
13	*N*-Phenyl-*p*-phenylenediamine	0.22
14	*N*-Phenyl-*N*′-cyclohexyl-*p*-phenylenediamine	0.24
15	*N*-*p*-Methoxyphenyl-*N*′-cyclohexyl-*p*-phenylenediamine	0.20
16	*N*-Phenyl-*N*′-isopropyl-*p*-phenylenediamine	0.26
17	*N*,*N*′-Diphenyl-*p*-phenylenediamine	0.35
18	*N*-*p*-Methoxyphenyl-*p*-phenylenediamine	0.20
19	*N*,*N*′-Di-*o*-nitrophenyl-*p*-phenylenediamine	0.84
20	*N*,*N*′-Di-*o*-tolyl-*p*-phenylenediamine	0.33
21	*N*,*N*′-Di-*p*-tolyl-*p*-phenylenediamine	0.30
22	*N*,*N*′-Diphenyltriphenylenediamine	0.35
23	*N*,*N*′-Di-2-naphthyl-*p*-phenylenediamine	0.40
24	Benzidine	0.18
25	*N*,*N*′-Dicyclohexylbenzidine	0.45
26	*N*,*N*′-Diphenylbenzidine	0.48
27	2,2,4-Trimethyl-1,2-dihydroquinoline	0.47
28	6-Ethoxy-2,2,4-trimethyl-1,2-dihydroquinoline	0.33

[a] These numbers identify the compounds in Figures 2.16 and 2.17.
[b] Decomposition potential.

of oxygen with radicals from azobisisobutyronitrile, (*c*) the relative ability to inhibit the oxidation of tetralin, and (*d*) the amount of ozone consumed by the amine in rubber solutions. The relative activities observed were compared with the oxidation potentials of the amine derivatives listed in Table 2.9.

Other workers[64–67] have observed correlation of oxidation potentials with activity of antioxidants and antiozonants. Good antioxidants were reported[64] to have potentials of 0.7 to 0.9 V, whereas effective antiozonants showed oxidation potentials lower than 0.4 V.[9] In the case of the compounds listed in Table 2.9, the antioxidant or antiozonant activity increases with oxidation potential as shown in Figure 2.16 and reaches a maximum at about 0.4 V for antioxidant effectiveness and at about 0.25 V for antiozonant activity. It is evident that some amines show activity both as antioxidants and as antiozonants for rubber. Their effectiveness in each case is clearly related to their oxidation potentials.

Oxidation potentials can also he correlated with the energy calculated for the highest occupied orbital by the Hückel method. The reactivity indices S^r and S^n for free radical and nucleophilic reactions can be calculated from

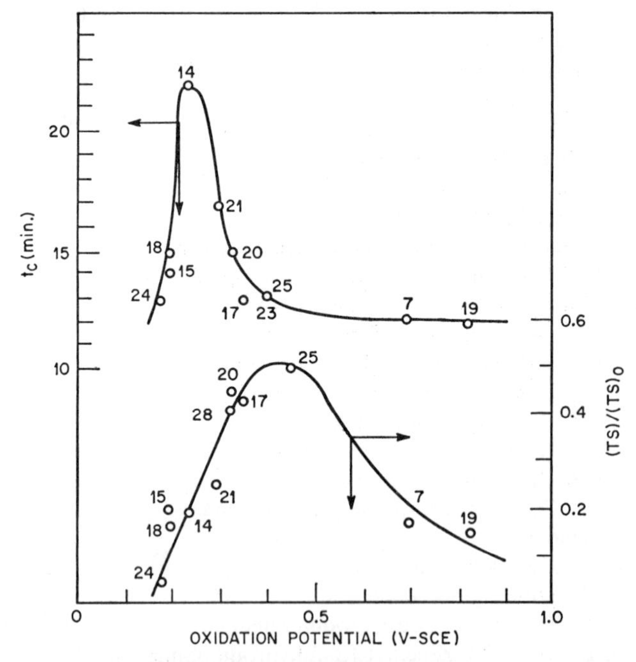

Fig. 2.16 Correlation between the relative tensile strength $(TS)/(TS)_0$ or the time for crack formation of NR gum vulcanizates, t_c, and the oxidation potential of the amine (see Table 2.9).[63]

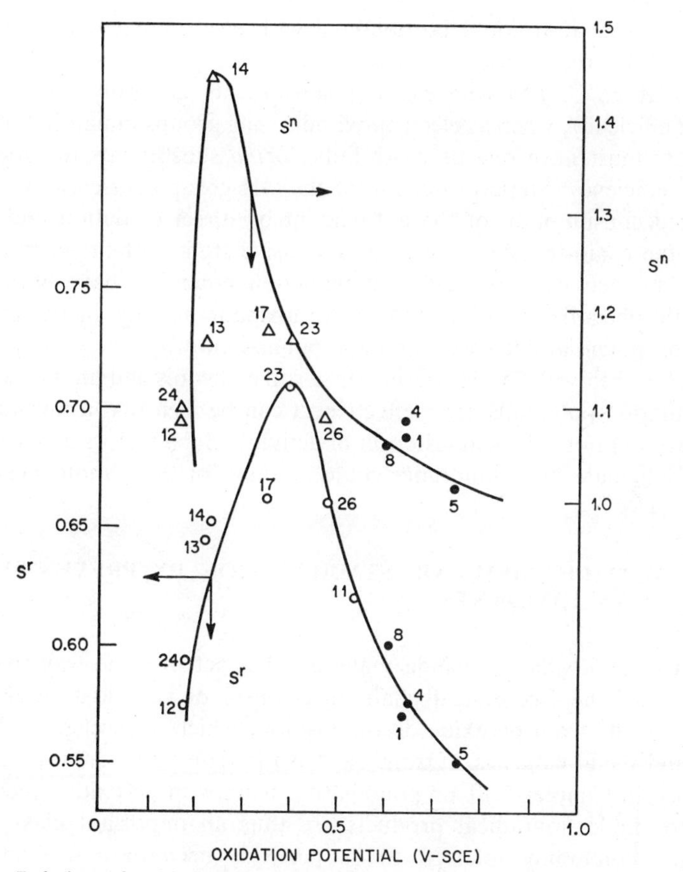

Fig. 2.17 Relation of oxidation potential of amine with its radical reactivity index S^r (open circles) and nucleophilic reactivity index S^n (triangles), oxidation half-wave potential (solid circles); decomposition potential (see Table 2.9).[63]

appropriate equations[63] involving summations of occupied and unoccupied orbitals, the calculated electron density on the nitrogen atom, and the energy levels of the orbitals. The relation of the oxidation potential of the amines in Table 2.9 to the calculated radical reactivity index S^r and nucleophilic reactivity index S^n is shown in Figure 2.17. Maximum values are observed at approximately 0.4 V for S^r and 0.25 V for S^n. Comparison of Figures 2.16 and 2.17 indicates that the mechanism of antiozonization is a nucleophilic polar process (see Chapter 5) and confirms the free-radical nature of the mechanism by which chain-breaking antioxidants stabilize against thermal autoxidation.

The relation of structure and antioxidant efficiency has been studied for a variety of phenolic antioxidants.[9,55] Howard and Ingold[62] showed that excellent correlation could be obtained with electrophilic substituent constants (σ^+) for the reaction of various substituted phenols with styrylperoxy radicals at 65°C. Electron-releasing substituents generally improve antioxidant efficiency, whereas electron-withdrawing groups impair it.[55] Phenolic inhibitors must have one or more bulky *ortho* substituents for good antioxidant efficiency. Steric protection of the OH group is essential to prevent too rapid consumption of the antioxidant by direct oxidation and also to reduce the chain-transfer reaction.[27] Extensive studies of the correlation of antioxidant activity with oxidation–reduction potentials have been carried out with phenolic inhibitors.[65,66,68] Antioxidant activity increases as the oxidation potential decreases, and compounds known to be good inhibitors have values below 0.7 V. However, since some phenols and amines with very low oxidation potentials are ineffective, it can be seen that a low value is a necessary but not sufficient criterion of activity.[9] Such factors as volatility or lack of adequate steric hindrance could account for these anomalies.

F. MECHANISMS OF STABILIZATION BY PREVENTIVE ANTIOXIDANTS

Preventive antioxidants include materials that act in some way to prevent initiation of the free-radical chain mechanism of autoxidation. Initiation usually results from peroxide decomposition, which is accelerated by heat, light, and metal catalysis. (Ozone can also initiate autoxidation, as will be described in Chapter 5.) Compounds that react with peroxides and convert them to stable nonradical products are thus an important class of antioxidants, commonly referred to as peroxide decomposers. Compounds that absorb ultraviolet light may be either photosensitizers or inhibitors depending upon their ability to dispose of the energy absorbed. A material such as carbon black is an effective light screen when incorporated into polymers, and is an example of the class of antioxidants called ultraviolet stabilizers, which are discussed in Chapter 4. Any agent that functions by removing sources of free radicals can thus be regarded as a preventive antioxidant. Only two types will be included in the present discussion: metal-ion deactivators and peroxide decomposers.

1. Metal-Ion Deactivators

The prooxidant effect of metal ions capable of undergoing one-electron oxidation–reduction reaction is well known. The evidence suggests that metal ions exert their catalytic effect upon autoxidation by forming unstable coordination complexes with alkyl hydroperoxides, followed by electron transfer to give free radicals.[9] Two oxidation–reduction reactions can be involved depending on the metal and its state of oxidation:

$$RO_2H + M^{n+} \rightarrow RO\cdot + M^{(n+1)+} + OH^-$$
$$RO_2H + M^{(n+1)+} \rightarrow RO_2\cdot + M^{n+} + H^+$$

The relative importance of the above reactions depends upon the relative strength of the metal ion as an oxidizing or reducing agent.[55] Thus a strong reducing agent such as ferrous ion will react to form RO· radicals according to the first reaction, whereas a strong oxidizing agent such as lead tetra-acetate will react with hydroperoxides to form $RO_2\cdot$ as in the second reaction. When the metal ion has two valence states of comparable stability (e.g., Co^{+2} and Co^{+3}), both reactions will occur and a trace amount of the metal can convert a large amount of peroxide to free radicals according to the sum of the two reactions

$$2ROOH \xrightarrow{Co^{+2}/Co^{+3}} RO\cdot + RO_2\cdot + H_2O$$

The effect of the metal ion is thus to speed up the rate of initiation by accelerating the rate of peroxide decomposition to free radicals. The relative effectiveness of a metal ion as an oxidation–reduction catalyst probably depends both on the ease of coordination and on the oxidation–reduction potential of the metal in the complex with the hydroperoxide. It follows that inhibition of metal-catalyzed autoxidation may be achieved either by strongly complexing the metal ion to its maximum coordination number or, in some cases, by stabilizing one valence state at the expense of the other.[9] Reaction to form an insoluble product (such as FeS in the case of iron present in rubber during sulfur vulcanization) will also inactivate the metal catalyst.

The effectiveness of various ligands as metal-ion deactivators depends in large measure upon the stability of the metal chelate formed. Derivatives of oxamide are used, for example, to deactivate copper in polymers.[69] Since the ligand functions as a Lewis base in the formation of the coordination complex, electron-releasing substituents would be expected to increase and electron-attracting substituents to decrease the stability. This has been observed in a series of copper chelates of Schiff bases formed from salicylaldehyde and substituted anilines.[70]

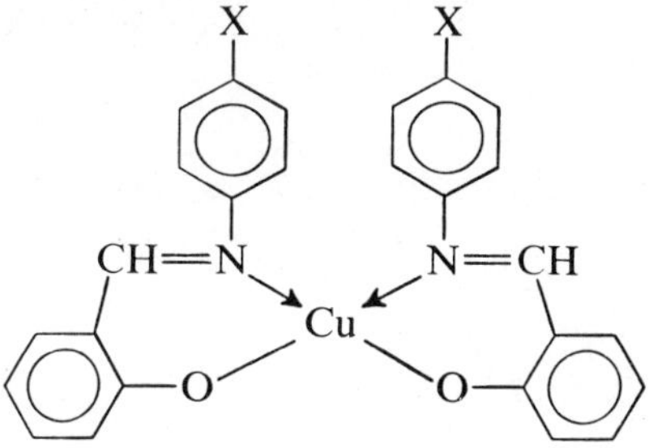

Also important to the problem of metal-ion inhibition is the formation of a favorable arrangement of the ligands around the metal ion.[9] The number and size of chelate rings that can be formed, the tying together of two chelating groups with bridges of varying length, and steric hinderance by alkyl substitution, have all been observed to affect the metal deactivating efficiency of various ligands.

Chelate stability alone is not a sufficient criterion of activity as a metal-ion deactivator. For example, Pedersen[71] has studied the effect of a variety of complexing agents upon the autoxidation of petroleum catalyzed by various transition metals, with the results summarized in Table 2.10. The efficiency,

Table 2.10 Catalytic and Deactivating Effects of Ligands on Metal-Catalyzed Autoxidation of Petroleum[71]

Metal Oleate (1.6×10^{-8} M)	Deactivating Efficiency, E_D				
Chelating Agent (0.002%)	Mn	Fe	Co	Ni	Cu
OH, HO; $CH{=}NCH_2CH_2N{=}CH$ **7**	−103	−43	−833	—	100
OH; $CH{=}N{-}OH$	—	0	−96	100	100
OH HO; $CH{=}N$	−84	100	96	−55	100
OH HO; $N{=}N$	−73	100	—	−124	100
[OH; $CH{=}N{-}CH_2{-}$]$_4$C **8**	100	100	100	100	100

E_D, of metal-ion deactivation is the percentage restoration of the original induction period observed with a control sample in the autoxidation of petroleum without metal or deactivator. Negative values indicate a pro-oxidant effect rather than deactivation. Thus in the case of N,N′-(disalicylidene)ethylenediamine (**7**), effective inhibition of copper ion catalysis is observed; but iron, manganese, and cobalt are activated by chelation with this compound. All the tetradentate ligands were found to be effective with copper, which forms planar four-coordination complexes; but transition metals with maximum coordination numbers greater than 4 were frequently converted to more efficient catalysts of autoxidation. The octadentate, *N,N′,N″,N‴*-tetrasalicylidenetetra(aminomethyl)methane (**8**), was the only compound in the group that was equally effective against copper, iron, nickel, cobalt, and manganese. Similar results have been reported by other workers,[72] who also found the hexadentate 1,8-bis(salicylideneamino)-3,6-dithiaoctane (**9**) to be equally effective against copper, iron, and cobalt ions.

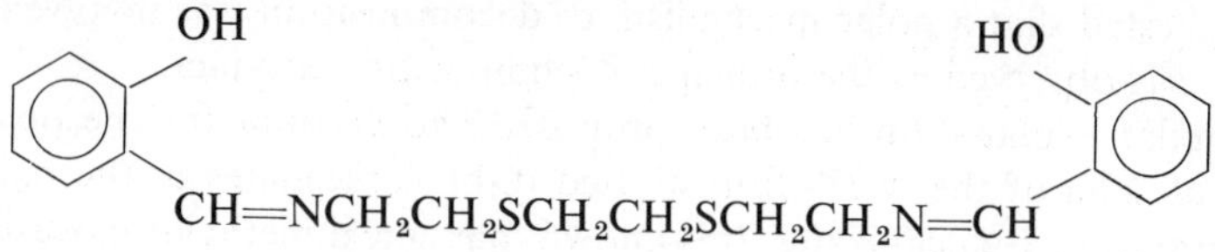

OH HO

$CH{=}NCH_2CH_2SCH_2CH_2SCH_2CH_2N{=}CH$

9

It would appear that the metal ion must be coordinated to its maximum in order to prevent complexing with hydroperoxide and the consequent electron-transfer reactions leading to the production of free radicals capable of initiating autoxidation. Incomplete coordination may in fact result in a prooxidant effect rather than deactivation as shown in Table 2.10. Complexing of the metal ion with hydroperoxide is thus not the only requirement for metal catalysis of autoxidation; it is also necessary to have a favorable oxidation–reduction potential for the electron-transfer reactions. Various ligands alter the relative strength of the metal ion as an oxidizing or reducing agent in a way that serves to enhance the catalytic effect in some cases and retard it in others.

Many amines, such as derivatives of *p*-phenylenediamines, are used as copper deactivators in polymers. Since these compounds also function as chain-breaking antioxidants and in part as peroxide decomposers, the observed activity probably reflects a combination of metal deactivation with these other modes of antioxidant action. The potential advantages of combining more than one mode of antioxidant action are discussed in Section 2-F.

The effect of sulfur containing ligands is also complex in that many of the ligands are also capable of exhibiting antioxidant activity by other mechanisms. Any S–H, N–H, or phenolic O–H group may function by hydrogen transfer to the $RO_2\cdot$ radical to stop the kinetic chain. Many of these compounds also appear to have activity as peroxide decomposers, as described in the following section. Indeed the metal complexes are frequently more effective in this regard than the original ligands. In many cases it is not possible to distinguish stabilization by metal deactivation from the effect of the stabilizer on hydroperoxide decomposition.[9] Both are preventive mechanisms that reduce the rate of peroxide initiation of the autoxidation reaction.

2. Peroxide Decomposers

Materials that promote the decomposition of organic hydroperoxides to form stable products rather than chain-initiating free radicals frequently function as preventive antioxidants in polymers. Kennerly and Patterson[73] demonstrated the ability of a number of compounds to destroy cumene hydroperoxide in a petroleum oil; these compounds included phenols, mercaptans, sulfides, sulfonic acids, and a zinc dialkyl dithiophosphate. It was suggested that a polar mechanism of decomposition was involved, since phenol was observed as the principal decomposition product.

A similar explanation has been proposed[74] to account for the protective action of salts of the *N,N′*-disubstituted dithiocarbamates in the oxidation of natural rubber vulcanizates. It is known that zinc dimethyldithiocarbamate (**10**) is formed in the vulcanization of rubber with tetramethylthiuram disulfide in the presence of zinc salts, and that the salt is extractable with acetone.[75] Since the superior aging resistance of these vulcanizates is destroyed by acetone extraction, it was proposed[76] that this salt was responsible for the original good aging resistance.

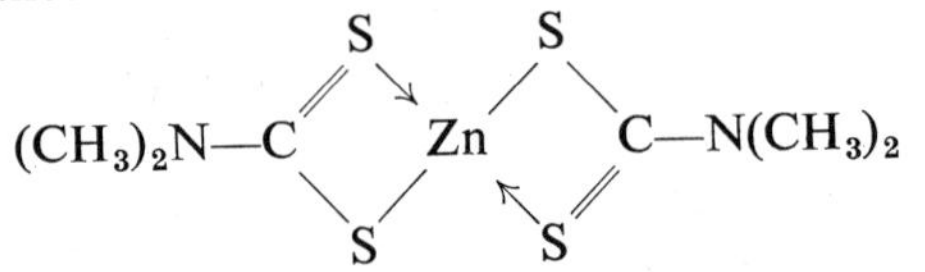

$$(CH_3)_2N{-}C{\overset{S}{\underset{S}{\lessgtr}}}Zn{\overset{S}{\underset{S}{\lessgtr}}}C{-}N(CH_3)_2$$

10

The mechanism of oxidation inhibition by zinc dialkyl dithiophosphates has been reviewed by Burn.[77] Although these compounds have been used as antioxidants for many years, the detailed mechanism of their action is still unknown. It is well established, however, that they are efficient peroxide decomposers and that they are effective as inhibitors of hydrocarbon autoxidation.[73] Studies of the oxidation of squalene[78] at 60°C initiated by azobisisobutyronitrile (AIBN) showed that the rate of oxidation was also retarded in this case by zinc dialkyl dithiophosphates and the related xanthate and dithiocarbamate salts. Since initiation by hydroperoxide would be negligible under these conditions, it was concluded that these salts also function as chain-breaking antioxidants. Since no active hydrogen comparable to those of the amine and phenolic antioxidants is present in these compounds, an electron-transfer process was suggested as illustrated below for a dialkyl dithiocarbamate salt.

$$[R_2NCS_2]_2Zn + RO_2\cdot \rightarrow (R_2NCS_2Zn\overset{+}{\dot{S}}CSNR_2)(RO_2:)^-$$

Burn[79] made a detailed study of the azonitrile-initiated oxidation of both squalane and cumene in the presence of dialkyl dithiophosphates and related compounds. Only compounds containing metal salts of thio anions were effective. A two-stage electron-transfer mechanism was proposed, involving a stabilized zinc salt–peroxy radical complex (**11**), which on attack by a second peroxy radical at the second thio anion of the salt would yield a disulfide (**12**) and two peroxy anions. The disulfide was isolated from a reaction of peroxy radicals and zinc diisopropyl dithiophosphate, lending support for the proposed mechanism summarized in the following reaction sequence:

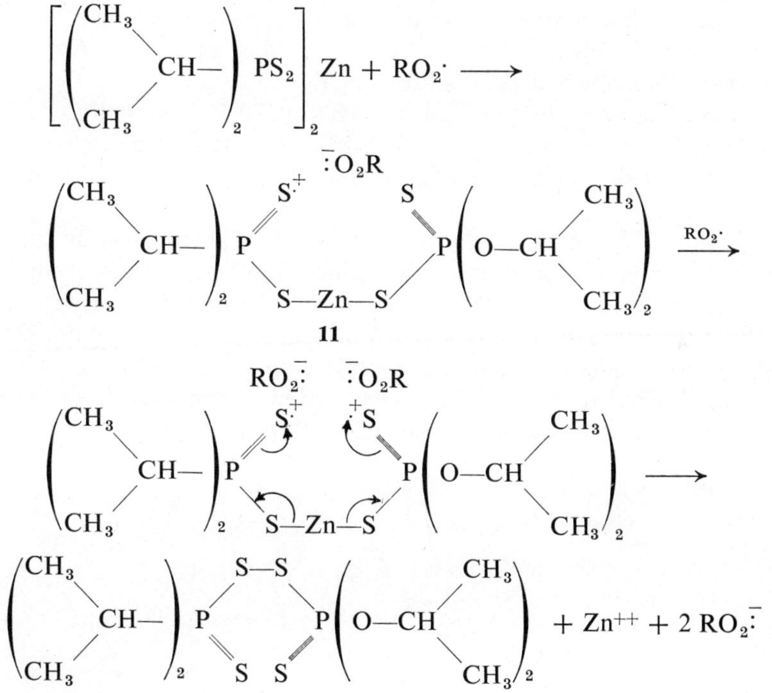

$$\left[\left(\begin{matrix}CH_3\\ \quad CH\text{—}\\ CH_3\end{matrix}\right)_2 PS_2\right]_2 Zn + RO_2\cdot \longrightarrow$$

$$\left(\begin{matrix}CH_3\\ \quad CH\text{—}\\ CH_3\end{matrix}\right)_2 P\begin{matrix}\overset{\overset{-}{:}O_2R}{\overset{+}{\dot{S}}} \qquad S\\ \\ S\text{—}Zn\text{—}S\end{matrix} P\left(O\text{—}CH\begin{matrix}CH_3\\ \\ CH_3\end{matrix}\right)_2 \xrightarrow{RO_2\cdot}$$

11

$$\left(\begin{matrix}CH_3\\ \quad CH\text{—}\\ CH_3\end{matrix}\right)_2 P\begin{matrix}\overset{RO_2\overset{-}{:}}{\overset{+}{\dot{S}}} \quad \overset{\overset{-}{:}O_2R}{\overset{+}{S}}\\ \\ S\text{—}Zn\text{—}S\end{matrix} P\left(O\text{—}CH\begin{matrix}CH_3\\ \\ CH_3\end{matrix}\right)_2 \longrightarrow$$

$$\left(\begin{matrix}CH_3\\ \quad CH\text{—}\\ CH_3\end{matrix}\right)_2 P\begin{matrix}S\text{—}S\\ \\ S \quad S\end{matrix} P\left(O\text{—}CH\begin{matrix}CH_3\\ \\ CH_3\end{matrix}\right)_2 + Zn^{++} + 2\,RO_2\overset{-}{:}$$

12

The author also included bonded radical structures as alternatives to the above ion-pair structures for the complex **11** and its reaction product with a second $RO_2\cdot$ radical. Comparison with the radical representations below shows the same electron pairing in each case, the only difference being the covalent or ion-pair nature of the bonding of RO_2 to sulfur. It seems likely that the true nature of the bond may be something between the two extremes, but with considerable ionic character involved.

```
        ROO                                   ROO      OOR
         |                                     |        |
         S·       S                            S·      ·S
        //         \\                         //         \\
(R'O)2P             P(OR')2         (R'O)2P               P(OR')2
        \          /                          \          /
         S—Zn—S                                S—Zn—S
```

The results of kinetic studies[79] of the AIBN-initiated oxidation of both cumene and tetralin inhibited by zinc diisopropyl dithiophosphate showed evidence of possible side reactions in which the zinc salt is consumed in addition to the chain-termination reaction. For example, the stoichiometric relation deduced from the observed induction period for the number of peroxy radicals reacting with each zinc salt molecule was 1.8, as compared to 2 predicted from the above reaction sequence. Furthermore, when the plots of initial oxidation rates as a function of substrate concentration were extrapolated to 0 hydrocarbon concentration, they gave an intercept on the rate axis. Experimental confirmation of the oxidation of a series of zinc dithiophosphates in the presence of AIBN at 70°C was obtained using *t*-butylbenzene as an inert solvent. No oxidation was observed in the absence of AIBN, but significant rates of oxidation were observed in the presence of the initiator.

The possibility of hydrogen abstraction was rejected in prior studies,[78] but abstraction of hydrogen from various alkyl phosphates by hydroxy radicals has been reported[80] to occur at room temperature in an aqueous medium. Electron-spin resonance (ESR) spectra of the phosphoralkyl radicals were obtained. Thus it is reasonable to consider the following additional reactions resulting from abstraction of hydrogen from the alkyl group of a zinc dialkyl dithiophosphate by $RO_2\cdot$ radical:[77]

```
       H
   \   |                                  \
     C—O      S                             Ċ—O      S
   /     \   //           RO2·             /     \   //
           P            ———————→                   P
         /   \                                   /   \
      RO      S—Zn—                           RO  |   S—Zn—
                                                  | O2
                                                  ↓
        O—H                                       O·
       /                                         /
      O                                         O
   \  |                                      \  |
     C—O      S             RH                 C—O      S
  + /    \   //          ←———————             /    \   //
R·         P                                         P
         /   \                                     /   \
      RO      S—Zn—                             RO      S—Zn—
```

If hydrogen abstraction from the zinc salt is able to compete with abstraction of hydrogen from the hydrocarbon substrate, the above reaction scheme represents a chain-transfer process. However, termination might also occur via an intramolecular complex (**13**) analogous to the complex **11** proposed to account for the chain-stopping activity of the zinc salt. Reaction of **13** with a peroxy radical in the other half of the molecule would give nonradical products similar to the reaction previously illustrated for **11**, leading to the disulfide **12**.

$$\begin{array}{c} \text{O} \\ \diagup\;\;\diagdown \\ \diagdown\text{C}\qquad\text{O:}^- \\ \diagup| \\ \text{O}\qquad\text{S}^{\cdot+}\qquad\text{S}\qquad\text{OR} \\ \diagdown\;/\!/\qquad\quad\backslash\!\backslash\;\;\diagup \\ \text{P}\qquad\qquad\qquad\text{P} \\ \diagup\;\;\diagdown\qquad\quad\diagup\;\;\diagdown \\ \text{RO}\qquad\text{S—Zn—S}\qquad\text{OR} \end{array}$$

13

Although peroxide decomposition is considered to be the major role of the metal dialkyl dithiophosphates and related compounds as preventive antioxidants, the mechanism has not yet been established. Many studies have been made of the effect of the metal and the structure of the alkyl groups; but interpretation of the observed results is complicated by prooxidant effects observed with certain metals and by the chain-breaking mechanisms of inhibition by both hydrogen and electron transfer, which are responsible for part of the observed effect of some of these compounds upon the rate of autoxidation. The catalytic nature of the peroxide decomposition has been clearly shown[81] for the decomposition of cumene hydroperoxide by metal xanthates, dithiocarbamates, and dithiophosphates. Complete decomposition was obtained with only 1 mole % of these compounds relative to the hydroperoxide. High yields of phenol and acetone indicate an electrophilic attack by the metal derivative or some oxidation or decomposition product according to the following scheme, in which E is the electrophilic reagent:

$$C_6H_5C(CH_3)_2OOH \xrightarrow{E} E\bar{O}H + C_6H_5C(CH_3)_2O^+ \longrightarrow$$

$$C_6H_5O\text{—}\overset{+}{C}(CH_3)_2 \xrightarrow[\text{[or } C_6H_5C(CH_3)_2OOH]]{EOH^-} C_6H_5OH + (CH_3)_2CO + E \quad [\text{or } C_6H_5C(CH_3)_2O^+]$$

Table 2.11 compares the effect of a series of metal dialkyl dithiophosphates upon the length of the induction period for the oxidation of squalane at 140°C. The superiority of the zinc salts is evident as well as the prooxidant effect of both nickel and iron salts. Variation of the alkyl groups for a given

Table 2.11 Effect of Metal Dialkyl Dithiophosphates, $[(RO)_2PS_2]_xM$(4×10^{-5} g-atoms P/liter), on the Oxidation of Squalane at 140°C[77]

Metal, M	x	Alkyl Group, R	Induction Period, (min)
Zinc	2	4-Methyl-2-pentyl	119 ± 9
		n-Hexyl	67 ± 5
		Isopropyl	80 ± 12
		n-Propyl	65 ± 10
		Ethyl[a]	70
Antimony	3	4-Methyl-2-pentyl	95 ± 11
Cadmium	2	4-Methyl-2-pentyl	79 ± 5
		n-Hexyl	30 ± 2
Lead	2	4-Methyl-2-pentyl	67 ± 6
		n-Hexyl	25 ± 0.5
Bismuth	3	4-Methyl-2-pentyl	45 ± 0.5
		n-Hexyl	49 ± 2
Iron	3	4-Methyl-2-pentyl	2.5
Nickel	2	4-Methyl-2-pentyl	2.5
		n-Hexyl	2.5
		Isopropyl	2.5
None (disulfide)	2	4-Methyl-2-pentyl	5 ± 0.5
		n-Hexyl	4 ± 0.5
		Isopropyl	4 ± 0.5
None	—	None	4 ± 0.5

[a] For $[(C_2H_5)_2PS_2]_2Zn$ (5×10^{-4} mole/liter) at 160°C.

metal shows that secondary alkyl esters of dithiophosphates are more effective than primary, as illustrated by 4-methyl-2-pentyl,

$$>P-OCH(CH_3)CH_2CH(CH_3)_2,$$

as compared to a *n*-hexyl ester.

Most workers have agreed that a reaction product of the metal salts is the active antioxidant. An acidic thermal decomposition product formed in a β-elimination reaction would account for the observed products according to the reactions illustrated above. However Burn[77] considers that such a decomposition product is unlikely to be the effective peroxide decomposer for the following reasons:

1. Peroxide decomposition by zinc salts has been observed at room temperature when the thermal decomposition of these salts could not be responsible.

2. The disulfide has comparable thermal stability and is inactive.

3. The dithiophosphinate, $[(C_2H_5)_2PS_2]_2Zn$, which cannot undergo β-elimination (since it is not an ester) still strongly inhibits the oxidation of squalane.

In view of the evidence of a catalytic decomposition of hydroperoxide, Burn[77] suggests a possible complexing of ROOH with the salt and the formation of a cyclic transition state (**14**) leading to the observed products in the case of cumene hydroperoxide:

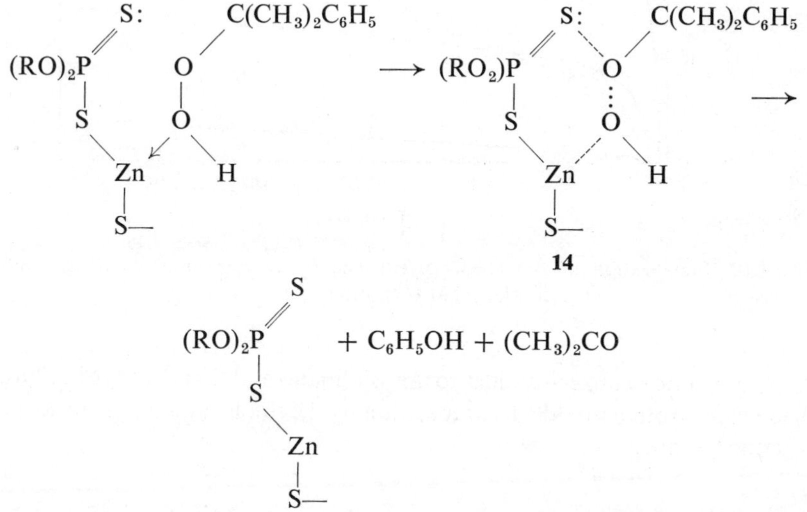

$$(RO)_2P(=S:)\text{—}S\text{—}Zn(\text{—}S\text{—}) \leftarrow O(H)\text{—}O\text{—}C(CH_3)_2C_6H_5 \longrightarrow (RO_2)P(=S:)\text{—}S\text{—}Zn(\text{—}S\text{—}) \cdots O(H)\cdots O\text{—}C(CH_3)_2C_6H_5 \;(\mathbf{14}) \longrightarrow$$

$$(RO)_2P(=S)\text{—}S\text{—}Zn\text{—}S\text{—} + C_6H_5OH + (CH_3)_2CO$$

No single mechanism of peroxide decomposition seems adequate to explain all the observed phenomena. Burn[77] concludes: "It is almost certainly a mistake to assume that the active intermediates involved in hydroperoxide decomposition by sulfur compounds are identical, even for such formally similar compounds as dithiophosphates and dithiocarbamates."

Many alkyl and aryl sulfides and disulfides, as well as elemental sulfur, have been shown to inhibit polymer oxidation. Aryl disulfides, for example, inhibit polyolefin oxidation.[82,83] Figure 2.18 shows effective inhibition of the thermal autoxidation of branched polyethylene at 140°C with 0.3% of 2-naphthyl disulfide. Lower concentrations of the inhibitor in this system were only effective after substantial reaction of the substrate with oxygen.[82] This suggests that it is not the disulfide, but rather oxidation products which are the effective inhibitors.[7] Studies of the Natural Rubber Producers' Research Association[84] (NRPRA) have shown that thiolsulfinates, which are formed by reaction of disulfides with hydroperoxides, immediately inhibit the oxidation of squalene in contrast to the delayed stabilization observed with the corresponding disulfides. It has been proposed[82,84] that peroxides formed

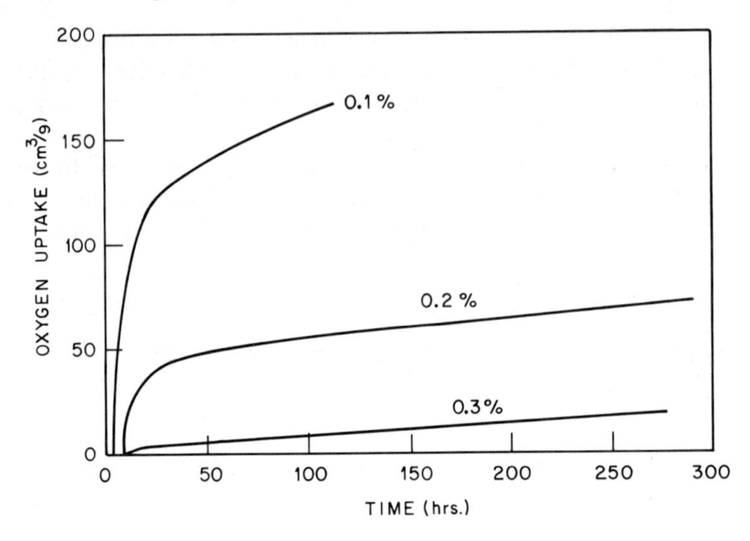

Fig. 2.18 Stabilization of branched polyethylene by varying amounts of 2-naphthyl disulfide (140°C, 1 atm O_2).[83]

in the substrate oxidize disulfides to thiolsulfinates and that effective inhibition then results from peroxide decomposition by the thiolsulfinate, or subsequent reaction products.

$$\text{RSSR} \xrightarrow{\text{R'OOH}} \text{R}\overset{\overset{\text{O}}{\uparrow}}{\text{S}}\text{SR}$$

$$\text{R'OOH} \xrightarrow{\text{R}\overset{\overset{\text{O}}{\uparrow}}{\text{S}}\text{SR}} \text{nonradical products}$$

Hawkins and Sautter[82] noted the thermal instability of phenyl benzenethiolsulfinate in the temperature range at which disulfides effectively inhibit the oxidation of polyethylene. They suggested that acidic decomposition products of the thiolsulfinate, including sulfur dioxide, may play an important role in the mechanism by which these inhibitors function. The relative contribution of the thiolsulfinate and its decomposition and oxidation products is probably a function of the temperature of oxidation.[7]

Figure 2.19 compares the effect of a peroxide decomposer with that of a chain-breaking hindered phenol when added to cumene undergoing autoxidation at 120°C. The hindered phenol (curve A) was effective immediately when added to the mixture after a substantial hydroperoxide concentration had been attained, but there was little change in hydroperoxide concentration during the period in which the stabilizer was effective. In contrast, oxidation

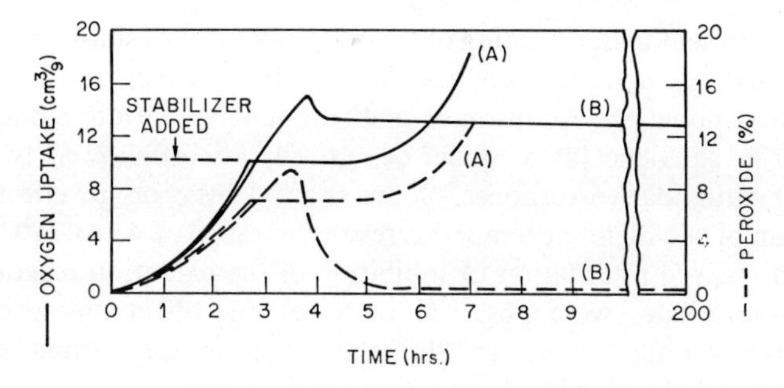

Fig. 2.19 Relation between peroxide content and stabilization in cumene oxidizing at 120°C: antioxidants: (*A*) 2,2′-methylenebis-(4-methyl-6-*t*-butylphenol); (*B*) phenyl disulfide.[83]

continued for some time after addition of the disulfide, (*B*) and hydroperoxide continued to accumulate. Eventually inhibition was observed, accompanied by complete decomposition of hydroperoxide.

A similar behavior was observed upon addition of the disulfide to solutions of cumene hydroperoxide or *t*-butyl hydroperoxide in an inert atmosphere at 100°C. A delayed decomposition occurred to form products from cumene hydroperoxide consistent with a heterolytic process as described above. The corresponding thiolsulfinate decomposed these hydroperoxides immediately at 100°C, but at a much slower rate at 40°C. Sulfur dioxide was even more effective, giving complete decomposition in minutes even at 0°C. One molar equivalent of the sulfur compound can decompose many moles of hydroperoxide consistent with a catalytic process. Other types of peroxide, such as the diperoxides formed in natural rubber and other polymers of 1,3 dienes, are resistant to decomposition by sulfur compounds.[7]

Activation of carbon black with sulfur yields a stabilizer that inhibits polyethylene oxidation for more than 2000 hours at 140°C. Hawkins[7] suggests that the sulfur-activated carbon black may also function as a peroxide decomposer, through reactions similar to those of other sulfur compounds. It is not certain whether carbon black functions as an inhibitor in sulfur-vulcanized rubbers, since a prooxidant effect is actually observed.[14] It has been suggested[85] that the accelerating effect may be caused by adsorption of antioxidant on the carbon surface, making the antioxidant unavailable as an inhibitor.

An extensive study of the autoxidation of organic sulfides, both alone and in the presence of unsaturated hydrocarbons, has been carried out in the laboratories of the Natural Rubber Producers' Research Association.[17] The nature of the oxidation products, like the observed effect of the various organic sulfur compounds upon the rate of autoxidation, has contributed much to our present understanding of the aging behavior of sulfur vulcanizates of natural and synthetic rubbers and the stabilizing effect of certain sulfur compounds in polymers. Addition of small amounts of various sulfides to an olefin such as squalene (as a model of natural rubber) showed two broad patterns of autoxidative response. Some showed little or no effect on the autoxidation of the olefin at temperatures in the range of 45 to 75°C. Others produced a marked retardation or inhibition of the oxidation reaction.

Most monosulfides were observed to have little effect unless one substituent was a *tert*-alkyl group and the other either another tertiary group or an allyl group alkylated in the 1- and 3 positions. A phenyl group was usually equivalent to a simple alkyl substituent, but increased activity was observed with a suitable substituted allylthio group. Examples of active and inactive sulfides included in the study are shown in Table 2.12. In contrast to monosulfides, simple alkyl disulfides showed considerable activity, which diminished in the case of the more complex alkyl and aryl compounds as shown in Table 2.13. It should be noted that the substituted allyl *tert*-alkyl sulfide grouping (which occurs in unaccelerated-sulfur vulcanizates of natural rubber) can confer marked antioxidant activity, whereas dialkenyl mono- and disulfides which are the characteristic crosslink structures in most accelerated-sulfur vulcanizates) are relatively inert in this respect.[17]

Table 2.12 Examples of Active and Inactive Sulfides as Inhibitors of Olefin Autoxidation[86]

"Inactive" Sulfide	Retardation Ratio[a]	"Active" Sulfide	Retardation Ratio[a]
n-Bu_2S	1.7[b]		
t-$BuSCH_3$	1.4	t-Bu_2S	256
t-$BuSC_6H_5$	1.4		
$CH_2{=}CHCH_2St$-Bu	1.5		
$MeCH{=}CHCHCH_3SCH_3$	3.1	$CH_3CH{=}CHCHCH_3St$-Bu	93,310[b]
$[(CH_3)_2C{=}CHCH(CH_3)]_2S$	4.4	$(CH_3)_2C{=}CHCHCH_3St$-Bu	276
$C_6H_5CH{=}CHCHC_6H_5SC_6H_5$	2.3	$CH_3CH{=}CHCHCH_3SC_6H_5$	270[b]

[a] Retardation ratio is a measure of inhibitory efficiency defined as the ratio of times for 1% (weight percent) oxygen uptake by olefin alone and by olefin plus sulfide (at 0.25 *M*).

[b] At 400-mm O_2 pressure, otherwise 760 mm.

Monosulfides with keto substituents in one alkyl side chain may be active or inactive depending upon the relative position of the carbonyl group

Table 2.13 Relative Activity of Disulfides as Inhibitors of Olefin Autoxidation[86]

Disulfide	Retardation Ratio	Disulfide	Retardation Ratio
$(i\text{-PrS})_2$	66	$(t\text{-BuS})_2$	1.6
$(n\text{-BuS})_2$	121	$[CH_3(CH_2)_4C(CH_3)_2S]_2$	1.7
$[CH_3(CH_2)_5S]_2$	51	$[(C_6H_5)S]_2$	2
$[(CH_3)_2C{=}CHCHCH_3S]_2$	9	$[(C_6H_5)CH_2S]_2$	8

relative to sulfur. Substantial retardation is observed if the sulfur is attached to a carbon which is β to the carbonyl group, provided the carbon β to sulfur (α to carbonyl) bears one or more hydrogen atoms as in the grouping $-S-CH_2-CH_2-\overset{|}{C}=O$. More branched alkyl substituents enhance this activity,[87] as illustrated by retardation ratios of 330 for $t\text{-BuCO(CH}_2)_2\text{S-}t\text{-Bu}$ as compared to 27 for $CH_3CO(CH_2)_2S-CH_3$ and the corresponding ratios in Table 2.12 for active and inactive sulfides.

Organic sulfides that were found to be active as stabilizers for thermal autoxidation of model compounds of polymers showed their inhibitory power only after the absorption of a measurable amount of oxygen, as shown in Figure 2.20.[17] This observation suggests that it is not the sulfides but rather certain of their oxidation products that function as the active inhibitors of autoxidation. The various possible oxygenated products of the active sulfides and disulfides were checked, and it was found that the only active oxygenated products formed by reaction with hydroperoxides are the sulfoxides derived from monosulfides and thiolsulfinates formed from disulfides. Consistent with these observations, addition of the active sulfoxides and thiolsulfinates showed inhibition at very much lower concentrations than the parent sulfides and showed this effect immediately, without requiring any oxygen uptake by the system as illustrated by the examples in Figure 2.21.[17]

Although the inhibitory effect of the sulfoxides and thiolsulfinates are observed at much lower concentrations than the sulfides and disulfides from which they are derived, the period of time in which they remain active is considerably reduced. This reflects the instability of these compounds, which appears to be related to the inhibitory action, especially in the case of the sulfoxides where a parallelism is evident in these properties.[88] Instability of the sulfoxide appears to be an important requirement for activity as an inhibitor, but there is a limit to the parallelism, since a compound that is too unstable will not exist long enough to provide adequate protection. It is for this reason that the parent sulfides and disulfides may be preferable as practical antioxidants; these compounds function as a reservoir from which

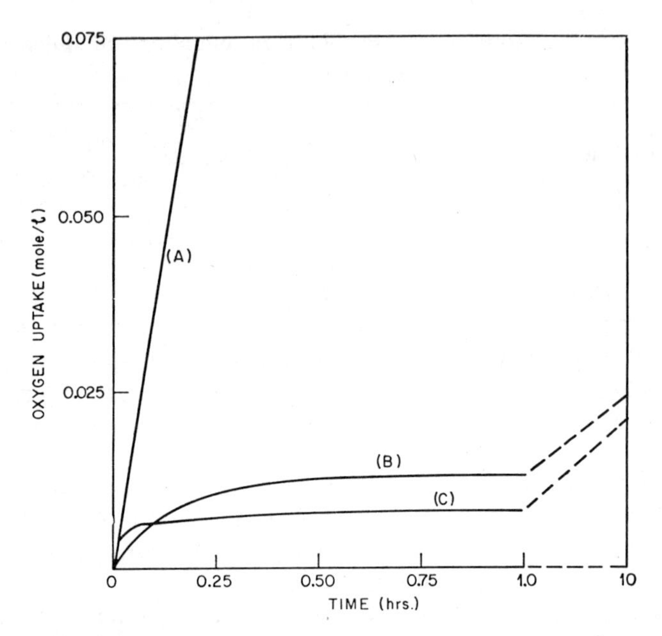

Fig. 2.20 Cooxidation of squalene and sulfides (0.25 *M*), (750°C, 760 mm O_2):[17,84] (*A*) squalene alone; (*B*) squalene + di-*t*-buty sulfide; (*C*) squalene + 1,3,3-trimethylallyl *t*-butyl sulfide.

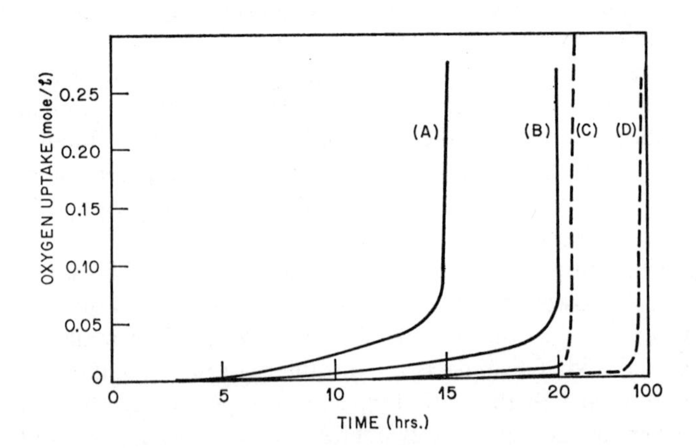

Fig. 2.21 Cooxidation of squalene with sulfoxides and thiolsulfinates (75°C, 760 mm O_2):[17,84] (*A*) squalene + *t*-Bu_2SO (0.01 *M*); (*B*) squalene + *n*-BuSSO*n*-Bu (0.005 *M*); (*C*) squalene + $(CH_3)_2C = CHCHCH_3SO$*i*-Pr (0.01 *M*); (*D*) squalene + *t*-BuSSO*t*-Bu (0.005 *M*).

the active inhibitor can be generated as required when hydroperoxides are formed in the autoxidation of the substrate.

Three factors apparently contribute to the observed behavior of organic sulfides as inhibitors in the systems studied.[17] The first is the ease of oxidation to the mono–oxy derivative. The second is the derivative's intrinsic inhibitory power. The third is an optimum stability for the derivative; because of this factor, the instability required for activity is balanced by stability to persist long enough to perform its function as a preventive antioxidant, presumably by decomposing hydroperoxide to inactive products. The interrelation of these three factors and the optimum stability of the oxyderivative vary with the nature of the substrate and the oxidation conditions encountered. Thus it has been observed[86] that different systems, or the same system at different temperatures, may respond differently to these inhibitors.

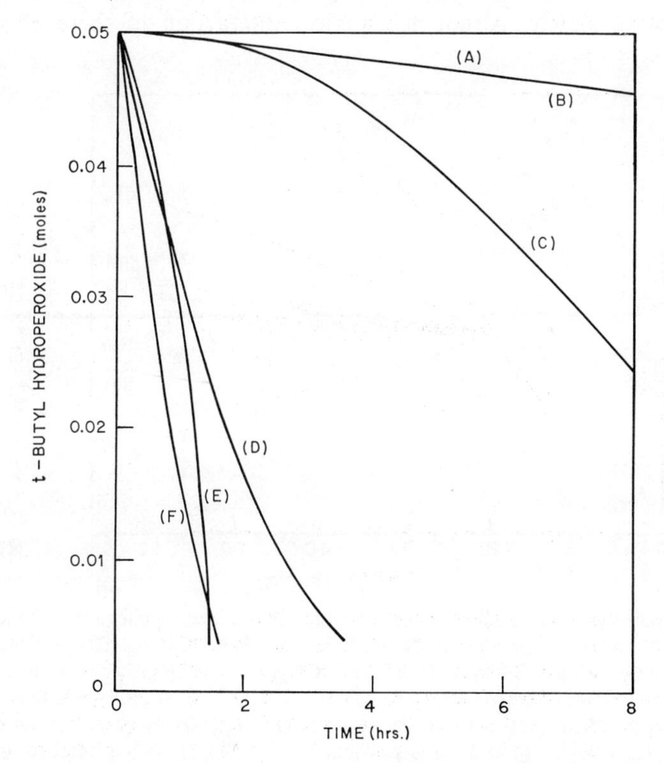

Fig. 2.22 Decomposition of *t*-butyl hydroperoxide by sulfoxides (0.05 *M*) and thiolsulfinates (0.05 *M*) in benzene at 75°C:[17,88] (*A*) no addition, $(CH_3)_2SO$, $(C_2H_5)_2SO$, *i*-Pr_2SO, *n*-Bu_2SO, $(C_6H_5CH_2CH_2)_2$-SO, or 2,6-di-*t*-butyl-*p*-cresol (0.05 *M*); (*B*) *t*-$BuCOC(CH_3)_2CH_2SOCH_3$, $CH_3CO(CH_2)_3SOCH_3$, or *t*-$BuCO(CH_2)_3SO$*t*-Bu; (*C*) $CH_3CO(CH_2)_2SOCH_3$; (*D*) *t*-$BuCO(CH_2)_2SO$*t*-Bu; (*E*) *t*-Bu_2SO; (*F*) *t*-BuSSO*t*-Bu.

Sulfoxides and thiolsulfinates derived from organic sulfides and disulfides are active as oxidation inhibitors only when hydroperoxide decomposition is involved in the autoxidation. They have little or no effect when initiation is caused by radicals produced from azobisisobutyronitrile, benzoyl peroxide, or perester decomposition. Thus they are clearly preventive antioxidants rather than chain-breaking antioxidants. Since these compounds have been shown to be effective hydroperoxide decomposers (Figure 2.22,[17]) their activity has frequently been attributed to conversion of hydroperoxides to inactive products without forming free radicals. This explanation is open to question, however, in view of the effects shown in Figure 2.23[17] of addition of various stabilizers to squalene as it undergoes autoxidation at 75°C after a 20-minute period during which a significant concentration of hydroperoxide had accumulated. Immediate retardation (or inhibition, in the case of the more effective materials) was observed; however the hydroperoxide concentration decreases rather slowly. Although chemical interaction of these effective

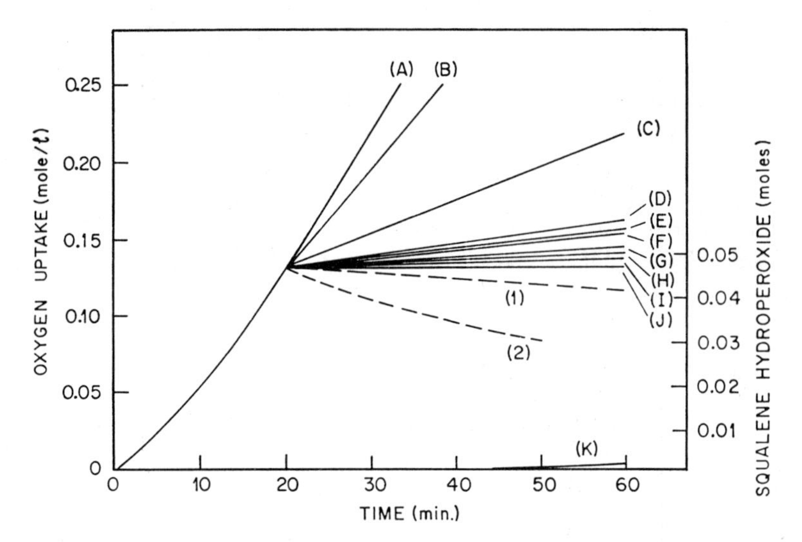

Fig. 2.23 Effect of various stabilizers when added to autoxidizing squalene after 20 min at 75°C.[17,88] (*A*) No addition, or *t*-BuSOCH$_3$ (0.25 *M*), or *t*-BuCOC(CH$_3$)$_2$CH$_2$SOCH$_3$ (0.01 *M*) added initially. (*B*) *t*-BuSSO*t*-Bu (0.005 *M*), (CH$_3$)$_2$C═CHCHCH$_3$SO*i*-Pr (0.005 *M*), or *t*-BuSO-*i*Pr (0.05 *M*) and stearic acid (0.05 *M*). (*C*) (CH$_3$)$_2$C═CHCHCH$_3$SO*i*-Pr (0.05 *M*) and stearic acid (0.25 *M*). (*D*) *t*-BuSO*i*-Pr, or *t*-BuCO(CH$_2$)$_2$SO*t*-Bu (0.05 *M*) and stearic acid (0.25 *M*). (*E*) *t*-Bu$_2$SO (0.05 *M*) and stearic acid (0.25 *M*). (*F*) β-Naphthol (0.005 *M*). (*G*) (CH$_3$)$_2$C═CHCHCH$_3$SO*i*-Pr (0.05 *M*). (*H*) *t*-Bu$_2$SO (0.005 *M*), or *t*-BuCO(CH$_2$)$_2$SO*t*-Bu (0.05 or 0.005 *M*). (*I*) *t*-BuSSO*t*-Bu (0.05 *M*). (*J*) *t*-Bu$_2$SO (0.05 *M*). (*K*) *t*-BuSO*i*-Pr (0.008 *M*), or *t*-Bu$_2$SO (0.008 *M*) both added initially; (1) variation of hydroperoxide content for *t*-BuSO*i*-Pr (0.05 *M*), or (CH$_3$)$_2$C═CHCH(CH$_3$)SO*i*-Pr (0.05 *M*); (2) variation of hydroperoxide content for *t*-BuCO(CH$_2$)$_2$SO*t*-Bu (0.05 *M*), or *t*-Bu$_2$SO (0.05 *M*).

inhibitors with hydroperoxide is clearly involved, some other mechanism must also be considered to account for the effective inhibition observed, even in the presence of undecomposed hydroperoxide. Some form of molecular association is proposed by Bateman and his co-workers at NRPRA[17,88] between the hydroperoxide and the SO-compound. This interpretation is supported by the observation that it is usually necessary for the SO-compound to be present in equivalent (or preferably in excess) concentration to the hydroperoxide to obtain immediate effective inhibition. The relation of activity of the sulfoxides to their proneness to undergo decomposition is also recognized as an important factor, but how this relates to the mechanism of inhibition remains to be determined.[17] The nature of the decomposition reaction and the identity of the initial products are discussed in the section of the chapter based on work in the present author's laboratory.

One of the most widely used sulfur compounds for the stabilization of polyolefins is dilauryl thiodipropionate, **15**, (DLTDP). The kinetics of the decomposition of tetralin hydroperoxide and of *t*-butyl hydroperoxide by DLTDP and related compounds has been studied by Marshall[89] to gain some insight into the mechanism by which sulfur compounds of this type function as stabilizers against thermal autoxidation. Rate studies in the temperature range of 30 to 90°C showed an initial slow reaction or induction period followed by a faster main reaction. The length of the induction period decreased as the initial concentration of hydroperoxide or sulfur compound was increased, and the subsequent rate of oxidation was first order in both reactants, at least until the amount of hydroperoxide decomposed approached a value equal to the initial concentration of sulfur compound. After this point the rate of hydroperoxide decomposition increased.

$$H_{25}C_{12}O{-}\overset{\overset{\displaystyle O}{\|}}{C}CH_2CH_2{-}S{-}CH_2CH_2\overset{\overset{\displaystyle O}{\|}}{C}{-}OC_{12}H_{25}$$

15

The addition of dilauryl β,β'-sulfinyldipropionate (DLTDP sulfoxide, **16**- had little effect on the length of the induction period, but the rate of the subsequent reaction was increased as shown in Figure 2.24.[89] The observed rate of hydroperoxide decomposition in the presence of the sulfoxide (curve *B*) was 8.6 times faster than the rate of decomposition in the presence of the corresponding sulfide (curve *A*). It is unlikely, however, that dilauryl β,β'-sulfinyldipropionate is the active species in this reaction, since addition of this sulfoxide failed to eliminate the induction period. Rather, the formation of some more reactive species from further oxidation and decomposition reactions is suggested to account for the catalytic nature of the overall

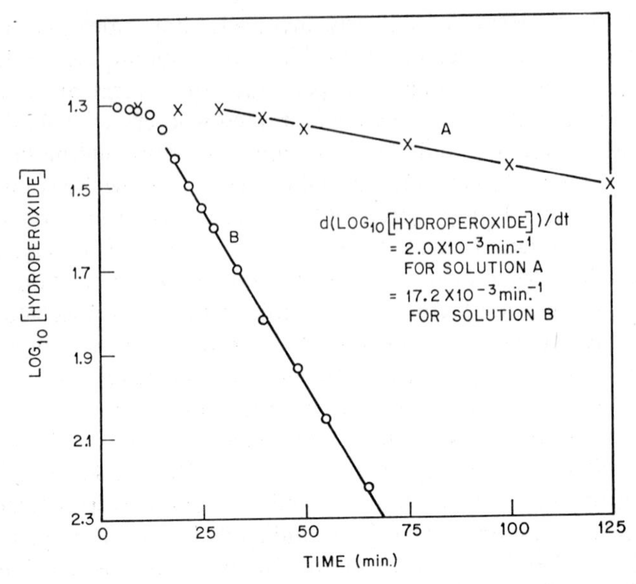

Fig. 2.24 Decomposition of tetralin hydroperoxide (5×10^{-2} *M*) by:[89] (*A*) dilauryl thiodipropionate (1.6×10^{-2} *M*); (*B*) dilauryl sulfinyldipropionate (1.6×10^{-2} *M*).

reaction in which at least 20 moles of hydroperoxide was eventually decomposed per mole of sulfur compound.

$$H_{25}C_{12}O{-}\overset{\overset{\displaystyle O}{\|}}{C}CH_2CH_2{-}\overset{\overset{\displaystyle O}{\uparrow}}{S}{-}CH_2CH_2\overset{\overset{\displaystyle O}{\|}}{C}{-}OC_{12}H_{25}$$

16

Further understanding of the mechanism by which organic sulfur compounds function as stabilizers against thermal autoxidation clearly requires additional information about the decomposition of sulfoxides and the kinds of products formed. It was with this objective in mind that we recently undertook a study of the thermolysis of a series of dialkyl and alkyl aryl sulfoxides. Previous workers[90–93] had shown that sulfoxides with one or more hydrogens on a carbon β to the sulfur atom undergo decomposition at moderate temperatures to give olefins by a stereospecific cis-elimination reaction.[90] Only the olefin product was isolated in most of these studies, but Colclough and Cunneen[91] also obtained *t*-butyl *t*-butanethiolsulfinate from the thermolysis of di-*t*-butyl sulfoxide at 75°C, and Neureiter and Bown[92] obtained several products, including thiolsulfinate along with acrylate ester from the decomposition of dilauryl β,β'-sulfinyldipropionate (16).

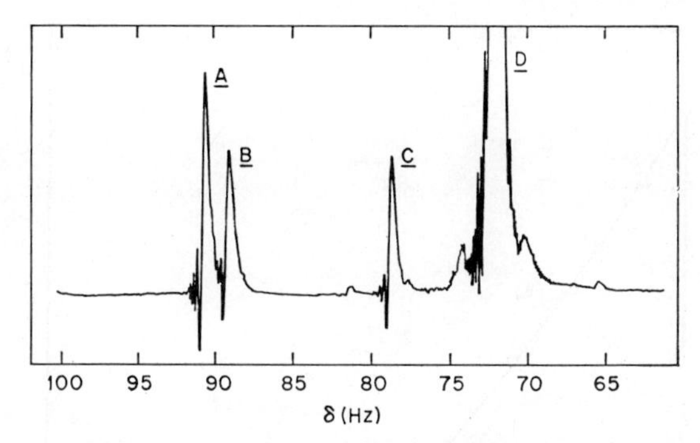

Fig. 2.25 NMR spectrum of reaction mixture of 0.2 *M* di-*t*-butyl sulfoxide in benzene after heating 1 hr at 80°C.[95]

Cunneen and co-workers[17,91] postulated the formation of a sulfenic acid, RSOH, in the cis-elimination reaction to form olefin from decomposition of di-*t*-butyl sulfoxide; they regarded this compound as an intermediate in the formation of the thiolsulfinate, although they were unable to demonstrate its presence in the reaction mixture. Our report[94] of the detection of *t*-butylsulfenic acid in solution by nuclear magnetic resonance (NMR) from the thermolysis of di-*t*-butyl sulfoxide, and the trapping of the compound by addition to certain electrophilic olefins and acetylenes was the first instance in which the existence of a simple aliphatic sulfenic acid was demonstrated. The NMR spectrum in Figure 2.25 shows the absorptions attributed to the hydrogens of the *t*-butyl groups in the sulfenic acid, *A*, and the thiolsulfinate, *B* and *C*, along with undecomposed sulfoxide, *D*. Subsequent condensation of two moles of sulfenic acid would account for the observed formation of thiolsulfinate:

$$(CH_3)_3C\overset{\overset{O}{\uparrow}}{S}C(CH_3)_3 \xrightarrow{65\text{ to }100°C} (CH_3)_3CSOH + (CH_3)_2C{=}CH_2$$

$$\underset{t\text{-Butanesulfenic acid}}{2(CH_3)_3CSOH} \xrightarrow{-H_2O} \underset{t\text{-Butyl }t\text{-butanethiolsulfinate}}{(CH_3)_3C\overset{\overset{O}{\uparrow}}{S}SC(CH_3)_3}$$

The formation of the sulfenic acid intermediate as a function of reaction time is shown in Figure 2.26 for the thermolysis of di-*t*-butyl sulfoxide in benzene at 80°C. The concentrations of the sulfoxide and reaction products at various reaction times were calculated by integration of the NMR peaks for the *t*-butyl hydrogens with a correction made for isobutylene formed and volatilized from the reaction mixture. The *t*-butanesulfenic acid formed from

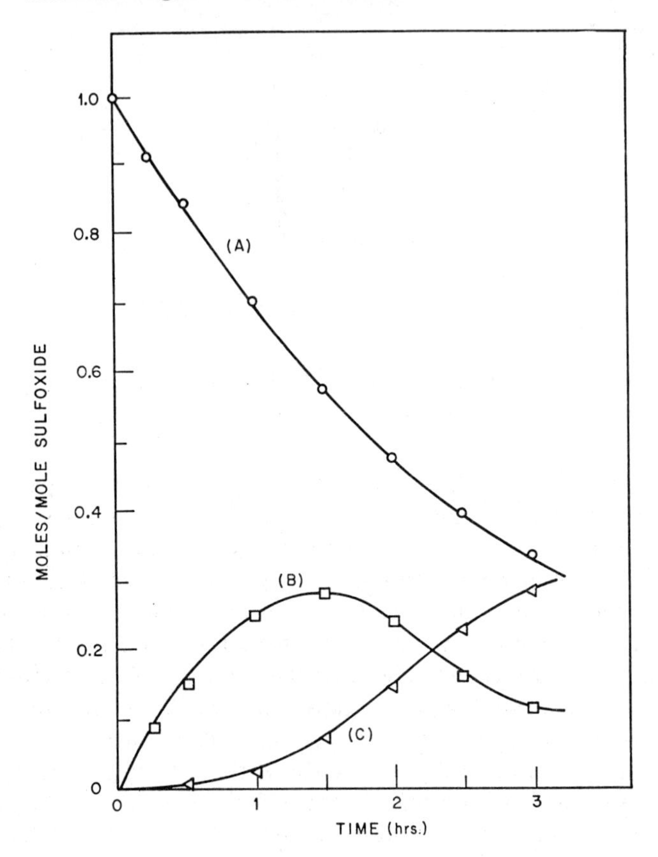

Fig. 2.26 Concentrations by NMR versus time of heating 0.5 *M* di-*t*-butyl sulfoxide in benzene at 80°C:[95] (*A*) sulfoxide; (*B*) sulfenic acid; (*C*) thiolsulfinate.

the sulfoxide decomposition reached a maximum concentration after 1.5 hours and then decreased as it was converted to the corresponding thiolsulfinate. It seems likely that the sulfenic acid and its oxidation products may play a major role in the peroxide decomposition reaction associated with inhibition by sulfoxides formed from sulfides used as preventive antioxidants in polymers.

Since the relative ease of decomposition of sulfoxides is an important aspect of their activity as oxidation inhibitors, it is of interest to compare their relative rates of decomposition reported in Table 2.14. The relative rates shown in the last column were calculated from the observed first-order rate constants at 100°C (as determined by gas–liquid partition chromatography measurements either of the residual sulfoxide or of the olefin formed) taking into

Table 2.14 Decomposition of Sulfoxides at 100°C[96]

R_1–S(→O)–R_2 R_1	R_2	First-Order Rate Constants[a] [$k(10^6)$/sec]	Relative k[b]
n-C_3H_7	n-C_3H_7	0.06 ± 0.01	0.02
i-C_3H_7	i-C_3H_7	6.0 ± 0.4	0.70
CH_3	$C(CH_3)_3$	6.3 ± 0.3	1.0
i-C_3H_7	$C(CH_3)_3$	46 ± 2	4.4
$C(CH_3)_3$	$C(CH_3)_3$	1170 ± 20	93
C_6H_5	i-C_3H_7	1.3 ± 0.1	0.31
C_6H_5	$C(CH_3)_3$	205 ± 8	33
n-C_7H_{15}	$CH_2CH_2CO_2C_2H_5$	390[c]	280
$CH_2CH_2CO_2C_2H_5$	$CH_2CH_2CO_2C_2H_5$	850[c]	300

[a] 0.05 M R_2SO in $C_6H_5CH_3$; rates by gas liquid partition chromatography.
[b] Corrected for number of β-Hs.
[c] Extrapolated from data at lower temperatures, 0.015 M in o-$C_6H_4Cl_2$.

account the number of hydrogens β to the SO-group in each case. A large steric acceleration effect is evident in the case of di-t-butyl sulfoxide, which shows a relative rate 93 times that of methyl t-butyl sulfoxide. The usual reactivity of the β-sulfinylpropionate esters is evident in the rate acceleration of approximately 15,000 relative to the thermolysis of di-n-propyl sulfoxide. This increased reactivity reflects the activating effect of the carbonyl in stabilizing the developing carbanion in the transition state as the hydrogen on the adjacent carbon is being transferred to the sulfinyl oxygen:

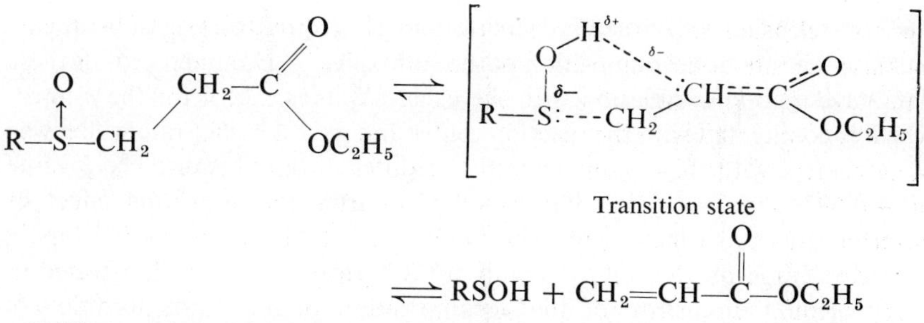

$$R{-}\overset{O}{\overset{\uparrow}{S}}{-}CH_2{-}CH_2{-}\overset{O}{\overset{\|}{C}}{-}OC_2H_5 \rightleftharpoons \left[R{-}\overset{\delta-}{S}(O{\cdots}\overset{\delta+}{H}){\cdots}CH_2{\cdots}\overset{\delta-}{C}H{=}C({=}O)OC_2H_5 \right]$$

Transition state

$$\rightleftharpoons RSOH + CH_2{=}CH{-}\overset{O}{\overset{\|}{C}}{-}OC_2H_5$$

Evidence of the reversibility of the above reaction in the case of the β-sulfinylpropionates was observed as a deviation from linearity in the first-order rate plots, which gave lower values as the reaction progressed. The reverse reaction was subsequently verified[94,95] by using ethyl acrylate as a trapping agent for several of the sulfenic acids formed in the thermolysis of the sulfoxides listed in Table 2.14. Simple olefins such as isobutylene formed in the decomposition of *t*-butyl sulfoxides do not react with sulfenic acids. Consequently the decomposition reaction is not reversible in those cases, consistent with the observation that all the compounds listed in Table 2.14 gave good first-order plots except for the β-sulfinylpropinate esters.

The effect of substituents on the benzene ring upon the rate of thermolysis of phenyl *t*-butyl sulfoxide is reflected in the rate constants observed at 101°C for a series of aryl *t*-butyl sulfoxides listed in Table 2.15. It is evident that

Table 2.15 Decomposition of Aryl *t*-Butyl Sulfoxides at 101°C[96]

R	Ar ($R-S(\rightarrow O)-Ar$)	First-Order Rate Constants[a] $[k(10)^4/\text{sec}]$
t-Bu	*p*-$CH_3OC_6H_4$	0.97 ± 0.04
t-Bu	*p*-$CH_3C_6H_4$	1.18 ± 0.08
t-Bu	*m*-$CH_3C_6H_4$	1.76 ± 0.02
t-Bu	C_6H_5	2.05 ± 0.08
t-Bu	*m*-$CH_3OC_6H_4$	2.27 ± 0.09
t-Bu	*p*-ClC_6H_4	2.42 ± 0.05
t-Bu	*m*-ClC_6H_4	3.22 ± 0.15
t-Bu	*m*-$NO_2C_6H_4$	4.97 ± 0.18
t-Bu	*p*-$NO_2C_6CH_4$	9.04 ± 0.26

[a] 0.10 *M* ArRSO in $C_6H_5CH_3$ (except *m*- and *p*-NO_2, 0.20 *M*).

electron-releasing substituents decrease and electron-attracting substituents increase the rate of decomposition of the sulfoxides. A Hammett plot of these data gave a good correlation with the usual σ values except for the *p*-nitro, which is conjugated with the reaction center and gave a higher rate which was in agreement with the σ value from the acidities of thiophenols. The ρ value of +0.695 observed (excluding *p*-NO_2) confirms the activating effect of electron-attracting substituents, which indicates a stabilization of a developing negative charge on the sulfinyl group in the transition state as illustrated in the preceding discussion of the decomposition of a β-sulfinylpropionate ester. The reaction is thus considered to proceed by a concerted polar mechanism in which the extent of carbon–sulfur and carbon–hydrogen bond breaking in the transition state varies with the nature of the alkyl and aryl groups involved,[96] as well as with temperature.

The mechanisms by which preventive antioxidants function to provide stabilization against thermal autoxidation are evidently complex, and they are not as well established as the mechanisms by which chain-breaking antioxidants interrupt the propagation cycle. Nevertheless considerable information is available to support the proposal that many organic oxygen, nitrogen, sulfur, and phosphorus compounds function as peroxide decomposers. There is good evidence that both phenols and aryl amines can act in part as peroxide decomposers.[14] Thus some compounds evidently function in more than one way with, for example, peroxide-decomposing ability superimposed upon activity as a chain-breaking antioxidant by a hydrogen-donation mechanism.

In the case of organic sulfides and disulfides, it is evident that reaction with hydroperoxides to form sulfoxides and thiolsulfinates is an essential feature of their activity. To be really effective in promoting the decomposition of more than a molar equivalent of peroxide, however, these SO-compounds must be capable of undergoing thermal decomposition to yield products that in some way contribute to the decomposition of many additional moles of hydroperoxide. It seems likely that the sulfenic acids, RSOH, formed as the first product of thermolysis of sulfoxides (and possibly RSSOH from thiolsulfinates), together with oxidation products of RSOH and its reaction products, play a major role in the mechanism of peroxide decomposition with this type of preventive antioxidant.

G. COMBINATIONS OF STABILIZERS TO PROTECT AGAINST THERMAL OXIDATION

The discussion of mechanisms of stabilization against thermal oxidation of polymers in the preceding sections of this chapter makes it clear that many different types of compounds may be involved and that they may function in many different ways. Thus it is possible to select stabilizers that function by inhibiting the initiation process (preventive antioxidants), or those that function by interrupting the propagation step (chain-breaking antioxidants) leading to termination of the chain reaction. Even within each of these broad classifications of antioxidants, there is evidence of a variety of different stabilizers and mechanisms of action. The availability of so many types of antioxidants and structural variations of each chemical type not only permits the selection of a stabilizer appropriate for a given polymer under the expected conditions of use, but also suggests the possibility of selecting combinations of two or more stabilizers to provide additional protection against a variety of environmental conditions.

The use of combinations of stabilizers against thermal oxidation can be beneficial in many instances, but the success of such a combination requires careful selection verified by actual testing. In many instances the observed protection is merely the additive effects of each used alone. It is also possible for each to affect the other in a way that either increases the effectiveness of each (synergism), or interferes with normal stabilizing ability (antagonism).

1. Additive Effects

One would logically expect to be able to combine stabilizers normally used to provide different kinds of protection and hope to retain the unique character of each. The combined effect should be at least additive and often synergistic. Thus combinations of conventional chain-breaking antioxidants of the aryl amine or phenolic type with preventive antioxidants such as a metal-ion deactivators, peroxide decomposers, and ultraviolet absorbers, should protect a polymer against both thermal and photodegradation by oxygen, and addition of an antiozonant would also provide stabilization against ozone attack. Compatibility of the particular stabilizer combination in a specific polymer must be verified under all the expected environmental conditions of manufacture, storage, and use. For example, an amine that is an effective stabilizer against thermal autoxidation may prove to be a photosensitizer, accelerating oxidative degradation when it is exposed to sunlight.

Combinations of antioxidants of the same type would normally be expected to give only additive effects, but they can sometimes be used to advantage. For example, two phenolic compounds differing in volatility and in the extent of steric hindrance by bulky alkyl groups may provide better protection over a wider range of temperatures than an equivalent amount of either one alone. The prooxidant effect observed with higher concentrations of most antioxidants may be avoided in some cases by using lower concentrations of two or more stabilizers that exhibit an additive effect of their normal activities when used in combination.

2. Antagonistic Effects

The influence of one inhibitor upon another when used in combination for the stabilization of polymers against oxidation, is of considerable importance, as is the interaction of inhibitors with other compounding ingredients. Although the interaction may be beneficial, as described in the next section of this discussion, many examples of detrimental effects have been noted.

Secondary aryl amines or alkylated phenols, which are effective stabilizers in most polymers, are appreciably less effective when added to polyethylene containing carbon black as an ultraviolet light screen as compared to the clear polymer without carbon.[7] It has been suggested that this loss in efficiency, which occurs even though carbon black is itself a moderately effective inhibitor against thermal oxidation,[97] may result from a catalytic effect of the carbon surface on the direct oxidation of the amine or phenol.[98] In a

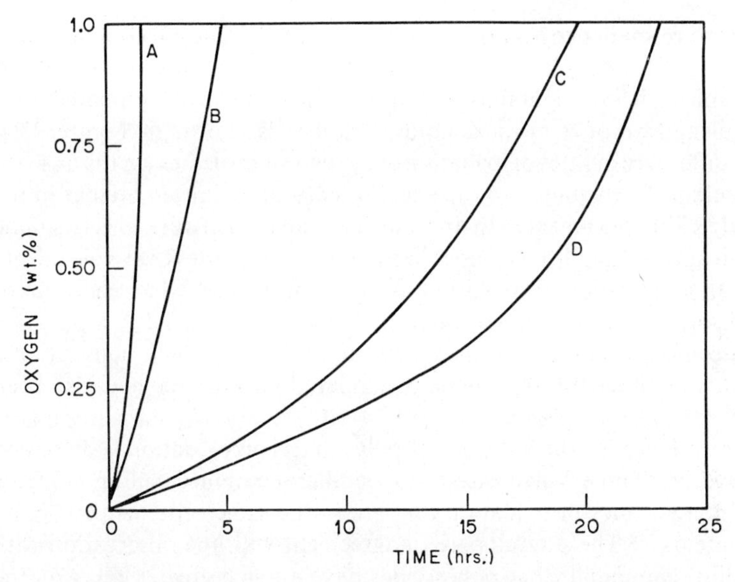

Fig. 2.27 Effect of an allylic dialkenyl trisulfide, $A_1S_3A_1$ (0.25 *M*), on the autoxidation of peroxide vulcanizates of natural rubber, RSSl, and purified natural rubber (100°C, 760 mm O_2):[19] (*A*) purified natural rubber, no addition; (*B*) purified natural rubber, $A_1S_3A_1$ added; (*C*) RSSl, $A_1S_3A_1$ added; (*D*) RSSl, no addition.

similar situation in elastomer oxidation, Kuzminskii and his co-workers[85] have attributed the accelerating effect of carbon black upon the rate of retarded autoxidation in the presence of amine antioxidants to adsorption of the antioxidant on the carbon surface, making it less available as an inhibitor. The activity of carbon black both as an inhibitor and as a promoter of oxidation has been confirmed by many independent studies in natural and synthetic rubber.[14,19] The loss of the inhibiting effect of the carbon black in the presence of antioxidants, together with decreased effectiveness of the antioxidants in the presence of carbon black, clearly indicates antagonistic effects; further study, however, is required to establish the nature of the interaction responsible for the observed behavior.

Cunneen[19] has reviewed the pieces of evidence of an antagonism between conventional antioxidants and certain sulfur compounds, particularly polysulfides. Parks and Lorenz[99] observed that sulfur vulcanizates of rubber with higher concentrations of polysulfides and cyclic sulfides exhibited a faster rate of oxidation. Possible ways in which these compounds might function as prooxidants were considered. However, studies of model compounds resembling the dialkenyl sulfide and polysulfide crosslinks (and some cyclic sulfides) in squalene have demonstrated that both polysulfides and cyclic sulfides act as antioxidants in this low-molecular-weight polyisoprene with a structure similar to that of natural rubber.[100,101] The effect of an allylic dialkenyl trisulfide, $A_1S_3A_1$, on the autoxidation of a peroxide vulcanizate of natural rubber, RSS1, containing natural antioxidants, as compared to a similar vulcanizate of a purified natural rubber is shown in Figure 2.27. The large difference in rates of oxidation of the two vulcanizates (in the absence of the trisulfide) is attributed to the effect of natural inhibitors present in the unpurified RSS1 vulcanizate. In the purified rubber network the trisulfide acted as an antioxidant just as it does in squalene, but in the RSS1 vulcanizate it caused an acceleration of oxidation. An even more marked acceleration of oxygen absorption was observed when the trisulfide was added to a peroxide vulcanizate made from the purified natural rubber to which conventional antioxidants of either the aryl amine or hindered phenol type had also been added.

Further evidence for the influence of polysulfides on oxidation is presented in Figure 2.28. Three vulanizates with similar combined sulfur contents showed different rates of oxidation that are in the same order as their polysulfide contents.[102] These results are in agreement with the observation with model sulfur compounds that polysulfides have a much greater effect on the

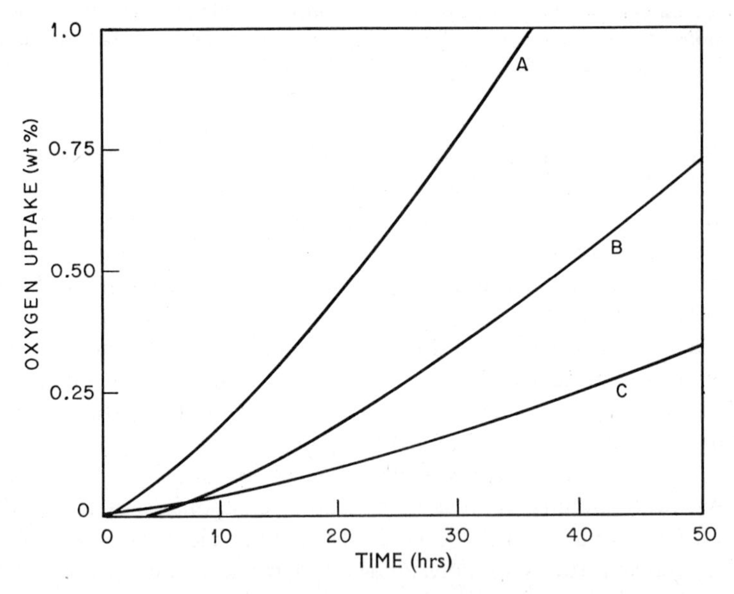

Fig. 2.28 Effect of polysulfide concentration on autoxidation of vulcanizates of natural rubber, RSSI, containing *N*-isopropyl-*N*′-phenyl-*p*-phenylenediamine (1%), (100°C, 760 mm O_2):[19] (*A*) zinc dimethyldithiocarbamate accelerated, high sulfur (combined sulfur 2.3%; polysulfide crosslinks, 6.05×10^{-5} moles/g of rubber network); (*B*) *N*-cyclohexyl benzothiazole-2-sulfenamide (CBS) accelerated sulfur (combined sulfur 2.1%; polysulfide crosslinks, 3.6×10^{-5} moles/g of rubber network); (*C*) high-modulus-efficient vulcanization with high CBS and low sulfur (combined sulfur 2.0%; no polysulfide crosslinks).

rate of oxidation of protected vulcanizates than do the other forms of combined sulfur. This antagonism toward antioxidants also occurs with monosulfides, but to a much lesser extent; presumably it is also found with unsaturated cyclic sulfides.[102]

The effect depends markedly on the structure of the sulfur compound as shown by the pronounced effect of the dialkenyl, $A_1S_xA_1$ compounds, whereas both di-*n*-butyl sulfide and di-*n*-hexyl trisulfide were inert. It is suggested that the antagonism is not caused by a molecular reaction between the sulfur compound and the antioxidant; rather, that "it appears to be due to reactions which involve the sulfur compound, peroxy radicals, and the conventional antioxidant."[19]

N-Phenylhydroxylamine is reported[101] to be a powerful catalyst of the autoxidation of squalene that has been inhibited by 2,6-di-*t*-butyl-*p*-cresol. Since the observed effect resembles that of added AIBN, the authors suggest that the increased rate results from initiation by reaction of oxygen with the hydroxylamine. In any event the net result is to repress the effectiveness of the antioxidant and thus it may be regarded as an example of antagonism.

3. Synergistic Effects

The phenomenon of synergism may be defined as a cooperative action such that the total effect is greater than the sum of the two or more individual effects taken independently. In Scott's discussion[9] of synergism between antioxidants, he credits Mattill[103] and co-workers with the original recognition of this greater than additive effect of a combination of antioxidants in their studies of the response of vegetable and animal fats to added antioxidants. The differences observed were attributed in some cases to the presence of natually occurring antioxidants, which augmented the action of other protective agents. Scott[9] considered two mechanistically distinct types of synergism: homosynergism, involving two compounds of unequal activity but operating by the same mechanism and heterosynergism, arising from the cooperative effect of two or more antioxidants acting by different mechanisms. In the latter category would be combinations of chain-breaking antioxidants with preventive antioxidants of various types. Both qualitative and quantitative aspects of the synergism of such combinations are reviewed in a recent book by Reich and Stivala.[55]

In the case of a combination of two different chain-breaking antioxidants that normally function by donation of hydrogen to a peroxy free radical, the most likely mechanism of synergism would involve transfer of hydrogen from one inhibitor molecule to the radical formed in the reaction of the other inhibitor with a peroxy radical. The more efficient free-radical chain stopper would be regenerated by this process from its oxidation products by a suitable hydrogen donor, which may also be a weak antioxidant when used alone.

Such compounds might be considered as a special type of antioxidant and classified as inhibitor regenerators.[6] The advantage of such combinations as two phenolic inhibitors differing in the degree of hindrance by bulky ortho substituents, or two amine-type antioxidants differing in structure and reactivity, or a combination of a secondary diaryl amine with a hindered phenol, is that the more reactive inhibitor will efficiently scavenge any oxy or peroxy radicals formed in the system and yet will not be depleted, since the less efficient hydrogen donor can still serve as a reservoir of hydrogen for regeneration of the more effective chain-breaking antioxidant. This mechanism would provide the efficiency of the more reactive stabilizer extended over a longer effective period of stabilization because of the contribution of the less reactive compound. If one or both of the compounds is also able to function as a peroxide decomposer, an even greater synergistic effect is possible.

Another factor in the synergism of a combination of phenolic inhibitors differing in the extent of bulky ortho substitution is the effect upon the chain-transfer reaction by which the oxy radical derived from the antioxidant may react with the substrate in the presence of oxygen to reinitiate the oxidation reaction. If the aryloxy free radical reacts more readily with the inhibitor regenerator than with oxygen or the substrate, chain transfer will be repressed. The following reactions illustrate the synergistic mechanism for a combination of phenols differing in the bulkiness of the ortho substituents:

$$RO_2\cdot + \text{(OH; } i\text{-Pr; } CH_3 \text{ phenol)} \longrightarrow RO_2H + \text{(O}\cdot\text{; } i\text{-Pr; } CH_3\text{)}$$

$$\text{(OH; } t\text{-Bu, } t\text{-Bu; } CH_3\text{)} + \text{(O}\cdot\text{; } i\text{-Pr; } CH_3\text{)} \longrightarrow \text{(O}\cdot\text{; } t\text{-Bu, } t\text{-Bu; } CH_3\text{)} + \text{(OH; } i\text{-Pr; } CH_3\text{)}$$

$$\text{(O}\cdot\text{; } t\text{-Bu, } t\text{-Bu; } CH_3\text{)} + RO_2\cdot \longrightarrow \text{(O; } t\text{-Bu, } t\text{-Bu; } RO_2\text{, } CH_3\text{ cyclohexadienone)}$$

Materials that are not effective inhibitors when used alone may nevertheless be able to function as synergists by reacting with an oxidized form of an antioxidant to regenerate it and thus prolong its effectiveness. Knapp and Orloff[104] have demonstrated synergistic effects with a number of hindered phenols and dialkylphosphonates in white mineral oil at 150°C. The following reactions illustrate regeneration of the oxidized forms by hydrogen transfer from the phosphonate ester:

$$R'O_2\cdot + \text{(2,6-R}_2\text{-4-R-phenol, OH)} \longrightarrow R'O_2H + \text{(2,6-R}_2\text{-4-R-phenoxy, O}\cdot\text{)}$$

$$\text{(2,6-R}_2\text{-4-R-phenoxy, O}\cdot\text{)} + H\text{—}\overset{O\uparrow}{P}(OR'')_2 \longrightarrow \text{(2,6-R}_2\text{-4-R-phenol, OH)} + \cdot\overset{O\uparrow}{P}(OR'')_2$$

$$2\ \text{(2,6-R}_2\text{-4-R-quinonoid, O}\cdot\text{)} + H\text{—}\overset{O\uparrow}{P}(OR'')_2 \longrightarrow \text{(2,6-R}_2\text{-4-R-phenol, OH)} + \text{(2,6-R}_2\text{-4-R-phenyl, O—}\overset{O\uparrow}{P}(OR'')_2\text{)} + \ldots$$

The mechanism of the synerigism observed with a wide variety of combinations of preventive antioxidants with chain-breaking antioxidants can be explained by consideration of the specific ways in which these different types of stabilizers function to retard or inhibit the oxidation reaction. The free-radical chain stopping action of an aryl amine or a hindered phenol will retard the formation of peroxides by terminating the kinetic chain promptly, thus effectively eliminating long kinetic chains. However at least one molecule of hydroperoxide will be formed in the transfer of hydrogen to a peroxy radical. If the small amount of hydroperoxide formed even in the presence of an efficient inhibitor of this type reacts with a peroxide decomposer to form nonradical products, rather than to decompose thermally to form chain-initiating free radicals, there will be a further reduction in the rate of oxidation. It is thus evident that the two antioxidants can act in a cooperative manner to give a synergistic effect.

The preventive antioxidant will reduce peroxide initiation to a negligible role as a source of free radicals so that the chain-breaking antioxidant will have fewer kinetic chains to stop and thus can more effectively inhibit the reaction over a much longer period of time. In fact, the main function of conventional antioxidants such as secondary aryl amines and hindered phenols in combination with a peroxide decomposer is to stop the chains resulting from initiation by direct attack of oxygen on the antioxidant itself. The preventive antioxidant is in turn protected by the chain-breaking inhibitor, since less peroxide is formed and the peroxide decomposer is consumed at a much slower rate and continues to inhibit peroxide initiation over an extended time period. Thus the two mechanisms can be expected to mutually augment each other to provide effective stabilization of the polymer for a useful life which is frequently far in excess of the additive effect of the same two antioxidants used separately.

Many examples of synergism involving combinations of stabilizers have been reported.[9] We have previously noted the effective combination of carbon black with thiols, disulfides, and elemental sulfur, even though these substances may be almost totally ineffective alone under comparable conditions.[7] Hawkins and Worthington[105] have reported synergistic combinations formed from a variety of chain terminators and sulfur compounds. A widely used combination of stabilizers for polyolefins is dilauryl thiodipropionate (DLTDP) with 2,6-di-*t*-butyl-4-methylphenol. This combination is of particular interest because both components are among the small group of stabilizers approved by the U.S. Food and Drug Administration for use in packaging materials for food products.

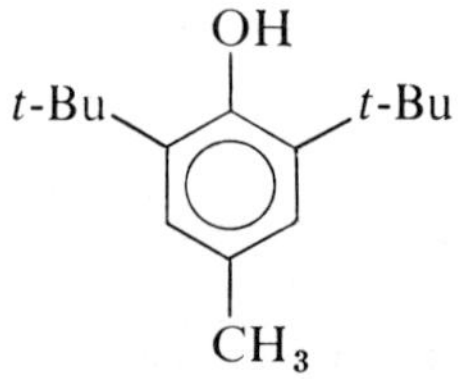

OH
t-Bu   *t*-Bu
CH$_3$

and $(H_{25}C_{12}O{-}\overset{\overset{\displaystyle O}{\|}}{C}CH_2CH_2)_2S$

DLTDP

Other synergistic combinations have been reported involving free-radical chain terminators used with either ultraviolet absorbers or metal deactivators as the preventive antioxidants.[7] Combinations of 2-hydroxybenzophenones as ultraviolet absorbers with either chain-breaking antioxidants or peroxide decomposers have been reported to give synergistic effects in polyethylene.[9] A combination of 2-mercaptobenzimidazole with both phenolic and amine antioxidants has been observed to give a synergistic protection in the presence of either copper or iron in both vulcanized rubber and the raw polymer.[106]

The differences in mechanism of action of these different types of stabilizers permits them to act independently but cooperatively to provide greater protection than would be predicted by the sum of their individual but separate effects. The prooxidant effect of ultraviolet light and of certain metal ions, which reflects their ability to decompose peroxides to form chain-initiating free radicals, can be nullified by ultraviolet absorbers and metal deactivators and thus can reduce the number of kinetic chains to be stopped by the conventional antioxidant. The ability of the latter component to efficiently stop the smaller quantity of kinetic chains reduces the amount of hydroperoxide formed and thus reduces the potential for initiation catalyzed by light or metals.

It would appear that a combination of three or more stabilizers might give tremendous synergistic effects in favorable situations. For example, an ultraviolet absorber, a metal deactivator, and a peroxide decomposer should act cooperatively to reduce peroxide initiation under a variety of environmental conditions. When these components are used in combination with an efficient free-radical chain stopper plus an inhibitor regenerator, we have the possibility of synergism involving the contributions of five different stabilizers to protect a polymer containing metal impurities and subject to both photodegradation and thermal degradation under the expected conditions of use. An antiozonant could of course be added to the above combination if the polymer is susceptible to this type of degradation and if exposure to ozone is anticipated. Synergistic interaction of the antiozonant might be expected with at least some of the stabilizers, since ozone is believed to contribute to free-radical initiation and many peroxide decomposers have the potential to decompose both ozone and ozonides.

Autosynergism is probably involved in the observed effectiveness of certain stabilizers that appear to have two kinds of antioxidant activity associated in the same molecule. For example, some amine antioxidants are known to have metal-complexing activity and are correspondingly more effective than phenols in this respect.[9] Kuzminskii et al.[107] have noted that many diarylamine antioxidants form stable complexes with copper and may act by a metal deactivation mechanism in addition to their known activity as chain-breaking antioxidants. Derivatives of dialkyldithiocarbamic acid may similarly function as metal deactivators as well as peroxide decomposers.[9] In our discussion of peroxide decomposers earlier in this chapter, it was pointed out that zinc dialkyl dithiophosphates and the related xanthate and dithiocarbamate salts appear to function as chain-breaking antioxidants in addition to their activity as preventive antioxidants. Carbon black is also an example of a multifunctional inhibitor showing evidence of reactivity as a free-radical trap and hydrogen donor in addition to its function as a light screen in the stabilization of polymers and its interaction with iron and other metal ions.[14]

Even phenolic and amine antioxidants show evidence of some activity as peroxide decomposers as well as the ability to complex metals; these properties, of course, are in addition to their main function as free-radical chain stoppers. The diarylamines show greater activity than the hindered phenols both as peroxide decomposers and as metal deactivators. The ability of phenyl-β-naphthylamine to reduce the contribution of peroxide to initiation was enhanced by the presence of carbon black in natural rubber vulcanizates.[21] Thus it appears that autosynergism may be involved in many situations in which stabilizers show outstanding activity as inhibitors against thermal autoxidation.

H. CONCLUSIONS

Much of our present knowledge of stabilization of polymers and related materials against autoxidation under various environmental conditions has been obtained by the empirical approach. Experience gained by trial and error gradually evolved a recognition of different types of antioxidants and mechanisms of stabilization and permitted a more efficient systematic approach to the problem. The continued development of improved antioxidants and synergistic combinations of stabilizers of different types will depend increasingly upon a better understanding of the mechanisms by which these materials function to inhibit or retard the autoxidation reaction.

It has been our purpose in this chapter to present a discussion of our present understanding of the fundamental nature of thermal autoxidation and the mechanisms of stabilization of polymers against degradation by reaction with oxygen under the various conditions encountered in synthesis, fabrication, storage, and use. Additional studies, both qualitative and quantitative, are needed to provide more completely information, particularly with regard to the detailed reactions involved in the action of the several types of preventive antioxidants, and the interactions responsible for the antagonism and synergism effects observed with many combinations of stabilizers.

Interest in the possibility of controlled destabilization of polymers in certain applications is now increasing. Materials used in packaging must ultimately be disposed of in some way. Polymers, which are being used in greater and greater amounts, thus contribute correspondingly to the problems of disposal both by incineration and by land-fill methods. Although collection and reuse of part of these materials may be possible in many situations, polymeric materials will inevitably continue to contribute to the accumulation of litter in our urban and recreation areas. Consequently the mechanisms of destabilization and accelerated degradation of polymers are also of interest. The discussion of mechanisms of oxidation and stabilization in this chapter provides the background needed for such studies designed to reduce the contribution of polymers to the pollution of our environment.

It is our hope that the present discussion, together with the other chapters of this book, will contribute to a better understanding both of the degradation processes and of the ways in which stabilizers function to control them. It has also been our objective to stimulate interest in additional fundamental research activity in the area of polymer degradation and the mechanisms of stabilization against various environmental conditions.

REFERENCES

1. J. L. Bolland, *Quart. Rev. (London)*, **3,** 1 (1949).
2. L. Bateman, *Quart. Rev. (London)*, **8,** 147 (1954).
3. N. Grassie, *Chemistry of High Polymer Degradation Processes*, Interscience, New York; Butterworth, London, 1956.
4. K. U. Ingold, *Chem. Rev.* **61,** 563 (1961).
5. N. Uri, "Mechanism of Antioxidation," in *Autoxidation and Antioxidants*, Vol. I, W. O. Lundberg, Ed., Interscience, New York, 1961, pp. 55, 133.
6. J. R. Shelton, *Offic. Dig., Fed. Soc. Paint Technol.* **34,** 590 (1962).
7. W. L. Hawkins, "Oxidative Degradation of High Polymers," in *Oxidation and Combustion Reviews*, Vol. I, C. F. H. Tipper, Ed., Elsevier, Amsterdam, 1965, p. 170.
8. A. S. Kuzminskii, "Aging and Stabilizating of Polymers," Consultants' Bureau, New York (1965).
9. G. Scott, *Atmospheric Oxidation and Antioxidants*, Elsevier, Amsterdam, 1965.
10. C. Dufraisse, "Autoxidation and Deterioration by Oxygen," in C. C. Davis and J. T. Blake Eds., *Chemistry and Technology of Rubber*, *ACS Monogr. No. 74*, Reinhold, New York, 1937.
11. J. O. Cole, "Aging and Stabilization of GR-S," in G. S. Whitby, *Synthetic Rubber*, Wiley, New York, 1954, p. 528.
12. E. M. Bevilacqua, *Rubber Age*, **80,** 271 (1956).
13. J. M. Buist, "Aging and Weathering of Rubber," *Monogr. Inst. Rubber Ind. (London)*, Heffer, Cambridge, 1956.
14. J. R. Shelton, *Rubber Rev. Rubber Chem. Technol.*, **30,** 1270 (1957).
15. E. M. Bevilacqua, "Oxidation and Antioxidants in Rubber," in *Autoxidation and Antioxidants*, Vol. II, W. O. Lundberg, Ed., Interscience, New York, 1962, p. 857.
16. J. C. Ambelang, R. H. Kline, O. M. Lorenz, C. R. Parks, C. Wadelin, and J. R. Shelton, *Rubber Rev. Rubber Chem. Technol.* **36,** 1497 (1963).
17. D. Barnard, L. Bateman, J. I. Cunneen, and J. F. Smith, in *Chemistry and Physics of Rubber-like Substances*, L. Bateman, Ed., Maclaren, London, 1963, p. 593.
18. P. M. Norling, T. C. P. Lee, and A. V. Tobolsky, *Rubber Rev. Rubber Chem. Technol.*, **38,** 1198 (1965).
19. J. I. Cunneen, *Rubber Rev. Rubber Chem. Technol.*, **41,** 182 (1968).
20. J. R. Shelton, *J. Appl. Polym. Sci.*, **2,** 345 (1959).
21. J. R. Shelton and W. L. Cox, *Ind. Eng. Chem.*, **46,** 816 (1954); *Rubber Chem. Technol.*, **27,** 671 (1954).

22. J. R. Shelton and W. L. Cox, *Ind. Eng. Chem.*, **45,** 392, 397 (1953); *Rubber Chem. Technol.*, **26,** 632, 643 (1953).
23. J. R. Shelton and E. T. McDonel, *J. Polym. Sci.*, **32,** 75 (1958).
24. J. R. Shelton, E. T. McDonel, and J. C. Crano, *J. Polym. Sci.*, **42,** 289 (1960).
25. J. R. Shelton and D. N. Vincent, *J. Amer. Chem. Soc.*, **85,** 2433 (1963).
26. Kah Ong, Ph.D. thesis, Case Western Reserve University, 1968.
27. A. F. Bickel and E. C. Kooyman, *J. Chem. Soc.*, **1953,** 3211; **1956,** 2215; **1957,** 2217.
28. K. U. Ingold, *Acct. Chem. Res.*, **2,** 1 (1969).
29. G. A. Russell, *J. Amer. Chem. Soc.*, **79,** 3871 (1957).
30. P. D. Bartlett and G. Guaraldi, *J. Amer. Chem. Soc.*, **89,** 4799 (1967).
31. J. A. Howard and K. U. Ingold, *J. Amer. Chem. Soc.*, **90,** 1056 (1968).
32. J. A. Kerr, *Chem. Rev.*, **66,** 465 (1966).
33. D. E. Van Sickle, F. R. Mayo, and R. M. Arluck, *J. Amer. Chem. Soc.*, **87,** 4824, 4832 (1965).
34. D. E. Van Sickle, F. R. Mayo, R. M. Arluck, and M. G. Syz, *J. Amer. Chem. Soc.*, **89,** 967 (1967).
35. D. E. Van Sickle, F. R. Mayo, E. S. Gould, and R. M. Arluck, *J. Amer. Chem. Soc.*, **89,** 977 (1967).
36. F. R. Mayo, *Acct. Chem. Res.*, **1,** 193 (1968).
37. L. Bateman, J. I. Cunneen, C. G. More, L. Mullins, and A. G. Thomas, in *Chemistry and Physics of Rubber-like Substances*, L. Bateman, Ed., Maclaren, London, 1963, p. 715.
38. W. L. Hawkins and F. W. Winslow in *Chemical Reactions of Polymers*, E. M. Fettes, Ed., Interscience, New York, 1964, p. 1055.
39. D. J. Metz and R. B. Mesrobian, *J. Polym. Sci.*, **16,** 345 (1955).
40. P. Kovacic and A. Kyriakis, *Tetrahedron Lett.*, **1962,** 467; *J. Amer. Chem. Soc.*, **85,** 454 (1963); P. Kovacic and R. M. Lange, *J. Org. Chem.*, **28,** 968 (1963).
41. P. Kovacic and F. W. Koch, *J. Org. Chem.*, **28,** 1864 (1963); P. Kovacic and L. Hsu *J. Polym. Sci., Part A-4*, **5** (1966).
42. H. Vogel and C. S. Marvel, *J. Polymer Sci.*, **50,** 511 (1961); *J. Polym. Sci., Part A*, **1,** 1531 (1963).
43. R. W. Lenz, *Organic Chemistry of Synthetic High Polymers*, Interscience, New York, 1967, p. 699.
44. R. D. Joyner and M. E. Kenney, *Inorg. Chem.*, **1,** 717 (1962); T. R. Janson, A. R. Kane, J. F. Sullivan, K. Knox, and M. E. Kenney, *J. Amer. Chem. Soc.*, **91,** 5210 (1969).
45. G. E. Meyer, W. E. Gibbs, F. J. Naples, R. M. Pierson, W. M. Saltman, R. W. Schrock, L. B. Tewksbury, and G. S. Trick, *Rubber World*, **136,** 529, 695 (1957); *Rubber Plastics Age*, **38,** 592, 708 (1957).
46. F. H. Winslow, C. J. Aloisio, W. L. Hawkins, and S. Matsuoka, *Chem. Ind.* (*London*), **1963,** 533.
47. J. R. Shelton and W. L. Cox, *Ind. Eng. Chem.*, **43,** 456 (1951); *Rubber Chem. Technol.*, **24,** 981 (1951).
48. J. R. Shelton, F. J. Wherley, and W. L. Cox, *Ind. Eng. Chem.*, **45,** 2080 (1953); *Rubber Chem. Technol.*, **27,** 120 (1954).
49. J. R. Shelton, W. L. Cox, and W. T. Wickham, *Ind. Eng. Chem.*, **47,** 2559 (1955); J. R. Shelton and E. T. McDonel, *J. Appl. Polym. Sci.*, **1,** 336 (1959).
50. M. E. Cain, G. T. Knight, P. M. Lewis, and B. Saville, *J. Rubber Res. Inst. Malaya*, (1968).

51. K. U. Ingold, "Inhibition of Autoxidation," in R. F. Gould, Ed., *Oxidation of Organic Compounds*, *Advan*, *Chem. Ser.*, **75-I,** American Chemical Society, Washington, D.C., 1968, p. 296.

52. M. J. Astle and J. R. Shelton, *Organic Chemistry*, 2nd ed., Harper, New York, 1949, p. 731.

53. C. J. Pedersen, *Ind. Eng. Chem.*, **48,** 1881 (1956).

54. M. S. Kharasch and A. Fono, *J. Org. Chem.*, **24,** 72 (1959).

55. L. Reich and S. S. Stivala, *Autoxidation of Hydrocarbons and Polyolefins*, Dekker, New York, 1969.

56. C. E. Boozer and G. S. Hammond, *J. Amer. Chem. Soc.*, **76,** 3861 (1954); G. S. Hammond, C. E. Boozer, C. E. Hamilton, and J. N. Sen, *J. Amer. Chem. Soc.*, **77,** 3238 (1955).

57. K. U. Ingold and J. E. Puddington, *Ind. Eng. Chem.*, **51,** 1319 (1959).

58. J. R. Thomas and C. A. Tolman, *J. Amer. Chem. Soc.*, **84,** 2930 (1962).

59. G. S. Hammond and U. S. Nandi, *J. Amer. Chem. Soc.*, **83,** 1217 (1961).

60. K. U. Ingold and J. A. Howard, *Nature*, **195,** 280 (1962).

61. O. Lorenz and C. R. Parks, *Rubber Chem. Technol.*, **34,** 816 (1961).

62. J. A. Howard and K. U. Ingold, *Can. J. Chem.*, **40,** 1851 (1962); **41,** 1744, 2800 (1963); **42,** 2324 (1964).

63. J. Furukawa, S. Yamashita, and T. Kotani, "Mechanism of Antioxidation and Antiozonization of Amines for Rubber," in R. F. Gould, Ed., *Stabilization of Polymers and Stabilizer Processes*, *Advan. Chem. Ser.*, **85,** American Chemical Society, Washington, D.C., 1968, p. 110.

64. C. D. Lowry, G. Egloff, J. C. Morrell, and C. G. Dryer, *Ind. Eng. Chem.*, **25,** 806 (1933).

65. J. L. Bolland and P. Ten Have, *Disc. Faraday Soc.*, **2,** 252 (1947).

66. G. E. Penketh, *J. Appl. Chem.*, **7,** 512 (1957).

67. W. G. Lloyd, R. G. Zimmerman, and A. J. Dietzler, *Ind. Eng. Chem.*, *Prod. Res. Develop.*, **5,** 326 (1966).

68. L. F. Fieser, *J. Amer. Chem. Soc.*, **52,** 5204 (1930).

69. R. H. Hansen, C. A. Russell, T. DeBenedictis, W. M. Martin, and J. V. Pascale, *J. Polym. Sci.*, *Part 2-A*, **587** (1964).

70. M. Calvin and R. H. Bailes, *J. Amer. Chem. Soc.*, **68,** 953 (1946).

71. C. J. Pedersen, *Ind. Eng. Chem.*, **41,** 924 (1949).

72. A. J. Chalk and J. F. Smith, *Nature*, **174,** 802 (1954); *Trans. Faraday Soc.*, **53,** 1214, 1235 (1957).

73. G. W. Kennerly, and W. L. Patterson, Jr., *Ind. Eng. Chem.*, **48,** 1917 (1956).

74. J. R. Dunn and J. Scanlan, *Trans. Inst. Rubber Ind.* (*London*), **34,** 228 (1958); *J. Polym. Sci.*, **35,** 267 (1959).

75. W. Scheele, O. Lorentz, and W. Drummer, *Kautschuk Gummi*, **7,** WT 273 (1954); *Rubber Chem. Technol.*, **29,** 1 (1956).

76. W. P. Fletcher and S. G. Fogg, *Rubber J.*, **134,** 16 (1958); *Rubber Chem. Technol.*, **31,** 327 (1958); *Rubber Age*, **84,** 632 (1959).

77. A. J. Burn, "Mechanism of Oxidation Inhibition by Zinc Dialkyl Dithiophosphates" in R. F. Gould, Ed., *Oxidation of Organic Compounds*, *Advan. Chem. Ser.*, **75-I,** American Chemical Society, Washington, D.C., 1968, p. 323.

78. T. Colclough and J. I. Cunneen, *J. Chem. Soc.*, **1964,** 4790.

79. A. J. Burn, *Tetrahedron*, **22,** 2153 (1966).

80. E. A. C. Lucken, *J. Chem. Soc.*, **1966,** 1354.

81. J. D. Holdsworth, G. Scott, and D. Williams, *J. Chem. Soc.*, **1964,** 4692.

82. W. L. Hawkins and Mrs. H. Sautter, *Chem. Ind.* (*London*), **1962,** 1825.

83. W. L. Hawkins and Mrs. H. Sautter, *J. Polym. Sci; Part 1A*, **3499** (1963).

84. D. Barnard, L. Bateman, E. R. Cole, and J. I. Cunneen, *Chem. Ind.* (*London*), **1958,** 918.

85. A. S. Kuzminskii, L. I. Lyubschanskyaya, N. G. Khitrava, and S. I. Bass, *Dokl. Akad. Nauk SSSR*, **85,** 131 (1952); *Rubber Chem. Technol.*, **26,** 858 (1953).

86. D. Barnard, L. Bateman, M. E. Cain, T. Colclough, and J. I. Cunneen, *J. Chem. Soc.*, **1961,** 5339.

87. M. E. Cain and J. I. Cunneen, *J. Chem. Soc.*, **1962,** 2959.

88. L. Bateman, M. E. Cain, T. Colclough, and J. I. Cunneen, *J. Chem. Soc.*, **1962,** 3570.

89. B. A. Marshall "Hydroperoxide Decomposition by Some Sulfur Compounds," in R. F. Gould, Ed., *Stabilization of Polymers and Stabilizer Processes, Advan. Chem. Ser.*, **85,** American Chemical Society, Washington, D.C., 1968, p. 140.

90. C. A. Kingsbury and D. J. Cram, *J. Amer. Chem. Soc.*, **82,** 1810 (1960).

91. T. Colclough and J. I. Cunneen, *Chem. Ind.*, **1960,** 626.

92. N. P. Neureiter and D. E. Bown, *Ind. Eng. Chem. Prod. Res. Develop.*, **1,** 236 (1962).

93. C. H. BePuy and R. W. King, *Chem. Rev.*, **60,** 431 (1960).

94. J. R. Shelton and K. E. Davis, *J. Amer. Chem. Soc.*, **89,** 718 (1967).

95. K. E. Davis, Ph.D. thesis, Case Western Reserve University, 1968.

96. J. R. Shelton and K. E. Davis, Division of Organic Chemistry, 159th National Meeting, American Chemical Society, Houston, February 1970.

97. W. L. Hawkins and F. H. Winslow, *Plastics Inst.* (*London*) *Trans. J.*, **29,** 82 (1961).

98. W. L. Hawkins and Mrs. M. A. Worthington, *J. Polym. Sci.*, **62,** S106 (1962).

99. C. R. Parks and O. Lorenz, *Ind. Eng. Chem. Prod. Res. Develop.*, **2,** 279 (1963).

100. M. E. Cain and J. I. Cunneen, *J. Chem. Soc.*, **1963,** 3323; C. L. M. Bell, M. E. Cain, D. J. Elliott, and B. Saville, *Kautschuk Gummi*, **19,** 133 (1966).

101. G. T. Knight and B. Saville, *Chem. Comm.*, **1969,** 1262.

102. C. L. M. Bell and J. I. Cunneen, *J. Appl. Polym. Sci.*, **11,** 2201 (1967).

103. H. S. Olcott and H. A. Mattill, *J. Amer. Chem. Soc.*, **58,** 2204 (1936); *Chem. Rev.*, **29,** 257 (1941).

104. G. G. Knapp and H. D. Orloff, *ACS, Gen. Papers*, **5** (1), 11 (April 1960).

105. W. L. Hawkins and Mrs. M. A. Worthington, *J. Polym. Sci.*, *Part 1-A*, **3489** (1963).

106. B. N. Leyland and R. L. Stafford, *Trans. Inst. Rubber Ind.* (*London*), **35,** 25 (1959).

107. A. S. Kuzminskii, V. D. Zaitseva, and N. N. Lexhnev, *Dokl. Akad. Nauk SSSR*, **125,** 1057 (1959).

3

THERMAL DEGRADATION AND STABILIZATION

L. D. LOAN AND F. H. WINSLOW

Bell Telephone Laboratories, Incorporated, Murray Hill, New Jersey

A. INTRODUCTION

Previous chapters have dealt with oxidative degradation in some detail; it is now necessary to consider a second major mode of degradation—that produced by heat in the absence of oxygen. Most polymers are used in air, and for these thermal degradation is of practical importance only when it, rather than oxidative degradation, leads to failure. Although this is not common a few special cases must be considered. First, many polymers are processed in ways that partially exclude oxygen, and if high processing temperatures are used, thermal degradation may ensue. This situation is perhaps best exemplified by poly(vinyl chloride). Another case has become more important with the advent of space age technology. Industry receives an increasing number of requests to produce polymers for exotic applications—some to be used in vacuum, some to be used at very high temperature—and, as a result, a new class of polymers, the so-called heat-resistant polymers, has appeared.

In this chapter we attempt to review the whole problem of stabilization against thermal degradation in the absence of oxygen; with the wide variety of polymers now available, however, it is obviously impossible to consider each separately. Luckily some classification of degradation mechanisms is possible, and our aim is rather to illustrate the major degradation and stabilization mechanisms, using as examples the most important polymers, with comparison with less common materials where this is possible. Some reader participation will be necessary in relating these examples to other less common examples with which they may be concerned.

The first widely available hydrocarbon polymer was probably natural rubber, and its thermal degradation was studied as long ago as 1883.[1] Even before this, in an attempt to determine its structure, Williams[2] heated rubber and succeeded in isolating a sample of isoprene, thus illustrating one important degradation mechanism—depolymerization to monomer.

The first systematic study of thermal degradation was probably the work of Grassie and Melville[3] on poly(methyl methacrylate), which again shows monomer as a main product. Subsequent work on a variety of polymers has, however, shown that monomer yield varies greatly from one polymer to another. A number of detailed studies indicate that a relatively small number of degradation mechanisms are found. These all involve molecular fragmentation with or without condensation to carbonaceous residues and are described in the following paragraphs.

One type of mechanism is depolymerization, in which, starting from a chain end or other weak point in the molecule, successive monomer units are lost to yield only one product, monomer. An example of this type of behavior is given by poly(methyl methacrylate)

$$\sim CH_2\underset{\displaystyle COOCH_3}{\underset{|}{C}}(CH_3)CH_2\underset{\displaystyle COOCH_3}{\underset{|}{\dot{C}}}(CH_3) \longrightarrow \sim CH_2\underset{\displaystyle COOCH_3}{\underset{|}{\dot{C}}}(CH_3) + CH_2{=}C(CH_3)COOCH_3$$

This type of degradation is characterized by rapid volatilization with little or no change in polymer molecular weight.

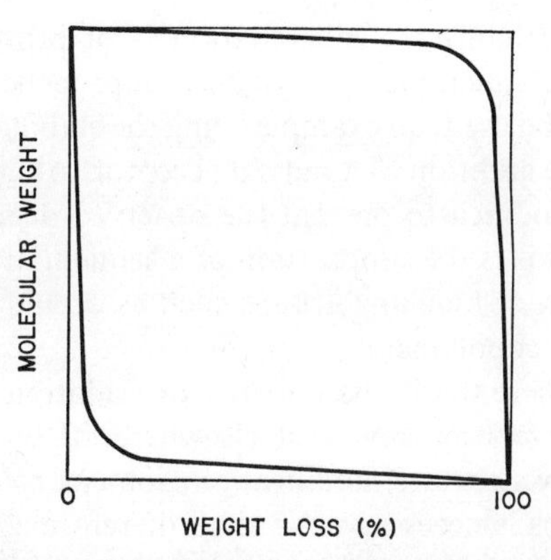

Fig. 3.1 Schematic representation of molecular weight of degrading polymer as a function of weight loss for depropagation (top) and random scission (bottom).

In random chain scission the production of lower molecular weight polymer is the main process. Examples of this type of degradation are given by polyethylene, polypropylene, and polystyrene.

$$\sim CH_2CH_2CH_2CH_2\sim \rightarrow \sim CH{=}CH_2 + CH_3CH_2\sim$$

At high extents of reaction some of the molecules produced will volatilize, but the main characteristic of this type of degradation is a rapid fall of molecular weight with little or no weight loss.

Practical polymers show all ranges of behavior between the first and second types, and consequently the molecular weight versus weight loss curves will fall between the two extremes sketched in Figure 3.1.

In addition to breakdown of the polymer backbone, a further type of degradation is often observed in which small molecules are eliminated without main-chain scission. A typical example is provided by poly(vinyl chloride) in which elimination of hydrogen chloride occurs and progresses along the polymer chain.

$$\sim CH_2CHClCH_2CHClCH_2CHClCH_2CHCl\sim \rightarrow$$
$$\sim CH{=}CHCH{=}CHCH{=}CHCH{=}CH\sim$$

A similar elimination, this time of water, occurs during the initial stage of degradation of poly(vinyl alcohol).[4] Such eliminations may also occur randomly; an example is the random elimination of isobutylene from poly(*t*-butyl methacrylate).

$$\sim CH_2{-}\underset{\displaystyle}{\overset{\displaystyle COOC(CH_3)_3}{\overset{|}{C}}}(CH_3)\sim \longrightarrow CH_2{=}C(CH_3)_2 + \sim CH_2{-}\overset{\displaystyle COOH}{\overset{|}{C}}(CH_3)\sim$$

Obviously the mechanism of degradation is of prime importance in developing more stable materials, and two basic approaches are available to this end. Additives may be used, an example being the stabilization of poly(methyl methacrylate) by the addition of a radical acceptor. Alternatively the polymer structure may be modified to prevent the observed degradation mechanism. A nice example of this is the suppression of a sequential elimination reaction by the incorporation of blocking groups such as occurs in methyl methacrylate/methyl acrylate copolymers.

It is not our aim here to discuss polymer degradation mechanisms in great detail, for these have been reviewed elsewhere,[5–7] but in any attempt to stabilize polymers by structure modification, some consideration of structure stability relationships is necessary. We shall therefore attempt to make some correlation between structure and stability.

Table 3.1 shows the major decomposition products for a number of polymers. It is clear that sequential elimination of monomer units (zipping) is favored when tertiary carbon atoms are present in the skeletal polymer structure. Thus polymethacrylates in general and poly(α-methyl styrene) degrade in this manner. Poly(*t*-butyl methacrylate) is an exception, but only because it contains a more unstable grouping and eliminates isobutene (some other higher esters also degrade differently giving less monomer); similarly poly(vinylidene chloride) eliminates hydrogen chloride. If no energetically favorable zipping type of degradation exists and if all side groups are stable, more complex reactions occur. Sometimes several reactions occur simultaneously and exact analysis of the reaction is difficult. This difficulty has been increased during recent years with the advent of specially synthesized, so-called thermally stable polymers.[8] These polymers are required to combat the severe environments often encountered in high-speed flight and in related areas. Their preparation may be regarded as the culmination of many stabilization steps, each resulting in a modification of polymer structure. These polymers thus contain only the most stable chemical bonds and are usually made up of heavily conjugated structures. As an example we may quote the polyimide structure that can lead to stability to above 500°C in

$$\left[-N\begin{matrix} CO \\ CO \end{matrix} C_6H_2 \begin{matrix} CO \\ CO \end{matrix} N - C_6H_4 - \right]_n$$

air and to 600°C in argon.[9]

In the remainder of this chapter we consider these matters in more detail and continue to discuss polymer stabilization according to degradation mechanism. We thus deal first with the polymers that degrade by sequential

Table 3.1 Summary of Polymer Degradation Data[5–7]

Polymer	Monomer Yield (%)	Energy of Activation (kcal/mole)	Temperature for Half-Life of about 30 Min[a] (°C)
Polyoxymethylene	100		
Poly(α-methylstyrene)	100		
Polytetrafluoroethylene	96	81	510
Poly(methyl methacrylate)	95	52	330
Polymethacrylonitrile	85		
Poly(*m*-methylstyrene)	45	56	360
Polystyrene	41	55	360
Polychlorotrifluoroethylene	26	57	380
Polyisobutene	20	49	350
Polybutadiene	20	62	410
Poly(ethylene oxide)	4	46	350
Poly(propylene oxide)	3	20	300
Poly(methyl acrylate)	1	34	330
Polyacrylonitrile	0	58	390
Polypropylene	0	63	400
Polyethylene	1		
Poly(butyl methacrylate)	0	High yield of isobutene	
Poly(vinyl chloride)	0[b]	32	260
Poly(vinyl acetate)	0[c]	17	270

[a] G. G. Achhammer, M. Tryon, and G. M. Kline, *Kunststoffe—Plastics*, **49**, 600 (1959).
[b] High yield of hydrogen chloride.
[c] High yield of acetic acid.

elimination of small molecules, including monomer, along the chain. This is followed by a discussion of polymers in which random main-chain scission is important. Finally a section is devoted to highly stable polymers.

In concluding this introductory section it is perhaps worth repeating that thermal degradation is normally of practical importance only in processing or in oxidation-resistant polymers. This is because plastics and rubbers are usually used in air and, under these conditions oxidation (see Chapter 2) is normally the cause of deterioration. However, oxygen is often also important in the initiation step of thermal degradation, since oxygenated structures, such as peroxides, often provide the initial bond-breaking step from which thermal degradation proceeds.

B. DEGRADATION BY SEQUENTIAL ELIMINATION (UNZIPPING) WITH MONOMER LOSS

This type of degradation is also called depolymerization or depropagation, since it is the reverse of the propagation step in addition polymerization. It most frequently occurs when the polymer chain contains a tertiary carbon atom. This is perhaps best illustrated by the polystyrenes shown in Table 3.2.

Table 3.2 Monomer Yields from Polystyrenes[10]

Compound	Monomer Yield (%)
Polystyrene	41
Poly(α-deuterostyrene)	68
Poly(β-deuterostyrene)	40
Poly(α,β,β-trifluorostyrene)	72
Poly(α-methylstyrene)	100

These results indicate the importance of a tertiary center in promoting depropagation, even when the substituent is deuterium.

Depropagation was probably first observed in linear polymers of formaldehyde, some of which smell strongly of formaldehyde even at room temperature. The first detailed study of this type of reaction was, however, the now classical work of Melville and Grassie[3] on the degradation of poly(methyl methacrylate). As outlined in Section A, they showed by simple molecular weight and weight loss measurements that the degradation was indeed a depropagation reaction and furthermore that it was free radical in nature.

The early work is reviewed in Grassie's book[5] and showed that radical acceptors were effective as stabilizers against degradation. In this type of unzipping reaction, however, a somewhat different and generally applicable stabilization mechanism seems attractive. In this method an inert blocking group is used to interrupt the chain reaction. The exact reactions involved are not certain, but as an example we may consider copolymers of methyl methacrylate and methyl acrylate. Whereas poly(methyl methacrylate) degrades by the unzipping mechanism outlined, poly(methyl acrylate) does not depolymerize and the yield of monomer from it is negligible. Thus a copolymer of these two might be expected to lose methyl methacrylate units along a chain until the terminal residue is methyl acrylate, whereupon the depolymerization should cease.

This copolymer system has been studied by Grassie and Torrance.[11,12] They investigated a range of polymer compositions using thermal volatilization to follow the degradation. Their results show that incorporation of methyl acrylate into the polymer appears to reduce the degradation initiated at unsaturated chain ends and also to move the peak characteristic of random-chain scission to higher temperatures. These results can be explained in either of two ways. The copolymer may contain less terminal unsaturation because of the copolymerization mechanism; that is, copolymer termination may be more by combination than is the homopolymerization. This effect is observed in methyl methacrylate–styrene copolymerization.[13] Alternatively the zip length may be substantially reduced by the methyl acrylate residues. The move to higher decomposition temperatures with increasing proportion of methyl acrylate shows the presence of the latter mechanism but does not preclude the former. Thus blocking does occur, but the presence of a methyl acrylate residue at the end of a depropagating chain apparently does not always suppress unzipping. This is shown in Table 3.3, which gives monomer

Table 3.3 Monomer Yield as a Function of Composition in the Degradation of Methyl Methacrylate–Methyl Acrylate Copolymers[11]

	Monomer Yield as Percentage of Total Volatiles	
Polymer[a]	Methyl Methacrylate	Methyl Acrylate
1/0	>96	—
112/1	>96	—
26/1	93	0.8
7.7/1	87	2.5
2/1	64	7.0
0/1	—	0.76

[a] Expressed as methyl methacrylate units/methyl acrylate units

yields from the degradation of the different copolymers. It is interesting to note that the yield of methyl acrylate from the 2/1 copolymer is several times greater than from the homopolymer.

Considerable care must be exercised in the choice of a suitable comonomer for blocking. Methyl acrylate is very well suited to the stabilization of poly-(methyl methacrylate) and is commercially used to stabilize this polymer.[11] Styrene and acrylonitrile also stablize against thermal degradation.[13] Other comonomers, however, may accelerate depropagation by acting as initiation centers. For example, when α-chloroacrylonitrile is copolymerized with methyl methacrylate, polymers of much lower thermal stability than the parent homopolymer result.[14] This is so even though poly(α-chloroacrylonitrile) degrades by elimination of hydrogen chloride rather than by depropagation. The copolymer apparently degrades by chain scission at the α-chloroacrylonitrile units after these have lost hydrogen chloride. Subsequent depropagation then occurs from these chain ends. Likewise in photodegradation, poly(methyl methacrylate) is made more unstable by the presence of acrylonitrile units in the chain.

Although from a historical point of view poly(methyl methacrylate) is perhaps regarded as the classical example of depropagation, having been the first system to be studied in detail, the polyaldehydes are probably more important commercially. Whereas poly(methyl methacrylate) in the absence of any stabilizing comonomer or other stabilizer is sufficiently stable for use, the polyaldehydes must be stabilized in order to obtain useful polymers.

The most important aldehyde polymer is polyformaldehyde or polyoxymethylene. The mechanism of the depropagation reaction that occurs is uncertain, but it probably does not involve free radicals.[15] The degradation has been studied in some detail by Kern and Cherdron,[16] who identified five processes:

1. Depropagation from chain ends.
2. Autoxidation.
3. Degradation by products of autoxidation.
4. Thermal degradation.
5. Hydrolysis.

In the unmodified polymer there is an hydroxy end group from which molecules of monomer are easily lost.

$$\sim CH_2OCH_2OCH_2OH \rightarrow \sim CH_2OCH_2OH + HCOH$$

This type of instability explains the observed inverse dependence of stability on molecular weight.[15] It is thus obvious that blocking of the terminal hydroxyl group will stabilize the resin, but there are difficulties in achieving sufficiently high extents of reaction. Suitable reaction processes have now been developed, and they form an important step in preparing commercial homopolymers.[17,18] The reactions most commonly used to replace the hydroxyl end groups are esterification and etherification, although a variety of others have been used.

Other degradative processes have also been observed. In the presence of oxygen, autoxidation occurs and leads to chain scission and subsequent monomer loss. This reaction may be minimized by the use of antioxidants; the usual classes of these materials have been used and some results are shown in Table 3.4.

Table 3.4 Effect of a Radical Acceptor in Stabilizing Acetylated Polyoxymethylene[19]

Di-β-naphthyl-p-phenylenediamine (%)	Weight Loss after 20 Min at 225°C (%)
0	80
0.005	45
0.01	23
0.05	4
0.1	4

Chain scission is also observed in the presence of acids. The effectiveness of acids correlates well with dissociation constant, as shown in Table 3.5. This acid-induced breakdown is of importance, since some antioxidants and some degradation products are or may lead to acids.

One other important stabilization mechanism remains to be discussed. In any of the degradation mechanisms mentioned, the loss of monomer may be minimized by interrupting the depropagation chain. This is quite logically achieved by introducing a second monomer. This approach has been followed in the preparation of the copolymers of trioxane that are now available commercially.

Table 3.5 Correlation between Dissociation Constant and Effectiveness of Acids in Degrading the Dimethyl Ether of Polyoxymethylene[16]

Acid	pK	Decomposition after 60 Min at 190°C (%)
Picric	0.4	87.6
Salicylic	3.0	36.2
p-Hydroxylbenzoic	4.5	17.8
β-Naphthol	8.0	10.8
Hydroquinone	10.0	1.3

C. DEGRADATION BY SEQUENTIAL ELIMINATION WITHOUT MONOMER LOSS

The outstanding example of a polymer showing this type of degradation is poly(vinyl chloride), one of the most important polymers in everyday use. Its main advantages are low cost and low flammability; its main disadvantage is low thermal stability. During processing, temperatures in the range 160 to

200°C are typically used, and stabilizers are required to prevent degradation. Where subsequent use conditions are prolonged or severe, some further stabilization may also be needed.

The typical thermal degradation of poly(vinyl chloride) leads to the formation of hydrogen chloride and a colored polymer containing conjugated double-bond sequences. The exact mechanism of this degradation is still unknown and the present state of knowledge has been recently reviewed by Geddes.[20] We shall not here repeat a detailed review but rather attempt to develop a background adequate for a detailed discussion of stabilization.

Other recent reviews of various aspects of dehydrochlorination and stabilization, differing somewhat in emphasis from our own, have been given by Chevassus and de Broutelles,[21] Neiman,[22] Gordon,[23] and Onozuka and Asahina.[24]

1. Mechanism of Thermal Dehydrochlorination

The basic structure of poly(vinyl chloride) would normally be considered to be quite stable as evidenced by the high stability of, for example 1,3,5-trichlorohexane.[25] Poly(vinyl chloride) is much less stable than this simple chloride, however; and in order to try and understand this it is useful to consider the stability of alkyl chlorides in general.

Barton, Onyon, and Howlett[26,27] developed a hypothesis, which has been widely substantiated, to predict the type of breakdown to be expected in simple alkyl chlorides. They assumed that two possibilities exist, a radical chain mechanism or a unimolecular evolution of hydrogen chloride. Their hypothesis is that where the first alternative is possible, it will occur. That is, if after initiation by a chlorine atom and abstraction of hydrogen to give hydrogen chloride, a polymeric radical capable of fragmentation to an olefin and a chlorine atom remains, then a radical chain will be observed. Otherwise unimolecular elimination of hydrogen chloride will take place. As examples, ethyl chloride with an hypothetical intermediate radical $CH_3\dot{C}HCl$, which cannot fragment, would decompose unimolecularly, whereas 1,2-dichloroethane, where the intermediate radical could fragment, would decompose by a radical chain.

$$\dot{C}HClCH_2Cl \rightarrow CHCl{=}CH_2 + Cl\cdot$$

On the basis of this hypothesis poly(vinyl chloride) may be understood to undergo either type of decomposition, depending upon which hydrogen atom is first extracted.

$$Cl\cdot + \sim CH_2CHCl\sim \;\nearrow\; \sim\dot{C}HCHCl\sim \;\longrightarrow\; \sim CH=CH\sim + Cl\cdot$$

$$Cl\cdot + \sim CH_2CHCl\sim \;\searrow\; \sim CH_2\dot{C}Cl\sim \;\longrightarrow\; \text{no propagation}$$

The results of Fredricks and Tedder[28] on 2-chlorobutane at 78°C indicate that abstraction of each type of hydrogen atom is equally probable. This suggests that a free-radical chain, being a more efficient process for hydrogen chloride elimination, will be the more important reaction.

One might think initially that the mere presence of long conjugated sequences in the degraded polymer suggests a chain reaction (either radical or ionic), but it is possible (although less likely) that a unimolecular elimination reaction might lead to such conjugation rather than isolated double bonds.

In addition to the molecular elimination and radical mechanisms mentioned, another possibility, an ionic reaction, should be considered. The case for this type of mechanism has been given by Marks, Benton, and Thomas.[29] It is perhaps fair to say that the most compelling piece of evidence for such a mechanism is the well-known catalytic effect of hydrogen chloride, which has never been satisfactorily explained on the basis of a free-radical reaction.

However the most widely accepted mechanism for poly(vinyl chloride) degradation is still one based on a free-radical chain. As evidence for such a mechanism we have the well-known acceleration by radical producers[30] and the detection of unpaired electron spins in degraded poly(vinyl chloride).[31] A variety of reaction sequences have been suggested and as examples we give those of Arlman[32] and Winkler.[33]

$$\sim\dot{C}HCHClCH_2CHCl\sim \rightarrow \sim CH{=}CH\dot{C}HCHCl\sim + HCl$$

$$\sim\dot{C}HCHClCH_2CHCl\sim \rightarrow \sim CH{=}CCH_2CHCl\sim + Cl\cdot$$

$$\sim CH{=}\overset{|}{C}\dot{C}HCHCl + HCl$$

In the latter scheme the preferential attack of the chlorine atom on the neighboring methylene group is explained on steric grounds.

In our discussion the initiation step has not yet been considered. Most investigations seeking an initiation mechanism have relied heavily upon free-radical propagation. Many possible initiation sites are apparent—chain ends, branch points, and oxygenated structures, to name just a few. The more obvious possibilities have been investigated. The dependence of dehydrochlorination rate upon molecular weight has been measured and shown, as in Figure 3.2, to follow an inverse relationship.[20,34] This indicates substantial involvement of end groups but fails to identify the group responsible. By investigating polymers prepared with a variety of initiators, the residues from these initiators have been classified according to their ability to promote degradation. Thus azo-initiated polymers appear to be the least stable, followed by peroxide- and gamma-ray-initiated materials.[36–38]

Other end groups, of course, exist as a result of various termination mechanisms;

$$\sim CH_2CHClCH{=}CHCl \quad \text{and} \quad \sim CH_2CHClCH_2CH_2Cl$$

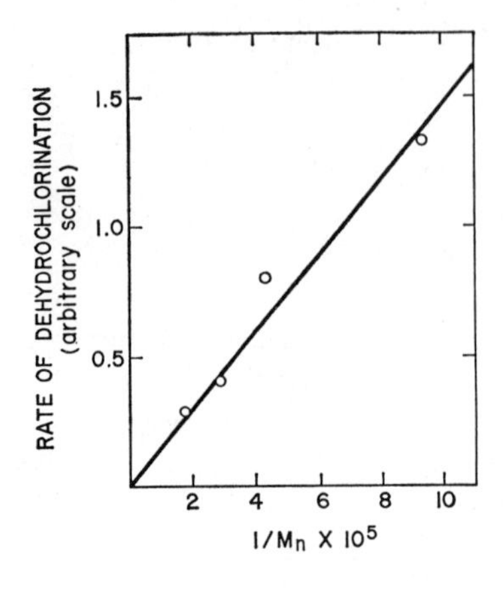

Fig. 3.2 Relation between dehydrochlorination rate and molecular weight.[32] Reprinted with permission from *Rubber Chemistry and Technology*.

result from disproportionation, whereas transfer to monomer gives also

$$\sim CHClCH_2CH{=}CHCl \quad \text{and} \quad \sim CHClCH_2CCl{=}CH_2$$

Of these end groups those having an allylic chlorine might be expected to show the highest activity, and such chlorine will also arise within the chain from any dehydrochlorination. Work with model compounds[39] has shown that the order of thermal stabilities of unsaturated chlorinated structures is

$$CH_2ClCH{=}CHCH_2CH_3,\quad CH_3CH_2{=}CClCH_2CH_3 > CH_3CHClCH_2CHClCH_3 > CH_2{=}CHCH_2CHClCH_2CH_3 > CH_2{=}CHCHClCH_2CH_3 > CH_3CH{=}CHCHClCH_2CH_3$$

In an attempt to measure the concentration of such labile groups Bengough and Onozuka have used an ester-exchange technique.[40] They have shown that there is a higher concentration of allylic chlorine in low-molecular-weight fractions than in unfractionated polymer.

Structural irregularities in the polymer chain may also provide the labile or reactive groups necesary for initiation. Possible abnormal structures are head-to-head linkages formed during termination by combination and tertiary chlorine structures produced by transfer to monomer. Bengough's work mentioned above has indicated the presence of the latter group, and it has been considered by many as an important initiator.

In this very brief survey we shall consider one more possible initiation site, the oxygenated structures likely to be introduced during polymerization and aging of poly(vinyl chloride). The most likely source of such groups is the unsaturated bonds of the parent polymer. These arise as mentioned above and also from any small amounts of dehydrochlorination. Oxidation of such structures is known to be easy and leads initially to the formation of hydroperoxides that, when decomposed homolytically, would lead to a rapid dehydrochlorination reaction.

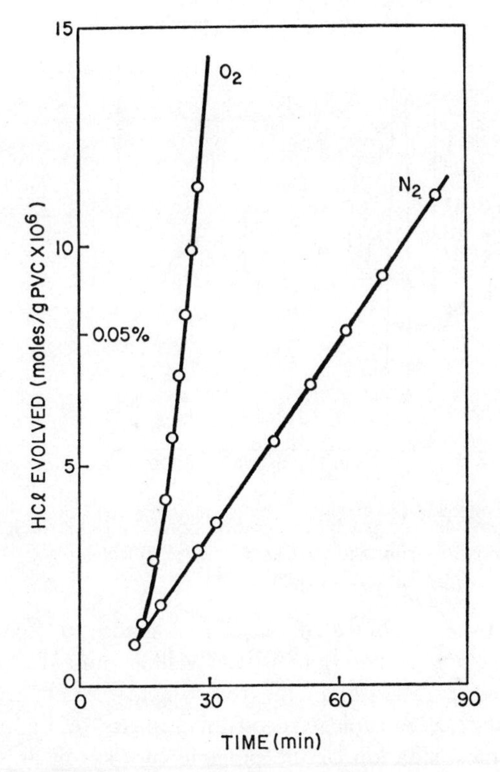

Fig. 3.3 Effect of oxygen atmosphere on dehydrochlorination rate.[41]

The effect of the atmosphere during dehydrochlorination is of importance. Three main environments have been used, nitrogen, oxygen, and hydrogen chloride. The effect of oxygen is large in increasing the rate of hydrogen chloride evolution, as shown in Figure 3.3, and in decreasing the rate of color development. These effects are perhaps best explained on the basis of an initiated oxidation that both limits the length of the conjugated sequences produced and leads to peroxy groups that give chain branching by decomposition.

$$\sim CH{=}CH\dot{C}HCHCl\sim + O_2 \rightarrow \sim CH{=}CHCH(OO\cdot)CHCl\sim$$

$$\downarrow + \text{polymer}$$

$$\sim CH{=}CHCH(OOH)CHCl\sim + \sim \dot{C}HCHCl\sim$$

$$\downarrow$$

$$\sim CH{=}CHCH(O\cdot)CHCl\sim$$

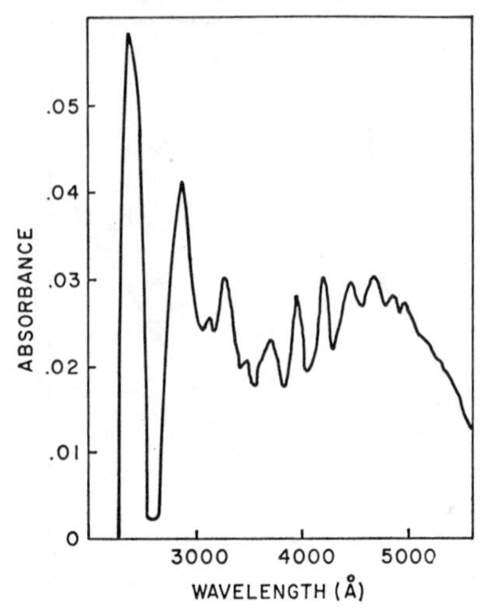

Fig. 3.4 Typical ultraviolet spectrum of degraded poly(vinyl chloride). Reprinted with permission from *Rubber Chemistry and Technology*.

The reduced length of conjugated sequences leads to light absorption in the ultraviolet region rather than in the visible, and degraded samples have a yellowish color rather than the red typical of dehydrochlorination in an inert atmosphere. A typical spectrum appears in Figure 3.4. There is also at least some oxidation of the polyene by oxygen as shown by the oxygen bleaching of degraded poly(vinyl chloride). Such oxidation, which also occurs during dehydrochlorination in oxygen, probably leads to crosslinks in the polymer (cf. oxidation of polybutadiene and styrene butadiene rubbers).

The effect of hydrogen chloride on the course of dehydrochlorination is more controversial. It was initially assumed that it catalyzed the reaction, and hence basic stabilizers were widely used. Later work on dehydrochlorination, however, showed no such catalysis,[42,43] and until quite recently this work has been accepted and widely quoted. More recent investigations have found a catalytic effect in inert[37,44,45] and oxidative atmospheres,[45] and such catalysis now seems well established. Its explanation is still unknown and perhaps is a strong piece of evidence supporting an ionic reaction.

2. Structure of Degraded Poly(vinyl chloride)

At the level of dehydrochlorination of interest, that is, up to about 1% loss of hydrogen chloride, our interest centers on the length of unsaturated sequences, polymer molecular weight, and the extent of crosslinking. There are two main approaches for measuring the first of these, analysis of ultraviolet spectra and chemical breakdown followed by molecular weight measurement.

The ultraviolet spectra of polymer with up to about 12 conjugated double bonds have been recorded.[46–51] Several equations have been used to correlate sequence length with absorption maxima[52–54] but none seems sufficiently accurate for use in the analysis of a degraded poly(vinyl chloride) spectrum such as that in Figure 3.4. Rough estimates ranging from 5 to 25 unsaturated bonds have been used as evidence for a very short chain in dehydrochlorination.[55] They may, of course, merely indicate a high transfer rate. It must also be remembered that these spectra detect only conjugated sequences, and much larger sequences of alternate double bonds may exist in which the conjugation is limited for steric reasons.

The alternative approach has been used by Geddes,[56] who measured the molecular weight change in degraded poly(vinyl chloride) after treatment with ozone to cut all double bonds. This method neglects chain end unsaturation, and values of about 25 double bonds per sequence were obtained.

The free radicals that are certainly present during poly(vinyl chloride) degradation may undergo reactions other than those associated directly with dehydrochlorination. Notably, crosslinking and scission reactions will occur, and these may have important effects on processibility. Crosslinking reactions have received most attention, and a number of possible mechanisms have been suggested. Combination of polymeric radicals is a probable mechanism,[57–59] but others, including copolymerization of polyene sequences,[43] Diels–Alder reactions, and intermolecular elimination of hydrogen chloride[34,59] have also been considered. In unstabilized polymer, discoloration usually leads to failure long before appreciable crosslinking has occurred. However, in stabilized compounds where color formation is intentionally suppressed, it is possible that crosslinking may lead to failure by producing too large a change in mechanical properties.

3. The Practical Effects of Degradation

As with all polymers, degradation of poly(vinyl chloride) means many things to many people (cf. Chapter 1). To the manufacturer of a white or pastel film the slightest coloration is detrimental, whereas to the maker of black flooring tiles the changes producing this coloration are unimportant. We are therefore faced with a large number of tests, each claiming to measure and indeed measuring degradation.

The main property changes occurring during dehydrochlorination are: hydrogen chloride evolution, coloration, and embrittlement. In the majority of cases color development is probably the most important sign of degradation, and thus some kind of color measurement would seem to be the most appropriate way of following degradation in practical applications. Although such tests have been and still are used, there is some difficulty in measuring color in a meaningful way, and since there is a rough correlation between color development and hydrogen chloride loss, many tests have relied upon measuring some aspect of this loss.

The most common test is probably the congo red test in which a polymer sample is heated and the time necessary for a congo red (or similar indicator) paper held in the effluent gas to change color is measured. In other tests samples are aged in a circulating air oven and color is assessed by comparison with standard color chips. Still others include some working of the polymer, such as milling, before color matching.

4. Stabilization

It is immediately obvious from the foregoing discussion that one useful function of a stabilizer is acid absorption. Rapid removal of hydrogen chloride is advantageous in two major ways, it prevents autocatalysis and it ensures that no catalytic metal chlorides are formed by reaction with metal equipment. An ancillary benefit is the absorption of traces of acid that might cause damage to surrounding materials. Many stabilizers are therefore merely acid absorbers; indeed some having other primary functions also act as acid absorbers. Since it is essential that the metal chlorides formed on reaction of the stabilizers be noncatalytic, lead salts are often used. Sodium, magnesium, and calcium salts have also been used, but less frequently.

Since the basic function of these materials is acid absorption, they may be classified according to the amount of acid with which they may safely react. In the most common case of lead salts, which are widely used in electrical applications, this is expressed by the effective percentage of lead (as litharge). Litharge itself has been used as a stabilizer, but it imparts an undesirable yellow color to the compound. It has now been superseded by basic salts such as basic lead carbonate and tribasic lead sulfate. A number of organic acid salts are also used, for example, maleates and phthalates. Lead stearate is often used, mainly in conjunction with other salts, but its action appears to be largely that of a lubricant.

Barium and cadmium salts of organic acids are also often used as stabilizers. They are, of course, acid acceptors but are more efficient than expected from this point of view and their mode of action has therefore been subjected to some investigation. The first detailed study of such systems was made by Frye and Horst.[60,61] They followed the attachment of ester residues to the polymer during heat treatment by both infrared absorption and isotope tracer methods. Both types of measurement showed that the hexanoate groups of cadmium, barium, and zinc 2-ethyl-hexanoates become chemically

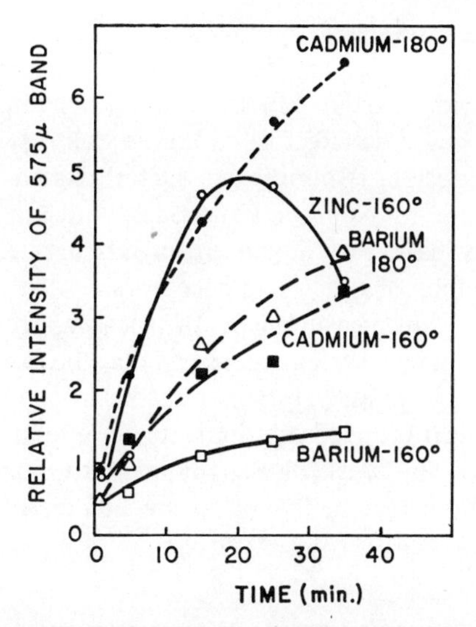

Fig. 3.5 Ester formation as a function of heating time for a variety of metal octoate–poly(vinyl chloride) mixtures.[60] Reprinted with permission from *Journal of Polymer Science.*

attached to the polymer chain during heat treatment. The extents of the attachment vary from salt to salt as indicated in Figure 3.5 (the 5.75-μ band is attributed to the carbonyl-stretching frequency of aliphatic esters of carboxylic acids).

Frye and Horst attribute this attachment to an ester-exchange reaction between the added salt and labile chlorine groups in the polymer. Such labile chlorine groups may reasonably be assumed to lead to initiation of dehydrochlorination chains upon heating, and thus such an exchange might be expected to result in stabilization. The type of stabilization expected from such a mechanism is true chemical stabilization with reduced coloration because of reduced hydrogen chloride loss. Hydrogen chloride evolution from a poly(vinyl chloride) metal carboxylate system is, of course, negligible owing to the acid acceptor properties of the salt; but measurements of dehydrochlorination have shown that the rate of chemical degradation of the polymer in the presence of these stabilizers is not decreased.[62,63] The coloration of the samples, however, is considerably lower than would be expected from the measured hydrogen chloride loss, which suggests that the stabilizer interferes with the formation of chromophores rather than with dehydrochlorination.

One complicating effect has, as yet, not been considered. This is the catalytic effect of the metal chloride formed from the carboxylate and hydrogen chloride. The use of such mixed stabilizers as barium-cadmium and zinc-calcium has been explained by postulating a minimization of the catalysis in such systems.[62,64] This explanation involves chloride ion exchange between two metals; in the barium-cadmium case the cadmium carboxylate is regenerated and the catalytic effect of cadmium chloride, which is higher than that of barium chloride, is minimized. Detailed measurements of this exchange have been made by Nagatomi and Saeki and are shown in Figure 3.6. Similar mechanisms have been advanced for mixtures of metal carboxylates and epoxy compounds or other acid acceptors.

In view of the more recent measurements of true dehydrochlorination rate in metal-carboxylate-stabilized systems, it seems that the original explanation of stabilization given by Frye and Horst cannot be true. Their experimental results are, of course, well founded and indicate definite attachment of ester groups to the polymer. The work of Bengough and Onozuka[40] mentioned earlier suggests that the attachment mechanism advanced by them is also

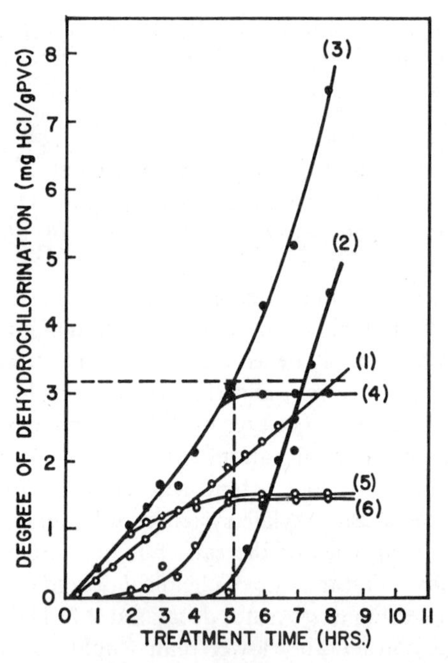

Fig. 3.6 Effect of cadmium and barium stearates on dehydrochlorination:[62] (1) hydrogen chloride loss in absence of stearates; (2) hydrogen chloride evolved as gas in presence of stearates; (3) hydrogen chloride lost from polymer in presence of stearates; (4) hydrogen chloride absorbed by stearates; (5) hydrogen chloride absorbed by barium stearate; (6) cadmium stearate. Dotted lines indicate points at which stearates are theoretically fully converted to chloride. Reprinted with permission from *Journal of Polymer Science.*

reasonable but that this reaction is not necessarily linked to stabilization. Stabilization of these materials appears to occur mainly by the prevention of discoloration, and it is therefore possible that double-bond sequences may be attacked to limit the light absorption in the visible region. One possible reaction would be the catalyzed addition of organic acid to the degraded molecule, but this is no more than speculation at this stage. It is also interesting to note that the degradation observed in the presence of chlorides of barium, cadmium, and copper leads to a lower degree of coloration than would be expected from the amount of hydrogen chloride evolved.

It thus seems that the stabilizing effect of these metal carboxylate systems is still not fully understood, although a number of the reactions have been well characterized.

The next class of stabilizers are the organotin compounds. We first consider those having the general formula

$$(C_4H_9)_2Sn(OCOR)_2$$

where the alkyl group R varies widely. In one of the early efforts to elucidate their mode of action and correlate this with a free-radical decomposition, Winkler[33] showed that several types of poly(vinyl chloride) stabilizers including epoxides, cadmium soaps, and dibutyl tin dilaurate were capable of retarding the cobalt-ion-catalyzed decomposition of *t*-butyl hydroperoxide.

In a later investigation of the action of the tin compounds, Frye, Horst, and Paliobagis[65–67] have followed the reaction between typical stabilizers and poly(vinyl chloride). They employed selective isotopic labeling of the butyl groups, the tin atom, and the carboxylate residue to show that, as with barium and cadmium carboxylates, the carboxylate residue becomes attached to the polymer chain during heat treatment (Fig. 3.7). Earlier work of Kenyon[68] had shown evidence for the attachment of butyl groups to the polymer during light irradiation. Repeated dissolution–precipitation cycles led to loss of radioactivity from the polymer stabilized with tin labeled stabilizer in most cases. This was used as evidence for a coordinate link between the stabilizer and polymer, with the subsequent possibility of

$$\begin{array}{ccc} \sim CH_2CHClCH_2\sim & & \sim CH_2\underset{|}{C}HCH_2\sim \\ & & Cl \\ + & \longrightarrow & | \\ & & Bu_2SnY_2 \\ Bu_2SnY_2 & & | \\ & & Cl \\ & & | \\ & & \sim CH_2CHCH_2\sim \end{array}$$

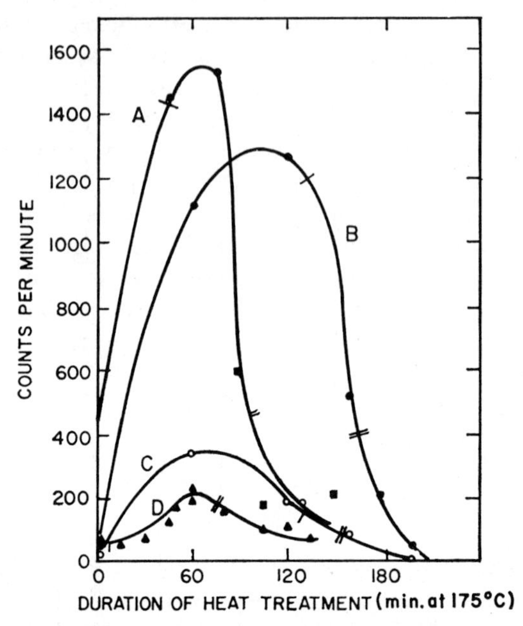

Fig. 3.7 Attachment of tin stabilizers to poly(vinyl chloride) as a function of heat treatment: (*A*) dibutyl tin bis(monomethyl maleate); (*B*) dibutyl tin β-mercaptopropanoate; (*C*) dibutyl tin bis(oxooctyl thioglycolate); (*D*) dibutyl tin bis(2-ethylhexanoate). Reprinted with permission from *Journal of Polymer Science*.

rearrangement and, especially in the presence of solvent, rupture of the coordinate bond. This cleaving reaction is accelerated by hydrogen chloride and thus would be increased during dehydrochlorination.

$$\begin{array}{ccc} \sim CH_2CHCH_2\sim & & \sim CH_2CHYCH_2\sim \\ | & & \\ Y & \longrightarrow & + \\ | & & \\ Bu_2SnClY & & Bu_2SnClY \\ | & & \\ Cl & & + \\ | & & \\ \sim CH_2CHCH_2\sim & & \sim CH_2CHClCH_2\sim \end{array}$$

These results, which show a chemical attachment of stabilizer residue to the polymer, were interpreted as an exchange between labile chlorine in the polymer and stabilizer ester groups. Such an exchange should lead to a true chemical stabilization, that is, a reduced dehydrochlorination rate; but, as in the case of heavy metal carboxylates, this is not experimentally confirmed. Results of Marks, Benton, and Thomas[29] reveal that the dehydrochlorination

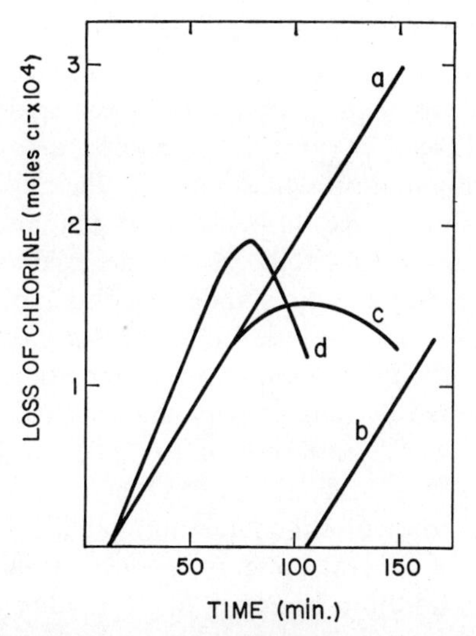

Fig. 3.8 Effect of tin stabilizers on dehydrochlorination in solution;[29] (*a*) hydrogen chloride loss without stabilizer; (*b*) loss with tin stabilizer; (*c*) and (*d*) chloride ion formed in the presence of R_2SnY_2 and RSY_3 type stabilizers (R—alkyl; Y—ester). Reprinted with permission from *Journal of the Society of the Chemical Industry, London.*

rate is unaffected by the presence of stabilizer (Figure 3.8). Thus, as with the heavy metal carboxylates, it seems that stabilization is achieved not by reducing dehydrochlorination but by reducing its effects, notably coloration.

In the case of dibutyl tin bis(monomethyl maleate) some residual activity remained after repeated dissolution and precipitation, and this may be accounted for by a different attachment reaction in this case. Possibly a Diels–Alder reaction takes place, between the maleic acid residue and polymer.

~CH=CHCH=CH~

\+

CH=CH (with $Bu_2YSnOCO$ and $COOCH_3$ substituents)

$\longrightarrow$

CH=CH / ~CH CH~ / CH—CH (ring), with $Bu_2YSnOCO$ and $COOCH_3$ substituents

Subsequent treatment with hydrogen chloride again leads to loss of activity, which may be explained by hydrolytic loss of the tin containing group to give

CH=CH / ~CH CH~ / CH—CH (ring), with COOH and $COOCH_3$ substituents $+ Bu_2SnCl_2 + HY$

The idea of a Diels–Alder addition in stabilization is an attractive one, since it would lead to a reduction of color development during dehydrochlorination. Active dienophiles are capable of bleaching degraded poly(vinyl chloride),[63] but they also act as powerful initiators. Less active dienophiles may well minimize this initiation while retaining the ability to remove discoloration. This possibility has yet to be fully substantiated experimentally.

Before leaving tin-containing stabilizers it is necessary to consider a few other types. First, there are a few tin stabilizers containing octyl rather than butyl groups. Octyl groups lead to lower stabilizer toxicity and sometimes to increased stabilization activity. A comparison of retention of octoate groups in the polymer as measured by radioactivity studies shows that, with certain heating conditions, the overall pattern is quite similar to that of compounds containing butyl groups.[66]

Another interesting group of tin-containing stabilizers is that including mercapto substituents. In general these are good thermal stabilizers and it seems probable that, in addition to acting like the other tin stabilizers, the sulfur, during processing, gives rise to peroxide decomposers that may act as stabilizers. The role of peroxide decomposers in stabilization is discussed in more detail below.

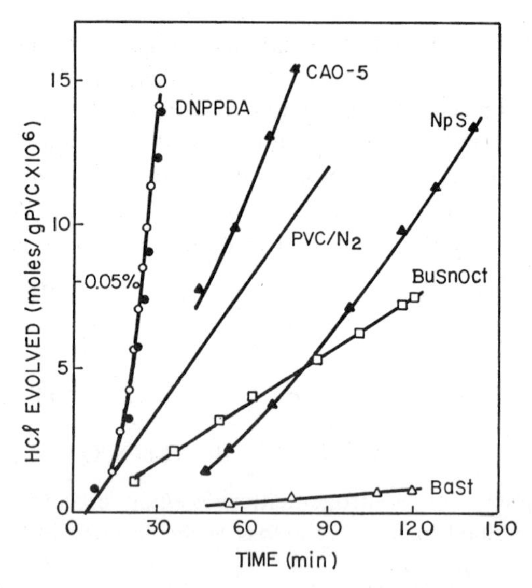

Fig. 3.9 Effects of various stabilizers on dehydrochlorination in oxygen: DNPPDA, di-β-naphthyl-p-phenylene diamine; CAO-5,2,2′-methylene bis(4-methyl-6-t-butyl phenol); NpS, naphthyl disulfide; BuSnOct, dibutyl tin octoate; BaSt, barium stearate. The curve labelled PVC/N_2 shows the dehydrochlorination rate of unstabilized poly(vinyl chloride) in nitrogen.

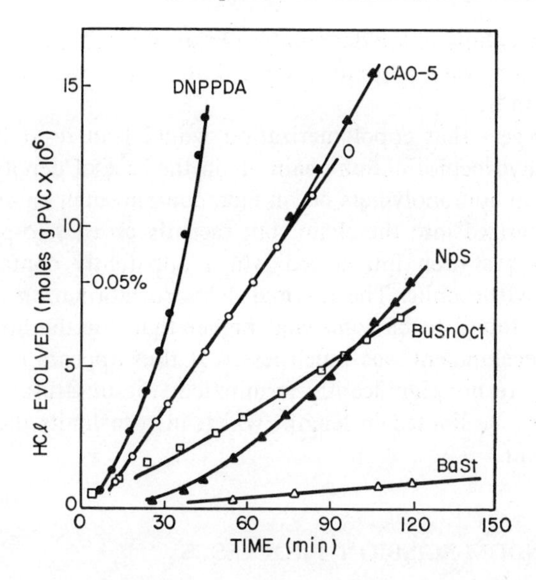

Fig. 3.10 Effects of various stabilizers on dehydrochlorination in nitrogen. Labels as in Figure 3.9.

In addition to the normally used classes of stabilizers mentioned above, a number of other compounds are used less commonly. First we might consider the chelating stabilizers, whose role is to remove metal ions that may be present in the resin and thus to prevent catalytic degradation. As previously explained, many metal ions do catalyze dehydrochlorination and the chelators, such as triphenyl phosphite, deactivate these in coordination compounds.

Conventional antioxidants are also sometimes used but are not outstandingly effective. They can help, however, in limiting degradation in an oxygen atmosphere.

One final class of materials will be mentioned, namely, peroxide decomposers. Although they are not currently used as such, their action may well be instructive in helping us to understand stabilization. The action of a typical peroxide decomposer, naphthyl disulfide, in degradation in oxygen appears in Figure 3.9 together with curves for a number of other stabilizers.[63] It may be seen that hydrogen chloride evolution is reduced, and in addition the coloration is decreased. It is somewhat surprising that similar results are obtained for degradation in nitrogen, as in Figure 3.10, but the reason for this becomes clear when it is remembered that some degradation chains may be initiated by peroxide groups in the polymers. The behavior of peroxide decomposers is perhaps best explained as the result of their ability to reduce or stop chain branching in the degradation in oxygen. This branching would be expected to occur via a peroxide intermediate, which would be vulnerable to these compounds.

One might expect that copolymerization would lead to stabilization by breaking the dehydrochlorination chain, as in the case of depolymerization. Many commercial homopolymers do, in fact, contain small amounts of vinyl acetate copolymerized into the chain, but recently an easy-to-process poly-(vinyl chloride) has been introduced which apparently contains a small number of propylene units. The thermal dehydrochlorination of this compound has been found to be somewhat higher than for the homopolymer, but the color development was much less.[41] It thus appears that, although reaction chains are not significantly terminated, the unsaturated sequences that develop may be limited in length, which in turn limits the amount of color development.

D. RANDOM-SCISSION PROCESSES

Many addition polymers and virtually all condensation polymers pyrolyze preponderantly by random-scission mechanisms. Two stages of breakdown have been recognized in polyolefins. An abrupt initial reduction in weight-average molecular weight at low temperatures has been ascribed to "weak links" in the main chain, and volatilization at higher temperatures has been attributed generally to depolymerization and hydrogen transfer reactions. Oakes and Richards[69] noted that the molecular weight of branched polyethylene decreased at temperatures above 290°C, whereas the rate of volatilization was insignificant below 370°C. They assumed that "weak links" responsible for initial degradation were associated with branch sites or with carbonyl, peroxide, or unsaturated groups. Wall and co-workers[70] pyrolyzed polyethylene under vacuum conditions and obtained a waxlike product containing only a trace of ethylene monomer. Infrared analysis of the polymeric residue showed that vinylene, vinylidene, and vinyl groups had formed during degradation. Since vinyl concentration approached a maximum in the final stage of breakdown of the linear polymer, it was proposed that transfer reactions of the following type were the main mode of polyethylene pyrolysis:

$$RCH_2\cdot + R'CH_2CH_2CH_2R'' \rightarrow RCH_3 + R'\dot{C}HCH_2CH_2R''$$
$$R'\dot{C}HCH_2CH_2R'' \rightarrow R'CH{=}CH_2 + R''CH_2\cdot$$

Bailey and Liotta[71] also pyrolyzed the polymer under reduced pressure in nitrogen at 415°C and obtained a variety of alkane and alkene products. As reaction temperature rose to 600°C the yield of ethylene monomer increased tenfold from 6 to 60%. Higher reaction temperatures promoted depolymerization because ethylene yield was independent of pyrolysis time. As a result, Bailey suggested that linear polyethylene, like all unsaturated compounds with at least one hydrogen in a gamma position with respect to unsaturation, would fragment by a unimolecular cyclic mechanism. He envisioned an initial reaction involving chain scission followed by inter- or intramolecular hydrogen abstraction. Tsuchiya and Sumi[72] decided that intramolecular radical transfer was the main mechanism of polymethylene volatilization. They suggested that the radical transfers successively from the first to the fifth, ninth, and thirteenth carbons in moving along the chain as follows:

$$\cdot CH_2(CH_2)_4CH_2\sim \rightarrow CH_3(CH_2)_3\dot{C}HCH_2\sim \rightarrow CH_3(CH_2)_7\dot{C}HCH_2\sim \ldots$$

Decomposition then proceeds by two routes,

$$R_2CH{-}\dot{C}HCH_2{-}R'$$

$$\swarrow \qquad\qquad \searrow$$

$$RCH_2CH{=}CH_2 + R'\cdot \qquad\qquad R\cdot + CH_2{=}CHCH_2R'$$

R represents the macromolecule; R′· signifies propyl, heptyl, or undecyl radicals; and $R'CH_2CH{=}CH_2$ can be 1-hexene, 1-decene, and so on.

Similar intramolecular radical transfer and propagation processes account for the volatile pyrolysis products from polypropylene.[73] Pyrolysis of polypropylene begins in the temperature range of 230 to 250°C, but catastrophic breakdown to volatile products is negligible below 300°C.[74] Davis[75] found that the weight-average molecular weight of polypropylene decreased rapidly and then leveled off as the ratio of weight-average to number-average molecular weight approached 2:1, the theoretical limit for a random distribution.[76]

Transfer occurs mainly in polymers such as polyethylene or polypropylene, which have readily accessible hydrogen atoms on secondary and tertiary alkyl carbons. Transfer is less likely in polystyrene because bulky phenyl groups shield the hydrogens bonded to main-chain carbon atoms. Styrene monomer yields of 40% from vacuum pyrolysis have been reported.[77] The volatilization rate reached a maximum at about 40% conversion and was independent of weight-average molecular weight in the range of 10^5 to 10^6. The degradation results indicate an end initiation mechanism with a superposed random-scission component.[78]

Branching, unsaturation, and oxygen-containing groups lower the stability of polyolefins, but processing procedures involve no serious problems, since nonoxidative degradation rates are relatively low at temperatures below 250°C. Familiar protectants such as radical scavengers and blocking groups are ineffective because breakdown is a random rather than a chain reaction. Even if they were effective, protectants would be hardly necessary, since polymethylene has greater thermal stability than any of the addition polymers except polytetrafluoroethylene and a few other highly fluorinated polymers.

None of the polymers described so far has useful properties at temperatures in the vicinity of 300°C, inasmuch as high heat resistance requires morphological (mechanical) as well as chemical (thermal) stability. The more important structure–stability relationships[79] have become apparent with the development of polymers with extreme heat resistance. In general, thermal stability depends on bond dissociation energies which, in turn, vary with the atomic species joined by the bonds and with the overall valence structure.

1. Bond Dissociation Energies

Maximum approximate energies required for homolytic cleavage of some common covalent bonds appear in Table 3.6. As might be expected, the bonds

Table 3.6 Typical Ultimate Bond Dissociation Energies[a]

Bond[b]	Energy	Bond[b]	Energy
C–F (al)	116	C–C (ar al)	93
C–N (ar)	110	C–O (al)	93
C–O (ar)	107	C–C (al al)	83
Si–O	106	C–N (al)	82
C–H (ar)	103	N–N	37
C–C (ar ar)	100	O–O	34

[a] S. W. Benson, *J. Chem. Educ.*, **42,** 502 (1965); C. E. Bawn, *Proc. Roy. Soc.* (*London*), **A282,** 91 (1964).
[b] ar = aromatic; al = aliphatic.

between aromatic carbon atoms are stronger than the corresponding aliphatic bonds. Heterobonds of oxygen or nitrogen atoms with silicon, boron, or carbon are especially strong. However several factors may increase reaction rates and therefore reduce pyrolysis temperatures. For example, although carbon–chlorine and carbon–oxygen side bonds have relatively high dissociation energies, they can rupture readily when formation of products with higher bond energies such as hydrogen chloride and water is possible (see Section C). In addition, resonance effects, bulky neighboring groups and free valences in the α-position also suppress bond energies, as shown in Table 3.7.

Table 3.7 Effects of Near-Neighbor Structures and Groups on Bond Dissociation Energies

Molecule Containing C–C	C–C Bond Energy (kcal/mole)	Molecule Containing C–H	C–H Bond Energy (kcal/mole)
$CH_3—CH_3$	88	$RCH_2—H$	98
$(CH_3)_3C—C(CH_3)_3$	68	$CH_2{=}CHCH_2—H$	85
$(C_6H_5)_3C—C(C_6H_5)_3$	15	$\cdot CH_2CH_2—H$	39

2. Thermal Stability and the Chain Unit

The relation between thermal stability and the repeat unit in the polymeric chain was illustrated in Table 3.1 by polymers that vaporize completely in a vacuum over a rather narrow temperature range. Note that unsaturation in the chain has a negligible effect on ultimate stability. Extensive replacement of hydrogen in polymethylene by fluorine raises the volatilization temperature, whereas steric effects of the type illustrated in Figure 3.11 decrease it.

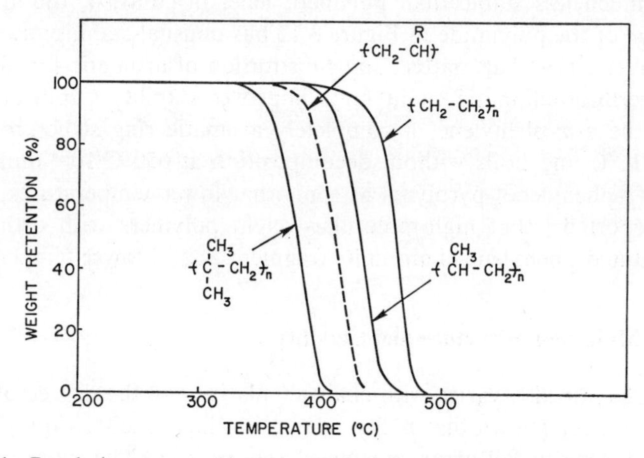

Fig. 3.11 Pyrolysis patterns of branched polyolefins. R signifies a cyclohexyl, a secondary propyl, or an isobutyl group in polymers represented by the dashed curve. Constant rate of temperature rise was 100°C/hr.

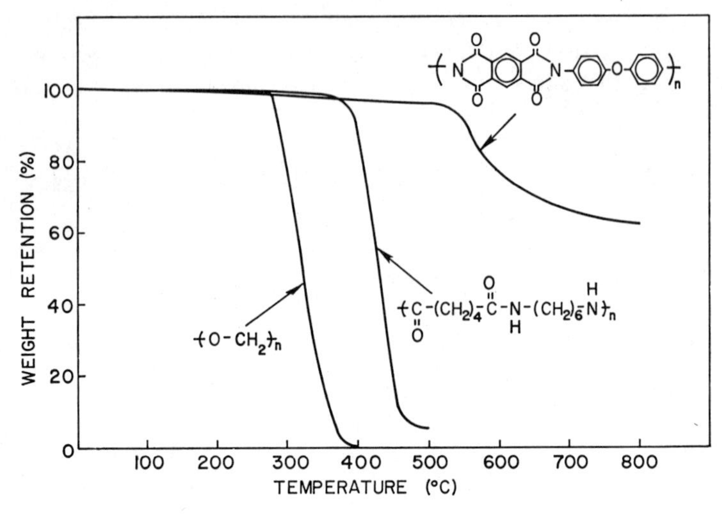

Fig. 3.12 Relative thermal stabilities of a polyether, a polyamide, and a polyimide; rate of temperature increase, 100°C/hr.

Evidently stereoisomerism has little effect, because isotactic and atactic forms of polypropylene, polystyrene, and poly(propylene oxide) have virtually identical decomposition temperatures characteristic of each polymer. On the other hand, substitution of oxygen for carbon in alkane chains reduces stability, since polyformaldehyde, poly(ethylene oxide), and poly(propylene oxide) are much less stable than polymethylene. In contrast, the aromatic ether linkage of the polyimide in Figure 3.12 has unusual stability in a nitrogen atmosphere. For that matter, any substitution of aromatic for aliphatic and even perfluoroaliphatic chain units improves stability significantly. In fact, cyclic hexa-*m*-phenylene, a completely aromatic ring structure, melts at 509 to 511°C and boils without decomposition at 650°C/10^{-4} mm.[80] The linear poly(*p*-phenylene) pyrolyzes at somewhat lower temperatures, and it has been reported[79] that high-molecular-weight polymers with ortho- and meta-substituted phenylene chain units fragment at still lower temperatures.

3. Molecular Structure and Stability

Pyrolytic behavior also varies with chain regularity and the degree of cross-linking. It has been shown that bulky side groups facilitate scission of chains consisting of head-to-tail arrangements of repeat units. The more crowded head-to-head structures generally reduce stability even further. Poly(vinyl chloride) and poly(vinylidene chloride) are noteworthy exceptions to the rule,

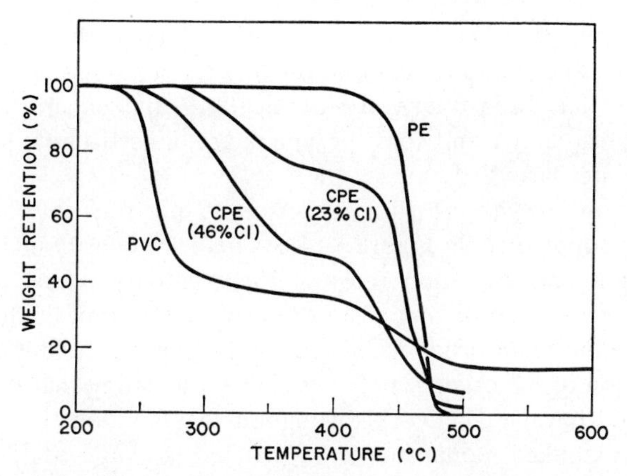

Fig. 3.13 Pyrolysis behavior of polyethylene, poly(vinyl chloride), and polyethylenes with hydrogen randomly replaced by chlorine; rate of temperature rise, 100°C/hr.

since hydrogen chloride is formed at a lower rate from the head-to-head than from the head-to-tail form of these polymers.[13,81] But, as might be expected, thermal stabilities of chlorinated polyethylenes are intermediate between those of polyethylene and poly(vinyl chloride), as revealed in Figure 3.13. Also introduction of ethylene oxide units into the polyoxymethylene chain augments stability by blocking depolymerization, as described in Section B.

4. Crosslinking and Thermal Stability

Rupture of a single covalent bond results in chain scission of a linear polymer. But the probability of molecular fragmentation is obviously less in the

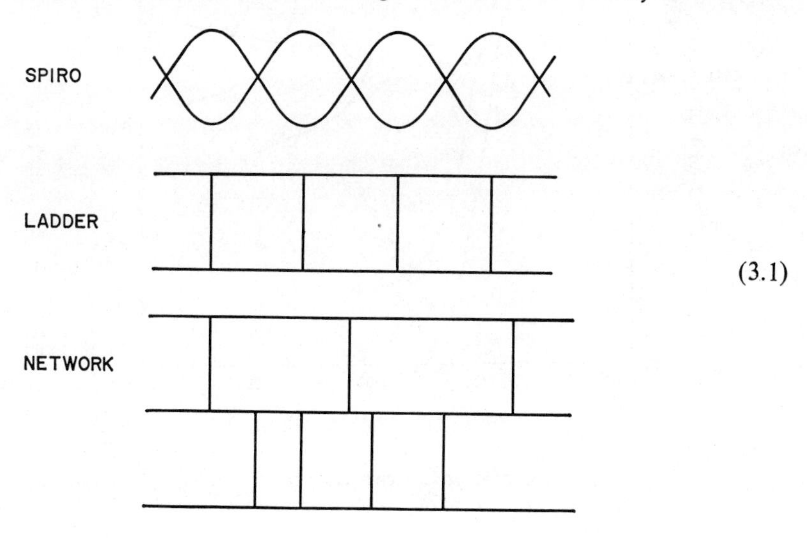

(3.1)

crosslinked systems of (3.1) because volatilization requires two or more bond breaks per ring. The random degradation of four- and six-membered-ring ladder polymers has been analyzed statistically,[82] and experimental work indicates that both ladder and spiro polymers are, indeed, more stable than their linear counterparts.[83,84]

Stability also varies with crosslink density,[85] as illustrated by the styrene–trivinylbenzene copolymers in Figure 3.14. As network density increased, the molecular weight and evolution rates of the volatile products decreased. Methane and a residue consisting of an extensive network of conjugated unsaturation were the main products. Carbonization of the residue proceeded with full retention of the original polymer shape, indicating that the polymer remained a rigid covalent network throughout degradation. Diamond, the ultimate in crosslinked nonaromatic structures, is quite stable in inert atmospheres at temperatures below 1300°C[86] but it graphitizes rapidly above 1500°C. The graphitic structure of fused aromatic rings is so stable that crystal moduli are nearly constant up to 2400°C where the vapor pressure is less than 10^{-3} mm Hg. Yet graphite and even its linear counterpart, poly(*p*-phenylene)[87,88] are too insoluble, infusible, and brittle for many applications for which some linear aromatic polymers are adequate.

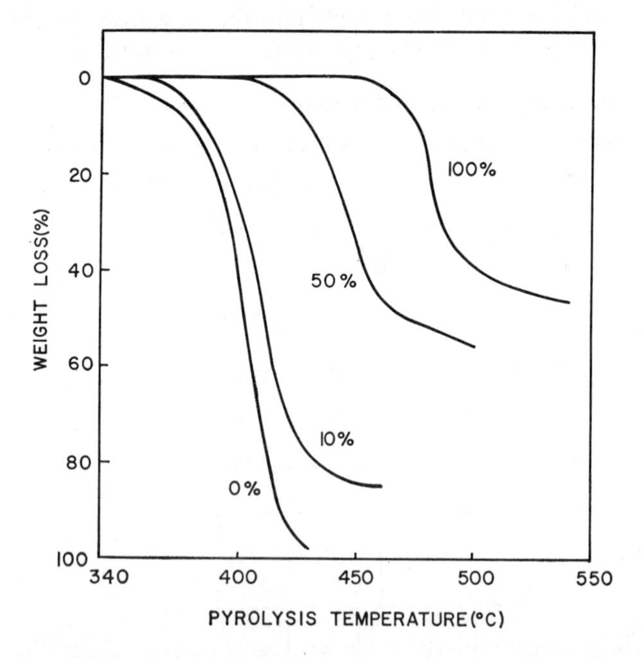

Fig. 3.14 Pyrolysis patterns of styrene–trivinylbenzene copolymers designated by the trivinylbenzene concentration; rate of temperature rise, 100°C/hr.

5. Heat-Resistant Polymers

Many commercial polymers, including some polyamides, polyesters, polycarbonates, polyphenylene oxides, and polysulfones, have melting or glass transition temperatures in the vicinity of 150°C or higher. But to qualify for reliable, long-term service at temperatures of 300°C or more a polymer must have (*a*) an extremely high melting point or glass transition temperature, (*b*) exceptional thermal and oxidative stability, and (*c*) properties of solubility, moldability, or both at least in an intermediate form. Most polymers satisfying these requirements consist of linear molecules with largely aromatic repeat units.[8] Planar aromatic ring structures lead to the stiff polymer chains associated with high melting and glass transition temperatures and with pyrolytic and oxidative stability.

Poly(*p*-xylylene) consisting of a chain of phenylene rings linked together by ethane units has excellent electrical properties and heat resistance. The polymer (parylene-*N*)[89] has been produced commercially from the cyclic dimer, di-*p*-xylylene [(2,2)-*p*-cyclophane], by a vapor deposition method in

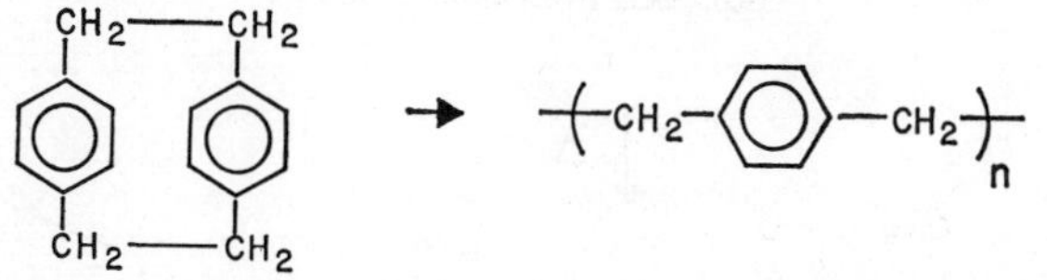

$$\text{[2.2]-}p\text{-cyclophane} \rightarrow \left(CH_2 - C_6H_4 - CH_2 \right)_n \qquad (3.2)$$

which dimeric pyrolysates were condensed and polymerized on a cool surface (3.2). The tetrafluoro derivative has also been synthesized by two different routes, and several other modifications of the polymer have been described.[90] Another commercial high-temperature polymer (Nomex or HT-1) has been reported to have the structure shown in (3.3). The polymer

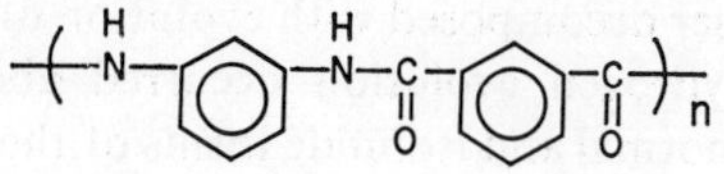

$$\left(NH - C_6H_4 - NH - C(=O) - C_6H_4 - C(=O) \right)_n \qquad (3.3)$$

has been made in the form of filaments and paper. Because it can withstand temperatures as high as 230°C for extended periods of time, it has been used for ironing board covers and braking parachutes for high-speed air- and spacecraft. Because of the high density of hydrogen bonds and rigid aromatic rings, the polymer does not melt.

Extraordinary heat resistance has been achieved with a large variety of polymers containing aromatic and heterocyclic rings.[91] To minimize processing problems, these polymers are first prepared as "prepolymers" and then subjected to a "post"-reaction during processing.[9,92–94] Aromatic poly-pyromellitimides are typical examples of such a class of materials. Soluble precursors are first prepared and then converted into intractable products by the procedure (3.4). At temperatures above 400°C and under nonoxidative

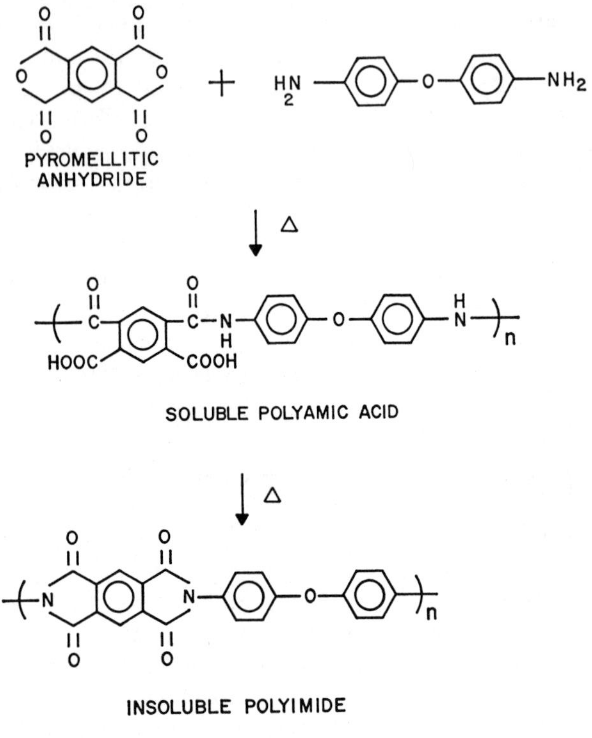

PYROMELLITIC ANHYDRIDE + $H_2N-C_6H_4-O-C_6H_4-NH_2$

↓ Δ

SOLUBLE POLYAMIC ACID (3.4)

↓ Δ

INSOLUBLE POLYIMIDE

conditions, the polymer decomposed with evolution of carbon monoxide and carbon dioxide.[95] Hydrogen evolution occurred above 525°C. A mobile equilibrium between normal and isoimide forms of the repeat units has been proposed to account for the products. The polymer is flame resistant and has excellent mechanical strength at liquid helium temperatures. Other properties are listed in Table 3.8. Ether, ester, carbonyl, or sulfone bridges between the phenylene groups provide more chain flexibility and, therefore, lead to greater polymer extensibility, solubility, and processibility without appreciable sacrifice of stability. Other high-temperature polymers have recurring units based on benzimidazole, benzothiazole, quinoxaline, and oxadiazole. Polybenzimidazole[96,97] has the distinct advantage over other aromatic heterocyclic polymers of being soluble in dimethyl sulfoxide and in phenols. Various

Table 3.8 Properties of Polypyromellitimide Films Based on *bis*(4-Aminophenyl) Ether

Stability (air)[a]		275°C	1 year
		300°C	1 month
		400°C	1 day
Melting point		>500°C	Decomposition
Glass-transition temperature		>400°C	
Density		1.42	
Tensile strength (psi)	25°C/25,000	200°C/17,000	300°C/10,000
Elongation (%)	70	90	120
Tensile modulus (psi)	400,000	260,000	200,000

[a] Retention of flexibility.

forms of the polymer have been made into cloth[98] and have also been marketed as high-temperature adhesives (Imidites).[99]

A few unusual polymers with all conjugated repeat units have been called "vat" polymers[100] because, like vat dyes, they can be reduced to a soluble leuco form and then oxidized back to an intractable colored form (3.5).

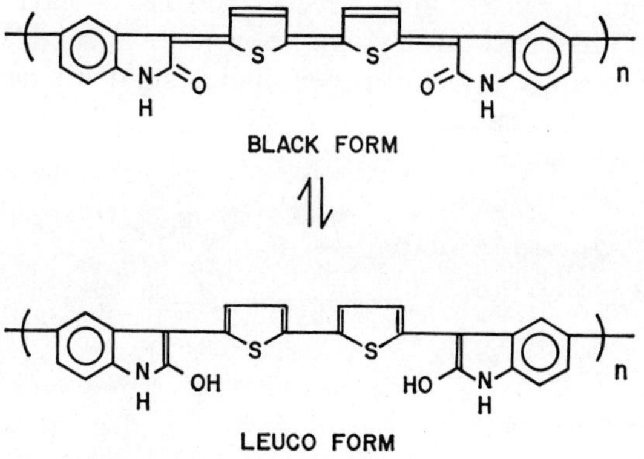

BLACK FORM

LEUCO FORM

(3.5)

6. Ladder and Spiro Polymers[101]

A well-defined "ladder" or "double-strand" polymer of equilibrated phenylsilsesquioxane has been synthesized.[102] It consisted of cis-anti-cis fused cyclotetrasiloxane rings as shown in (3.6), was soluble in benzene and other

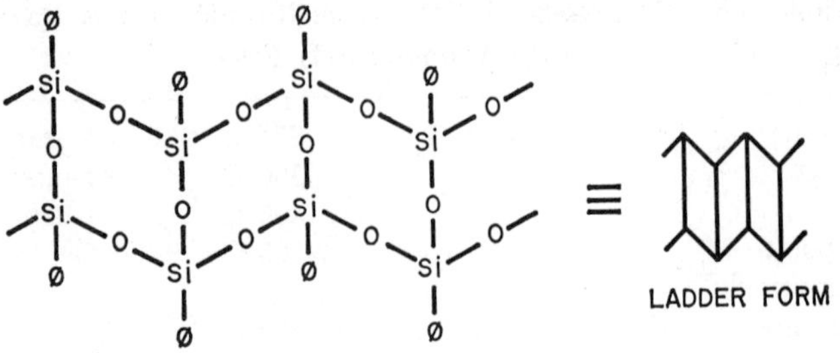

(3.6)

organic solvents, and had improved thermal stability over the corresponding linear siloxane. Although the polymer had a high molecular weight it was brittle and hydrolytically unstable. The nonequilibrated trifunctional silicone polycondensates evidently have interesting "bead-chain" structures[103] linked together in the main chain by single bonds as in (3.7).

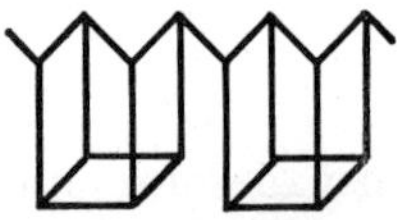

(3.7)

PHENYLSILSESQUIOXANE

Most ladder polymers have been made by a two-step process similar to that used for polyimides. For example, a ladder polymer of vinyl isocyanate has been prepared[104] by the two routes shown in (3.8). A soluble product with unreacted side groups was first obtained, indicating that a partial or "step" ladder structure was formed.

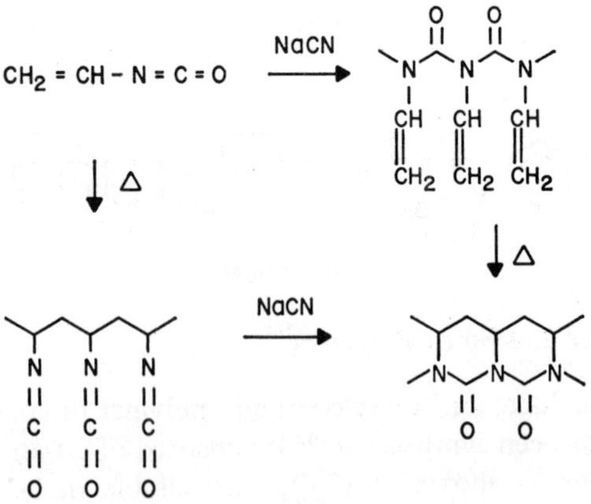

(3.8)

Several inorganic polymers[105] have been assigned a spiro structure, and a high-molecular-weight polyspiroketal[106,107] has been reported, but so far these polymers have attracted only minor interest.

In addition a number of hybrid polymers such as those of (3.9) have been

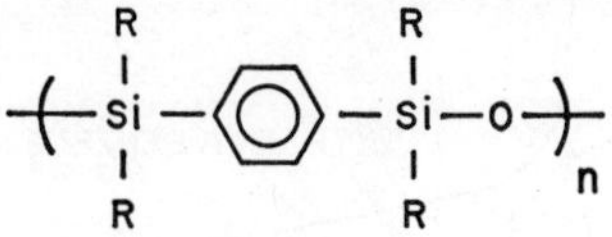

$$\left(-\underset{R}{\overset{R}{Si}}-C_6H_4-\underset{R}{\overset{R}{Si}}-O- \right)_n \quad \text{POLYSILPHENYLENESILOXANE}$$

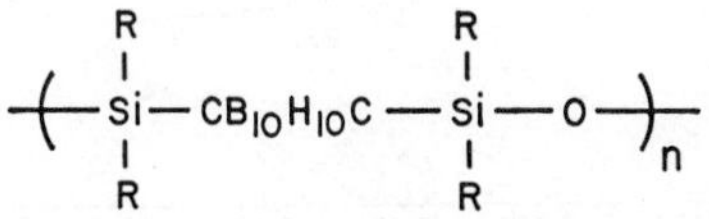

$$\left(-\underset{R}{\overset{R}{Si}}-CB_{10}H_{10}C-\underset{R}{\overset{R}{Si}}-O- \right)_n \quad \text{POLYCARBORANESILOXANE} \tag{3.9}$$

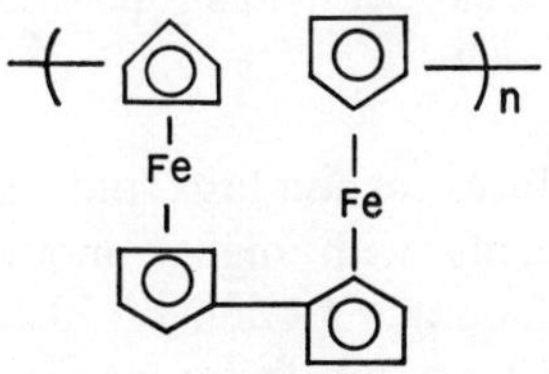

$$\text{POLYFERROCENYL}$$

prepared and studied. Although these polymers have good heat resistance, their cost at present precludes commercial development.

7. Pyrolytic Polymers

The highest level of heat resistance has been achieved with polymer carbons. Degradation of several char-forming polymers (see Figure 3.15) begins at relatively low temperatures. For example, loss of the first hydrogen chloride per repeat unit in polyvinylidene chloride becomes rapid above 150°C. It has been proposed that the polyene product cyclizes,[108] with rising temperature and further loss of hydrogen chloride, to form fused aromatic structures. Nearly all the original carbon skeleton is retained in a highly disordered carbon, which resists reordering at temperatures well above 2000°C. Numerous studies have indicated that polyacrylonitrile pyrolyzes to a ladder structure,[109–112] which converts to an aromatic system above 500°C. Further heating to 800°C leads to loss of up to one-quarter of the nitrile groups as

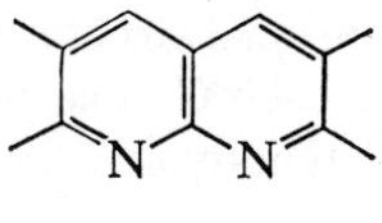

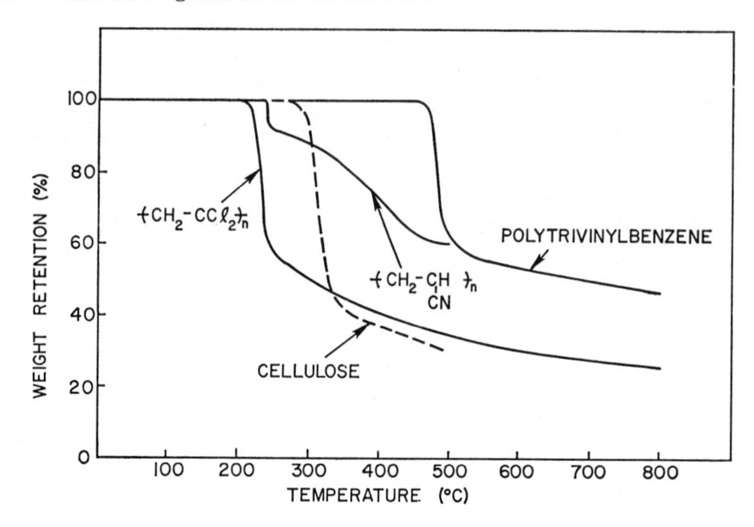

Fig. 3.15 Degradation characteristics of some char-forming polymers; heating rate, 100°C/hr.

hydrogen cyanide.[113] Preoxidation reduces carbon loss, and stress orientation during carbonization leads to filaments with higher moduli and tensile strengths. But Watt has reported[114] Young's moduli of 30 to 60 $\times$ 10^6 psi and tensile strengths of 300 to 430 $\times$ 10^3 psi, which are far less than the values of 130 $\times$ 10^6 and 3 $\times$ 10^6 psi, respectively, calculated for graphite whiskers.[115] He has shown that modulus increases directly with heat treatment temperature, but tensile strength reaches a maximum at a heat treatment temperature of 1500°C. Carbon filaments with comparable properties have been made from rayon and aromatic polyimides.[116] There is considerable evidence that formation of high-strength carbons requires extensive rearrangement of the carbon network during the pyrolytic condensation process.[117] Loss of volatile carbon compounds promotes reordering, but lack of carbon loss during polyvinylidene chloride pyrolysis leads to a highly porous, brittle product.

Thermal stability depends mainly on the energy necessary to break bonds in the main chain. It has been proposed that spiro and ladder polymers should show much higher thermal stabilities than their simple linear counterparts, because a chain break should require cleavage of at least two bonds. However, at degradation temperatures the rate of bond scission becomes very large and consequently it is doubtful that pyrolysis of any entirely aliphatic ladder or three-dimensional structure other than diamond requires temperatures much above 400°–500°C. The two factors having the greatest effect on pyrolytic stability are rigidity and aromaticity. The effect of rigidity is exemplified by a comparison of polystyrene and polytrivinylbenzene, whereas the combined effect of rigidity and aromaticity is evident in the comparison of Nylon 66 with polypyromellitimide.

8. Ablative Materials

Since organic polymers have low thermal conductivities and high specific heats, they are superior ablative materials for protecting space vehicles from temperatures up to 5000°C or more during reentry. For example, the thermal erosion rate of a nylon–phenolic composite[118] is lower than the rates for most metals and ceramics and is comparable with the rates for silicon carbide and graphite. When a vehicle enters the earth's atmosphere, the surface of the polymer coating heats up and chars. Gas evolved from the degrading polymer cools the char somewhat, and as the char thickens the charring rate decreases. It may seem odd that phenolics are such fine ablative materials since they begin to pyrolyze well below 400°C at moderate heating rates, but at rapid heating rates they act as superb thermal insulators.

E. EPILOGUE

Having dealt in some detail with a number of important polymers it is now perhaps worthwhile to review the basic considerations that determine the effectiveness of various types of stabilization. Although we shall consider mainly thermal stabilization, similar considerations will be relevant to any type of stabilization.

The easiest method available is the simple use of a stabilizing additive. Such an additive may act in either of two ways or in both of them. It may increase the stability of the most reactive bond in the molecule by reacting preferentially with it to produce a stronger bond. Examples of this type of behavior are the use of esterifying or etherifying materials to remove the unstable hydroxyl end groups in polyacetals and the use of ester-containing compounds to replace labile chlorine in poly(vinyl chloride). Additives may also be effective by interrupting a degradative chain reaction. Such a mechanism would explain the reported stabilization of poly(methyl methacrylate) by the use of radical traps. It should be noted, however, that such stabilization is only likely to be effective where the initiation rate is low. When the energy of activation for initiation is very high, the degradation is likely to become catastrophic because of a very high initiation rate. The use of additives to limit chain length may then be ineffective. This is probably the reason for the relative ineffectiveness of radical traps in thermal degradation, as compared with their effectiveness in oxidation. Furthermore radical traps such as hindered phenoxy radicals, which act as chain stoppers in low-temperature oxidations, may be lost by evaporation, they may function as radical initiators at the much higher temperatures encountered during pyrolysis, or they may combine these behaviors.

Where stabilization by additives is inappropriate, stabilization by structural modification becomes the logical choice. For reactions that progress along the polymer chain, the introduction of occasional blocking groups is effective, as evidenced by the stability of methyl methacrylate–methyl acrylate copolymers and of formaldehyde–ethylene oxide copolymers. Where such a chain reaction does not occur, notably where a radical backbiting is observed as in the case of polyethylene, the readily abstractable groups or atoms may be replaced by more firmly bound ones to provide stabilization. A result of this approach is the "conversion" of polyethylene to polytetrafluoroethylene.

These stabilization methods may be extended to give the very stable, highly aromatic compounds discussed at the end of the chapter, since here all weak bonds have been eliminated, thus limiting initiation and transfer reactions.

In concluding we should perhaps ask ourselves the question, What next? The field of inorganic polymers is being investigated. Apart from the temperature resistant, naturally occurring carbons and silicates, the only commercially available inorganic polymers are the silicones; and these, of course, have inorganic backbones. Ionic bonds are in general stronger than the single covalent type, and it is therefore logical to consider inorganic structures as forming the basis for a further generation of even higher temperature polymers. However tangible justification for such logic has yet to be found.

REFERENCES

1. R. Burghart, *J. Soc. Chem. Ind.*, **2,** 119 (1883).
2. C. G. Williams, *Proc. Roy. Soc. (London)*, **10,** 516 (1860).
3. N. Grassie and H. W. Melville, *Proc. Roy. Soc. (London)*, **A199,** 1, 14, 24, 39 (1949).
4. Y. Tsuchiya and K. Sumi, *J. Polym. Sci. Part A-1*, **7,** 3151 (1969).
5. N. Grassie, *Chemistry of High Polymer Degradation Processes*, Butterworth, London, 1956.
6. S. L. Madorsky, *Thermal Degradation of Organic Polymers*, Interscience, New York, 1964.
7. N. Grassie, in *Encyclopedia of Polymer Science and Technology*, Interscience, New York, 1966.
8. J. I. Jones, *J. Macromol. Sci., Rev. Macromol. Chem.*, **C2,** 303 (1968); *Chem. Brit.*, **6,** 251 (1970).
9. J. I. Jones, F. W. Ochynski, and F. A. Rackley, *Chem. Ind.*, **1686** (1962).
10. Ref. **6,** p. 301.
11. N. Grassie and B. J. D. Torrance, *J. Polym. Sci., Part A-1*, **6,** 3303 (1968).
12. N. Grassie and B. J. D. Torrance, *J. Polym. Sci., Part A-1*, **6,** 3315 (1968).

13. N. Grassie and E. Farish, *Eur. Polym. J.*, **3,** 305, 619 (1967).
14. N. Grassie and E. M. Grant, *Eur. Polym. J.*, **2,** 255 (1966).
15. N. Grassie and R. S. Roche, *Makromol. Chem.*, **112,** 16 (1968).
16. W. Kern and H. Cherdron, *Makromol. Chem.*, **40,** 101 (1960).
17. C. E. Schweitzer, R. N. MacDonald, and J. O. Punderson, *J. Appl. Polym. Sci.*, **1,** 158 (1959).
18. T. A. Koch and P. E. Lindvig, *J. Appl. Polym. Sci.*, **1,** 164 (1959).
19. R. N. MacDonald and M. J. Roedel, U.S. Patent 2,920,059 (Jan. 5, 1960); through J. Furukawa and T. Saegusa, *Polymerization of Aldehydes and Oxides*, Interscience, New York, 1963, p. 268.
20. W. C. Geddes, *Rubber Chem. Technol.*, **40,** 177 (1967).
21. F. Chevassus and R. de Broutelles, *The Stabilization of Poly(vinyl chloride)*, St. Martins, New York, 1963.
22. M. B. Neiman, *Aging and Stabilization of Polymers*, Consultants Bureau, New York, 1965.
23. G. Ya. Gordon, *Stabilization of Synthetic High Polymers*, Israel Program for Scientific Translations, Jerusalem, 1964.
24. M. Onozuka and M. Asahina, *J. Macromol. Sci., Rev. Macromol. Chem.*, **C3,** 235 (1969).
25. J. J. P. Staudinger, *Plastics Progr.*, 9 (1953).
26. D. H. R. Barton and P. F. Onyon, *Trans. Faraday Soc.*, **45,** 725 (1949).
27. D. H. R. Barton and K. E. Howlett, *J. Chem. Soc.*, 155, 165 (1949).
28. P. S. Fredricks and J. M. Tedder, *J. Chem. Soc.*, 144 (1960).
29. G. C. Marks, J. L. Benton, and C. M. Thomas, *Soc. Chem. Ind. (London), Monograph*, **26,** p. 204 (1967).
30. W. C. Geddes, *Eur. Polym. J.*, **3,** 733 (1967).
31. I. Ouchi, *J. Polym. Sci., Part A*, **3,** 2685 (1965).
32. E. J. Arlman, *J. Polym. Sci.*, **12,** 547 (1954).
33. D. E. Winkler, *J. Polym. Sci.*, **35,** 3 (1959).
34. W. I. Bengough and H. M. Sharpe, *Makromol. Chem.*, **66,** 31 (1963).
35. A. Cittadini and R. Paolillo, *Chim. Ind. (Milan)*, **41,** 980 (1959).
36. C. Corso, *Chim. Ind. (Milan)*, **43,** 8 (1961).
37. A. Talamini, G. Cinque, and G. Palma, *Materie Plastiche*, **30,** 317 (1964).
38, R. R. Stromberg, S. Straus, and B. G. Achhammer, *J. Polym. Sci.*, **35,** 355 (1959).
39. M. Asahina and M. Onozuka, *J. Polym. Sci., Part A*, **2,** 3515 (1964).
40. W. I. Bengough and M. Onozuka, *Polymer*, **6,** 625 (1965).
41. L. D. Loan, unpublished results.
42. E. J. Arlman, *J. Polym. Sci.*, **12,** 543 (1954).
43. D. Druesedow and C. F. Gibbs, *Natl. Bur. Std. (U.S.) Circ. 525* (1953) p. 69.
44. A. Crosato-Arnaldi, A. Palma, and A. Talamini, *Materie Plastiche*, **32,** 50 (1966).
45. A. Rieche, A. Grimm, and H. Mucke, *Kunstoffe–Plastics*, **52,** 265 (1962).
46. F. Sonheimer, D. A. Ben-Efraim, and R. Wolovsky, *J. Amer. Chem. Soc.*, **83,** 1675 (1960).
47. G. F. Woods and L. H. Schwartzman, *J. Amer. Chem. Soc.*, **70,** 3394 (1948).
48. A. D. Mebane, *J. Amer. Chem. Soc.*, **74,** 5227 (1952).
49. P. Naylor and M. C. Whiting, *J. Chem. Soc.*, 3037 (1955).
50. F. Bohlmann and H. Mannhardt, *Chem. Ber.*, **89,** 1307 (1956).
51. C. W. Spangler and G. F. Woods, *J. Org. Chem.*, **30,** 2218 (1965).
52. G. N. Lewis and M. Calvin, *Chem. Rev.*, **25,** 237 (1939).
53. M. J. S. Dewar, *J. Chem. Soc.*, 3544 (1952).

54. K. Hirayama, *J. Amer. Chem. Soc.*, **77,** 373 (1955).
55. A. Guyot, P. Roux, and P. Quang Tho, *J. Appl. Polym. Sci.*, **9,** 1823 (1965).
56. W. C. Geddes, *Eur. Polym. J.*, **3,** 747 (1967).
57. C. F. Bersch, M. R. Harvey, and B. G. Achhammer, *J. Res. Natl. Bur. Std., U.S.*, **60,** 481 (1958).
58. A. Guyot and J. P. Benevise, *J. Appl. Polym. Sci.*, **6,** 103 (1962).
59. B. Baum, *SPE (Soc. Plastics Engrs.) J.*, **17,** 71 (1961).
60. A. H. Frye and R. W. Horst, *J. Polym. Sci.*, **40,** 419 (1959).
61. A. H. Frye and R. W. Horst, *J. Polym. Sci.*, **45,** 1 (1960).
62. R. Nagatomi and Y. Saeki, *J. Polym. Sci.*, **61,** S60 (1962).
63. L. D. Loan, *ACS Polym. Preprints*, **11,** 224 (1970).
64. M. Onozuka, *J. Polym. Sci.* **5,** Part A-1, 2229 (1967).
65. A. H. Frye, R. W. Horst, and M. A. Paliobagis, *J. Polym. Sci., Part A*, **2,** 1765 (1964).
66. A. H. Frye, R. W. Horst, and M. A. Paliobagis, *J. Polym. Sci., Part A*, **2,** 1785 (1964).
67. A. H. Frye, R. W. Horst, and M. A. Paliobagis, *J. Polym. Sci., Part A*, **2,** 1801 (1964).
68. A. S. Kenyon, *Natl. Bur. Std. (U.S.) Circ. 525* (1953) p. 81.
69. W. G. Oakes and R. B. Richards, *J. Chem. Soc.*, 2929 (1949).
70. L. A. Wall, S. L. Madorsky. D. W. Brown, S. Straus, and R. Simha, *J. Amer. Chem. Soc.*, **76,** 3430 (1954).
71. W. J. Bailey and C. L. Liotta, *ACS Polym. Preprints*, **5,** 333 (1964).
72. Y. Tsuchiya and K. Sumi, *J. Polym. Sci., Part B*, **6,** 357 (1968).
73. Y. Tsuchiya and K. Sumi, *J. Polym. Sci., Part A-1*, **7,** 1599 (1969).
74. J. van Schooten and P. W. O. Wijga, in *Thermal Degradation of Polymers, Soc. Chem. Ind. (London) Monogr.* **13,** 433–452 (1961).
75. T. E. Davis, R. L. Tobias, and E. B. Peterli, *J. Polym. Sci.*, **56,** 485 (1962).
76. P. J. Flory, *Principles of Polymer Chemistry*, Cornell University Press, Ithaca, N.Y., 1953, p. 325.
77. L. A. Wall and S. Straus, *ACS Polym. Preprints*, **5,** 325 (1964).
78. M. Gordon, *Trans. Faraday Soc.*, **53,** 1662 (1957).
79. V. V. Korshak and S. V. Vinogradova, *Russ. Chem. Rev.*, **37,** 885 (1968).
80. H. A. Staab and F. Binning, *Tetrahedron Lett.*, **319** (1964).
81. N. Murayama and I. Amagi, *J. Polym. Sci. Part B*, **4,** 115 (1966).
82. M. M. Tessler, *J. Polym. Sci., Part A-1*, **1,** 2521 (1966); *ACS Polym. Preprints*, **8,** 152 (1967).
83. W. J. Bailey and B. D. Feinberg, *ACS Polym. Preprints*, **8,** 165 (1967).
84. M. Okada and C. S. Marvel, *ACS Polym. Preprints*, **8,** 220 (1967).
85. F. H. Winslow and W. Matreyek, *J. Polym. Sci.*, **23,** 315 (1956).
86. J. J. Lander and J. Morrison, *Surface Sci.*, **4,** 241 (1966).
87. C. S. Marvel and G. Hartzel, *J. Amer. Chem. Soc.*, **81,** 448 (1959).
88. P. Kovacic and F. W. Kosh, *J. Org. Chem.*, **28,** 1864 (1963).
89. W. F. Gorham, *J. Polym. Sci., Part A-1*, **4,** 3027 (1966).
90. S.-W. Chow and L. A. Pilato, British Patent 1,067,156 (1967); S.-W. Chow, U.S. Patent 3,268,599 (1966).
91. A. H. Frazer, *Sci. Amer.*, **221,** 96 (1969).
92. C. E. Sroog, S. V. Abramo, C. E. Berr, W. M. Edwards, A. L. Endrey, and K. L. Olivier, *ACS Polym. Preprints*, **5,** 132 (1964).
93. J. I. Jones, *Rept. Progr. Appl. Chem.*, **49,** 621 (1964).
94. G. M. Bower and L. W. Frost, *J. Polym. Sci. Part A-1*, **1,** 3135 (1963).
95. F. P. Gay and C. E. Berr, *J. Polym. Sci., Part A-1*, **6,** 1935 (1968).
96. K. C. Brinker and I. M. Robinson, U.S. Patent 2,895,948 (1959).

97. H. Vogel and C. S. Marvel, *J. Polym. Sci.*, **50,** 511 (1961).
98. A. B. Conciatori, E. C. Chenevy, T. C. Bohrer, and A. E. Prince, Jr., *J. Polym. Sci., Part C*, **19,** 49 (1967).
99. S. Litvak, *Appl. Polym. Symp.*, **3,** 279 (1966).
100. A. Schopov, *J. Polym. Sci., Part B*, **4,** 1023 (1966).
101. W. DeWinter, *J. Macromol. Sci.: Rev. Macromol. Chem.*, **1,** 329 (1966).
102. J. F. Brown, Jr., *J. Polym. Sci., Part C*, **1,** 83 (1963).
103. J. F. Brown, Jr., and G. M. J. Slusarczuk, *ACS Polym. Preprints*, **8,** 157 (1967).
104. C. G. Overberger, S. Ozaki, and H. Mukamal, *J. Polym. Sci., Part B*, **2,** 627 (1964).
105. S. H. Rose and B. P. Block, *J. Amer. Chem. Soc.*, **87,** 2076 (1965).
106. W. J. Bailey and A. A. Volpe, *ACS Polym. Preprints*, **8,** 292 (1967).
107. W. J. Bailey, in *Encyclopedia of Polymer Science and Technology*, Vol. 8, Interscience, New York, 1967, p. 97.
108. F. H. Winslow, W. O. Baker, and W. A. Yager, *Proc. 1st, 2nd* Conf. Carbon (1956), p. 93.
109. R. C. Houtz, *Text. Res. J.*, **80,** 786 (1950).
110. W. G. Vosburgh, *Text. Res. J.*, **30,** 882 (1960).
111. N. Grassie and J. N. Hay, *J. Polym. Sci.*, **56,** 189 (1962).
112. C. S. Marvel, *ACS Polym. Preprints*, **5,** 167 (1964).
113. W. N. Turner and F. C. Johnson, *J. Appl. Polym. Sci.*, **13,** 2073 (1969).
114. R. Moreton, W. Watt, and W. Johnson, *Nature*, **213,** 690 (1967).
115. R. Bacon, *J. Appl. Phys.*, **31,** 283 (1960).
116. H. M. Ezekiel and R. G. Spain, *J. Polym. Sci., Part C*, **19,** 249 (1967).
117. F. H. Winslow, W. Matreyek, and W. A. Yager, *Ind. Carbon Graphite*, **190** (1958).
118. I. J. Gruntfest, in *Encyclopedia of Polymer Science and Technology*, Vol. 1, Interscience, New York, 1964, p. 1.

4

STABILIZATION AGAINST OXIDATIVE PHOTODEGRADATION

ANTHONY M. TROZZOLO

Bell Telephone Laboratories, Incorporated, Murray Hill, New Jersey

A. INTRODUCTION

The deleterious effect of sunlight on polymeric materials has been ascribed to a complex set of reactions in which both the absorption of ultraviolet light and the presence of oxygen are participating events. As a result, the process has been termed oxidative photodegradation (in this chapter, the term is used interchangeably with photooxidation), and it has been the object of many studies and review articles.[1–8] Also important, from the practical standpoint, have been efforts to prevent the occurrence of this degradative process. Recent dramatic developments in the area of organic photochemistry[9–15] have given impetus to attempts to interpret oxidative photodegradation processes in terms of the newly acquired knowledge and to searches for novel methods of stabilization. This chapter reviews photodegradation mechanisms in some detail. Although these discussions relate to photostabilization indirectly, they emphasize the reactions responsible for initiation of photooxidation and thus serve as a broad directive for stabilization. Furthermore, an understanding of the photodegradation mechanisms provides an essential background for a scientific review of the current status of photostabilization mechanisms.

It is obvious that, since polymeric materials are derived from a large number of monomers whose chemistry is quite diverse, the effects of oxidative photodegradation vary from polymer to polymer. Also, the mechanism by which photodegradation occurs depends upon the particular polymer used. However several general observations may be made regarding most of the polymers that are normally used in commercial applications.

In this chapter we begin by considering the effects of oxidative photodegradation. As will be seen, these depend upon several factors, probably the most important of which is the sunlight that is responsible for the initial or primary photo process in polymer degradation. Next comes a discussion of the mechanisms that have been suggested for the degradative process in various polymers. Then follows a description of general methods of stabilization, and finally we take a look at the perspectives of stabilization against photooxidation.

B. ULTRAVIOLET RADIATION

1. Effects of Oxidative Photodegradation

The exposure of polymers to ultraviolet light in an oxygen-containing atmosphere gives rise to a variety of physical and chemical effects. Although it is difficult to detect photooxidation in its early stages (see Chapter 10) the almost imperceptible chemical changes that occur in the polymer slowly accumulate and lead to visible physical effects such as discoloration, surface cracking, or deterioration of mechanical and electrical properties. These visible results of oxidative photodegradation naturally cause some concern to the plastics consumer.

a. Mechanical and Optical Properties. Visual evidence of photodegradation varies with the polymer type. For example, the oxidative photodegradation of polystyrene leads to a yellowing, whereas poly(vinyl chloride) develops a yellow → red → black reddish coloration after exposure. In some instances, oxidation products of stabilizers or other additives either discolor or are extracted, resulting in distortion or shrinkage.

Surface cracking or embrittlement often accompanies the photochemical discoloration. This leads to drastic reductions in toughness and tensile

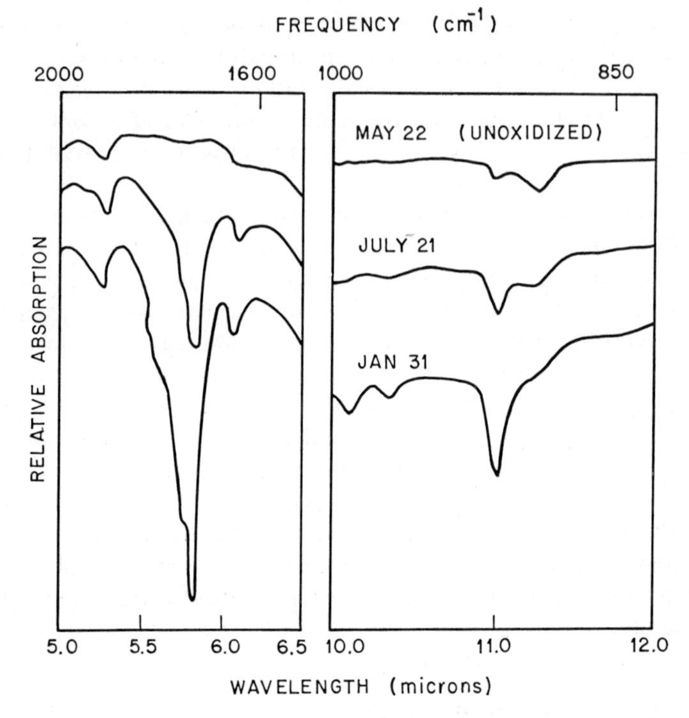

Fig. 4.1 Development of carbonyl and vinyl bands in the infrared transmission spectrum of polyethylene during outdoor exposure.[16] (Reproduced, with permission, from *ACS Polymer Preprints*.)

strength and ultimately to mechanical failure. Many of these changes in physical properties can be used to measure the extent of oxidative photodegradation, and the procedures are discussed in Chapter 10.

b. Chemical Changes. Photooxidation very often leads to chain scission or crosslinking, with concomitant formation of oxygen-containing functional groups such as ketones, carboxylic acids, peroxides, and alcohols. These changes can be detected by using infrared spectroscopy. Figure 4.1 shows the changes in the infrared spectrum of polyethylene during photooxidation.[16] The oxidative reactions are discussed in detail in the section on mechanisms, but it should be mentioned that the chemical changes also serve as convenient methods for monitoring the extent of oxidative photodegradation (see Chapter 10).

c. Electrical Properties. Polyolefins, as well as many other polymers, have very good electrical properties[17] and are used widely as dielectrics. Since photooxidation leads to an accumulation of polar groups, drastic

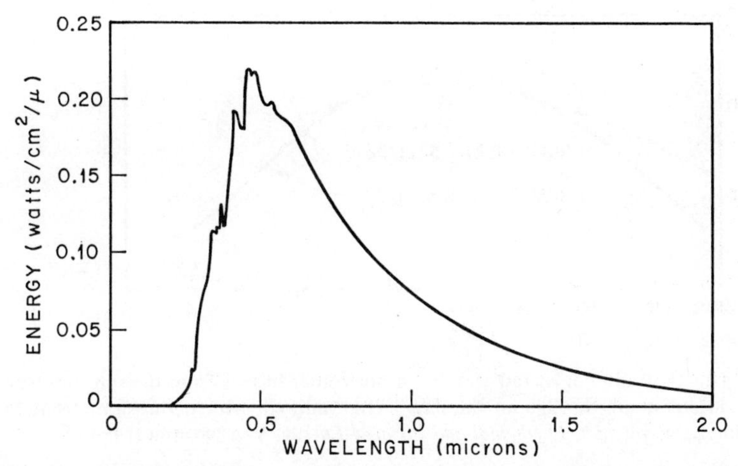

Fig. 4.2 Energy distribution in the solar spectrum outside the earth's atmosphere.[18] (Reproduced, with permission, from *Journal of Meteorology*.)

changes in dielectric constant and surface resistivity may occur, leading to electrical failure.

2. Solar Radiation

Although the weathering of plastics is dependent upon many environmental factors, it is generally accepted that only a relatively narrow band of the electromagnetic spectrum of sunlight is responsible for the primary photochemical processes in the oxidative photodegradation of polymers. Although the energy distribution in the solar spectrum in space (Figure 4.2) extends to wavelengths below 200 nm,[18,19] almost all the radiation of wavelength less than 290 nm is absorbed in the earth's atmosphere; therefore very little of the shorter wavelength radiation reaches the earth's surface[19,20] (Figure 4.3).

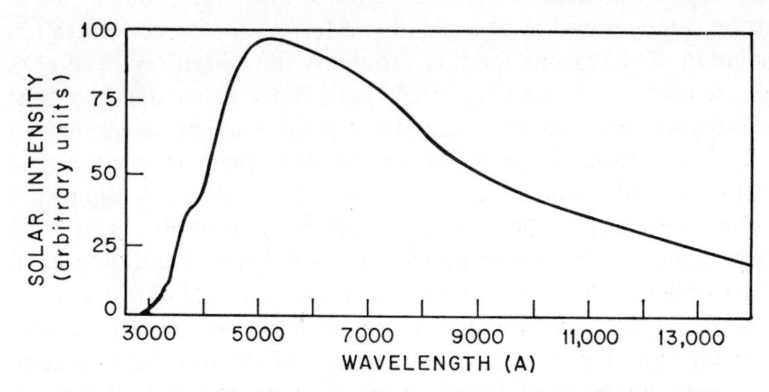

Fig. 4.3 Solar energy distribution for midsummer noonday sun in Washington, D.C.[20] (Reproduced, with permission, from L. R. Koller, *Ultraviolet Radiation*, Wiley, 1965.)

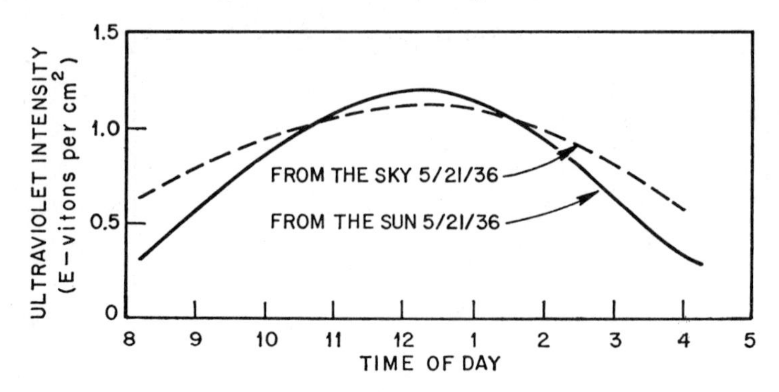

Fig. 4.4 Intensity of ultraviolet radiation on a horizontal surface from the sun and from the sky at various hours throughout the day.[21] (Reproduced, with permission, from M. Luckiesh, *Germicidal, Erythemal and Infrared Energy*, Van Nostrand, 1946.)

Most of the absorption at the shorter wavelengths is caused by a layer of ozone (of thickness equivalent ~0.25 cm at standard temperature and pressure) which exists at higher altitudes. Thus ozone, which has been implicated in air-pollution processes such as smog in the lower atmosphere, constitutes a sun screen essential to maintain life on earth as we know it.

a. Variations in Solar Spectrum. Although the general features of the solar radiation spectrum at the earth's surface (Figure 4.3) do not vary significantly with location, the ultraviolet end of the spectrum ($\lambda < 313$ nm) is strongly dependent upon several factors such as altitude, latitude, time of day, season, and local weather conditions. Since the shorter wavelength radiation is most effective in promoting photooxidation, it is of interest to note these variations. (Some of these variations and the procedures to determine their effect in weathering tests are discussed in Chapter 10.)

With the exception of local weather conditions, most variations in ultraviolet radiation can be traced to changes in two factors: ozone concentration and sky radiation. The reasons for this variability have been reviewed elsewhere[20] and are commented on only briefly here. Since ozone absorbs most of the solar radiation between 200 and 300 nm, it follows that the amount of ultraviolet light of wavelength near 300 nm reaching the earth's surface as direct radiation depends strongly upon the ozone layer above. A significant amount of ultraviolet light reaches the earth's surface as the result of scattering in the sky (sky light) as well as directly from the sun. On a clear day, in fact, there are two periods when the amount of ultraviolet light directed on a horizontal surface from sky light is greater than the amount that comes directly from the sun (Figure 4.4). In the middle of the day the two components are approximately equal, but in the morning and late afternoon,

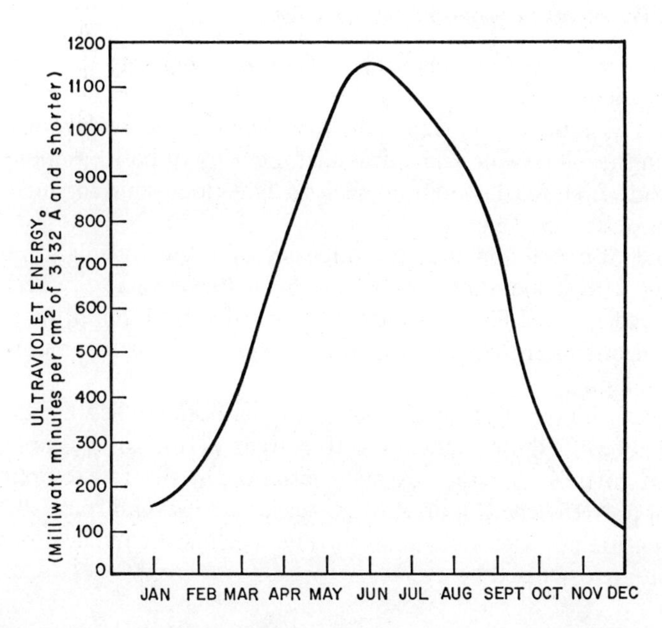

Fig. 4.5 Integrated monthly ultraviolet radiation of wavelength less than 313.2 nm.[23] (Reproduced, with permission, from *Journal of Research*, National Bureau of Standards.)

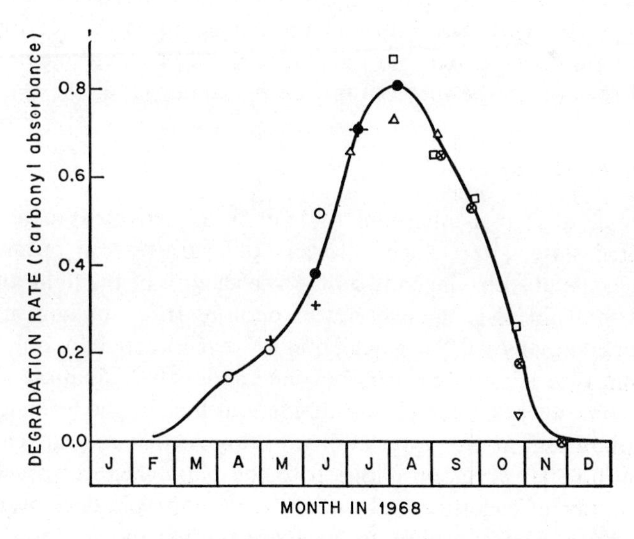

Fig. 4.6 Rate of increase in carbonyl absorbance during weathering of branched polyethylene films at Murray Hill, N.J., during 1968.[16] (Reproduced, with permission, from *ACS Polymer Reprints*.)

the sky component is considerably larger. Thus it is possible to obtain a severe sunburn even when one is shielded from the sun if a large part of the sky is visible. The ratio of sky–sun radiation shows a seasonal variation, being greater in the winter when the ultraviolet intensity of both components is lowest.[22] In addition, local conditions such as light clouds and fog increase the proportion of scattered light.

Coblentz and Stair[23] found that the intensity of ultraviolet light varies during the day with a monotonic dependence on the altitude of the sun. They also showed that the ultraviolet intensity less than 313 nm in the early morning and in late afternoon is very small compared with the amount in the middle of the day.

The seasonal variation of ultraviolet intensity has been studied by several groups.[21,23] Figure 4.5 shows data for a three-year period in Washington, D.C.[23] It is of interest that the seasonal variation in the rate of photodegradation of polyethylene (Figure 4.6) as measured in several studies[16,24,25] correlates well with this seasonal variation. The mechanistic implications of this correlation are discussed below.

C. MECHANISMS OF OXIDATIVE PHOTODEGRADATION

In the previous section we saw that the absorption of sunlight can cause a variety of changes in the physical and chemical properties of polymers. This section reviews the details of what happens on a molecular scale after the initial absorption of light. Some of the underlying principles of organic photochemistry are briefly given, but the reader should consult one or more textbooks and reviews on the subject for a deeper understanding.[9–14]

1. Photoactivation

When a molecule absorbs a quantum of light, it is activated to an *electronically* excited state, after which a variety of processes can occur. The energy the molecule absorbs depends on the wavelength of the light and, as can be seen from Table 4.1, the energies of photons from sunlight may be sufficient to break many of the single bonds[26] encountered in polymeric systems. Chemical reaction thus can be one mode of dissipation of the absorbed electronic energy. Such chemical reactions include the formation of free radicals, photoionizations, cyclizations, intramolecular rearrangements, and fragmentations. However, in addition to the photochemical processes, there are a number of radiative and nonradiative photophysical processes that do not lead to a net chemical reaction yet are alternate modes for dissipation of the absorbed energy (Figure 4.7). Since a major photostabilization process is essentially that of ensuring that all the excitation energy is dissipated

Table 4.1 Strength of Chemical Bonds

Chemical Bond	Bond Energy (kcal/mole^{-1})[26]	Wavelength of Corresponding Energy (nm)
O—H	110.6	259
C—F	105.4	272
C—H	98.8	290
N—H	93.4	306
C—O	84.0	340
C—C	83.1	342
C—Cl	78.5	364
C—N	69.5	410

in only photophysical processes, it is important to consider these in more detail.

a. Photophysical Processes. The photophysical processes which an excited molecule A* can undergo fall under three different categories:

$$A^* \xrightarrow{\text{emission}} A_0 + h\nu$$

$$A^* \xrightarrow[\text{conversion}]{\text{radiationless}} A_0 + \text{heat}$$

$$A^* + B_0 \xrightarrow[\text{transfer}]{\text{energy}} A_0 + B^*$$

Figure 4.7 shows schematically the photophysical processes that may occur

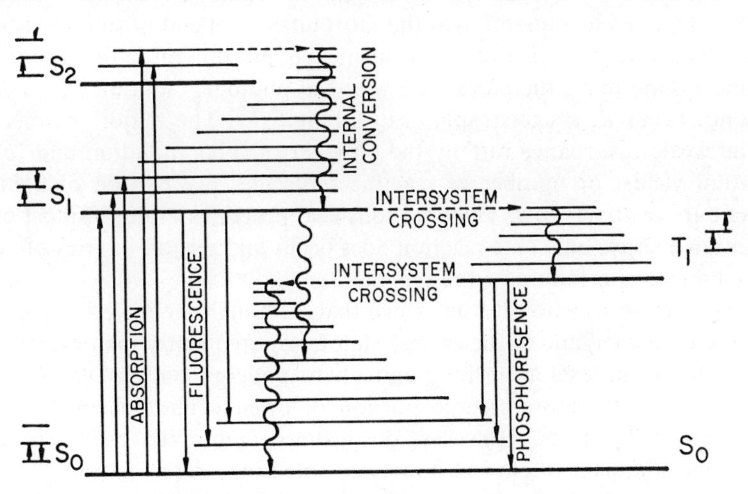

Fig. 4.7 Photophysical processes of molecules.

after an organic molecule has absorbed a quantum of light. Since it has been found that the lowest-excited-singlet (S_1) and triplet (T_1) states are the starting points for most organic photochemical processes, the discussion will center on these states. Other higher singlet and triplet states will not be described.

After a molecule has been raised (by light absorption) from its ground state S_0 to the first-excited-singlet state S_1, it can relax back to S_0 by a number of different routes. It can emit a photon of light (fluorescence) or it can transform the excitation into vibrational energy (internal conversion). Another alternate is to change the spin of one of the electrons in the half-filled orbitals and thereby become a triplet state T_1 (intersystem crossing). The triplet state T_1 can revert back to S_0 either by emitting a photon (phosphorescence) or by intersystem crossing ($T_1 \rightsquigarrow S_0$) to a vibrationally excited level of S_0. Another important photophysical process involves energy transfer to another molecule from either S_1 or (usually) T_1. Energy transfer in polymeric systems has received increased study recently, particularly as a source of photodegradation.[15] As will be seen later, the process also offers a method of stabilization, particularly where T_1 is a reactive species undergoing photochemical reaction.

Associated with each of these modes is a "lifetime" that may be viewed in terms of a first-order decay process. Fluorescence lifetimes of S_1 vary from 10^{-9} to 10^{-6} sec for most organic molecules. In contrast, phosphorescence lifetimes vary from 10^{-3} to 20 sec. The longer lifetime of phosphorescence is the result of a relatively slow spin inversion.

b. Photochemical Processes. The data in Table 4.1 indicate that the energy contained in a quantum at the short ultraviolet end of sunlight is more than sufficient to break many of the chemical bonds present in the commercial polymers. One may, therefore, ask why the photodegradation of polymers does not occur at a catastrophic rate in sunlight. The major reasons are: (*a*) the weak absorbance rate by the polymer of solar radiation and (*b*) the quantum yields, or number of reacting molecules per photon of light absorbed, are relatively low. The photophysical processes drain off most of the excitation energy, but some reaction does occur and sets off a series of events that culminates in failure of the polymer.

In the previous section it was shown that, although the lowest triplet state T_1 is the least energetic of the excited states, it is frequently the only one that retains electronic excitation long enough to undergo bimolecular chemical reactions, and therefore a large portion of organic photochemistry is the chemistry of T_1 states. However, one cannot ignore the chemistry of S_1 states, which appear to play an important role in the deterioration of several polymer systems.

It was noted previously that energy transfer is a photophysical process by which an excited molecule A* can dispose of its excitation energy without undergoing a chemical reaction. However, it is quite possible that the acceptor of this energy B* can undergo reaction. This is the case in photosensitized reactions:

$$A \xrightarrow{h\nu} A^*$$

$$A^* + B \rightarrow A + B^*$$

$$B^* \rightarrow \text{products}$$

In general, the longer the lifetime of an excited state, the greater is the probability that it will undergo energy-transfer processes. For this reason triplet states are much more likely than singlet states to take part in energy-transfer processes.

c. "Hot" Ground-State Reactions. One of the photophysical processes mentioned earlier is internal conversion, whereby an excited molecule, S_1, converts its excitation energy into vibrational energy as it returns to the ground state, S_0. However the vibrationally excited ground state is capable of undergoing the same types of chemical reactions that would occur if the ground state were thermally excited. It is therefore possible that some reactions that may appear to be photochemical (reaction from electronically excited states) are in fact photoinduced thermal reactions (reaction from vibrationally excited or "hot" ground state).

2. General Mechanisms for Oxidative Photodegradation

The mechanism for oxidative photodegradation of a polymer may be divided into three types of processes: primary photochemical initiation, secondary photochemical initiation, and dark reactions.

a. Primary Photochemical Initiation. Primary photochemical initiation is the set of chemical reactions that occur when the ultraviolet radiation is absorbed by the "pure" polymer, by possible oxygen–polymer complexes, by ozone or ozone–polymer complexes, or by molecular oxygen itself.

Although Table 4.1 reveals that the short ultraviolet end of sunlight contains enough energy to break single bonds in most polymers, we have seen that, because of the very low quantum yields of these fragmentations, direct absorption by the "pure" polymer containing only single bonds appears to play a relatively minor role in initiation. However polymers that contain multiple bonds (carbonyl groups, in particular) can initiate fragmentation reactions by direct absorption; and obviously these reactions will occur even in the absence of oxygen.

It is also conceivable that electronically excited oxygen molecules may be involved in the initiation step, and therefore the spectroscopy of oxygen must be considered.[27] Almost all the absorption bands of molecular oxygen occur in the vacuum ultraviolet region. The longest wavelength absorption in the ultraviolet region is the forbidden Herzberg band at 2454 Å, in which the excited state quickly dissociates to form oxygen atoms:

$$O_2(^3\Sigma_g^-) + h\nu \rightarrow O_2(^3\Sigma_u^+) \rightarrow 2\,O(^3P) \tag{4,1}$$

Since solar radiation at the earth's surface does not extend down to these wavelengths, the process shown in (4.1) cannot be important as an initiation step. The weak absorption bands in the visible region (the Fraunhofer lines of the solar spectrum at 7600–7650 and 6870–6910 Å) corresponding to the strongly forbidden transitions also would not be expected to lead to initiation, since light of these wavelengths would be filtered out by atmospheric oxygen molecules.

$$O_2(^3\Sigma_g^-) + h\nu \rightarrow O_2(^1\Sigma_g^+) \tag{4.2}$$

The role of ozone in initiating oxidative photodegradation merits more attention than it has received in the past (see Chapter 5). Ozone is photolyzed by ultraviolet and visible light to form oxygen atoms.[28] Electronically excited oxygen molecules are also formed:

$$O_3 \xrightarrow{h\nu} O_2^* + O(^3P). \tag{4.3}$$

Both these products are known to react with hydrocarbon substrates. In addition, ozone may react by nonphotochemical modes to initiate degradation processes (Chapter 5).

Polymer–oxygen interactions in the excited states may be responsible for initiation in some systems. New ultraviolet absorption bands have been reported for hydrocarbon–oxygen complexes which are not present in the hydrocarbon alone.[29–32] These absorptions have been identified[31] as contact charge–transfer absorptions. Several authors have suggested that the excited charge–transfer states could initiate photooxidation.

$$RH + O_2 \xrightarrow{h\nu} (RH^+\text{—}O_2^-) \rightarrow \text{reactive intermediates} \tag{4.4}$$

Chien[33] found that certain alkanes and alkenes, when saturated with oxygen, exhibited ultraviolet bands with tails beyond 300 nm, whereas the oxygen-free hydrocarbons are transparent in that region. Chien suggested the intermediate shown in (4.4) to explain the new absorption band and hypothesized that it may be responsible for the initiation step in photooxidation. Carlsson and Robb[34] proposed that the same process is responsible for the oxygen–tetralin and oxygen–indene spectra. A recent study[35] of the *p*-cymene–oxygen system led to similar conclusions.

Still another possible polymer–oxygen interaction is the oxygen-perturbed singlet → triplet transition, which has been observed in olefinic and aromatic hydrocarbons.[36] However little is known about the triplet-state energy and singlet → triplet transitions for alkanes. Chien[33] suggested that the oxygen-perturbed $S_0 \rightarrow T_1$ excitation found in alkene–oxygen systems might initiate photooxidation according to the scheme shown in (4.5) and (4.6):

$$^3A + O_2 \rightarrow \text{peroxide} \tag{4.5}$$

$$A + O_2 + h\nu \rightarrow {}^3A + O_2(^1\Delta_g) \tag{4.6}$$

where A is the alkene. However, since oxygen molecules are excellent triplet quenchers,[37] it seems probable that the production of peroxide proceeds via attack of $O_2(^1\Delta_g)$ produced by quenching of the olefin:

$$^3A + {}^3O_2 \rightarrow A + {}^1O_2 \tag{4.7}$$

$$O_2(^1\Delta_g) + A \rightarrow AO_2 \tag{4.8}$$

b. Secondary Photochemical Initiation. A variety of adventitious agents generally can be incorporated into the polymer during the processing of the polymer. Included in this group are catalytic residues, or traces of hydroperoxide and carbonyl moities. Absorption of light by these substances may lead to initiation of photooxidation. The photochemistry of additives that are introduced intentionally also must be considered as a possible source of initiation.

Once polymer photooxidation has started, a new series of photochemical initiation processes may replace the original initiation processes, since oxidation products such as hydroperoxides, ketones, carboxylic acids, and aldehydes may absorb to give initiation.

(1) Initiation by Catalyst Residues. It is well known that transition metal ions act as sensitizers for the photooxidation of polyolefins.[38] It also is generally recognized that polyolefins obtained by polymerization with Natta–Ziegler catalysts contain transition metal residues. Kujirai, Hashiya, Shibuya, and Nishio[39] recently found that the oxidative photodegradation of polypropylene in ultraviolet light is dependent on the concentrations of oxygen and of the catalyst residue. They concluded that the oxidative photodegradation is sensitized by the initiator residues.

The photochemistry of transition metal ions[40] will not be discussed in detail here, but it is worthwhile mentioning that the initiation process probably involves the production of radical ions via excited-state–oxidation reduction reactions.

(2) Initiation by Hydroperoxide and Carbonyl Groups. It is very likely that hydroperoxide or carbonyl groups can be found in the unexposed polymer, in addition to the metallic impurities that may be present. These groups can

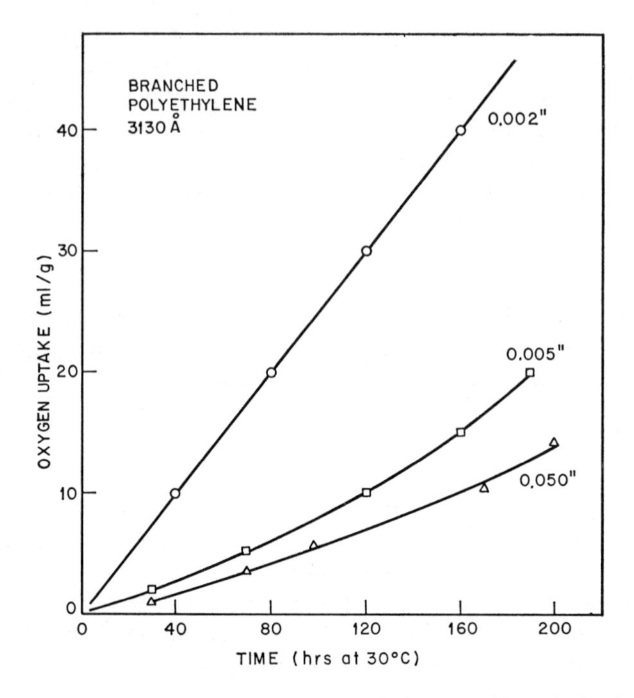

Fig. 4.8 Effect of film thickness on the photooxidation rate of branched polyethylene under RPR 3000-Å lamps at 30°C.

be formed by thermal oxidation during polymerization or processing. Carbonyl groups also could result from copolymerization with carbon monoxide[41] or by reaction with ozone during outdoor exposure. For example, the data in Figure 4.8 are consistent with an ozone-initiated degradation at the surface of polyethylene films, since oxidation was more rapid at the surface than in the bulk.[16] Hydroperoxides and carbonyl groups also are formed during the oxidative photodegradation process, and therefore they are able to initiate further reactions. Hence, in order to understand the secondary initiation processes more fully, it is necessary to review briefly the photochemistry of these groups.

Although the absorption maximum of alkyl hydroperoxides is around 210 nm, the tail of this band extends beyond 300 nm; thus absorption of solar radiation can occur. The photochemical reaction brought about by this absorption involves cleavage of the O–O bond:[42]

$$\mathrm{ROOH} \xrightarrow{h\nu} \mathrm{RO\cdot} + \mathrm{\cdot OH} \tag{4.9}$$

Quantum yield measurements[42] indicate a quantum efficiency near unity. The fragments produced in (4.9) can undergo the same general types of reactions outlined in Chapter 2 on thermal oxidation, and these processes are considered to make an important contribution to the oxidative photodegradation of polypropylene.

Before considering the photochemistry of carbonyl compounds, it is worthwhile describing the spectroscopy of this functional group, since the photoprocesses are related intimately to the descriptions of the excited states.[9–14] Aliphatic aldehydes and ketones show a relatively weak absorption band with a maximum between 270 and 290 nm and extending into the region beyond 300 nm. The absorption is the result of a "forbidden" electronic transition in which an electron is promoted from a nonbonding, *n*-orbital localized on the oxygen atom to a delocalized antibonding π^*-orbital that is distributed over the entire carbonyl group. The excited-singlet state, $^1(n, \pi^*)$ is capable of intersystem crossing (see Section C.1.a) to form the triplet state, $^3(n, \pi^*)$. Both these excited states are capable of undergoing a variety of cleavage reactions.

The known photochemistry of aliphatic ketones[9–14] suggests that the principal routes available for chain scission are the Norrish type I reaction

$$R\overset{\overset{\displaystyle O}{\|}}{C}CH_2R' \xrightarrow{h\nu} R\overset{\overset{\displaystyle O}{\|}}{C}\cdot + \cdot CH_2R' \tag{4.10}$$

and the Norrish type II reaction

$$R\overset{\overset{\displaystyle O}{\|}}{C}{-}CH_2{-}CH_2{-}CH_2{-}R' \xrightarrow{h\nu} R\overset{\overset{\displaystyle O}{\|}}{C}CH_3 + CH_2{=}CHR' \tag{4.11}$$

The quantum yields of these reactions have been measured for a variety of aliphatic ketones,[43–45] and where both processes are possible (ketones that have at least one γ-hydrogen atom), the type I cleavage appears to occur less readily than the type II process.

Rice and Teller[46] and Noyes and Davis[47] have considered alternative mechanisms for the type II cleavage of ketones, but present evidence[48] strongly favors the existence of a biradical intermediate that quickly fragments or returns to the starting ketone. The nature of the excited states that

$$\begin{array}{c} \text{H} \\ \diagdown \\ \overset{\overset{\displaystyle O}{\|}}{RC} \quad \overset{CHR'}{\underset{CH_2}{|}} \\ \diagdown \;\; \diagup \\ CH_2 \end{array} \underset{}{\overset{h\nu}{\rightleftharpoons}} \begin{array}{c} \overset{\overset{\displaystyle OH}{|}}{RC\cdot} \quad \overset{\cdot CHR'}{\underset{CH_2}{|}} \\ \diagdown \;\; \diagup \\ CH_2 \end{array} \longrightarrow \begin{array}{c} \overset{\overset{\displaystyle OH}{|}}{RC} + \overset{\overset{\displaystyle CHR'}{\|}}{CH_2} \\ \diagdown\!\!\diagdown \;\; \\ CH_2 \\ \downarrow \\ \overset{\overset{\displaystyle O}{\|}}{RCCH_3} \end{array} \tag{4.12}$$

participate in the type II cleavage has been a subject of considerable controversy.[48–53] However it has been shown recently[54] that both the excited $^1(n, \pi^*)$ state and the $^3(n, \pi^*)$ state take part in the type II cleavage of several aliphatic ketones.

In addition to the type I and type II reactions of ketones, a third process has been observed that leads to cyclobutanol derivatives:[55]

$$\mathrm{RC(=O)—CH_2—CH_2—CH_2R'} \xrightarrow{h\nu}$$

$$\mathrm{R\dot{C}(OH)—CH_2—CH_2—\dot{C}H—R'} \longrightarrow \mathrm{RC(OH)—CHR'—CH_2—CH_2\ (cyclobutanol\ ring)} \qquad (4.13)$$

This reaction is considered to involve a biradical intermediate similar to that of the type II process. This reaction is not considered to occur to any significant extent in polymeric systems, compared with the type II process.

The relative contributions of the above three processes in the photochemistry of aliphatic ketones have been reviewed by Hartley and Guillet.[43] They studied the dependence of quantum yield of the type I and type II processes on the chain length of a variety of di(n-alkyl) ketones. Their results are given in Figures 4.9 and 4.10. Although the quantum yield of the type II process appeared to decrease quite regularly with increasing chain

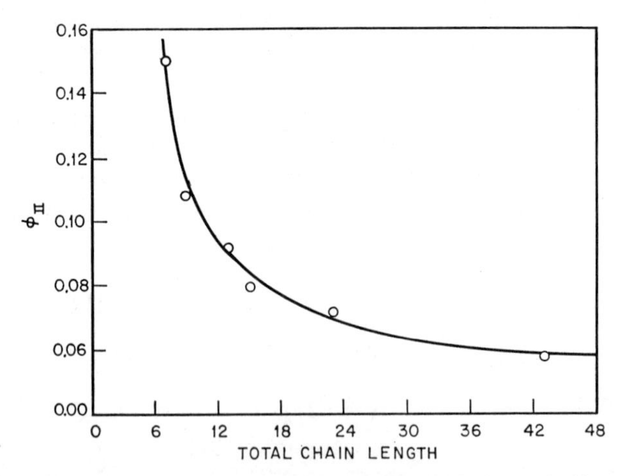

Fig. 4.9 Type II quantum yield (ϕ_{II}) as a function of chain length.[43] (Reproduced, with permission, from *Macromolecules*.)

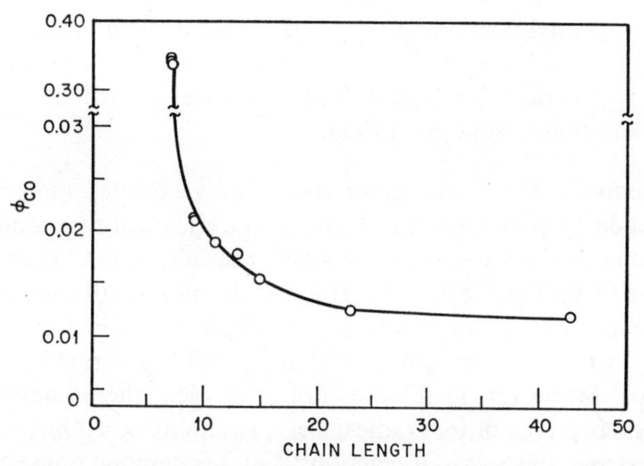

Fig. 4.10 Quantum yield of carbon monoxide evolution (ϕ_{CO}) as a function of chain length.[43] (Reproduced, with permission, from *Macromolecules*.)

length, it remained as high as 0.06 in tritetracontan-22-one. As can be seen in Figure 4.9, the quantum yield had not yet reached a limiting value at chain lengths of more than 40 carbon atoms and was still decreasing slowly. It would appear, therefore, that the value of the quantum yield, $\phi_{II} = 0.025$, reported by Hartley and Guillet[43] for a carbon monoxide–ethylene copolymer (1 mole % CO) is in the expected range. On the other hand, the type I reactions of the ketones showed a precipitous drop in quantum yield (Figure 4.10) with increasing chain length and reached a limiting value of about 0.012. These authors also found that although the type II process appeared to be insensitive to changes in temperature or solution viscosity, the type I quantum yield was temperature dependent and affected by solution viscosity.

Aldehydes are also known to undergo the cleavage reactions cited above, but the quantum yields have not been investigated extensively. In the few reported cases,[10] it appears that the type I process is relatively more important than it is in the photolysis of ketones.

The importance of both the type I and type II processes in the photodegradation of polymeric systems containing carbonyl groups was first pointed out by Guillet and Norrish[56] in the photolytic reactions of poly(methyl vinyl ketone). Later studies[57] confirmed these results and the type II process was again observed in the photolysis of polyesters.[58] In addition, the elegant studies of Hartley and Guillet on ethylene–carbon monoxide copolymers clearly indicates the importance of the type II process as a main source of chain scission in model systems related to polyethylene. Moreover, Guillet and co-workers[59] have reinforced Osborn's suggestion[58] that carbonyl photocleavage reactions may be the major cause of photodegradation in a wide variety of carbonyl-containing polymers with carbonyl groups such as polyesters, polyamides, polyureas, polyurethanes, and cellulose esters, as well as in the polymers that always contain at least traces of carbonyl functions as a result of oxidation during processing.

c. Dark Reactions. If the excitation energy of an excited molecule has not been dissipated by photophysical paths, then a chemical reaction occurs, which often results in the formation of reactive radicals. These intermediates can initiate a variety of processes: (*a*) elimination of a small molecule, (*b*) further cleavage into smaller fragments, (*c*) "unzipping" or depolymerization, (*d*) crosslinking with adjacent polymer chains, and (*e*) formation of unsaturated groups. Moreover, in the presence of oxygen, these intermediates can undergo the same type of free-radical chain reactions noted in Chapter 2. In polymeric systems, this series of reactions is quite complex, thus making it difficult to separate thermal- from photooxidative ones. Evidence for this complexity has been derived from the relatively high temperature coefficients for polymer oxidative photodegradations.[60] The temperature dependence of a single primary photochemical process would be expected to be small.

d. Model System–Polymer Comparisons. Since most of the model system photochemistry is in fluid solution, one must be somewhat cautious about applying those results to polymeric systems. There are several significant differences between the two types of experiments, the most obvious being the difference in rigidity. The consequences of this are likely to be exhibited in variation in oxygen diffusion, energy-transfer processes, excited-state lifetimes, and the quantum yields of alternative photochemical paths.

3. Photodegradation of Polyolefins

Although all the members of this group are hydrocarbon polymers, there appear to be a variety of mechanisms for the oxidative photodegradations,[61] and each polymer will be considered separately.

a. Polyethylene. The rapid disintegration of polyethylene in sunlight came as a surprise to early workers, since the very low absorption of the paraffins in the near-ultraviolet range[62] would have predicted good weathering resistance for polyethylene. However, commercially prepared samples display an ultraviolet absorption spectrum with a weak intensity at the long-wavelength end, but which does overlap the ultraviolet spectrum of solar radiation. Pross and Black[63] suggested that this small absorption was caused by traces of carbonyl groups formed by oxidation during the preparation and processing of the polymer. Burgess[41,64] elaborated this hypothesis by proposing that the Norrish type I and type II photofragmentation reactions of carbonyl compounds could account for the products observed in the oxidative photodegradation of polyethylene:

$$\sim CH_2{-}CH_2{-}CH_2{-}\overset{\overset{\displaystyle O}{\|}}{C}{-}CH_2{-}CH_2{-}CH_2\sim \xrightarrow[\text{type I}]{h\nu}$$

$$\sim CH_2{-}CH_2CH_2\overset{\overset{\displaystyle O}{\|}}{C}\cdot + \cdot CH_2{-}CH_2{-}CH_2\sim$$

$$\sim CH_2{-}CH_2{-}CH_2\cdot + CO \qquad (4.14)$$

$$\sim CH_2{-}CH_2{-}CH_2{-}\overset{\overset{\displaystyle O}{\|}}{C}{-}CH_2{-}CH_2{-}CH_2{-}CH_2\sim \xrightarrow[\text{type II}]{h\nu}$$

$$\sim CH_2{-}CH_2{-}CH_2{-}\overset{\overset{\displaystyle O}{\|}}{C}{-}CH_3 + CH_2{=}CH{-}CH_2\sim \qquad (4.15)$$

Burgess measured the development of carbonyl, hydroxyl, and vinyl groups that are produced during the oxidative photodegradation of polyethylene in 313-nm light. The results, shown in Figure 4.11, suggest that the degradation becomes autocatalytic at long exposure times. It is also of interest to note that formation of hydroxyl groups does not occur appreciably until the autocatalytic stage commences. In support of his conclusion that carbonyl impurities were the chief cause of the oxidative photodegradation, Burgess determined the quantum efficiency of carbonyl formation in three wavelength regions (Table 4.2). The results showed that the efficiency was greatest in the

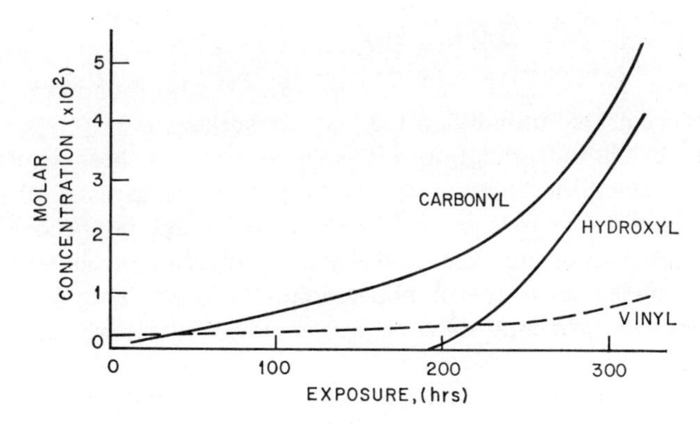

Fig. 4.11 Photooxidative products of polyethylene.[41] (Reproduced, with permission, from National Bureau of Standards.)

Table 4.2 Photodegradation of Polyethylene[41]

Wavelength (nm)	Quantum Efficiency of Carbonyl Formation
254	0.04
250–350	0.11
350–450	0

near-ultraviolet region (250–350 nm), which corresponds to the carbonyl-absorption band.[65] The activation spectrum obtained by Hirt, Searle, and Schmitt[66] showed a peak at 300 nm for polyethylene oxidative photodegradation. Winslow, Matreyek, and Trozzolo[16] found that radiation of wavelength greater than 340 nm was ineffective in photodegrading polyethylene. The rate of degradation of polyethylene films was found to increase with increasing temperature, ozone concentration, surface–volume ratio, and molecular disorder, but was virtually independent of relative humidity and low concentrations of thermal antioxidants. It has been pointed out (Section B.2) that the seasonal variation of the weathering rate of polyethylene was found[16,24] to be remarkably similar to the seasonal variation in ultraviolet radiation—less than 313 nm, measured by Coblentz and Stair[23] in Washington, D.C.

Of particular interest is the effect of film thickness in the photooxidation of polyethylene shown in Figure 4.8.[16] It can be seen that the very thin sample showed no evidence of autocatalysis, but the thicker samples exhibited a slight acceleration in rate at longer exposure times. The results have been interpreted in terms of a distinct difference between the reaction rates at the surface and in the bulk. This view is supported by the internal reflection infrared spectroscopic work of Chan and Hawkins,[67] who found that aging effects were concentrated initially on the polymer surface.

The oxidative photodegradation of polyethylene has been explained recently[68] by a mechanism based on (*a*) the known photochemistry of mode systems resembling polyethylene, (*b*) the known products of the oxidative photodegradation of polyethylene, and (*c*) the formation of electronically excited oxygen molecules in the photodegradative process. The mechanism and the relevant experimental data have been reviewed elsewhere[61,68] and will be described only briefly here.

The proposed mechanism has four steps: (*a*) absorption of light by carbonyl groups, (*b*) Norrish type II cleavage involving the n–π^* excited states of carbonyl groups, (*c*) formation of *singlet* oxygen molecules by quenching of the (n, π^*) triplet state of the carbonyl groups, and (*d*) reaction of singlet oxygen molecules with vinyl groups formed in the type II cleavage. The steps are outlined in (4.16 to 4.18):

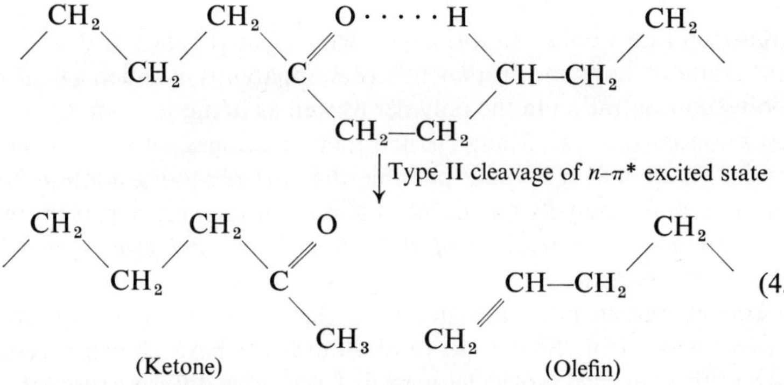

```
  CH2         CH2      O · · · · · H                CH2
 /   \       /   \   //             \              /   \
       CH2         C                  CH—CH2
                    \                /
                     CH2——CH2
                        |  Type II cleavage of n–π* excited state
                        ↓
 CH2         CH2        O                          CH2
/   \       /   \     //                          /   \
      CH2         C                    CH—CH2                  (4.16)
                   \                  //
                    CH3         CH2
       (Ketone)                           (Olefin)
```

$$(\text{ketone})^{3}{}_{n-\pi^*} + O_2 \longrightarrow \text{ketone} + {}^1O_2 \tag{4.17}$$

$${}^1O_2 + RCH_2CH{=}CH_2 \longrightarrow RCH{=}CHCH_2OOH \longrightarrow \text{further reactions} \tag{4.18}$$

Further reactions of the hydroperoxide would lead to additional carbonyl groups that could undergo type II cleavages. The formation of the hydroperoxide by singlet oxygen attack is also attractive from a steric point of view, since the ketone and vinyl groups in the polymer system are separated only slightly after type II cleavage.

Additional support for the mechanism is given by data obtained in the photolysis of polyethylene, which indicate that little autocatalysis occurs and that chain lengths are quite short. This behavior would be expected if the rate-determining step were the diffusion of oxygen. Although the mechanism suggests the formation of hydroperoxides, no hydroperoxides were detected by Winslow, Matreyek, and Trozzolo[16] in branched polyethylene that was photooxidized either outdoors or under sunlamps. However the authors found that when hydroperoxide was deliberately formed in the polymer by autoxidation (in a dark reaction) at 110°C and the oxidized polymer irradiated with 300-nm light, very little hydroperoxide remained after a three-hour exposure. These results are consistent with those of Bamford and Dewar,[69] who found that cyclohexene hydroperoxide absorbed 330-nm radiation to form α,β-unsaturated ketone. Additional evidence for the process in (4.18) has been obtained recently by Kaplan and Kelleher,[70] who treated *n*-docosene-1 adsorbed on alumina with ${}^1O_2$ generated from a microwave discharge and obtained hydroperoxides as products, presumably by decomposition of the initially formed hydroperoxide. This model compound was chosen because it contains a terminal vinyl group, and it had been reported previously[71] that terminal vinyl groups were relatively inert to singlet molecular oxygen. The above evidence, therefore, suggests that the rate of hydroperoxide decomposition is much greater than the rate of formation during photolysis.

There are, however, several unanswered questions regarding the mechanism for oxidative photodegradation of polyethylene. If carbonyl groups act as sensitizers for the photooxidation of polyethylene, one would expect that the reaction rate would rise with increased carbonyl concentration. However recent studies[16] have shown that this rate apparently is independent of bulk carbonyl concentration in the polymer as well as in model systems containing known concentrations of long-chain aliphatic ketones. In view of the results given in Figure 4.8, it is quite possible that the photodegradation begins at the surface and gradually moves into the bulk of the polymer. If this were the case, then *a large proportion of the internal carbonyl groups would be ineffective as initiators.*

If singlet oxygen plays a major role, the sensitization process should be diffusion dependent. Schnuriger and Bourdon[72] have shown recently that only half these excited molecules were deactivated in diffusing through a 10-nm soap film. Some singlet oxygen molecules actually penetrated 50-nm films and presumably could also diffuse from carbonyl to vinyl groups in polyethylene. It is still not known whether the carbonyl groups are the effective primary sensitizers or whether precursors of carbonyls exist that are important in the initial steps.

b. Polypropylene. Polypropylene has found extensive commercial use in films and fibers despite its greater susceptibility than polyethylene to oxidative photodegradation. This lower photooxidative stability is generally attributed to the presence of a labile hydrogen on the tertiary carbon in each repeating unit. As in the case of polyethylene, the "pure" polymer would not be expected to absorb ultraviolet radiation beyond 200 nm[73] and should therefore, be inert toward sunlight. However the processing of polypropylene often involves temperatures at which thermal oxidation of the polymer[74–76] (see Chapter 2) would be expected to give hydroperoxides as initial products. A variety of carbonyl products is also formed by secondary processes.[74] Both the hydroperoxides and carbonyl groups might act as initiators in the photooxidation.

Carlsson and Wiles[77] prepared melt-oxidized polypropylene films in order to identify the "impurities" that might be generated during production of polypropylene fibers. Three distinct types of ketones were identified:

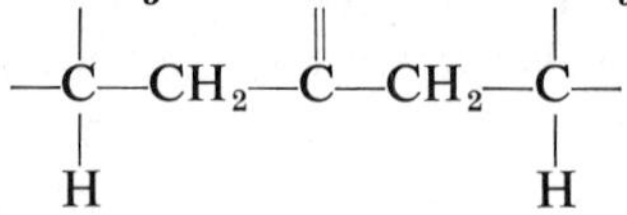

$$-\underset{\text{H}}{\overset{\text{CH}_3}{\text{C}}}-\text{CH}_2-\overset{\text{O}}{\overset{\|}{\text{C}}}-\text{CH}_2-\underset{\text{H}}{\overset{\text{CH}_3}{\text{C}}}- \qquad \textbf{1}$$

$$-\underset{\text{H}}{\overset{\text{CH}_3}{\text{C}}}-\text{CH}_2-\overset{\text{O}}{\overset{\|}{\text{C}}}-\text{CH}_3 \qquad \textbf{2}$$

$$-\text{CH}_2-\underset{\text{H}}{\overset{\text{CH}_3}{\text{C}}}-\overset{\text{O}}{\overset{\|}{\text{C}}}-\underset{\text{H}}{\overset{\text{CH}_3}{\text{C}}}-\text{CH}_2- \qquad \textbf{3}$$

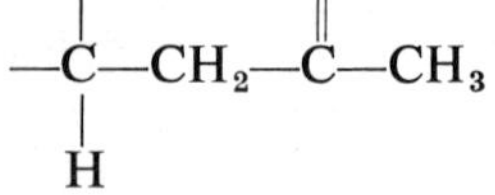

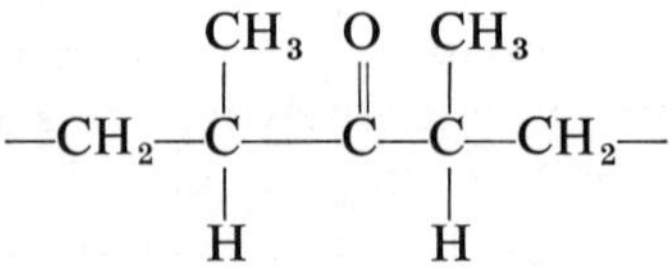

$$\text{—CH}_2\underset{\text{H}}{\overset{\text{CH}_3}{\text{C}}}\text{CH}_2\overset{\text{O}}{\overset{\|}{\text{C}}}\text{CH}_2\underset{\text{H}}{\overset{\text{CH}_3}{\text{C}}}\text{—}\ \ (\mathbf{1}) \xrightarrow{h\nu}$$

$$\xrightarrow{\geqslant 90\%} \text{—}\underset{\text{H}}{\overset{\text{CH}_3}{\text{C}}}\text{CH}_2\cdot + \text{O}{=}\dot{\text{C}}\text{CH}_2\underset{\text{H}}{\overset{\text{CH}_3}{\text{C}}}\text{—} \longrightarrow \text{CO} + \cdot\text{CH}_2\underset{\text{H}}{\overset{\text{CH}_3}{\text{C}}}\text{—} \tag{4.19a}$$

$$\xrightarrow{\leqslant 10\%} \left\{ \begin{array}{l} \text{—CH}{=}\text{C}(\text{CH}_3)\text{H} \\ \text{—CH}_2\text{C}({=}\text{CH}_2)\text{H} \end{array} \right\} + \text{CH}_2{=}\overset{\text{OH}}{\text{C}}\text{CH}_2\underset{\text{H}}{\overset{\text{CH}_3}{\text{C}}}\text{—} \longrightarrow \text{CH}_3\overset{\text{O}}{\overset{\|}{\text{C}}}\text{—CH}_2\underset{\text{H}}{\overset{\text{CH}_3}{\text{C}}}\text{—} \tag{4.19b}$$

$$\text{—CH}_2\underset{\text{H}}{\overset{\text{CH}_3}{\text{C}}}\text{CH}_2\text{C}({=}\text{O})\text{CH}_3\ \ (\mathbf{2}) \xrightarrow{h\nu}$$

$$\xrightarrow{\sim 15\%} \text{—CH}_2\underset{\text{H}}{\overset{\text{CH}_3}{\text{C}}}\text{CH}_2\cdot + \text{CH}_3\text{CO}^{\cdot} \longrightarrow \text{CH}_3\text{CHO};\quad \text{CH}_3\text{CO}^{\cdot} \longrightarrow \text{CH}_3\cdot + \text{CO} \tag{4.20a}$$

$$\xrightarrow{\sim 85\%} \left\{ \begin{array}{l} \text{—CH}{=}\text{C}(\text{CH}_3)\text{H} \\ \text{—CH}_2\text{C}({=}\text{CH}_2)\text{H} \end{array} \right\} + \text{CH}_2{=}\overset{\text{OH}}{\text{C}}\text{CH}_3 \longrightarrow \text{CH}_3\text{COCH}_3 \tag{4.20b}$$

A study of the changes in the infrared spectrum and the composition of the volatile products produced during irradiation of the melt-oxidized polypropylene films indicated that the type **1** ketones photolyze primarily by a Norrish type I cleavage to give carbon monoxide and two monoradicals, whereas ketone **2** is photolyzed by a Norrish type II scission to give acetone and an unsaturated polymer chain end.

The Norrish type II scission of ketone **2** from either the excited-singlet or triplet level would result in the loss of a small molecule and generate unsaturation. The type II cleavage of ketone **1** would result in chain scission yielding unsaturation and a ketone **2**. In the presence of oxygen, it is conceivable that energy transfer might occur from the ketone triplet to oxygen to form singlet oxygen, $^1O_2$.[68] The singlet oxygen would then react with neighboring vinyl groups to generate allylic hydroperoxides, which rapidly photolyze and would continue the photodegradation. Carlsson and Wiles[78] found a high quantum efficiency (~1) for polypropylene hydroperoxide photolysis, which led them to suggest that tertiary hydroperoxide photolysis is a key step in the acceleration and possibly also in the initiation of the oxidative photodegradation of polypropylene. Their mechanism is summarized in the scheme:

$$-\underset{H}{\overset{CH_3}{|}}{C}-CH_2-\underset{H}{\overset{CH_3}{|}}{C}- \xrightarrow[\text{processing}]{O_2} \text{ketones } \mathbf{1+2} \xrightarrow{h\nu} \left(\,>C{=}O\,\right)^{*} \longrightarrow \text{type II cleavage} \longrightarrow \text{molecular products}$$

$$\left(\,>C{=}O\,\right)^{*} \longrightarrow \text{Type I cleavage} \longrightarrow R\cdot \xrightarrow{O_2} RO_2\cdot \xrightarrow{RH} ROOH \xrightarrow{h\nu} RO\cdot + \cdot OH \longrightarrow \text{ketones } \mathbf{1+2} \qquad (4.21)$$

However, in comparing the oxidative photodegradation of polypropylene, polyethylene, and poly(4-methylpentene-1), Winslow and Matreyek found little evidence of autocatalysis (Figure 4.12). This implies that the chain lengths of degenerate branching reactions in (4.21) are relatively short.

*c. **Polystyrene.*** The oxidative photodegradation of polystyrene results in a yellowing of the polymer and a deterioration of mechanical properties. The activation spectrum[66] shows that in sunlight 318-nm radiation is most effective in causing degradation. The reaction is accelerated by traces of monomer and sulfur compounds.[80] In addition to the yellowing effect, a yellow-green fluorescence appears,[66] and the infrared spectrum shows carbonyl and hydroxyl bands.[81] The volatile products of the photooxidation are water and carbon dioxide.[82,83]

The yellowing is confined to the illuminated surface, and absorption of ultraviolet light in the yellow layer can protect the remainder of the polymer. Tryon and Wall[84] followed the oxidation by measuring the absorbance at 340 nm and found that the rate was halved by deuteration in the α position but was unaffected by deuteration in the β position. They therefore concluded that the rate-controlling step in the degradation is hydrogen abstraction specifically from the α position.

It has been suggested[85] that the color of oxidized polystyrene is caused by quinomethanes produced by the reaction of the polystyryl radical through

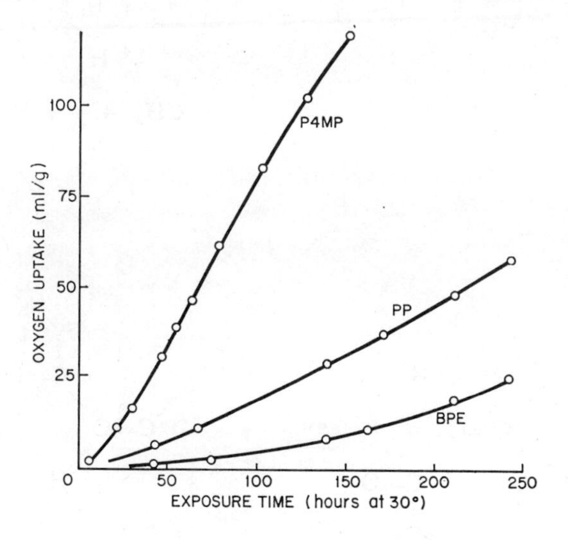

Fig. 4.12 Oxidative photodegradation of branched polyethylene (BPE), polypropylene (PP), and poly(4-methylpentene-1) (P4MP).[79]

one of the other resonance forms, as below:

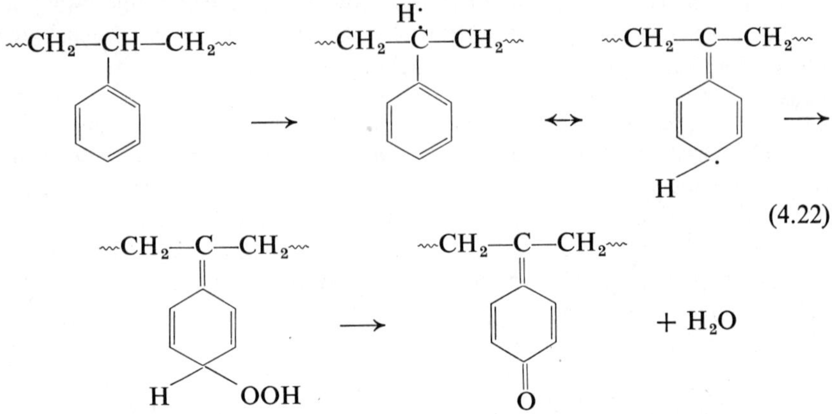

(4.22)

Tryon and Wall[86–88] suggested an alternative mechanism to account for the changes in the visible and ultraviolet spectra of photooxidized polystyrene as well as for certain post-irradiation effects. The post-irradiation

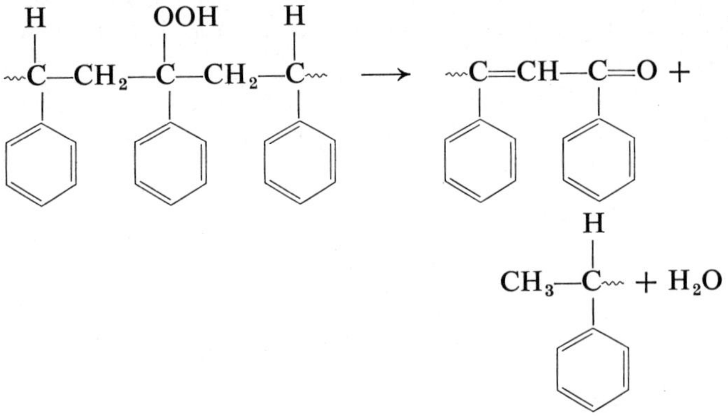

(4.23)

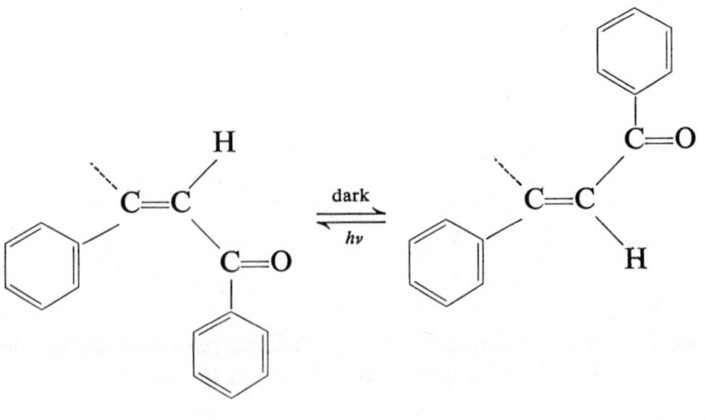

dark

$h\nu$

reaction appears to comprise two first-order reactions that take place at widely differing rates. The fast reaction is presumably the decomposition of hydroperoxides, known to be present[84] because of their ability to initiate polymerization of methyl methacrylate. They suggested that the slow reaction is cis–trans isomerization of a benzalacetophenone formed by the decomposition of the hydroperoxide. However, Grassie and Weir[89] found that the yellow coloration develops in polystyrene under 254-nm irradiation at comparable rates in the presence or absence of oxygen. Thus they concluded that the coloration is not a result of oxidation. They suggested that the color is caused by conjugated carbon–carbon unsaturation in the polystyrene backbone, but that the lack of mobility of the molecules within the rigid polymer framework prevents the occurrence of a coplanar sequence of double bonds with concomitant longer wavelength absorption bands. Also they proposed that the decomposition of the hydroperoxide occurs via a six-membered transfer state:

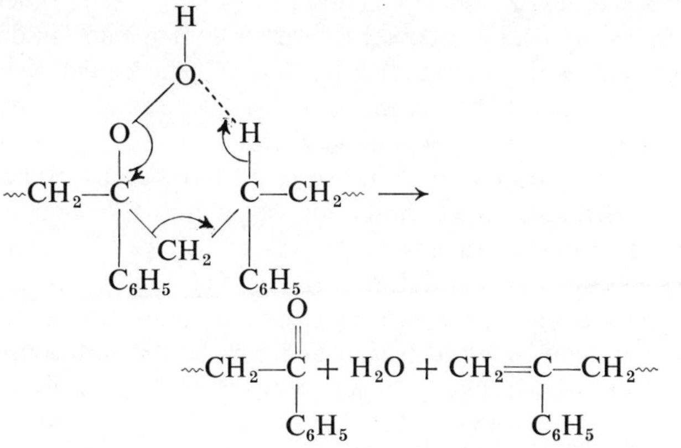

$$\sim CH_2-\overset{\overset{\overset{H}{|}}{\overset{O}{|}}}{\underset{|}{\overset{O}{|}}}\!\!C(C_6H_5)-CH_2-C(H)(C_6H_5)-CH_2\sim \;\longrightarrow$$

$$\sim CH_2-\overset{O}{\overset{\|}{C}}(C_6H_5) + H_2O + CH_2{=}C(C_6H_5)-CH_2\sim \qquad (4.24)$$

If the above scheme is valid, then subsequent oxidation via the intermediacy of singlet oxygen is conceivable, since it is known[90] that aryl ketone triplet states are quenched by oxygen, resulting in the formation of singlet oxygen molecules, and that the olefin produced in (4.24) would be quite reactive toward singlet oxygen attack.[71]

d. Rubber. In spite of a considerable amount of literature on the thermal aging of rubber and elastomers, relatively little work has been done on the photodegradative processes. This is largely because elastomers are used primarily in black formulations, which are affected only slightly by solar radiation. However, since clear rubber materials have been developed, it is worthwhile considering the photodegradation process in some detail. The relevant research has been reviewed by J. Morand;[91] therefore this section will deal only briefly with the more pertinent results as well as current developments.

The photodegradation of rubber involves two types of photoreactions: those that take place in the absence of oxygen (vacuum photolysis, isomerization) and those that require the presence of oxygen. In the oxidative photodegradation of crude elastomers, hydroperoxides have been detected at the beginning of the process, and they develop at an increasing rate before reaching a limiting concentration.[92,93] Carbonyl groups appear after a relatively long induction period (one to several hours).[94]

In the photooxidation of nonconjugated olefins (which serve as model systems for rubber), the quantum yield was found to depend strongly on the wavelength of the light employed, being close to unity for 254-nm light and dropping to 0.25 for 366-nm light.[95–98] Since the quantum yield for the decomposition of the hydroperoxides was found to be near unity with 254-nm light, Bateman and Gee[99] concluded that the olefinic photooxidation at this wavelength proceeds by a radical chain-reaction mechanism, which would also account for the autocatalysis observed earlier by Bateman[97] in the photooxidation of rubber. However Hart and Matheson[96] believed that the reaction is autocatalytic only at the shorter wavelengths. They found a constant rate when light of wavelength near 400 nm was used.

Irradiation in the absence of oxygen can result in cis–trans isomerization. Golub[97] showed that ultraviolet irradiation of poly(*cis*-1,4 butadiene) converted the polymer into the trans form. However this type of transformation has not been detected in natural rubber. In addition to isomerization, irradiation under vacuum results in the evolution of gases, which, in the case of natural rubber, consist of hydrogen and simple hydrocarbons (mostly methane).[101] It is presumed that hydrogen results from the photolytic scission of the C–H bond, and methane forms by combination of methyl radicals (resulting from cleavage of the $C\text{–}CH_3$ group) with hydrogen atoms. The quantum yield for formation of these gaseous products is 10^{-3} to 10^{-4},[101,102] a value several orders of magnitude smaller than that for oxygen absorbed.[97] Morand[95] has summarized the above effects and has also suggested a role for singlet oxygen in the photodegradation of elastomers.

Kaplan and Kelleher[103] have shown that singlet oxygen molecules generated from the triphenylphosphite–ozone adduct[104] are able to oxidize cis- and trans-polybutadiene as well as acrylonitrile-butadiene-styrene (ABS) polymers. They suggest that photooxidation of ABS probably occurs through initial oxidation of the polybutadiene portion of the polymer.

Morand studied the effects of additives on the photodegradation of rubber.[95] Mercaptobenzthiazole and sulfur were found to have a protective effect in solar radiation, but zinc oxide was found to be a sensitizer in the region of 400 nm, although it retarded degradation when the irradiation was below

370 nm. It is believed that this effect is due to singlet oxygen formation which occurs via a complex of ZnO and physically adsorbed O_2:

$$ZnO + O_2 \xrightarrow{h\nu} ZnO^{+} \cdots O_2^{-}$$
$$O_2^{-} \rightarrow {}^{1}O_2 + e$$
$$ZnO^{+} + e \rightarrow ZnO \qquad (4.25)$$

This suggestion has also been made by Egerton[105] to explain the photosensitivity of dyes and certain white pigments (zinc oxide, zinc sulfate, and titanium dioxide) in promoting the phototendering of cotton.[106]

4. Photodegradation of Other Polymers

Since most of the other commonly used organic polymers share to some degree the photooxidation reactions of the polyolefins, the preceding discussion is of some general value. However most other polymers contain one or more additional characteristic functional group whose contribution to the overall photooxidation process must be emphasized.

a. Chlorinated Hydrocarbon Polymers. The relatively rapid deterioration of poly(vinyl chloride) (PVC) and poly(vinylidene chloride) that occurs on outdoor exposure has been known for some years. Whereas the photodegradation and aging of most polymers result in a deterioration in physical properties, PVC decomposes with loss of hydrogen chloride. Although the accompanying discoloration is aesthetically displeasing, the yellowing or browning may be accompanied by negligible changes in mechanical properties. The effects of photochemical degradation are similar to those found in thermal degradation (see Chapter 3, Section C), but oxidative and crosslinking processes play a relatively more important role.[107,108]

Sunlight at wavelengths of 300 to 320 nm is reported as being the most effective in causing the dehydrochlorination.[66,108] The mechanism of the photodegradation almost certainly involves a radical chain, and this related work has been reviewed by Geddes.[109] The exact nature of the initiation step is uncertain, however. Kenyon[110] found that simple alkyl chlorides are stable even at wavelengths of 220 to 230 nm, but they decompose in the presence of carbonyl groups,[107] which absorb at higher wavelengths (280–300 nm) and could act as sensitizers. Unsaturated structures formed during thermal treatment may also act as sensitizers.[107,108]

Photodegradation of PVC is usually associated with oxidative attack,[108,110] which probably is similar to thermal oxidation (Chapter 2). PVC becomes more susceptible to thermal degradation after exposure to light. This leads one to believe that the oxidative photodegradation causes an increase in the number of thermally labile groups in the polymer. That oxidation of unsaturation is an important factor in the total reaction is shown in the bleaching of discolored samples by oxygen,[110] with the final color depending on the balance of dehydrochlorination and bleaching.

Thus the overall mechanism for the oxidative photodegradation would seem to involve two processes: dehydrochlorination by a radical-chain process and oxidation resulting in hydroperoxides and ultimately carbonyl groups. Both these processes can become an autocatalysis effect, but the effects are somewhat mutually antagonistic; that is, the double bonds activating adjacent chlorine atoms are removed by oxidation, but the photolysis of oxidation products may in fact produce intermediates that are initiators for dehydrochlorination.

b. Polyesters. Poly(methyl methacrylate) (PMMA) is extraordinarily resistant to oxidative photodegradation, primarily because of the absence of any absorption in the "pure" polymer down to 285 nm. This resistance is a significant factor in the use of PMMA in a variety of applications, however catalyst residues and other impurities may cause yellowing and therefore should be removed for improved light resistance. When degradation does occur, it is probably very similar in mechanism to the thermal degradation (Chapter 3, Section B).

Poly(ethylene terephthalate) is very resistant to photooxidation, but polycarbonates yellow rapidly under the influence of light. However, the yellow layer usually protects the rest of the polymer.

c. Polyamides. The presence of the amide group leads to the possibility of hydrolysis reactions[111] that may greatly affect the nature of the oxidative photodegradation of these polymers. However the hydrocarbon segments linked by the amide groups in forming the polymer are susceptible to the same types of oxidative processes described for hydrocarbon polymers.

Very few product studies have been carried out on photodegraded polyamides. Ford[112] showed that there was a marked increase in the absorption band at 290 nm when nylon was irradiated with 254-nm light in air. A "post-irradiation" effect was also observed; that is, the absorbance at 290 nm continues to increase even after irradiation has been stopped. On re-irradiation the absorbance fell rapidly to a value corresponding to that prior to the dark period and then proceeded to increase again with continued irradiation. The above observations are consistent with a scheme proposed by Kroes[113] and Sharkey and Mochel[114] in which the primary step of photolysis is the following:

$$\text{—}\overset{\overset{\displaystyle O}{\|}}{C}\text{—}\overset{\overset{\displaystyle H}{|}}{N}\text{—}CH_2\text{—} \xrightarrow{h\nu} \text{—}\overset{\overset{\displaystyle O}{\|}}{C}\text{—}\overset{\overset{\displaystyle H}{|}}{N}\cdot + \cdot CH_2\text{—} \qquad (4.26)$$

Evidence supporting this scheme was obtained by Heuvel and Lind,[115] who detected free-radical formation in irradiated samples of nylon 6.

Studies of the photolysis of simple amides[10] shows that the Norrish type I and type II cleavages are observed but that the type I process appears to predominate. The quantum yield of type II cleavage in the photolysis of poly(*t*-butyl *N*-vinylcarbamate) was found to be solvent dependent,[116] and

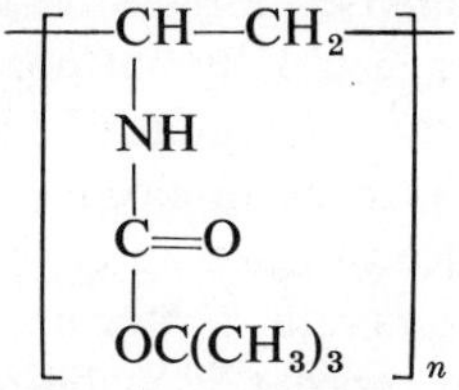

$$\left[\begin{array}{l} -CH-CH_2- \\ \quad | \\ \quad NH \\ \quad | \\ \quad C{=}O \\ \quad | \\ \quad OC(CH_3)_3 \end{array} \right]_n$$

the results were interpreted in terms of changes in the optical transition. Implications of this behavior in stabilization processes are discussed in the next section on stabilization mechanisms.

d. Miscellaneous Polymers. The completely halogenated polymers, poly(tetrafluoroethylene) and poly(chlorotrifluoroethylene) are essentially totally resistant to photooxidation, but they are very expensive and are seldom employed for purposes where only outdoor weatherability is required. Polyacrylonitrile also is quite resistant to oxidative photodegradation, provided high temperatures are not involved. Where these polymers undergo degradation, it usually involves "unzipping" processes of the type described in Chapter 3, Section B.

Cellulose is exceptionally unstable to ultraviolet irradiation, particularly in the presence of dyes.[106] A comprehensive review of its degradation has been given by McBurney.[117] The mechanism of degradation appears to be quite dependent on the presence of moisture[106] or trace contaminants that are readily absorbed and retained. The intermediacy of oxygen species such as $^1O_2$ and $O_2^{\dot{-}}$ in the photooxidation has been discussed by Egerton[106] and Schnuriger and Bourdon.[118]

Polyacetals oxidize rapidly and have very low weathering resistance. It is probable that the mechanism for photodegradation involves a combination of "unzipping" reactions and oxidation. Hydrolysis may also be important.

D. GENERAL METHODS OF STABILIZATION

In practice, polymers are usually protected against photooxidation by the addition of stabilizers. These additives are of three general types: light screens, ultraviolet absorbers, and quenching compounds. By strict definition, a light screen is a material that is interposed as a shield between the polymer and the radiation source. A typical example is the coating of polyethylene with aluminum powder.[119] However pigments dispersed in a polymer matrix are usually classified as light screens, since particles concentrated at the surface do filter out radiation from the polymer bulk.

Ultraviolet absorbers function by absorbing and dissipating ultraviolet irradiation that would otherwise initiate degradation of the polymer. Stabilizers of this type, although usually not opaque to visible light, are by definition screens in the ultraviolet region of the spectrum. Thus the distinction between light screens, as exemplified by pigments, and ultraviolet absorbers is one of convenience rather than of basic mechanism.

The use of quenching agents in protecting polymers against photooxidation is a relatively recent concept.[120] Stabilizers of this type function by dissipation of the excited state energy from polymer molecules (Section D.2.d) Quenching agents could be particularly effective in combination with ultraviolet by absorbers.

The modification of polymer structure to improve the resistance to photodegradation is an important although somewhat neglected method of stabilization. Polymer modification encompasses both the purification of existing polymers and the development of new polymers with inherent photostability.

1. Light Screens

Light screens function either by absorbing damaging radiation before it reaches the polymer surface or by limiting its penetration into the polymer bulk. Although protection is essentially a physical phenomenon, we will see in the following sections that chemical reactions of light screens can be important.

Light screens include exterior coatings such as paints, protective films that are exuded from within the polymer bulk, and additives, notably pigments, that are dispersed throughout the polymer.

Although carbon black and other pigments are usually classified as light screens, they must also absorb damaging radiation. An external coating of a pigment is a light screen in the strictest sense. Within a polymer matrix, however, pigments protect primarily by absorption.

a. Coatings. The classical example of a protective coating for polymers is the painting of cellulosic materials, particularly wood. Conventional paints, however, have not been widely used to protect synthetic polymers against photodegradation. We should recognize that painting of wood is but one example of the use of a polymer coating to protect another polymeric material. Extension of this ancient technology, employing polymer films designed for maximum light screening, has considerable potential in the stabilization of polymers against photooxidation.

In theory, at least, any polymer having good photostability and capable of absorbing damaging radiation could be used as a coating to protect a more vulnerable polymer. It is obvious that the coating must be of sufficient depth to prevent radiation from reaching the base polymer. An important factor, of course, is the extent to which the coating absorbs ultraviolet radiation. Poly[2-hydroxy-4-(2-acryloxyethoxy)benzophenone] has good photostability and a high level of ultraviolet absorbance, and its use as a protective ultraviolet coating has been reported.[121] Much remains to be accomplished, however, in the synthesis of specialized polymers whose composition is designed to afford the maximum protection as light screens. Although such polymers may be relatively expensive, their use as coatings to extend the useful life of less expensive and less stable polymers could be economically feasible.

Another type of protective coating consists of compositions in which a highly efficient ultraviolet absorber is dispersed in a polymer capable of retaining the additive in reasonably high concentration. The polymer in these compositions need not have inherent stability to ultraviolet irradiation. Its primary role would be to retain an adequate concentration of the ultraviolet absorber, resisting loss by evaporation, leaching, or migration of the stabilizer into the base polymer. The analogy to conventional paints is apparent—the polymer is functioning as the vehicle and the ultraviolet absorber acts as the pigment.

Protective coatings can also be formed as a result of migration of incompatible ultraviolet absorbers out of the polymer bulk to the surface. This migration results in a concentration gradient with the maximum concentration of absorber at the surface. Unfortunately, the process rarely if ever stops at this point. Additives that migrate to the surface are invariably lost by evaporation, leaching, or both.

In the use of coatings to inhibit photooxidation, it cannot be assumed that no interactions will occur between the coating and the base polymer. Migration of additives across the interface has been mentioned. There is also the possibility that energy absorbed in the coating could be transferred to the base polymer. Perhaps even more significant is the possibility that products formed in deterioration of the polymer coating or its additives could diffuse across the interface and initiate deterioration of the base polymer.

An interesting modification of photostabilization by surface coating is based on changes in polymer morphology. Since photooxidation is dependent on the diffusion of oxygen into the polymer, formation of a less permeable surface layer would be an effective route to stabilization. This could be accomplished by increasing the density at the surface. Luongo and Schonhorn[122] have reported a higher density modification of polyethylene, formed when the polymer is crystallized against a high-energy surface such as gold. Dense surface layers such as this could increase stability to the extent that they restrict diffusion of reactants into the polymer.

Modification of either the physical or the chemical structure of the surface could contribute to stabilization by reducing the permeability to reactants. Rogers and co-workers[123] have shown that crosslinking of polyethylene to the extent of one crosslink per 50 carbon atoms in the chain reduces the permeability to nitrogen by a factor of approximately 2. However adequate stabilization would require a much higher crosslink density, perhaps resulting in embrittlement of the surface layer.

b. Pigments. The addition of pigments to polymers is a very effective, practical method for increasing resistance to photodegradation. Although coating polyethylene with aluminum powder has been shown to impart excellent stability to this polymer,[119] the practical problems of pigment adhesion to a polymer surface severely limit the use of pigments as coatings. Pigments are usually dispersed throughout the polymer. Their effectiveness is determined by their ability to shield inner layers of the polymer from damaging radiation and by their proficiency in absorbing ultraviolet radiation.

Carbon black is by far the most effective pigment. In general, the efficiency of pigments decreases from the darker to the lighter colors. Several dark pigments give some protection, but none is comparable to carbon black.[124] Pigments that scatter absorbed radiation throughout the polymer matrix can increase the rate of photodegradation.[125] Carbon black was discussed in Chapter 2 as a thermal antioxidant. Its effectiveness in protecting polymers against outdoor weathering, particularly at high temperatures, can be attributed in part to its ability to inhibit thermal oxidation which often occurs concurrently with photodegradation. In addition, carbon black is a very effective light screen and functions as an ultraviolet absorber through energy level transitions in its polynuclear aromatic structure. It is effective as a stabilizer in a variety of polymers, as shown in Table 4.3. Its effectiveness, coupled with low cost, accounts for the wide use of this pigment in protecting polymers against photodegradation. However it is obvious that its use is limited to applications in which black compositions are acceptable.

The effectiveness of carbon black as a light screen depends on several additional factors, for example, concentration, particle size, and dispersion. The relation between these variables and the protective effect has been described by Wallder et al.[124] and reviewed in several recent publications.[126,127] Both particle size and concentration influence dispersion of the pigment in the polymer matrix, and the ultimate dispersion of carbon black determines its effectiveness. This can be interpreted as an indication of the

Table 4.3 Effect of Carbon Black on the Outdoor Weathering of Typical Polymers[126]

Polymer	Years Required to Reach Visible Surface Failure
Polyethylene–natural	1–1½
Polyethylene + 1 pph channel black	>25
Plasticized poly(vinyl chloride)–natural	1–2
Plasticized poly(vinyl chloride) + 10 pph channel black	>15
Neoprene–natural	½–1
Neoprene + 40 pph SRF black	>20

relative importance of light screening in comparison to the other mechanisms by which carbon black can function as a stabilizer.

Until recently carbon blacks prepared by the channel process were the only effective light screens. However new modifications in the furnace process have provided furnace blacks of particle size comparable to that of the channel blacks previously demonstrated to be effective. These small-particle-size furnace blacks appear to be as effective light screens as channel blacks of comparable particle size. Wallder et al.[124] established an optimum particle size of 15 to 25 nm for adequate stabilization and a concentration between 2 and 5%. Higher concentrations, although effective in stabilization, usually have an adverse effect on physical properties of the polymer. Within the particle size and concentration limits described, carbon black provides excellent protection, but only when adequately dispersed in the polymer matrix. Figure 4.13 compares a good and a poor dispersion of the same concentration of a carbon black. At 100× magnification, a good dispersion is described as one exhibiting a uniform dark background with a minimum of discrete particles or agglomerates visible. In contrast, the poor dispersion shows many agglomerates and large areas having little or no pigmentation.

The effectiveness of a good dispersion of carbon black is evident from the data shown in Table 4.3. As further evidence, polyethylene samples containing only 1% of a channel black show no visible change and no significant loss in mechanical properties after outdoor exposure in Florida for more than 30 years. In summary, then, excellent protection against photodegradation can be obtained through proper pigmentation with carbon black. Other pigments, particularly those that are dark, protect, but to a much lesser extent. Carbon black still ranks as the most effective protection against photodegradation for most polymers, limited in application only by the color it imparts to the formulation.

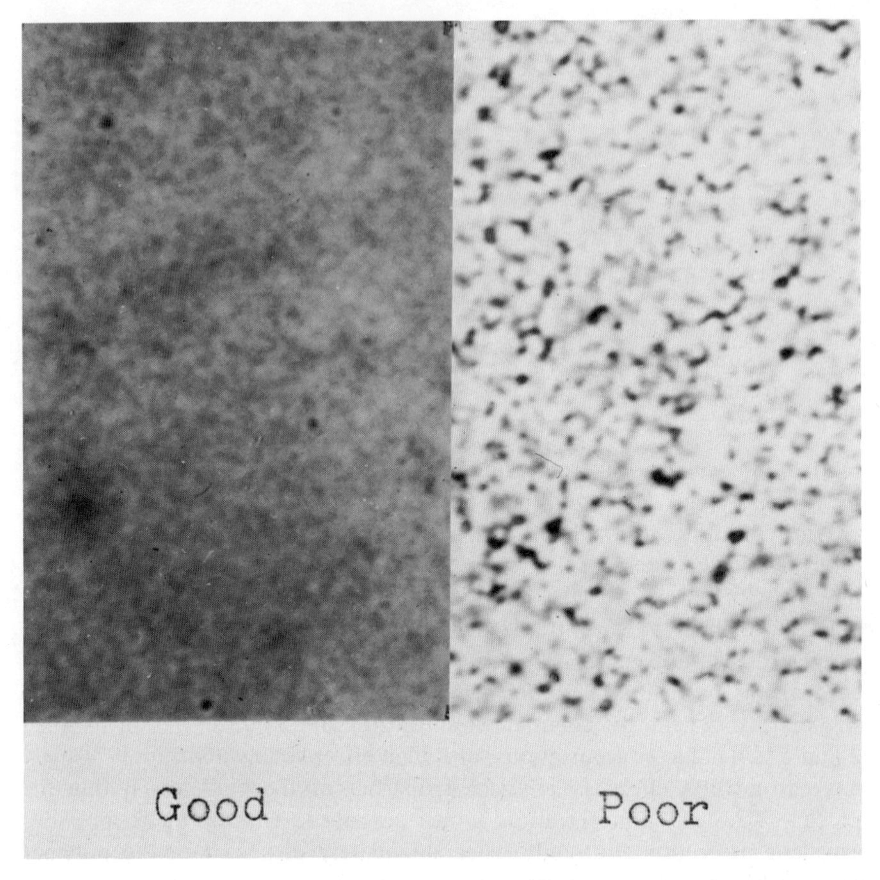

Fig. 4.13 Comparison of dispersions of carbon black:[126] (*A*) good; (*B*) poor. (Reproduced from *Reinforcement of Elastomers*, Interscience, 1965.)

2. Photophysical Mechanisms of Stabilization: Ultraviolet Absorbers and Quenchers

One might inquire why, in a chapter devoted to stabilization against oxidative photodegradation, there has been so much discussion of the mechanisms of the photodegradative processes. The answer is fairly straightforward: the more one is able to understand the degradative processes at a molecular level, the better one is able to take appropriate preventive measures. The same holds true for the photostabilization process. If one can achieve an understanding of these reactions on a molecular basis, then it should be possible to "tailor-make" stabilizers for a specific polymer. This section deals with recent developments in photochemical concepts as applied to the harmless deactivation of excited states.

The deactivation of excited states can occur by any of the photophysical processes outlined in Section C.1.a and shown in Figure 4.7. These are (*a*) radiative processes such as fluorescence ($S_1 \rightarrow S_0$) and phosphorescence ($T_1 \rightarrow S_0$), (*b*) intersystem crossing ($S_1 \rightsquigarrow T_1$), (*c*) internal conversion ($S_1 \rightsquigarrow S_0$, $S_2 \rightsquigarrow S_1$), and (*d*) quenching by energy transfer. Each of the above paths could conceivably function as a photostabilization process.

In addition to the stabilization mechanisms involving photophysical processes, it is possible to have a photoexcited species that undergoes a molecular rearrangement to generate either a more photostable product or one that reverts back to the original material in the dark. Examples of both types will be discussed.

The range of stabilization obtainable by the incorporation of ultraviolet absorbers in plastics is illustrated in Table 4.4, which gives data for cellulose acetate butyrate.

Table 4.4 Stabilization of Cellulose Acetate Butyrate[128]

	Stabilizing Effectiveness[b]	
Additive[a]	Hours in Modified Weather-Ometer	Years Outdoors in Kingsport, Tenn.
None	200	1
Phenyl salicylate	1,000	5
Resorcinol monobenzoate	1,800	7
2,4-Dihydroxybenzophenone	2,400	8
4-(4-Nitrophenylazo)phenol	>8,000	—
6,13-Dichloro-3,10-diphenyltriphenodioxazine	>22,000	—
Carbon black (channel)	>30,000	>10

[a] 1% concentration.
[b] Exposure time required to cause 25% loss of flexural strength.

a. Radiative Processes. Radiative processes in general involve the emission of light at longer wavelengths than those of the light absorbed. It is the mechanism by which optical brighteners operate, and at first glance fluorescence would seem to offer an ideal mechanism for deactivation. However it is conceivable that the emitted light would be reabsorbed by the polymer and degradation would occur. Also most efficient fluorescing molecules will photodegrade slowly to nonfluorescent products; thus many of these compounds, such as phenylpyrazolines, benzoxazoles, and fluoranthene, give good protection only until they are destroyed by the ultraviolet irradiation. However, one light-stable fluorescent compound, 6,13-dichloro-3,10-diphenyltriphenodioxazine, was found[128] to be an excellent stabilizer (see Table 4.4), approaching carbon black in effectiveness.

b. Intersystem Crossing. Intersystem crossing ($S_1 \rightsquigarrow T_1$) may result in the production of triplet states that are relatively inert to further photochemical reactions. If the intersystem-crossing efficiency is near unity, then it is possible to achieve a photostabilizer system by this mechanism. An interesting example of this technique is illustrated in the type II cleavage of substituted butyrophenones. It is generally agreed[129] that the cleavage of these

$$X\text{-}C_6H_4\text{-}C(=O)CH_2CH_2CH_2\text{-}H \xrightarrow{h\nu} X\text{-}C_6H_4\text{-}C(=O)CH_3 + CH_2{=}CH_2 \quad (4.27)$$

aryl ketones occurs via the lowest triplet state, T_1, and it is also known that the intersystem crossing efficiency of butyrophenone is near unity. However, the data in Table 4.5 show that the quantum yield for scission is very sensitive to the nature and position of the ring substituent. The results have been interpreted as follows: the absorption spectra of these compounds indicate that only butyrophenone and 4-methylbutyrophenone possess $S_1(n\text{–}\pi^*)$ states, whereas for the other derivatives listed in the table, the $^1(\pi, \pi^*)$ state merges with the $^1(n, \pi^*)$ state, and it is difficult to determine which state is

Table 4.5 Quantum Yields of Ethylene Formation in the Photolysis of Butyrophenone Derivatives in Benzene Solution at Room Temperature[129]

Compound	$\phi_{CH_2=CH_2}$
Butyrophenone	0.40
4-Methylbutyrophenone	0.39
4-Aminobutyrophenone	0
4-Hydroxybutyrophenone	0
2-Hydroxybutyrophenone	0

lower in energy. There is a corresponding change in the triplet states, and it has been suggested that the unreactive ketones (with the exception of the 2-hydroxy derivative, which is discussed in the next section) possess a lowest triplet state, $T_1(\pi, \pi^*)$. A long lived phosphorescence has been observed in the unreactive ketones, which is in accordance with the assignment of the $T_1(\pi, \pi^*)$ state, since it is generally agreed that $^3(\pi, \pi^*)$ states have longer lifetimes than $^3(n, \pi^*)$ states. The above type of mechanism for photostabilization has not received much attention.

c. Internal Conversion. Internal conversion is the process that converts electronic energy into vibrational energy by a radiationless route without a change in spin multiplicity ($S_1 \rightsquigarrow S_0$, $S_2 \rightsquigarrow S_1$). It is the major mechanism for photostabilization in the case of the *o*-hydroxybenzophenones, **4**, *o*-hydroxyphenylbenzotriazoles, **5**, and salicylates **6**. All these compounds have

4 **5** **6**

a common structural feature, the intramolecular hydrogen bond. It is generally accepted that the presence of this structural grouping is responsible for the

Table 4.6 Correlation Between NMR Shift and Stabilizing Effectiveness[128]

Additive[a]	NMR Hydroxy Proton Shift (C/sec)	Stabilizing Effectiveness,[b] Hours in Modified Weather-Ometer
None	—	200
2,6-Dihydroxybenzophenone	−160	600
Phenyl salicylate	−220	1000
2,2′-Dihydroxybenzophenone	−220	1000
2,4-Dihydroxybenzophenone	−280	2400
2-Hydroxy-4,4′-dimethoxybenzophenone	−310	>2600
3-Benzoyl-2,4-dihydroxybenzophenone	−340	>4000

[a] 1% concentration.

[b] Exposure time required to cause 25% loss of flexural strength.

efficient deactivation of the electronically excited states of the ultraviolet absorber. In fact, the strength of the hydrogen bond (as measured by NMR spectroscopy) of a variety of substituted *o*-hydroxybenzophenones has been correlated with their relative efficiencies as ultraviolet stabilizers (Table 4.6). However, it would be of interest to interpret what is the peculiar function of this structural grouping in the excited state, since it is the *excited* state that undergoes deactivation.

In attempting this interpretation, one should first discuss the absorption process involved for the long-wavelength band of *o*-hydroxybenzophenone and for the absorption of solar radiation. This transition is very probably an intramolecular charge–transfer transition.

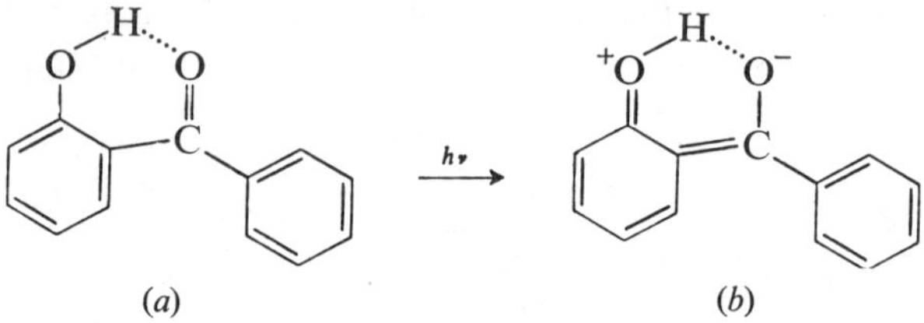

(*a*) $\xrightarrow{h\nu}$ (*b*)

Structure (4.28*b*) consists of an acid–base pair in which a facile proton transfer should occur. In other words, the phenolic portion becomes more acidic in the excited state and the carbonyl function becomes more basic. The heightened acidity of phenols and basicity of carbonyl groups in the excited-singlet state is a well-documented phenomenon and has been reviewed by Weller.[130] Heller[131] has suggested that photochemical reactions involving hydrogen transfer are best viewed as controlled radiationless relaxation processes. Therefore the photostabilization mechanism in *o*-hydroxybenzophenones is considered to be the result of a rapid tautomerism in the excited state which may be facilitated by the strong intramolecular hydrogen bond. (This view has also been suggested recently by O'Connell[131a] who studied the photoreactivity of 2-hydroxy-4-*t*-butylbenzophenone.

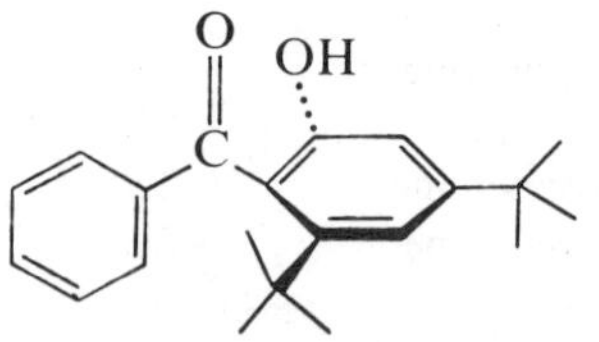

Although geometrical restrictions presumably preclude the existence of an intramolecular hydrogen bond in the ground state of this molecule, it was found to be photostable in hydrocarbon media.) The reverse step is associated with a simultaneous internal conversion process, $S_1 \rightsquigarrow S_0$.

Spectroscopic studies support this view. Beckett and Porter[132] reported that 2,4-dihydroxybenzophenone gives no luminescence in a rigid hydrocarbon matrix at 77°K, whereas 4-hydroxybenzophenone (which does not possess the intramolecular hydrogen bond) exhibits an intense phosphorescence. Hammond, Turro, and Leermakers[133] observed that salicylaldehyde, 2-hydroxy-4-methoxybenzophenone, and 2,4-dihydroxybenzophenone do not sensitize the cis–trans isomerization of the piperylenes (1,3-pentadienes) in benzene solution, whereas 4-hydroxybenzophenone is a highly efficient sensitizer. Lamola and Sharp[134] have shown that the spectroscopic and photochemical properties of the *o*-hydroxy aromatic carbonyl compounds are strongly dependent on the nature of the solvent. In polar solvents and particularly in hydrogen-bonding solvents (e.g., those containing alcohols), the internal hydrogen bond is broken in some of the molecules; the rapid tautomerism is prevented and phosphorescence is observed. However, in nonpolar solvents (e.g., 3-methylpentane), essentially all the solute molecules are intramolecularly hydrogen bonded and no emission is observed. Another effect of this rapid and efficient $S_1 \rightsquigarrow S_0$ internal conversion is that intersystem crossing may not take place.

A similar process is found in the *o*-hydroxyphenylbenzotriazoles. Here the spectroscopic evidence has been obtained by Merrill and Bennett[135] on the analogous isomers of 2-(amidophenyl)-2,1,3-benzotriazole. From the infrared spectrum of the ortho isomer they deduced that it contained an intramolecular hydrogen bond. The quantum yields for fluorescence and chemical reaction of the ortho isomer were very much lower than those of the nonhydrogen-bonded meta and para isomers and the internal conversion $S_1 \rightsquigarrow S_0$ was found to be one hundred-fold more rapid in the hydrogen-bonded species. As a consequence, most of the photoexcitation energy is thermally dissipated in the singlet manifold. It may be assumed that the 2-(2-hydroxyphenyl)-2,1,3-benzotriazole and the salicylates function by a similar process.

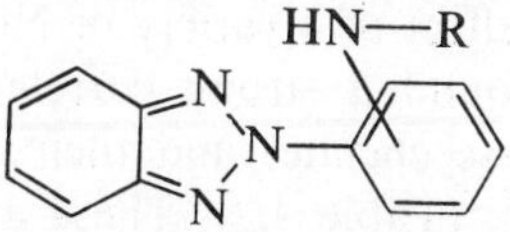

d. Quenching. Quenchers need not possess a very high absorbance at the wavelengths at which polymer degradation is most efficient, since, unlike absorbers, their function is to remove energy by an intermolecular process. The quenching may occur in either of two ways: (*a*) energy transfer resulting in a nonreactive excited quencher molecule:

$$A^* + Q \rightarrow A + Q^* \rightsquigarrow Q \qquad (4.29)$$

(*b*) formation of an excited-state complex that undergoes other photophysical processes:

$$A^* + Q \rightarrow [A \cdots Q]^* \rightarrow \text{photophysical processes (fluorescence, internal conversion, etc.)} \tag{4.30}$$

The use of stabilization mechanisms involving energy transfer from the polymer to the stabilizer has received greater attention in recent years.[136–138] For example, Heskins and Guillet[139] showed that it was possible to use a known triplet quencher, cyclooctadiene (COD), to quench about 45% of the Norrish type II cleavage in the photolysis of the ethylene–carbon monoxide copolymer. On the other hand, the use of this quencher in the photodegradation of polyphenylvinylketone reduced the rate of degradation to 7% of the rate in the absence of COD. The authors suggested that, although the triplet-quenching mechanism would not be an effective means of photostabilization in aliphatic ketone-containing polymers (only 40% quenching), this mechanism could be an effective method of stabilizing the polymers that degrade by paths involving the photolysis of phenylketone groups.

Another set of triplet quenchers that has been studied extensively is the Ni(II) chelates. Briggs and McKellar[140] have used the flash-photolysis technique to elucidate the mechanism of ultraviolet stabilization by these chelates. They studied the effect of a variety of Ni(II) chelates on the triplet state of anthracene and found a strong correlation between the triplet-quenching efficiency of these chelates and their effectiveness as ultraviolet stabilizers in polypropylene (Table 4.7). These authors suggested that the protective action in polypropylene involved quenching by the chelate of *triplet* carbonyl groups in the polymer. However, in view of the known photochemistry of aliphatic ketones involving the excited-singlet state,[53,54] it would appear that the quenching of the excited-singlet states might also occur.

e. Molecular Rearrangement. Another type of stabilization mechanism involves the dissipation of energy by a molecular rearrangement (the rapid tautomerism of the *o*-hydroxybenzophenones is a trivial case) following the absorption of ultraviolet light.[128] As stated earlier, the rearrangement should be reversible in the dark. An example of such a phototropic stabilizer is the azo dye 4(4′-nitrophenylazo) phenol, which is a very good stabilizer for cellulose acetate butyrate (see Table 4.4).

In contrast to the above rearrangement by a radiation-absorbing molecule is the use of a "nonabsorbing" stabilizer that undergoes a molecular rearrangement. This species is illustrated by resorcinol monobenzoate (Table

Table 4.7 Triplet-Quenching Effect and Ultraviolet Stabilizing Activity of Chelates[140]

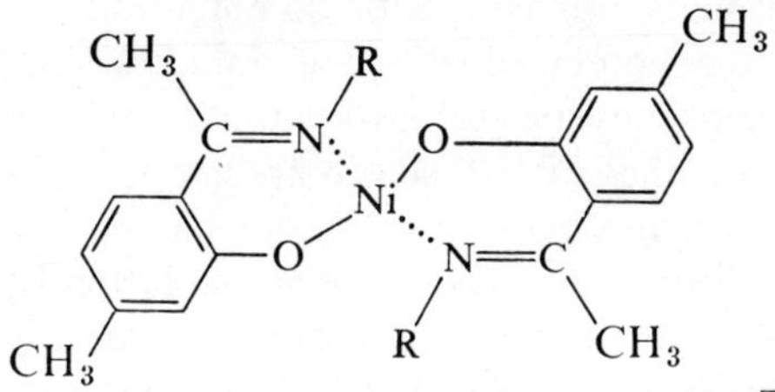

Stabilizer	Quenching Effect on Triplet Anthracene	Toughness of Yarn[a] Retained After Exposure in Weather-Ometer (%)		Time for Polypropylene Plastic Films[c] to Absorb 0.06% O_2 as Carbonyl on Exposure to Xenotest, (hr)
		500 hr[b]	600 hr[b]	
R = OH	Strong	52	53	2000
R = n-C_4H_8	Moderate	47	23	825
R = C_6H_5	None	17	8	470
2-Hydroxy-4-n-octyl-oxybenzophenone	None	—	20	650
None (same antioxidants)	—	—	3	—

[a] Profax 4/160 (a thermally degraded polypropylene supplied by Hercules Company)· Spinning temperature 225°C, draw ratio 4:1; 0.5% ultraviolet stabilizer, 0.1% Topanol CA, 0.5% Negonox DLTP, 0.4% calcium stearate. Toughness of yarn = tenacity (g/den) × extension at break (%) × 2 × 10^{-2}.
[b] Separate exposures.
[c] Polypropylene (MFI = 3) 0.02-in. thick films; compression-molded at 190°C for 2 min.; 0.5% ultraviolet stabilizer, 0.25% Negonox DLTP.

4.4), which is almost as effective as 2,4-dihydroxybenzophenone. The comparable effectiveness, however, is not accidental, since resorcinol monobenzoate is converted by sunlight into 2,4-dehydroxybenzophenone by a photo-Fries rearrangement.[141]

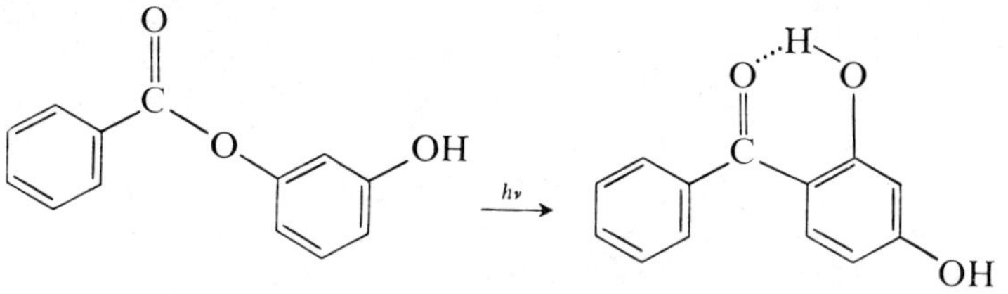

(3.41)

3. Combinations of Photostabilizers and Antioxidants

Since photooxidation and thermal oxidation often take place simultaneously during weathering, effective protection is obtained with combinations of photostabilizers and antioxidants. The combined effect of an ultraviolet stabilizer and an antioxidant has been shown[128] to exceed the protection anticipated from the effectiveness of the individual components. This phenomenon, referred to as synergism, is discussed in detail in Chapter 2.

Since most antioxidants alone have little or no effect in photooxidation, it is perhaps logical to assume that their contribution to outdoor protection is simply the result of inhibition of the concurrent thermal degradation. However, to the extent that free-radical reactions contribute to the overall mechanism of photooxidation, antioxidants could provide photostabilization in the absence of any thermal degradation.

The exceptional effectiveness of carbon black in protecting polymers against photooxidation has been emphasized, and it was shown in Chapter 2 that this pigment is also a good thermal antioxidant. Thus carbon black probably functions much like the combination of a photostabilizer and an antioxidant.

4. Polymer Purification and Modification

Impurities that reduce the photostability of polymers can be foreign materials incorporated during synthesis or fabrication, or they can be an integral part of polymer molecules. The first type of impurity includes metals and their various compounds, many of which are catalysts in photooxidation. The effect of traces of manganous stearate on the photooxidation of polyethylene containing carbon black (Table 4.8) is typical. The catalytic effect of copper

Table 4.8 Photooxidation of Black[a] Polyethylene Films[128]

Added Concentration of Manganous Stearate (%)	Embrittlement Time[b] (Weeks)
0	50
0.05	12
0.1	9
0.2	3

[a] Films contained 3% carbon black (Witco 100).
[b] Outdoors in summer at Kingsport, Tenn.

and other transition metals in thermal oxidation has been discussed in Chapter 2. It has been suggested[142] that metal catalysis results from an increase in the rate at which hydroperoxides are decomposed into radicals. Thus the removal of reactive metals can increase oxidative stability either by suppressing photooxidation or by reducing the rate of concurrent thermal degradation.

Even the complete removal of foreign impurities does not always result in a chemically pure polymer. Many polymers contain chemical groupings that should not be in the ideal structure. Traces of these foreign groups can reduce the resistance of many polymers to photooxidation. The effect of carbonyl groups in sensitizing polyethylene to photooxidation, discussed in Section C.3.a, is the classical example.

Elimination or reduction in the amount of impurities, either foreign or as part of polymer molecules, is a practical approach to stabilization. In some polymers this requires increased control over the synthetic process, but most reactions leading to polymer sensitization occur during fabrication. Thus the stability of many commercial polymers could be improved by more careful control during these processes. For example, the exclusion of oxygen would minimize the formation of carbonyl and other sensitizing groups. Reduction of the contamination with metal particles, which occurs during some fabrication steps, is also important in attaining the maximum level of stability. It is not the intent of this review to discuss details of possible purification methods for individual polymers, but rather to point out areas within which effective improvement could be attained.

There is a continuing effort to synthesize polymers whose structures impart a high degree of stability to photooxidation. Obviously these would have structures as free as possible of chemical groups that absorb damaging radiation. Within this concept, one should include polymers in which stabilizers are incorporated in the structure during polymerization or are grafted to the polymer in a subsequent reaction.

E. FACTORS GOVERNING CHOICE OF STABILIZER

The basic requirements for ultraviolet stabilizers that operate as absorbers are the following: (*a*) ability to absorb ultraviolet light and dissipate the energy in a nondegradative form, (*b*) stability of the compound itself toward ultraviolet light, and (*c*) a relatively low absorption in the visible region if color is undesirable. In addition to these requirements, the choice of a suitable stabilizer depends upon a complex interaction of the above criteria and other factors such as type of polymer, use of polymer, compatibility of stabilizer and polymer, and stabilization mechanism.

It becomes apparent quickly to the worker in the field that, although our knowledge of the basic mechanisms of photodegradation and stabilization has improved considerably in recent years, the selection of a particular stabilizer for a specific polymer application still is very likely to be an educated guess based on a consideration of the factors described here.

1. Type of Polymer

Because of the diversity of polymer degradation and stabilization mechanisms, it is not altogether surprising that a given stabilizer, effective in one polymer, may be ineffective or even a sensitizer in another polymer system. For example, a nickel complex of 2,4-pentanedione was reported to increase the outdoor stress-crack life of polypropylene while reducing that of polyethylene.[142] Since these complexes presumably function as triplet quenchers,[140] it is possible that the difference in effectiveness in the two polymers is a reflection of the variation in the mechanism of oxidative photodegradation. Some typical cases are given here.

a. Polyolefins. Because the polyolefins are so light sensitive, until recently only carbon black could protect them adequately. Of course there is one important drawback to its use, namely, the black color in the product. Now, however, hydroxybenzophenones, such as 2-hydroxy-4-octoxybenzophenone provide substantial protection while imparting relatively little color. In addition to the hydroxybenzophenones, the benzotriazoles, substituted acrylonitriles, and salicylates are useful in stabilization of the polyolefins. The metal-chelated nickel complexes also seem particularly useful for dyed polypropylene.

As mentioned in the section on mechanisms of photodegradation, the yellowing of polystyrene after exposure to ultraviolet light appears to be a surface phenomenon. To eliminate it, one needs an absorber with a high extinction coefficient. Otherwise a higher concentration of stabilizer is required, which may be uneconomical. Recent studies[143] have shown that a combination of the ultraviolet absorber with a phenol may solve this problem.

b. Poly(vinyl chloride). The benzophenones (especially the substituted alkyl types) benzotriazoles, and acrylonitriles also are used to stabilize poly(vinyl chloride). However, because of the interplay between the thermal, oxidative, and photooxidative processes, other stabilizers also are useful, particularly in combination with one of the above stabilizers. For example, although tin compounds are usually employed as heat stabilizers, they exhibit some light stabilization, particularly in rigid poly(vinyl chloride).

c. Cellulosics. Resorcinol monobenzoate is perhaps the most widely used stabilizer for cellulosics, although carbon black, the benzotriazoles, and benzophenone also are effective. McBurney and Evans[144,145] have shown that phenols (chain breakers) as well as sulfur-containing compounds (peroxide decomposers) are effective against both thermal and ultraviolet degradation. Other studies[146] show that combinations of ultraviolet absorbers and antioxidants enhance greatly the resistance to oxidative photodegradation.

2. Thickness of Fabricated Polymer Product

It may not be generally recognized that the best photostabilizer for a thick article is not necessarily the most suitable one for a thin product of the same polymer. The latter case often requires a stabilizer whose most important characteristic is not its inherent protective efficiency but rather its immobility in the polymer, which prevents its volatilization, migration, or leaching. Thus it would appear that a high-molecular-weight stabilizer is desirable for thin films and fibers. In fact, a polymeric analog of phenyl salicylate, poly-(4-acetyl-β-resorcylic acid) has been reported to be superior to the simple ester for stabilizing cellulose ester films.[147]

Migration of the stabilizer may be a desirable property in thick articles, however, since stabilizer in the surface layers is exhausted during the exposure and therefore additional stabilizer can reach the surface and provide further protection. The data in Table 4.9 illustrate the effect of thickness on the

Table 4.9 Effect of Sample Thickness on Ultraviolet Degradation of Cellulose Acetate Butyrate Protected with Phenyl Salicylate[148]

Sample Thickness (in.)	Outer Layer Degradation (% molecular weight decrease)
0.125	20
0.100	30
0.075	46
0.050	85
0.025	70

stability of the top layer of a sample of cellulose acetate butyrate, which is stabilized with 1% of phenyl salicylate.

3. Color

When absence of color or of color development during exposure is important in a particular application, then some compromise with maximum possible stabilization efficiency generally is necessary. There are no known colorless or noncoloring stabilizers that are even half as effective as the best colored stabilizers. For example, cellulose acetate butyrate that was stabilized with 1% of 6,13-dichloro-3,10-diphenyl-triphenodioxazine (reddish-brown) lasted

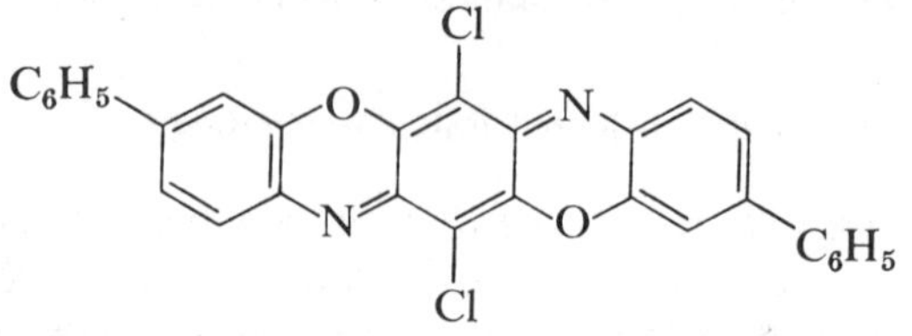

more than nine times as long (Table 4.4) as a sample that was stabilized with one percent of the colorless 2,4-dihydroxybenzophenone.[128]

4. Compatibility

Unlike the ester-containing polymers, such as poly(methyl methacrylate), or the cellulosics, polyethylene and polypropylene have low solvating power for stabilizers such as the benzophenones. For example, it can be seen from the data in Table 4.10 that the more symmetrical compounds with high melting

Table 4.10 Solubilities and Compatibilities of Ultraviolet Absorbers[149]

Substituents in Benzophenone	Solubility in Hexane (weight %)	Melting Point (°C)	Rate of Migration (mg/day)	Compatibility in Polyethylene (at 0.2%)
2-Hydroxy-4-methoxy	4.4	62	—	No bloom
2,2′-Dihydroxy-4-methoxy	2.3	71	0.02	No bloom
2,2′-Dihydroxy-4,4′-dimethoxy	0.1	136	0.08	Bloom
2,2′,4,4′-Tetrahydroxy	0.01	200	—	Slight haze (no ultraviolet absorption)

points have the greater tendency to bloom to the surface of polyethylene. The importance of a solubilizing group is clearly demonstrated by the studies of Coleman[149] on the 2,2′-dihydroxybenzophenones in polyethylene (Table 4.11). Although the 4-methoxy derivative is virtually ineffective, the high homologs are excellent stabilizers even though the change in alkyl group has little effect on the absorption spectrum.

Although exudation of the stabilizer during fabrication, storage, or exposure is usually considered to be an undesirable characteristic, sometimes exudation actually may be advantageous in order to provide maximum protection. Such apparently is the case in polyethylene aerial cables containing

Table 4.11 Effect of Varying the Alkoxyl Group in 2,2′ Dihydroxy-4-Alkoxybenzophenones[148]

OH HO

CO OR

R—	Concentration (%)	Carbonyl (%) Fadeometer 500 hr		Arizona 2 months		Retained Elongation[a] (%) 2 months	
		500 hr	1000 hr	2 months	4 months	2 months	4 months
CH_3	0.1	0.22	0.54	0.41	0.6	7	0
C_2H_2	0.1	0.07	0.37	0.25	0.6	20	5
C_3H_{17}	0.1	0.04	0.19	0.08	0.6	44	4
$C_{12}H_{25}$	0.1	0.04	0.17	0.33	0.6	61	3
No additive	—	0.30	0.64	0.33	0.6	9	0

[a] Arizona.

exuding concentrations of ultraviolet absorber, which were reported[150] to give improved performance by virtue of the self-renewing surface protectant.

F. PROSPECTS

In concluding this chapter, it is worthwhile to discuss the importance of photostabilization and to look toward the future of this very active area of polymer science.

1. Importance of Photostabilization

The boom in ultraviolet stabilizers that began in the mid-1960s continued to new sales records in 1968. It is estimated that 2.5 million lb of ultraviolet stabilizers (Table 4.12) were sold in 1968. Of this total, more than 60% went into polyolefins, particularly polypropylene. This increase in sales reflects the fact that about one-fourth of all plastics produced in the U.S. have outdoor uses. It also reflects a certain degree of success by ultraviolet stabilizer producers in improving the efficiency of their products, with accompanying greater acceptability to the consumer. As a result of the increased volume, prices of the major-volume stabilizers have continued their downward trend.[151]

Table 4.12 Plastics' Use of Ultraviolet Stabilizers (in thousand lb)[151]

Markets	1964	1965	1966	1967	1968
Polyolefins	185	400	700	950	1550
PVC	180	210	250	245	265
Polyesters	125	135	150	165	165
Cellulosics	110	130	145	150	155
Polystyrenes	100	110	125	135	140
Coatings, and so on	50	80	100	115	125
Other resins	40	50	55	70	100
Totals[a]	790	1115	1525	1830	2500

[a] Totals for 1967 and 1968 include salicylates.

Optical brighteners are not classified as ultraviolet stabilizers (although some may function as such—see Section D.2.a), but they overcome some of the yellowing effects of photodegradation by acting as "aesthetic" stabilizers. In 1958, the total sales of these additives was about $4 million, which tripled the sales recorded as recently as 1965.[151]

2. Future Trends

Much of recent polymer research has been directed toward the development of new polymers having excellent strength, toughness, and flexibility, but only a modest effort has been directed toward achieving outdoor durability. With the better understanding of the mechanisms of photodegradation and photostabilization, there undoubtedly will be an increased effort to produce polymers with increased photoresistance and also more effective stabilizers. However more knowledge is needed about the primary and secondary reactions involved in oxidative photodegradation. It should be possible to measure the quantum yields of primary reactions with the advanced techniques of photochemistry, and although model systems already have received increased attention, their use will facilitate the separation of the factors involved in the complex photodegradative process. There is, of course, much to be done in the development of new stabilizers. Perhaps one of the future trends will be in the development of excited-singlet-state quenchers.

Although this chapter has emphasized the undesirable aspects of oxidative photodegradation, some mention should be made of the positive side of these processes. As discussed earlier, one of the overall effects of oxidative photodegradation in some polymers is crosslinking. It may be possible to acquire some control over this process, with a resulting crosslinked product that is more photostable than the original polymer.

A novel, potentially useful photodegradative process has been reported by Guillet[152] and Scott[153] who suggested the fabrication of articles using easily photodegradable polymers in order to expedite their disposal after use.

Finally it is clear that the research field of oxidative photodegradation offers ample opportunity for the organic photochemist and the polymer chemist to collaborate in work that is not only intellectually stimulating but of practical importance as well.

ACKNOWLEDGMENTS

The writing of this chapter was facilitated greatly by the use of unpublished manuscript material supplied by Dr. J. W. Tamblyn. The author is indebted to Dr. Tamblyn for making this work available. In addition, Dr. W. L. Hawkins made several substantive contributions to the chapter, which are gratefully appreciated. Finally, the author thanks several of his colleagues, Drs. F. H. Winslow, L. D. Loan, A. A. Lamola, E. D. Feit, and G. N. Taylor for helpful comments.

REFERENCES

1. H. H. G. Jellinek, *Degradation of Vinyl Polymers*, Academic Press, New York, 1955.
2. W. L. Hawkins and F. H. Winslow, "Degradation and Stabilization," in R. A. V. Raff and K. W. Doak, Eds., *Crystalline Olefin Polymers*, Pt. II, Interscience, New York, 1964, Chapter 8.
3. S. H. Pinner, Ed., *Weathering and Degradation of Plastics*, Gordon and Breach, London, 1966.
4. M. R. Kamal, Ed., "Weatherability of Plastic Materials," *Appl. Polym. Symp.*, No. **4,** Interscience, New York, 1967.
5. M. B. Neiman, Ed., *Aging and Stabilization of Polymers*, (English transl.), Consultants' Bureau, New York, 1965.
6. G. Scott, *Atmospheric Oxidation and Antioxidants*, Elsevier, New York, 1965.
7. W. L. Hawkins, "Oxidative Degradation of High Polymers," in *Oxidation and Combustion Reviews*, C. F. H. Tipper, Ed., Vol. I, Elsevier, New York, 1965, p. 169.
8. Yu. A. Ershov, S. I. Kuzina, and M. B. Neiman, *Russ. Chem. Rev.*, **38,** 147 (1969) [*Usp. Khim.*, **38,** 289 (1969)].
9. W. A. Noyes, Jr., G. S. Hammond, and J. N. Pitts, Jr., Eds., *Advances in Photochemistry*, Interscience, New York, Vols. 1–7, 1963–1969.
10. J. G. Calvert and J. N. Pitts, Jr., *Photochemistry*, Wiley, New York, 1966.
11. N. J. Turro, *Molecular Photochemistry*, Benjamin, New York, 1965.
12. R. O. Kan, *Organic Photochemistry*, McGraw-Hill, New York, 1966.
13. D. C. Neckers, *Mechanistic Organic Photochemistry*, Reinhold, New York, 1967.
14. A. A. Lamola and N. J. Turro, Eds., *Energy Transfer and Organic Photochemistry*, Interscience, New York, 1969.

15. R. B. Fox and R. F. Cozzens, *Macromolecules*, **2,** 181 (1969), and references therein.
16. F. H. Winslow, W. Matreyek, and A. M. Trozzolo, *ACS Polym. Preprints*, **10** (2) 1271 (1969).
17. V. T. Wallder, J. B. DeCoste, J. B. Howard, and W. J. Clarke, *Ind. Eng. Chem.*, **42,** 2320 (1950).
18. F. S. Johnson, *J. Meteorol.*, **2,** 431 (1954).
19. R. E. Barker, Jr., *Photochem. Photobiol.*, **7,** 275 (1968).
20. L. R. Koller, *Ultraviolet Radiation*, 2nd ed., Wiley, New York, 1965, Chapter 4.
21. M. Luckiesh, *Germicidal, Erythemal and Infrared Energy*, Van Nostrand, New York, 1946, p. 52.
22. P. Beuer, Contract AF 61(052)-618 Tech. Note 2.
23. W. W. Coblentz and R. Stair, *J. Res. Natl. Bur. Std.* (*U.S.*), **33,** 21 (1944).
24. M. G. Chan and W. L. Hawkins, *ACS Polym. Preprints*, **9** (2), 1638 (1968).
25. R. C. Hirt and N. Z. Searle, in *Appl. Polym. Symp.*, *No.* **4,** M. R. Kamal, Ed., Interscience, New York, 1967, p. 61.
26. L. Pauling, *Nature of the Chemical Bond*, 3rd ed., Cornell University Press, Ithaca, N.Y., 1960.
27. J. R. McNesby and H. Okabe, *Advan. Photochem.*, **3,** 157 (1965).
28. R. P. Wayne, *Advan. Photochem.*, **7,** 400 (1969).
29. A. V. Munck and J. R. Scott, *Nature*, **177,** 587 (1956).
30. D. F. Evans, *J. Chem. Soc.*, **345** (1953).
31. H. Tsubomura and R. S. Mulliken, *J. Amer. Chem. Soc.*, **82,** 5966 (1960).
32. G. J. Hoijtink, *Acct. Chem. Res.*, **2,** 114 (1969).
33. J. C. W. Chien, *J. Phys. Chem.*, **69,** 4317 (1965).
34. D. J. Carlsson and J. C. Robb, *Trans. Faraday Soc.*, **62,** 3403 (1966).
35. J. C. Betts and J. C. Robb, *Trans. Faraday Soc.*, **64,** 2402 (1968).
36. D. F. Evans, *J. Chem. Soc.*, **1735** (1960).
37. K. Kawaoka, A. U. Khan, and D. R. Kearns, *J. Chem. Phys.*, **46,** 1842 (1967).
38. F. H. Winslow and W. L. Hawkins, in *Appl. Polym. Symp.*, *No.* **4,** M. R. Kamal, Ed., Interscience, New York, 1967, p. 29.
39. C. Kujirai, S. Hashiya, K. Shibuya, and K. Nishio, *Chem. High Polymers* (*Tokyo*) **25,** 193 (1968).
40. N. Uri, *Advan. Chem. Ser.*, **36,** 102 (1962).
41. A. R. Burgess, *Natl. Bur. Std.* (*U.S.*), *Circ.*, **525,** 149 (1953).
42. R. G. W. Norrish and M. H. Searby, *Proc. Roy. Soc.* (*London*), **A237,** 464 (1956).
43. G. H. Hartley and J. E. Guillet, *Macromolecules*, **1,** 165, 413 (1968).
44. T. J. Dougherty, *J. Amer. Chem. Soc.*, **87,** 4011 (1965) and references therein.
45. P. J. Wagner and G. S. Hammond, *Advan. Photochem.*, **5,** 21 (1967).
46. F. O. Rice and E. Teller, *J. Chem. Phys.*, **6,** 489 (1938).
47. W. Davis, Jr., and W. A. Noyes, Jr., *J. Amer. Chem. Soc.*, **69,** 2153 (1947).
48. N. C. Yang, S. P. Elliott, and B. Kim, *J. Amer. Chem. Soc.*, **91,** 7551 (1969).
49. P. J. Wagner, *J. Amer. Chem. Soc.*, **89,** 5898 (1967).
50. E. J. Baum, J. K. S. Wan, and J. N. Pitts, Jr., *J. Amer. Chem. Soc.*, **88,** 2652 (1966).
51. P. Ausloos and R. E. Rebbert, *J. Amer. Chem. Soc.*, **4512** (1964).
52. G. R. McMillan, J. G. Calvert, and J. N. Pitts, Jr., *J. Amer. Chem. Soc.*, **86,** 3602 (1964).
53. D. R. Coulson and N. C. Yang, *J. Amer. Chem. Soc.*, **88,** 4511 (1966).
54. P. J. Wagner and G. S. Hammond, *J. Amer. Chem. Soc.*, **87,** 4009 (1965); **88,** 1245 (1966).
55. N. C. Yang and D. D. H. Yang, *J. Amer. Chem. Soc.* **80,** 2913 (1958).

56. J. E. Guillet and R. G. W. Norrish, *Proc. Roy. Soc.* (*London*), **A233,** 153 (1955).
57. K. F. Wissbrun, *J. Amer. Chem. Soc.*, **81,** 58 (1959).
58. K. R. Osborn, *ACS Div. Org. Coatings Plast. Chem., Gen Papers*, **21,** 411 (1961).
59. J. E. Guillet, J. Dhanraj, F. J. Golemba, and G. H. Hartley, *Advan. Chem. Ser.*, **85,** American Chemical Society, Washington, D.C., 1968, p. 272.
60. F. H. Winslow, W. Matreyek, A. M. Trozzolo, and R. H. Hansen, *ACS Polym. Preprints*, **9,** 377 (1968).
61. O. Cicchetti, *Advan. Polym. Sci.*, **7,** 70 (1970).
62. F. A. Hessel, *Matier. Grasses*, **16,** 7010 (1924); *Chem. Abstr.*, **19,** 888 (1925).
63. A. W. Pross and R. M. Black, *J. Soc. Chem. Ind.* (*London*), **69,** 113 (1950).
64. A. R. Burgess, *Chem. Ind.* (*London*), 78 (1952).
65. H. H. Jaffe and M. Orchin, *Theory and Applications of Ultraviolet Spectroscopy*, Wiley, New York, 1962.
66. R. C. Hirt, N. Z. Searle, and R. G. Schmitt, *SPE Trans.*, **1,** 21 (1961).
67. M. G. Chan and W. L. Hawkins, presented at the 9th National Meeting of the Soc. of Applied Spectroscopy, Oct. 9, 1970, New Orleans, La.
68. A. M. Trozzolo and F. H. Winslow, *Macromolecules*, **1,** 98 (1968).
69. C. H. Bamford and M. J. S. Dewar, *Proc. Roy. Soc.* (*London*), **A198,** 252 (1949).
70. M. L. Kaplan and P. G. Kelleher, *ACS Polym. Preprints*, **12,** 2, in press, 1971.
71. K. R. Kopecky and H. J. Reich, *Can. J. Chem.*, **43,** 2265 (1965).
72. B. Schnuriger and J. Bourdon, *Photochem. Photobiol.*, **8,** 361 (1968).
73. R. H. Partridge, *J. Chem. Phys.*, **45,** 1679 (1966).
74. Y. Kato, D. J. Carlsson, and D. M. Wiles, *J. Appl. Polym. Sci.*, **13,** 1447 (1969).
75. L. Dulog, E. Radlman, and W. Kern, *Makromol. Chem.*, **60,** 1 (1963).
76. C. R. Boss and J. C. W. Chien, *J. Polym. Sci., Part A-1*, **4,** 1543 (1966).
77. D. J. Carlsson and D. M. Wiles, *Macromolecules*, **2,** 587 (1969).
78. Ref. 77, p. 597 (1969).
79. F. H. Winslow and W. Matreyek, unpublished results, 1970.
80. L. A. Matheson and R. F. Boyer, *Ind. Eng. Chem.*, **44,** 867 (1952).
81. B. G. Achhammer, M. J. Reiney, and F. W. Reinhart, *J. Res. Natl. Bur. Std.* (*U.S.*), **47,** 116 (1951).
82. B. G. Achhammer, M. J. Reiney, L. A. Wall, and F. W. Reinhart, *J. Polym. Sci.*, **8,** 555 (1952).
83. N. Grassie and N. A. Weir, *J. Appl. Polym. Sci.*, **9,** 987 (1965).
84. M. Tryon and L. A. Wall, in W. O. Lundberg, Ed., *Autoxidation and Antioxidants*, Vol. II, Interscience, New York, 1961, p. 919.
85. B. G. Achhammer, M. J. Reiney, L. A. Wall, and F. W. Reinhart, *Natl. Bur. Std.* (*U.S.*) *Circ.*, **525,** 205 (1953).
86. L. A. Wall, M. R. Harvey, and M. Tryon, *J. Phys. Chem.*, **60,** 1306 (1956).
87. L. A. Wall and M. Tryon, *Nature*, **178,** 101 (1956).
88. L. A. Wall and D. W. Brown, *J. Phys. Chem.*, **61,** 129 (1956).
89. N. Grassie and N. A. Weir, *J. Appl. Polym. Sci.*, **9,** 999 (1965).
90. A. M. Trozzolo and S. R. Fahrenholtz, Abstracts of Papers, 155th Meeting, American Chemical Society, San Francisco, March 1968, p. 138.
91. J. Morand, *Rubber Chem. Technol.*, **39,** 537 (1966).
92. A. F. Postovskaya and A. S. Kuzminskii, *Rubber Chem. Technol.*, **29,** 598 (1956).
93. E. H. Farmer and A. Sundralingarn, *J. Chem. Soc.*, **125** (1943).
94. A. Thac and V. Kellö, *Rubber Chem. Technol.*, **28,** 383, 968, 989 (1955).
95. J. Morand, *Rev. Gen. Caoutchouc Plast.*, **45,** 615, 999 (1968).
96. E. J. Hart and M. S. Matheson, *J. Amer. Chem. Soc.*, **70,** 784 (1948).

97. L. Bateman, *Trans. Faraday Soc.*, **42,** 267 (1946).
98. L. Bateman and G. Gee, *Proc. Roy. Soc.* (*London*), **A195,** 376, 391 (1948).
99. L. Bateman and G. Gee, *Trans. Faraday Soc.*, **47,** 155 (1951).
100. M. A. Golub, *J. Polym. Sci.*, **25,** 373 (1957).
101. L. Bateman, *J. Polym. Sci.*, **2,** 1 (1947).
102. A. F. Postovskaya and A. S. Kuz'minskii, *Rubber Chem. Technol.*, **25,** 872 (1952).
103. M. L. Kaplan and P. G. Kelleher, *J. Polym. Sci.*, *part A-1*, *8*, 3163 (1970).
104. R. W. Murray and M. L. Kaplan, *J. Amer. Chem. Soc.*, **90,** 4161 (1968).
105. G. S. Egerton, *Nature*, **204,** 1153 (1964).
106. G. S. Egerton, *J. Soc. Dyers Colour.*, **65,** 764 (1949).
107. W. Jasching, *Kunstoffe—Plastics*, **52,** 458 (1962).
108. G. P. Mack, *Mod. Plastics*, **31,** 150 (1953).
109. W. C. Geddes, *Rubber Chem. Technol.*, **40,** 177 (1967).
110. A. L. Scarborough, W. L. Kellner and P. W. Rizzo, *Natl. Bur. Std.* (*U.S.*) *Circ.*, **525,** 95 (1953).
111. S. Strauss and L. A. Wall, *J. Res. Natl. Bur. Std.* (*U.S.*), **69,** 39 (1958).
112. R. A. Ford, *Nature*, **176,** 1023 (1955).
113. G. Kroeş, *Rec. Trav. Chim. Pays-Bas*, **82,** 979 (1963).
114. W. Sharkey and W. Mochel, *J. Amer. Chem. Soc.*, **81,** 3000 (1959).
115. H. M. Heuvel and K. C. J. B. Lind, *J. Polym. Sci.*, *Part A-2*, **8,** 401 (1970).
116. A. R. Monahan, *Macromolecules*, **1,** 408 (1968).
117. L. F. McBurney, in E. Ott, H. M. Spurlin, and M. W. Grafflin, Eds., *Cellulose and Cellulose Derivatives*, Interscience, New York, 1954, pp. 140, 168.
118. J. Bourdon and B. Schnuriger, "Photosensitization of Organic Solids," in D. Fox, M. M. Labes, and A. Weissberger, Eds., *Physics and Chemistry of the Organic Solid State*, Vol. III, Interscience, New York, 1967, Chapter 2.
119. V. T. Wallder, *AIEE Elec. Eng.*, (*Soc. Plastics Engrs.*), **71,** 59 (1952).
120. F. J. Golemba and J. E. Guillet, *SPE* (*Soc. Plastics Engrs.*) *J.*, **26,** 88 (1970).
121. R. L. Horton and H. G. Brooks, Jr., U.S. Patent 3,313,866, April 11, 1967; *Chem. Absts.*, **67,** 12143v (1967).
122. J. P. Luongo and H. Schonhorn, *J. Polym. Sci.*, *Part A-2*, **6,** 1649 (1968).
123. C. E. Rogers, J. A. Meyer, V. Stannett and M. Szwarc, *TAPPI Monogr. Ser.*, **23,** 12 (1962).
124. V. T. Wallder, W. J. Clarke, J. B. DeCoste, and J. B. Howard, *Ind. Eng. Chem.*, **42,** 2320 (1950).
125. F. H. McTigue and M. Blumberg, "Factors Affecting Light Resistance of Polypropylene," in M. R. Kamal, Ed., *Appl. Polym. Symp.*, No. **4,** Interscience, New York, 1967, p. 175.
126. W. L. Hawkins and F. H. Winslow, "Antioxidant Properties of Carbon Black," in G. Kraus, Ed., *Reinforcement of Elastomers*, Interscience, New York, 1965, p. 563.
127. W. L. Hawkins, M. A. Worthington, and F. H. Winslow, *Proc. 4th Conf. Carbon*, 63 (1960).
128. J. H. Chaudet, G. C. Newland, H. W. Peters, and J. W. Tamblyn, *SPE* (*Soc. Plastics Engrs.*) *Trans.*, **1,** 26 (1961).
129. E. J. Baum, J. K. S. Wan, and J. N. Pitts, Jr., *J. Amer. Chem. Soc.*, **88,** 2652 (1966).
130. A. Weller, in G. Porter, Ed., *Progress in Reaction Kinetics*, Vol. 1, Pergamon, New York, 1961, p. 196.
131. A. Heller, *Mol. Photochem.*, **1,** 257 (1969).
131a. E. J. O'Connell, *J. Amer. Chem. Soc.*, **90,** 6550 (1968).
132. A. Beckett and G. Porter, *Trans. Faraday Soc.*, **59,** 2051 (1963).

133. G. S. Hammond, N. J. Turro, and P. A. Leermakers, *J. Phys. Chem.*, **66,** 1144 (1962).
134. A. A. Lamola and L. J. Sharp, *J. Phys. Chem.*, **70,** 2623 (1966).
135. J. R. Merrill and R. G. Bennett, *J. Chem. Phys.*, **43,** 1410 (1965).
136. J. A. Melchore, *Ind. Eng. Chem. Prod. Res. Div.*, **1,** 232 (1962).
137. R. G. Schmitt and R. C. Hirt, *J. Appl. Polym. Sci.*, **7,** 1565 (1963).
138. H. H. Hormann, *Ind. Eng. Chem. Prod. Res. Div.*, **5,** 92 (1966).
139. M. Heskins and J. Guillet, *Macromolecules*, **3,** 224 (1970).
140. P. J. Briggs and J. F. McKellar, *J. Appl. Polym. Sci.*, **12,** 1825 (1968).
141. V. I. Stenberg, "Photo-Fries Reaction and Related Rearrangements," in O. L. Chapman, Ed., *Organic Photochemistry*, Vol. 1, Dekker, New York, 1967, p. 127.
142. G. C. Newland and J. W. Tamblyn, *J. Appl. Polym. Sci.*, **9,** 1947 (1965).
143. C. Savides, J. A. Stretanski, and L. R. Costello, *Advan. Chem. Ser.*, **85,** American Chemical Society, Washington, D.C., 1968, p. 287.
144. E. F. Evans and L. F. McBurney, *Ind. Eng. Chem.*, **41,** 1256 (1949).
145. L. F. McBurney, *Natl. Bur. Std.* (*U.S.*) *Circ.*, **525,** 191 (1953).
146. J. B. Batdorf and G. M. Gantz, *Offic. Dig.*, **28,** 65 (1965).
147. D. J. Shields, C. J. Kibler, and R. M. Schulken, U.S. 2,856,305 (October 14, 1958); *Chem. Abstr.* **53,** 2689 (1959).
148. G. C. DeCroes and J. W. Tamblyn, *Mod. Plastics*, **29,** No. 4, 127 (1952).
149. R. A. Coleman, *ACS Chem., Papers*, **19,** 45 (1959).
150. R. C. Harrington, Jr., presented at IEEE Summer Power Meeting, New Orleans, La., July 1966.
151. *Mod. Plastics*, **46** (9), 87 (1969).
152. *Chem. Eng. News*, **48** (20), May 11, 1970, p. 61.
153. *Chem. Eng. News*, **48** (30), July 27, 1970, p. 11.

5

PREVENTION OF DEGRADATION BY OZONE

R. W. MURRAY

University of Missouri, St. Louis, Missouri

A. INTRODUCTION

Any discussion of the stabilization of polymers against ozone deals mainly with the stabilization of polymers containing appreciable amounts of residual unsaturation. From a polymer deterioration viewpoint, such materials are generally confined to the class of polymers known as elastomers.

Ozone has been used in a positive sense in the area of saturated polymers most notably in graft polymerization[1,2] and to initiate crosslinking.[3,4] There are, however, two closely allied areas in which ozone does appear to play a role in the deterioration of saturated polymers. In both cases the chemistry involved appears ultimately to be largely free radical in character. Stabilization against such degradation is then attempted, using antioxidants of the type described in Chapter 2. For the sake of completeness in considering ozone reactions with polymers, these deterioration pathways are mentioned briefly here.

Ozone is known to react with virtually all organic materials. The rate of attack of ozone on saturated materials is considerably slower than on unsaturated materials, but such reactions probably do occur to such an extent that they can serve as the initiation reaction in autoxidation. Ozone-initiated autoxidation has probably not received the attention it deserves. Although it is true, as discussed below, that ozone normally occurs in small concentrations in the atmosphere, such concentrations are probably sufficient to cause initiation reactions. In addition, polluted atmospheres, particularly those subject to photochemical smog, have considerably higher ozone concentrations. Likewise, since oxidative photodegradation studies are frequently carried out under conditions where ozone concentrations can be expected to be relatively high, ozone-initiated reactions should not be neglected in interpreting the results of such studies.

As early as 1939 Durland and Adkins[5] described the oxidation of *cis*- and *trans*-decalin with ozone. Schubert and Pease[6] studied the gas-phase oxidation of alkanes with ozone. Ozone oxidation of alkanes has also been reported by Long and Fieser.[7] More recently ozone oxidations of other saturated hydrocarbons, including adamantane[8] and cyclohexane[9] have been investigated in detail. Such reports suggest that ozone-initiated autoxidation should be considered when the mechanism of saturated polymer degradation is discussed.

The second area in which ozone may play a role in polymer degradation, even though it is not directly involved in a reaction with unsaturation, has only very recently been discussed. It now seems quite likely that certain reactions of ozone with organic materials, including polymers, can serve as sources of singlet oxygen.[10–14] This species is electronically excited, long-lived, and capable of undergoing very specific chemical reactions. Thus ozone may be responsible for a secondary degradation reaction in addition to the primary reaction of ozone with the polymeric substrate. A discussion of the chemistry of singlet oxygen is not within the scope of this chapter, but the reader is referred to some recent reviews for a further description of this subject.[15–16]

The possible role of singlet oxygen in one of the fundamental processes involved in oxidative photodegradation of polyethylene has recently been discussed by Trozzolo and Winslow.[17] In this case the singlet oxygen is postulated as arising from the quenching of n–π^* triplet states of ketone groups by molecular oxygen (Chapter 4).

1. Ozone in the Exposure Environment

The observation of ozone in the environment has an early history. When Homer, in the *Iliad*[18] and the *Odyssey*,[19] referred to a freshness after a storm, he was describing what today we call the odor of ozone. The first to undertake a study of the chemical properties of ozone was Schönbein.[20] In fact it was Schönbein who gave the name ozone from the Greek, to smell, to the peculiar odor that had been detected much earlier by so many others.

Ozone is a natural constituent of the atmosphere in concentrations varying approximately from 0 to 10 pphm (parts per hundred million). The local concentration varies over the surface of the earth and is a function of meteorological conditions as well as the makeup of the environment. Ozone is produced in the upper atmosphere by ultraviolet photolysis of oxygen. The oxygen atoms thus produced combine with oxygen molecules to give ozone. The wavelengths required are in the 1100 to 2200 Å region. This process has produced an ozone layer in the upper atmosphere at altitudes between 12 and 22 miles, and concentrations as high as 500 pphm occur in this layer. This concentrated layer acts as an efficient absorber of solar short-wave ultraviolet radiation ($<$3000 Å) and so serves to protect life at the earth's surface from the harmful effects of this radiation. In addition the absorption of these wavelengths in the ozone layer makes this a warm layer, for both the absorption and the exothermic decomposition of ozone occur.

Surface ozone is the result of winds which bring ozone, which has diffused to the troposphere, down to the earth's surface.[21] The observed surface concentration is then the result of the balance attained between this input of ozone and that which is destroyed by reaction with materials in the environment. These destructive reactions normally occur with organic materials such as trees and houses. The local concentration can vary widely and weather disturbances can lead to unusually high local concentrations. Ozone concentrations are also normally higher in coastal areas, presumably because the environment contains fewer materials that can decompose ozone.

Even these rather low normal concentrations of ozone cannot be ignored as far as stabilization of elastomers is concerned, but a far greater problem is posed by the alarmingly high ozone concentrations being attained more and more frequently in polluted atmospheres. Perhaps the most extreme example is in the Los Angeles area, where ozone concentrations reach 40 to 100 pphm[22] during periods of severe photochemical smog. There are now indications that this kind of pollution is occurring in more and more urban areas. Such high ozone concentrations have added greater emphasis to the problem of stabilization against ozone deterioration.

In polluted areas ozone concentrations are not reported as such but are included in a total oxidant level. Ozone frequently represents some 90% of such concentrations, however. It is also important to note that peroxidic fractions of photochemical smog do not, themselves, cause cracking in rubber. Such materials have been trapped out of smog and have failed to cause cracking in separate tests.[23]

The mechanism of production of ozone in polluted atmospheres has been a matter of much study. Haagen-Smit has demonstrated that mixtures of low-molecular-weight hydrocarbons and nitrogen dioxide can generate ozone when photolyzed.[24,25] It seems likely that the process of photolysis of nitrogen dioxide to nitric oxide and oxygen atoms with subsequent combination of these oxygen atoms with molecular oxygen to give ozone is the additional source of ozone in polluted atmospheres:

$$NO_2 + h\nu \rightarrow NO + O$$
$$M + O + O_2 \rightarrow O_3 + M$$

The nitrogen dioxide in turn has its origin in nitric oxide emitted in automobile exhaust. The exact mechanism of the conversion of nitric oxide to nitrogen dioxide has also received a great deal of attention. The generally accepted scheme[26,27] for this conversion calls for the reaction of an acyl radical with molecular oxygen to give a peroxyacyl radical. This species then combines with nitric oxide to give a peroxyacyl nitrite, which is then photolyzed further to an acyloxy radical and nitrogen dioxide. This nitrogen dioxide then serves as the precursor to increased ozone concentrations as described above.

$$R{-}\overset{\overset{\displaystyle O}{\|}}{C}\cdot + O_2 \longrightarrow R{-}\overset{\overset{\displaystyle O}{\|}}{C}{-}O{-}O\cdot$$
$$R{-}\overset{\overset{\displaystyle O}{\|}}{C}{-}O{-}O\cdot + NO \longrightarrow R{-}\overset{\overset{\displaystyle O}{\|}}{C}{-}O{-}O{-}N{-}O$$
$$R{-}\overset{\overset{\displaystyle O}{\|}}{C}{-}O{-}O{-}N{-}O \longrightarrow R{-}\overset{\overset{\displaystyle O}{\|}}{C}{-}O\cdot + NO_2$$

The origin of the acyl radicals in the above scheme has been a subject of separate study. They may arise from a series of photooxidation reactions.[26] More recently it has been suggested[28] that they may arise in atmospheres containing olefinic hydrocarbons through a reaction of singlet oxygen with the olefinic materials.

The combination of normal ozone concentrations with other sources of ozone found in polluted atmospheres continues to represent a problem to the chemist or technologist concerned with stabilization of elastomers. In the next section we consider this deleterious ozone attack.

2. The Deterioration Process

a. General Observations. Although today it is widely recognized that the unique form of degradation of elastomers under stress, in which cracks are formed perpendicular to the direction of stress, is caused by ozone, this realization has not always been the case. In fact recognition that ozone was the culprit in such cases came only slowly. This slow recognition was probably caused by several factors, the most important being the extremely low concentrations of ozone necessary to cause the observed deterioration. Although the presence of ozone as a minor constituent of normal atmospheres was recognized, it was generally felt that the levels present could not be responsible for the severe cracking observed. Thompson[29] first reported that ozone generated in the laboratory could crack stretched, vulcanized rubber. On the basis of extensive work on the problem of cracking, Newton[30] finally concluded that the essential factors are ozone and strain in the sample.

Part of the difficulty in recognizing ozone as a separate problem in deterioration was the necessity to recognize and distinguish the physical manifestation of the problem. As mentioned above, cracks caused by ozone occur perpendicular to the direction of stress, and on that basis they can be distinguished from other kinds of oxidative degradation. These latter types of deterioration are variously described as sun checking, sun cracking, light-oxidized cracking, alligatoring, crazing, and mud cracking. In general, these phenomena are the result of photooxidation, perhaps sometimes occurring along with ozone cracking. It has been shown that the small square crack patterns on tire sidewalls can be reproduced by ozone when a rubber sample is simultaneously stressed in two directions.[31] Thus it is apparent that ozone can cause deterioration, with a physical appearance that might have led to its classification as mud cracking, with photooxidation assumed as the source.

Another factor that slowed the recognition of ozone as a separate deterioration source is the effect of light. Newton[30] has indicated that many workers concluded that light was essential to the cracking process when in fact the experimental procedures that led to this conclusion excluded both light and ozone. In one case it was concluded that ultraviolet light caused the observed cracking because a sample that was covered with black paper did not crack.[32] Williams had shown that light can actually decrease the extent of cracking, presumably by forming an oxidized layer on the rubber surface that protects the underlying material.[33] Likewise, Zuev[34] had found that prior exposure to ultraviolet irradiation can increase the ozone resistance of rubber samples. Perhaps the most important evidence obtained against the light-effect idea was that of Van Rossem and Talen[35] and Potter,[36] who found that samples exposed only at night cracked in the same way as those exposed during daylight.

Today the separate effect of ozone is well recognized and the general phenomenon is known as ozone cracking, following the suggestion of Newton.[30] Other kinds of oxidative degradation are generally referred to as crazing. Cracks caused by ozone are usually deeper than those assigned to crazing and, as mentioned earlier, the former always occur perpendicular to the direction of stress. Also, crazing can occur even in the absence of stress, but both stress and ozone are necessary for ozone cracking. Ozone certainly must react with unstressed rubber, but no cracking occurs. Likewise, a rubber sample that has been exposed to ozone while unstressed will not crack when subsequently subjected to stress.[35]

b. The Ozonolysis Reaction. Since the cracking problem is associated only with elastomers containing some degree of residual unsaturation, it seems reasonable to assume that the process is somehow related to the attack of ozone on the unsaturation. The general process of attack of ozone on unsaturation is referred to as ozonolysis, as distinguished from ozonization or ozonation. The ozonolysis reaction, and particularly the mechanism of the reaction, has received considerable study.

It is felt that the cracking phenomenon as well as measures taken to prevent cracking are both intimately related to an understanding of the mechanism of ozonolysis, and that area is reviewed here in some detail. The study of this reaction apparently began with the work of Schönbein[37] who reported in 1855 that ozone reacts with ethylene to give carbonic acid, formaldehyde, and formic acid. This work ultimately led to the wide use of ozone in structure determination. A number of mechanisms for the reaction have been proposed over the years. All these mechanisms, including those prevalent today, accept the general idea first suggested by Staudinger[38] in 1925 that ozone adds to the double bond to give a first adduct, which is usually very unstable, and which Staudinger called the molozonide. In the usual case this molozonide or initial ozone–olefin adduct then goes on to give a new structure called the normal ozonide.

The general question of whether the ozonolysis reaction is free radical or ionic has been at least partially answered. In this connection it is important to emphasize that ozone is a powerful oxidant and will react with virtually all organic materials. In many of these reactions one can expect the formation of peroxidic materials that may subsequently become involved in free-radical processes. Our attention here is on the ozonolysis reaction and whether free-radical chemistry is associated with it.

When it was observed that ozonolysis of styrene gave the usual ozonolysis products but no polystyrene, Criegee[39] concluded that the reaction was probably not free radical in nature. This observation is certainly a powerful argument against the intervention of free radicals, but it may not be the final answer on this point.

Ozonolysis studies are usually carried out at very low temperatures under conditions possibly causing the suppression of free-radical reactions. At any rate, the chemistry of the reaction is generally interpreted in terms of one or more ionic mechanisms. This point is important to an understanding of the problems of crack formation and growth, and stabilization. If free-radical contributions were nonexistent or minimal, then antioxidant-type inhibitors that are designed for use in free-radical situations would not be expected to be useful as stabilizers against ozone cracking. In general this has been found to be the case; that is, materials that are good antioxidants are not usually useful as antiozonants, although there are some exceptions to this general observation.

The addition of ozone to the double bond appears to be electrophilic in nature as first suggested by Wibaut[40,41] and since confirmed by a number of observations. Huisgen[42] has suggested that the addition is of the general form of a 1,3-dipolar cycloaddition reaction leading to a 1,2,3,-trioxolane structure for the initial adduct. There is growing evidence that in many cases the initial adduct is formed in a one-step addition process and that it does have the 1,2,3-trioxolane structure. Thus Criegee and Schröder[43] showed that the initial adduct in the case of *trans*-di-*t*-butylethylene is a crystalline material that can be reduced to the racemic diol. This means that the addition was a one step process and that the unstable adduct still had one carbon–carbon bond intact. Similar observations have been made by Greenwood[44,45] in the cases of the initial adducts from a number of other *trans* olefins, but to date no such reduction of an initial adduct has been completed in the case of a *cis* olefin.

An important experiment relative to the structure of the initial adduct was that initially carried out by Bailey and co-workers[46] in which NMR evidence was used to demonstrate that the adduct in the case of *trans*-di-*t*-butylethylene did indeed have the 1,2,3-trioxolane structure. Durham and Greenwood[47] have also used NMR evidence to come to the same conclusion regarding the structure of the initial adduct for a number of other *trans* olefins and also for several *cis* olefins.[48] There are a number of cases where it appears that the initial adduct does not have the 1,2,3-trioxolane structure (**1**) but instead had the so-called σ structure (**2**).[49–51] Structure **2** seems

particularly likely in those cases of a terminal olefin, which is heavily substituted on one side of the double bond. In such cases the predominant ozonolysis product is the epoxide.[50,52,53]

A large body of experimental data was accounted for by the mechanistic proposal for the ozonolysis reaction given by Criegee.[39,54] According to this proposal the initial adduct, probably with structure **1**, undergoes a concerted decomposition to give a zwitterion (**3**) and a carbonyl compound (**4**). In most cases fragments **3** and **4** will recombine to give the normal ozonide (**5**) but the extent of this recombination reaction will depend upon the structures of **3** and **4**. If **4** is a particularly unreactive carbonyl compound, the zwitterion **3** may dimerize to a diperoxide (**6**) or give a higher peroxide (**7**). Depending upon the reaction medium the recombination of **3** and **4** may not be the favored reaction pathway; and in a reactive solvent such as methanol, a new product, a methoxyhydroperoxide (**8**) will be formed exclusively. It is also possible for the fragments **3** and **4** to recombine in a more random matter to give a polymeric ozonide. The essential elements of the Criegee proposal are summarized in (5.1.)

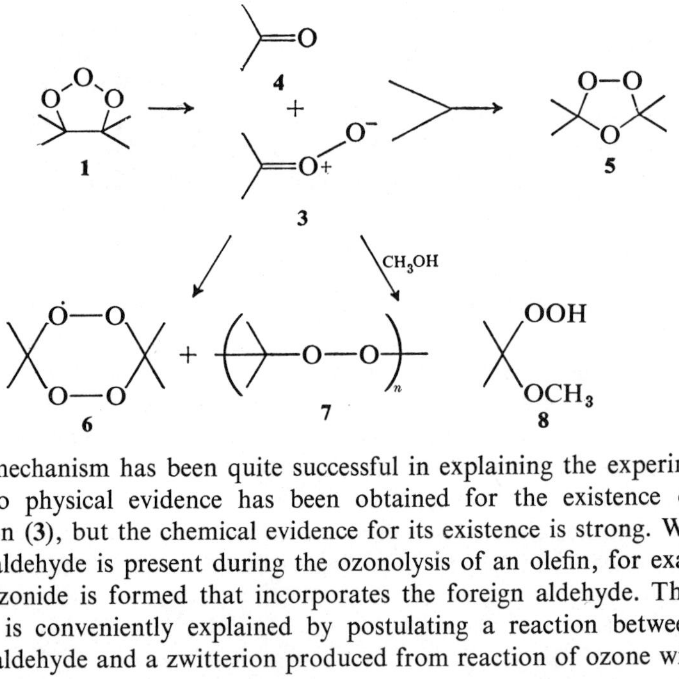

(5.1)

This mechanism has been quite successful in explaining the experimental data. No physical evidence has been obtained for the existence of the zwitterion (**3**), but the chemical evidence for its existence is strong. When a foreign aldehyde is present during the ozonolysis of an olefin, for example, a new ozonide is formed that incorporates the foreign aldehyde. The new ozonide is conveniently explained by postulating a reaction between the foreign aldehyde and a zwitterion produced from reaction of ozone with the olefin.[39]

The recombination reaction of fragments **3** and **4** would seem to be quite dependent upon their structures. The recombination of an aldehyde and an aldehydic zwitterion proceeds quite smoothly, and the ozonide **5** is the usual product in such cases. For some time it was felt that simple ketones would not react with zwitterions.[39,55] Although it is certainly true that ketones are not as reactive as aldehydes in this reaction, it has been shown that acetone, for example, will react with aldehydic zwitterions when the acetone is present in great excess.[56] Likewise the ketonic zwitterion, derived from tetraphenylethylene, will react with acetone to give the ozonide of 1,1-diphenyl-2,2-dimethyl ethylene.[57]

If a simple ketone contains an electron-withdrawing substituent so that its reactivity is enhanced, then it will react smoothly with a zwitterion. Thus *trans*-1,4-dibromo-2,3-dimethylbutene-2 reacts with ozone to give a good yield of the normal ozonide.[58]

According to the Criegee meachanism, an unsymmetrical olefin should give two different zwitterions and two different carbonyl compounds. Random recombination of these fragments ought to give not one but three different ozonides. One of these ozonides is the normal ozonide of the parent olefin. The other two are symmetrical ozonides and are referred to as cross ozonides.

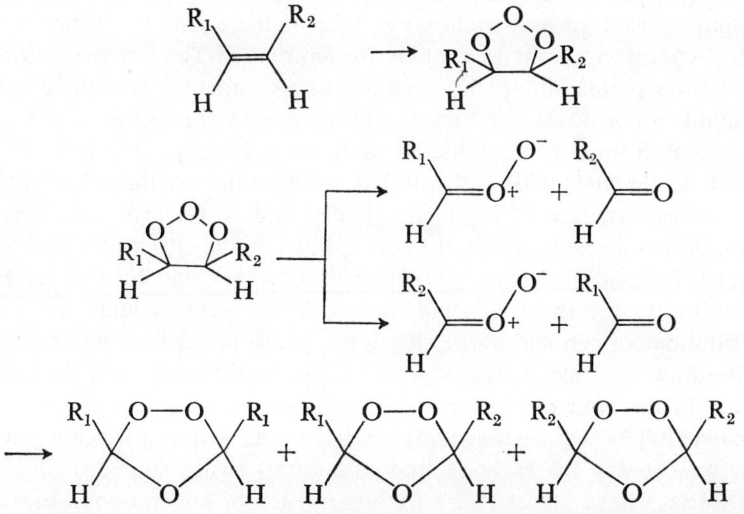

In addition, in the case of a 1,2-disubstituted olefin, as illustrated, each of the three ozonides produced is capable of existing as a cis–trans pair. When he found that the unsymmetrical olefin 3-heptene, gave only 3-heptene ozonide, Criegee[39] postulated that the zwitterion and carbonyl fragments are indeed formed but undergo further reaction in a solvent cage and so are not free to recombine randomly. It has been shown subsequently that the prediction of the Criegee mechanism can be realized under the proper reaction conditions. The ozonolysis of methyl oleate was reported to give all three expected cis–trans pairs of ozonides,[59,60] and in 1965 it was shown that the simple olefin 2-pentene[61] will give all three predicted ozonide pairs. Since that time there have been more and more similar cases.

The Criegee mechanism has been quite successful in correlating a large amount of experimental data on the ozonolysis reaction, but there is a growing body of experimental data that are not adequately explained by the mechanism as presented. These data include the observations (*a*) that cis and trans isomers of 1-arylpropenes do not form the same proportions of ozonides, aldehydes, and peroxides;[62–65] (*b*) that cis and trans olefins give different ozonide cis–trans ratios both in the normal[49,58,66–72] and the cross ozonides;[49,71] and (*c*) that ozonide cis–trans ratios and yields are a function of olefin steric requirements,[49] olefin concentration,[73] and the nature of the solvent.[73] It was quite startling to find, for example, that *cis* and *trans*-4-methyl-2-pentene gave different ozonide cis–trans ratios in all three of the ozonides obtained.[71] This result would not have been expected from the simple zwitterion–carbonyl recombination reaction.

Several attempts have been made to bring some of the more recent data into a general mechanistic scheme. All these attempts use the Criegee mechanism as a general framework. One of these newer schemes includes several competing paths to ozonide formation.[49,74] The Criegee zwitterion–carbonyl recombination path is retained as a competing path along with the possibility of a direct reaction of aldehyde with the initial adduct and a direct path from a σ initial adduct to ozonide. Clearly the extent of the influence of the latter pathways will depend upon the reaction temperature as well as the structure of the olefin. Bailey and co-workers have devised a mechanistic scheme based on the concept of syn and anti zwitterions.[72] Such a possibility had also been considered by Criegee[54] and by Murray et al.[73] According to the Bailey scheme, cis and trans olefins could give different distributions of syn and anti zwitterions. The subsequent reactions of these zwitterions with aldehydes is seen as giving different ozonide cis–trans ratios. Fliszár and co-workers have suggested an alternative form of the aldehyde initial olefin–ozone adduct scheme[75] as well as a different mode of reaction between the carbonyl compound and zwitterion.[76] Finally Fliszár and Carles[76] have suggested a variation of the syn–anti proposal in which a zwitterion–olefin interaction could lead to preferential stabilization of syn and anti zwitterions.

Attempts have been made to gain further information on the relative importance of the various reaction pathways discussed. Several such attempts have used an ^{18}O tracer technique and are based on the prediction that the original Criegee scheme[39] would place such a label in the ether bridge of the ozonide when it is introduced via an added labeled aldehyde, whereas a contribution of aldehyde–initial-adduct reaction proposed by Murray, Story, and Youssefyeh[71] would predict some ^{18}O label in the peroxide bridge. Using such an approach and using various ozonide reduction techniques in conjunction with mass spectrometry, Story et al.[77] have found evidence for considerable peroxide bridge labeling. On the other hand, using mass spectrometry directly on the ozonides, Fliszár and co-workers[76,78] found only ether bridge labeling in the cases of some aromatic ozonides. It should be pointed out that the latter work was carried out under temperature conditions wherein the aldehyde–initial-adduct reaction would not be expected to make a significant contribution. Also using mass spectrometry directly on the ozonides to determine ^{18}O distribution, Hagen and Murray[79] have found considerable evidence for peroxide bridge labeling and furthermore found that such labeling was more important at lower temperatures as predicted by the alternative pathway proposed by Murray, Story, and Youssefyeh.

At the moment, our understanding of the mechanism of the ozonolysis reaction must be regarded as incomplete, but considerable progress has been made. It is hoped that the intensive efforts being made by several groups will lead to further improvements in this understanding.

c. The Physics and Chemistry of Cracking. The data arising from the work on the mechanism described above have important implications for the ozone cracking problem. It is clear, for example, that the stereochemistry of the elastomer and the nature and bulk of substituents can be expected to affect the ozonolysis reaction and, as argued below, the susceptibility to cracking.

The cracking process can be regarded as the result of one of two general proposals for the chemistry of the process. In the first of these it is felt that the crack has its origin in molecular scission; that is, that ozone literally cuts the elastomer chain.[30,80,81] One possible chemical understanding of this process would be that the zwitterion and carbonyl fragments arising from decomposition of the adduct formed between ozone and surface unsaturation are prevented from recombining to give ozonide by the strain on the elastomer. The idea has a certain amount of attraction. It does account for the dual requirement of ozone and strain. Likewise the critical strain observation—that is, the observation that a certain strain is required to initiate and perpetuate cracking—would be translated to the strain required to prevent most recombinations of zwitterion and carbonyl fragments. In the last analysis, however, it is felt that this proposal must be rejected in favor of that which follows. It seems likely that, even at the critical strain, the elastomer chains are sufficiently coiled and the proximity and rate of reaction of the reactive fragments are such that normal ozonide formation is probably not deterred to any large extent.

In the second general proposal the surface must undergo a physical change in the direction of embrittlement so that stress or flexing causes cracking of this surface. In fact, in the view of the writer, such a physical change might be expected to accompany the chemical change occurring at the surface. When subjected to ozone, the surface unsaturation should react rapidly with ozone to give mostly normal ozonide. This means that the surface is being rapidly converted from one with a varying amount of carbon–carbon unsaturation to one containing many units of the somewhat strained ozonide ring. Under this view the stress is required to elongate the elastomer chains, thus exposing more unsaturation to be converted to the more brittle ozonide backbone. A second possible chemical explanation for the postulated physical change is that ozonidelike linkages are formed between successive layers of elastomer chains and that stress enhances the opportunity for this to take place. Such chemical reactions are actually crosslinking type reactions and should lead to embrittlement of the surface. It may be that some of each type of reaction takes place. The first kind of reaction is certainly an expected one and intuitively one would expect it to lead to a more brittle surface.

This description of the chemistry of cracking is also related to the stabilization problem to be discussed later. Suffice it to say here that according to this explanation the elastomers that resist ozonide formation would be expected to be less susceptible to cracking. An elastomer like polychloroprene, which involves an acid chloride as the carbonyl fragment after cleavage of the initial adduct, would not be expected to form much ozonide. In addition to the generally reduced reactivity of halogen-bearing unsaturation, the reactive acid chloride would be expected to be diverted to nonozonide-forming reactions such as reaction with atmospheric or surface moisture. This idea is also consistent with the general observation of a development of an oily film on exposed polychloroprene.

The amount of research reported that might support or refute the above ideas is minimal, partly because of the complex chemical composition of the usual rubber mix. Even when the formulation used reduces the number of chemical components to a minimum, there is still the problem of measuring chemical changes in a sample of ozone-exposed elastomer. This kind of measurement is made difficult since it appears that only a very small portion of the rubber needs to be affected chemically in order for the accompanying physical changes, including cracking, to occur. These problems have led workers to undertake studies of the reaction of ozone and elastomers and antiozonants in solution. It is dangerous, however, to extrapolate the results obtained under these conditions to the conditions normally present when a solid, strained elastomer is subject to the exposure. Indeed, a comparison of the results of accelerated tests (which are performed under conditions specifically tailored to closely match those in the exposure environment) to results obtained under service conditions suggests that even such tests are not always reliable.

Ozone has been used extensively in studies to determine the structure of elastomers. This work, beginning with the classical studies of Harries,[85] suggests that ozone reacts with elastomer unsaturation in the usual way; that is, to give a normal ozonide that is usually converted in structure determination studies, into the corresponding carbonyl compounds. On the other hand, Kendall and Mann[86] have used infrared spectroscopy to show that ozonolysis of thin elastomer films gives structures other than the normal ozonide; these authors have suggested a diperoxide structure as one possibility. Current understanding of the ozonolysis reaction indicates that a variety of compounds can be expected in a normal ozonolysis. The distribution and nature of these products will depend upon the stereochemistry and substituents at the unsaturation as well as the concentration of other reactive species in the immediate environment of the double bond.

In contrast to the situation with respect to the chemistry of the ozone cracking problem, extensive work has been done on the physical aspects of the problem. A number of research groups[30,87–90] have found that there is a critical stress at which the rate of cracking reaches a maximum, and this has spurred efforts aimed at understanding the phenomenon. A variety of methods have been devised to evaluate cracking, particularly as a function of strain and ozone concentration. These have been reviewed extensively[91–94] and will not be discussed in detail here.

The effect of the degree of vulcanization of an elastomer on the susceptibility to cracking has been shown to be related to mechanical properties,[35] particularly the modulus. The general conclusion of the effect of strain studies is that cracking reaches a maximum at some relatively low value of strain, 3 to 5% has been reported for a large number of cases.[33] Newton[30] has pointed out that it is important to distinguish between the rate of formation and the rate of growth of cracks. The rate of formation of cracks appears to increase up to about 70% strain, whereas, as indicated above, the rate of growth of cracks is a maximum at much lower strains. Zuev and Pravednikova[34] have carried out extensive studies of the effect of strain and have demonstrated that the rate of crack growth is a maximum at what is termed the critical elongation. Likewise, the time required to completely rupture a sample was found to be a minimum at the critical elongation. Vodden and Wilson[90] have carried out similar studies using the stress-relaxation technique.

A rather extensive study of the physics of ozone cracking has been carried out by Braden and Gent.[95–97] These workers suggest that ozone cracking can be regarded as having two stages, the crack initiation stage and the crack growth stage. It had been recognized by many groups that a threshold stress has to be applied in order to initiate cracking. This threshold stress varies as the square root of the modulus of the elastomer. The rate of growth of the initiated crack was studied by Braden and Gent by following the rate of growth of a single cut made with a razor.[95–97] The rate of crack growth was found to be constant for a given elastomer, and this rate paralleled the known resistance to cracking. It was also concluded that the rate of crack growth is inversely proportional to the number of network chains per unit volume.

By studying the rate of crack growth as a function of several other variables, some additional conclusions could be reached. The rate of crack growth rises with temperature and plasticizer content until a rather high maximum value is reached. More recently it has been shown that the dependence of the rate of crack growth upon temperature is the same as the dependence of segmental motion upon temperature.[98–99] This suggests that the tendency of a crack to grow depends upon the ability of the broken polymer chains to pull apart, which will be determined by the internal viscosity of the elastomer. Consistent with this view is the observation that the rate of crack growth also depends inversely on the crosslink density of the elastomer.[100]

B. STABILIZATION BY PHYSICAL METHODS

1. Waxes

The use of wax to inhibit ozone deterioration probably goes back to 1881,[101] when wax was added to rubber compounds to prevent deterioration. It seems unlikely that the users recognized its capacity against ozone attack, but the wax no doubt performed this function among others. A great variety of waxes and combinations of waxes have been used to provide protection against ozone cracking. The use of pure paraffin wax proved early to be ineffective, for this wax tends to flake off the rubber surface, particularly under conditions of constant flexing. As a result, combinations of waxes are generally used today. The use of pure microcrystalline waxes suffers from the shortcoming of slow blooming.[102] When combined with a paraffin wax, however, the microcrystalline waxes migrate to the surface more rapidly and the desired surface bloom is attained.[102]

The manner in which waxes provide protection has been well studied,[103] and it is believed that they function by providing a layer on the rubber surface that is not itself susceptible to ozone attack. When using wax to protect against ozone cracking, it is important to realize that, unlike oxygen, ozone reacts only at the surface; it does not diffuse into the sample. Thus a surface layer of relatively unreactive wax presents an impervious surface to ozone. This explains the attention paid to blooming characteristics in blending paraffin and microcrystalline waxes. Even when the desired blooming rate has been achieved, the use of waxes alone to provide protection against ozone attack is rather well restricted to static conditions of service.

Whenever constant flexing is present, even the more strongly held microcrystalline waxes flake off and protection is lost. This problem has led to the wide use of combinations of waxes and chemical antiozonants to provide protection under both static and dynamic conditions of service. In fact, it is felt that waxes can aid in the diffusion of the chemical antiozonant to the rubber surface. The subsequent action of the antiozonant on the surface is discussed in Section C.

When wax alone is used to provide ozone protection, particular care must be taken to use a sufficient quantity. When insufficient wax is used, then the usual surface condition of many fine cracks, observed in the absence of protection, is converted to one of a few very large cracks that can grow to the point of rupture.[95,96] On the other hand, use of too much wax will lead to a surface coat that is more susceptible to the flaking problem referred to earlier. Also, Zuev and Zaitseva have shown that higher concentrations of wax in samples under high deformations leads to lower durability of elastomers.

Since it is generally agreed that wax needs to bloom to the surface in order to afford protection, a great deal of attention has been paid to the factors affecting migratory aptitude. This emphasizes the significance of the degree of solubility of the various waxes in the particular elastomer to be protected, as well as possible interference by fillers, pigments, and other rubber additives. Van Pul[105] has suggested that there is a reasonable correlation between the protective power of a particular wax and its melting point, refractive index, and degree of branching. Waxes with good protective power have melting points between 65 and 72°C, refractive indices in the range of 1.432 to 1.438, and branching to the extent of 30 to 50% side chains.[105] A summary of the characteristics of the two general categories of waxes, paraffin and microcrystalline, has been given by Bennet.[106] A study by Levitin et al.[107] again confirmed that paraffin waxes were satisfactory only under static conditions. A synthetic ceresin wax was found to be the best under dynamic conditions. This wax had the highest viscosity of all those tested. Satisfactory protection could only be obtained when waxes were used in conjunction with antiozonants, however.

Not all authors agree on what constitutes desirable physical properties in a wax as far as ozone protection is concerned. Winkelman[102] argues that a linear-chain wax is better than a branched-chain wax. Likewise he argues that a blend of low-melting microcrystalline wax and high-melting paraffin wax is a desirable combination. Ferris et al.[108] have concluded that it is not possible to arrive at any reasonable correlation between protective ability and physical properties. Part of the difficulty here may be a variation of testing and service conditions. Various wax blends give quite different results depending upon whether one uses indoor accelerated tests or actual outdoor exposure. Crabtree and Kemp[109] found that the performance of various waxes was quite dependent upon the exposure temperature. Higher exposure temperatures gave poorer protection. The authors suggest that at higher temperatures some resorption of the surface wax into the elastomer may occur. Also at lower temperatures the waxes are more inclined to embrittlement and flaking off. Sharp[110] has also indicated that accelerated testing can lead to questionable results when waxes are being evaluated.

In general, blends of waxes can provide reasonably good protection under conditions of static service. There is considerable disagreement about whether waxes alone can provide protection under dynamic conditions. If only moderate flexing is encountered, then a carefully designed wax blend can be useful. Where considerable flexing is present, then a combination of wax and chemical antiozonant or one of the better antiozonants used alone should be employed.

2. Other Physical Methods

The premise underlying the use of waxes to provide protection against ozone attack—namely, the creation of a surface that is itself not attacked by ozone—has led to the development of a number of other similar methods. Some of these methods have been known for some time and were used to prevent deterioration even before ozone was widely recognized to be a separate deterioration factor. Williams,[33] for example, had found that oxidized rubber was less susceptible to ozone attack than fresh rubber. On this basis Williams suggested treatment of the surface with copper chloride to promote oxidation as a protective measure.

Various methods have been suggested that would remove surface unsaturation and, presumably, reduce ozone attack. Hydrogenation of such unsaturation has been patented as a protective method.[111] Crabtree and Kemp[87] have suggested a number of similar methods including air oxidation, exposure to bromine or nitrogen peroxide, and bomb aging. In some of these cases the surface becomes so brittle that it suffers from its own form of cracking, however.

The general idea of a separate surface coating has also been tried. Norton[112] has suggested that alkyd resins be used for this purpose; Newton[113] and Buist[114] have used a polyurethane coating for protective purposes. Patents have been issued for a phenol–formaldehyde resin[115] as a surface coating and for the addition of hydrogen sulfide adducts.[116] Poly(vinyl chloride) paints[117] or a complete film of cellulose[118] have also been advocated as surface coatings against ozone. Likewise the modification of acrylonitrile–butadiene polymers with poly(vinyl chloride) resins to improve ozone resistance has been described by several groups.[119–122]

Perhaps this section should not be concluded without at least brief mention of another more recent method of providing stabilization, namely, the use of EPT (ethylene propylene termonomer) elastomers. In such elastomers the termonomer is generally a diene of such structure that the terpolymer contains residual unsaturation only in the side chains. Since the polymer backbone is largely free from unsaturation, it is less susceptible to attack by ozone. At the same time, the polymer contains sufficient unsaturation to permit vulcanization using the usual recipes. We may see increasing use of such types of elastomers as one way of providing stabilization against deterioration by ozone.

C. STABILIZATION BY CHEMICAL MEANS—ANTIOZONANTS

The failure of waxes to provide protection against ozone degradation under most conditions of dynamic service led to a further search for agents that could give this protection. Many of the chemicals first tried were those that had been found to be effective as antioxidants. In general these materials were not effective as antiozonants.[123] In one such study Tuley[123] found, however, that some of the poorer antioxidants were more effective as antiozonants than other materials tested. Included in this group were the primary diamines *p*-phenylenediamine, benzidine, and 4,4′-diaminodiphenylmethane. Derivatives of these materials are still among the most effective antiozonants available today. Other groups attempted to use antioxidants as antiozonants.[125–128] As a result of his studies, Thompson[125] classified the antioxidants as good, ineffective, or harmful as antiozonants; the antioxidants that did display any activity as antiozonants were amines, and this discovery led to further examination of a variety of amines. In 1943 Barton[129] received a patent for the use of 1-(*p*-aminophenyl)-2,5-dimethylpyrrole as a protective additive. Nickel dibutyldithiocarbamate was also described early as an antiozonant.[130] In 1953 it was reported that 6-ethoxy-1,2-dihydro-2,2,4-trimethylquinoline performed well as an antiozonant.[131] The condensation products between aldehydes and amines and ketones and amines were also found to be slightly effective as antiozonants.[128]

At this point the *p*-phenylenediamine derivatives still seemed the most promising, and these were extensively investigated, particularly at the Rock Island, Ill., Arsenal. In 1954 the Rock Island group published a paper describing the results of a major investigation of amines as potential antiozonants.[132] This paper disclosed the high effectiveness of *N*,*N*′-di-*sec*-butyl-*p*-phenylenediamine and some related compounds as antiozonants. The best antiozonants were all *N*,*N*′-dialkyl-*p*-phenylenediamines, and these materials were even more effective when used along with waxes. On the other hand,

Table 5.1 Examples of Materials Used as Antiozonants

Amines	
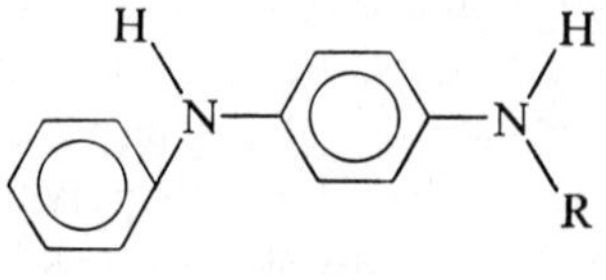	R = alkyl
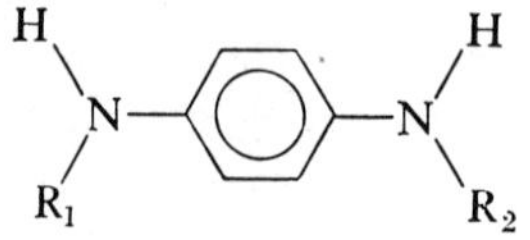	R_1, R_2 = alkyl
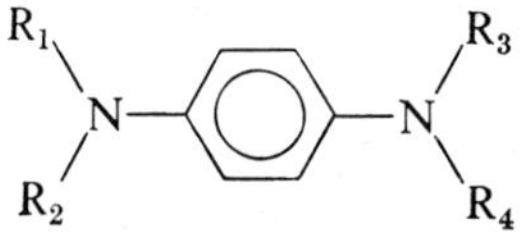	R_1, R_2, R_3, R_4 = alkyl
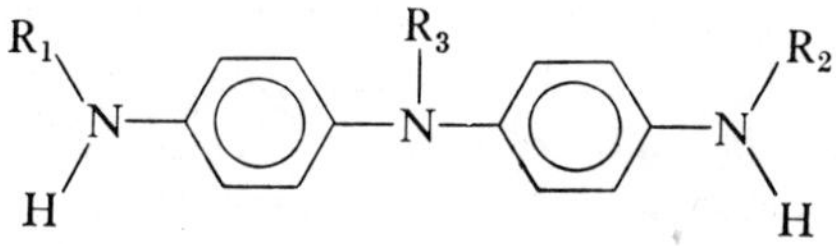	R_1, R_2, R_3 = alkyl
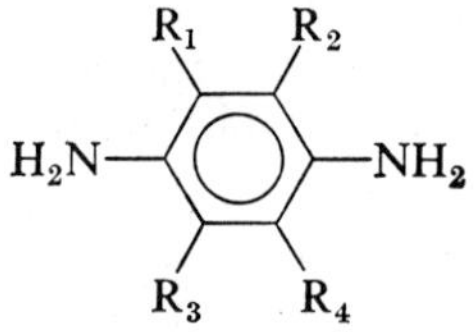	R_1, R_2, R_3, R_4 = alkyl
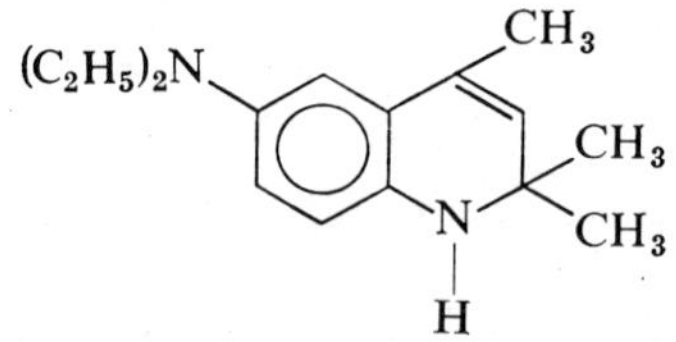	
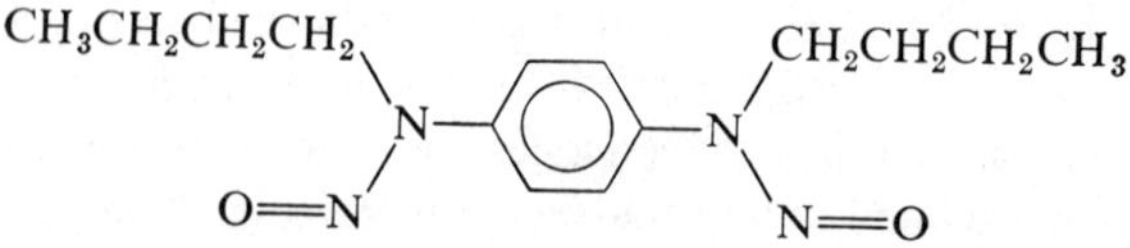	

Table 5.1 Continued

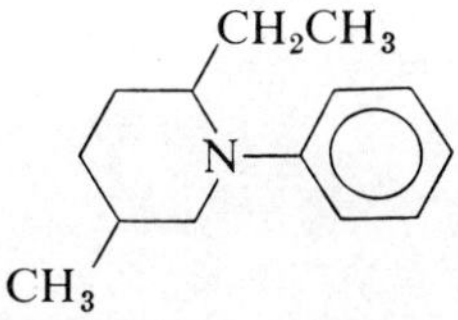

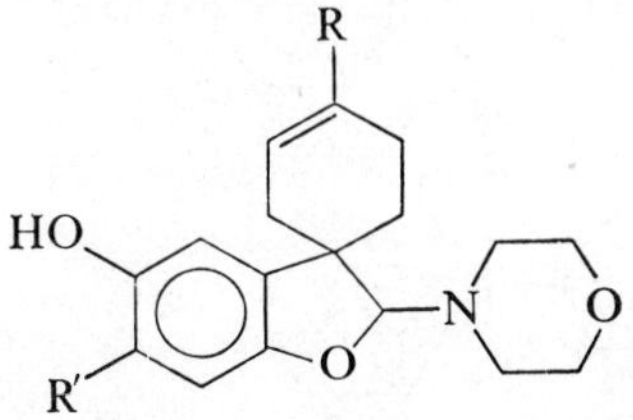

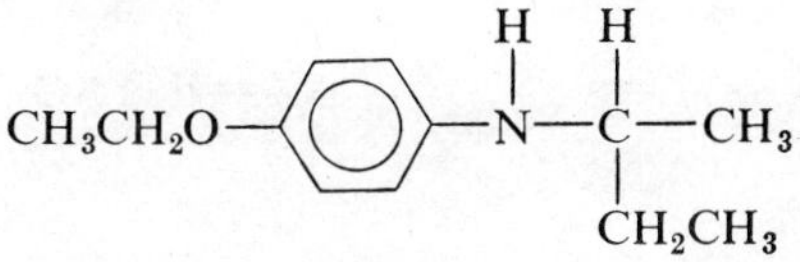

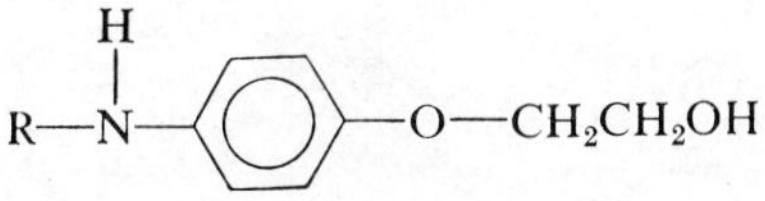

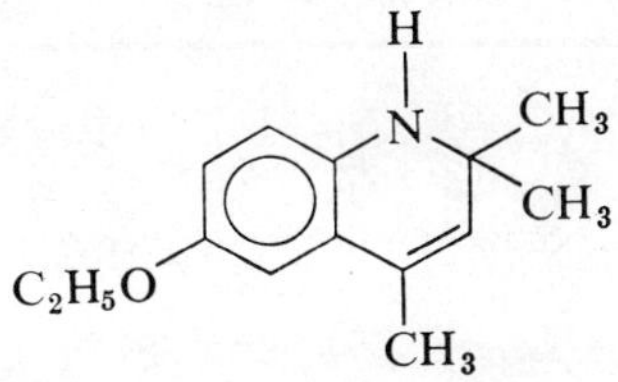

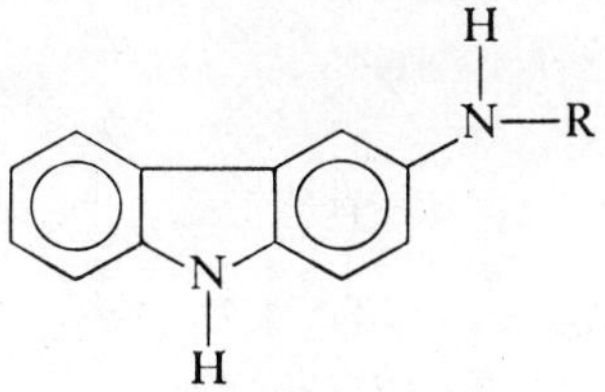

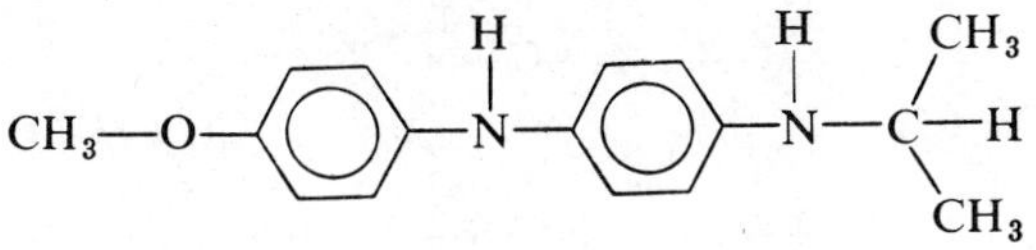

CH_2CH_3

N

CH_3

R

HO

N

O

O

R′

H H

CH_3CH_2O—⟨⟩—N—C—CH_3

CH_2CH_3

H

R—N—⟨⟩—O—CH_2CH_2OH

H

N

CH_3

CH_3

C_2H_5O

CH_3

H

N—R

N

H

H H CH_3

CH_3—O—⟨⟩—N—⟨⟩—N—C—H

CH_3

Table 5.1 Continued

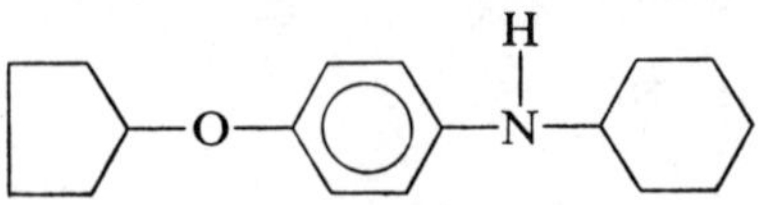

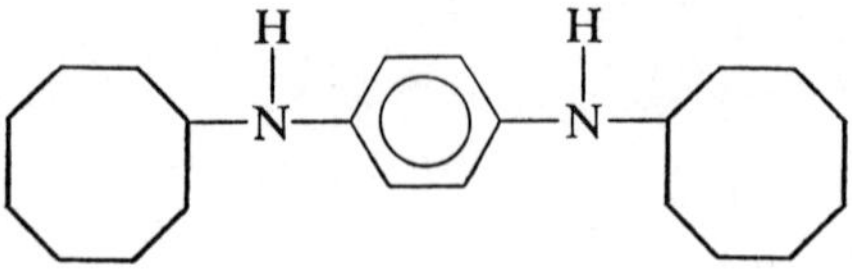

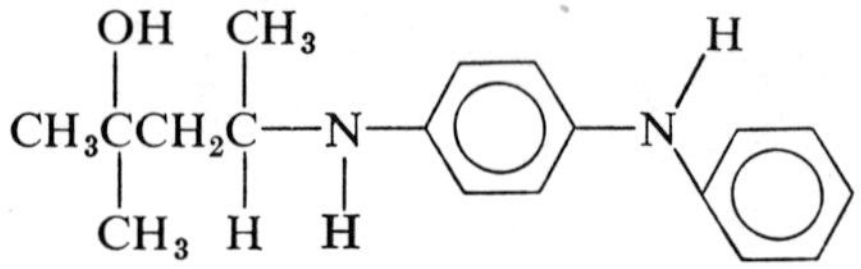

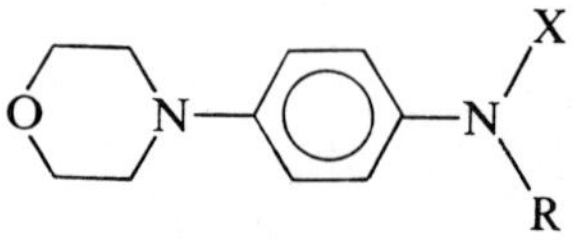

$X = H$ or NO

Sulfur Compounds

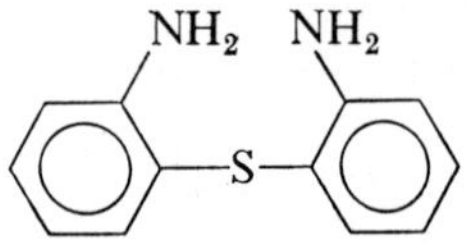

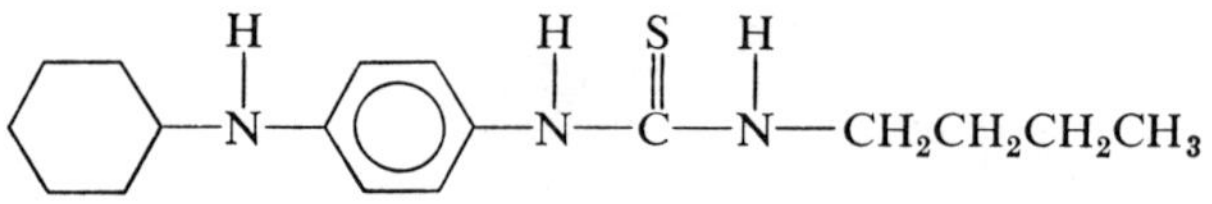

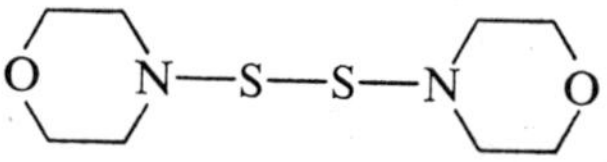

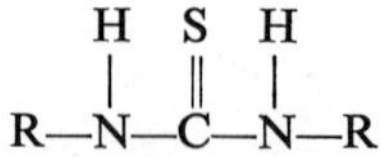

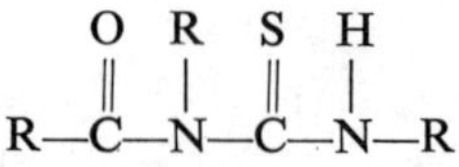

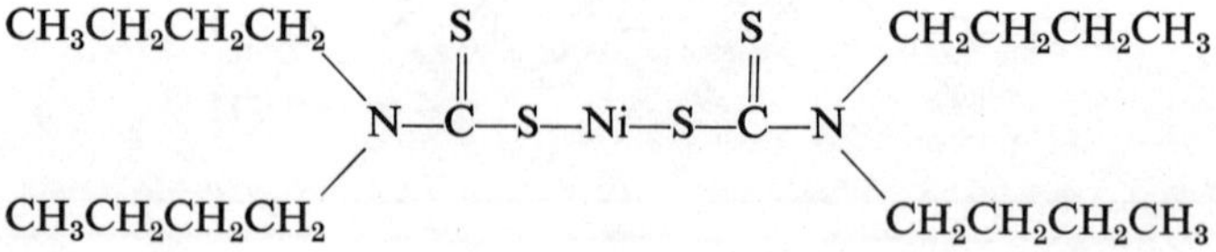

$CH_2CH_2CH_2CH_3$

$R{-}NH{-}C(=S){-}NH{-}R$

$R{-}C(=O){-}N(R){-}C(=S){-}NH{-}R$

$(CH_3CH_2CH_2CH_2)_2N{-}C(=S){-}S{-}Ni{-}S{-}C(=S){-}N(CH_2CH_2CH_2CH_3)_2$

Table 5.1 Continued

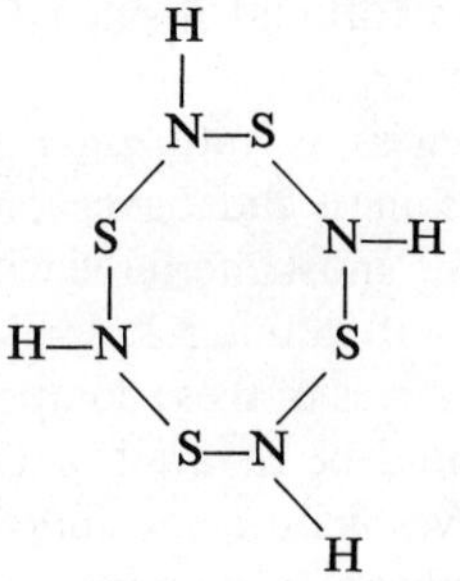

Compounds Claimed to be Nonstaining

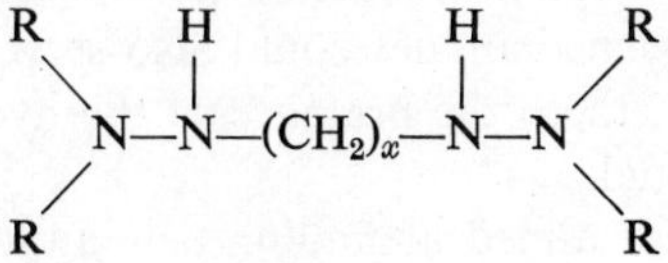

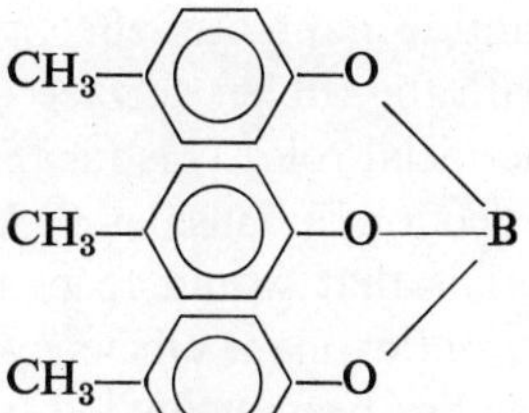

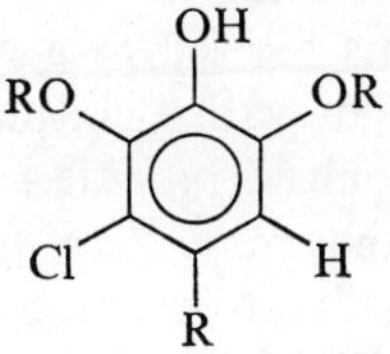

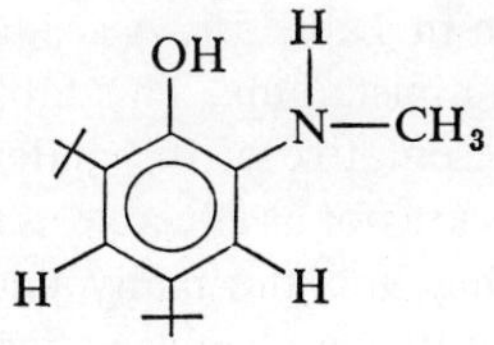

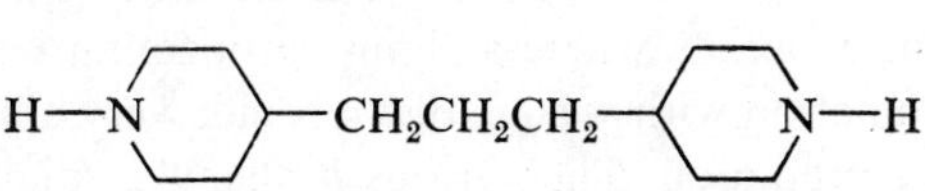

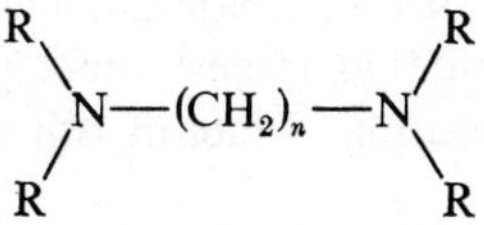

Van Pul[126–128] and Bergstrom[133] found *N*-phenyl-*N'*-cyclohexyl-*p*-phenylenediamine to be the most effective.

Although high in effectiveness as antiozonants, the class of compounds, *N*,*N'*-dialkyl-*p*-phenylenediamines, did have some drawbacks. Many were toxic, some were too volatile,[124] most interfered with the vulcanization process, and all had severe problems of staining and discoloration. This has led to extensive efforts to find derivatives of these compounds or related compounds in which these difficulties could be reduced or eliminated. Based on results from many sources, some investigators concluded that the hydrogen-bearing nitrogen in the secondary *p*-phenylenediamines was essential for effectiveness as an antiozonant. However, Bruce and co-workers[135] found that some tetrasubstituted *p*-phenylenediamines could also serve as very effective antiozonants. At this point, then, the presence of the hydrogen appeared to be beneficial but not essential.

A portion of the work aimed at finding new antiozonants was based in part on the idea that the antiozonant was effective because ozone reacted more rapidly with it than with the rubber surface. Thus it was believed that the antiozonant played a sacrificial role. This idea is discussed further below when the mechanism of protection is considered. With this idea in mind, a search was begun for materials that would react more rapidly with ozone than rubber did. Indeed the antiozonant efficiency of a large number of *p*-phenylenediamine derivatives has been evaluated on the basis of the rate of the reaction of these compounds with ozone.[134]

The search for new antiozonants has led to a wide variety of materials being patented for this purpose. The varied chemical composition of these materials makes it even more difficult to provide a mechanism for their protection behavior. It seems likely that more than one mechanism should be considered. Nevertheless it is possible to see a common characteristic for most of them. A summary of the various kinds of compounds that have been used as antiozonants is given in Table 5.1. A separate section is included for those that are claimed to be nonstaining. This table is not meant to be comprehensive, but merely representative of the materials available. In general, the nonstaining antiozonants are not as effective as the best staining materials. Despite the extensive searches and the many patents that have appeared, certain of the *N*,*N'*-disubstituted *p*-phenylenediamines still seem to be the best-performing antiozonants where staining antiozonants can be tolerated. When used in conjunction with waxes, they provide very efficient systems for preventing ozone degradation. The various *p*-phenylenediamines derivatives have been extensively investigated, and Biggs[136] has published a table in which the protective power of these materials has been rated.

Choice of the antiozonant that is best for a particular elastomer or use clearly must involve consideration of factors other than the ability to prevent cracks. Among these are volatility, compatibility, toxicity, and possible interference with other materials present, particularly the cure system. The volatility and compatibility problem is frequently overcome by including long alkyl chains. Many of the amine compounds are known to be scorchy. Sometimes this problem can be solved by altering the amount of accelerator or activator used. At any rate, the rubber compounder can draw from a variety of antiozonant materials to achieve his particular specifications.

D. THE MECHANISM OF STABILIZATION BY ANTIOZONANTS

Although there appears to be general agreement that stabilization against ozone degradation by physical means, especially by waxes, involves the creation of a surface layer that is not susceptible to ozone attack, there is far less agreement about the mechanism by which chemical antiozonants impart protection.

It does appear quite clear that the mechanistic function of antiozonants is not like that of antioxidants. The very best antioxidants are only poor antiozonants, but some very poor antioxidants are reasonably good antiozonants. There are some indications that a small part of the antiozonant present may function like an antioxidant, namely, inhibiting radical-chain processes. This is not inconsistent with our understanding of the reactions of ozone with organic compounds, including the ozonolysis reaction. Although the greater part of the chemistry of the ozonolysis reaction has been interpreted in ionic rather than free-radical terms, the production and involvement of several different kinds of peroxidic species would certainly indicate that some free-radical chemistry could take place. In addition, of course, one usually expects to have some peroxidic material present as a result of thermal or photochemical oxidation processes. In all these cases a portion of some antiozonants could play an antioxidant-type role. The major role of most antiozonants, however, does not appear to be the inhibition of free-radical processes.

Another important difference between the functioning of antioxidants and antiozonants is that oxygen diffuses into the bulk of the sample being protected, but the ozone reactions appear to be only surface reactions. An important consequence of this difference is the expectation that an antiozonant must be at the surface to be effective. This means that rates of diffusion become important to mechanistic considerations. One of the reasons given for the high effectiveness of some antiozonant–wax combinations is the ability of the wax to assist in the diffusion of the antiozonant to the surface.

One of the most widely held views of the mechanism of antiozonant action is the so-called scavenger theory. According to this view antiozonants are effective because they diffuse to the surface and there scavenge ozone because of their greater reactivity.[136–138] Numerous efforts to find new antiozonants are still based on this theory, but it has a number of serious drawbacks. Many materials that are extremely active toward ozone are not effective as antiozonants. Thus dilauryl selenide, which is very reactive toward ozone, showed no effectiveness as an antiozonant.[97]

The scavenger theory requires a stoichiometric relation between the antiozonant and ozone. Considering the extended periods of protection provided by some materials, this would not seem to be a realistic requirement. Such a view requires antiozonant to diffuse to the surface at a rate far more rapid than the rate of diffusion measured for one *p*-phenylenediamine antiozonant,[139] for example.

Another argument against the scavenger mechanism is the observation made by a number of groups that increasing the concentration of antiozonant does not always increase the degree of protection. In some cases increasing the concentration above only 1% resulted in no additional protection.[128,140] In another case, an optimum concentration of antiozonant was found, above which the level of protection began to fall off.[141]

It is certainly true that correlations exist between the effectiveness of antiozonants and their ability to bloom to the surface,[142] but one should not conclude that this requires a scavenger mechanism for protection. Many of the *p*-phenylenediamines are made more effective by a variety of agents besides waxes, which appear to assist them in reaching the surface. Again this observation would also be consistent with nonscavenger mechanisms of protection.

The measurements carried out by Erickson and co-workers[143] can also be used to argue against the scavenger mechanism. These workers measured the rate of ozone uptake in both protected and unprotected samples of SBR vulcanizates. It was found that the rate of ozone absorption in the substituted *p*-phenylenediamine protected sample although initially much higher than an identical unprotected sample, dropped to a much lower level in a very short time (40 min.). If the scavenger mechanism were operative one would expect the initially higher rate presumably resulting from reaction with the *p*-phenylenediamine. However, once the antiozonant available on the surface was depleted, the rate in the protected sample should have fallen off to the same rate observed in the unprotected sample. The lower rate suggests a more complex mechanism of protection.

A second mechanism for antiozonant action has been termed the protective film mechanism; here it is believed that the antiozonant diffuses to the surface and then spreads out over the surface to form a protective film.[134,143,144] A modification of this mechanism suggests that it is the ozonized antiozonant that makes up the protective film.[145] This proposal also would not seem to be consistent with the work of Erickson et al.[143] referred to earlier. If the protective film were to be composed only of antiozonant or ozonized antiozonant, then some rather unrealistic rates of diffusion for the antiozonant would have to be assumed in order to form such a film in the time period observed. Nor would the explanation seem to be consistent with the optimum concentrations of antiozonant sometimes found.

In order to attempt to elucidate the mechanism of ozonolysis, a number of groups have studied the basic chemistry of antiozonants. As a result of such a study, Lorenz and Parks[146] have concluded that antiozonants may (*a*) react directly with ozone, (*b*) react with ozonides and other peroxidic ozonolysis products, or (*c*) react with aldehydic ozonolysis products. The latter kind of reaction led to an additional proposal for the mechanism of antiozonant action, namely, a relinking of severed chain ends by the difunctional *p*-phenylenediamines. A similar proposal has been made by Braden and Gent[97] based on their observation that many *p*-phenylenediamine derivatives increase the critical energy required for cracking, whereas some of the less efficient antiozonants such as 6-ethoxy-2,2,4-trimethyl-1,2-dihydroquinoline confer protection without increasing the critical stress. The work of Lorenz and Parks[146] also showed that a portion of the antiozonant became attached to the rubber network.

The relinking mechanism appears to be an attractive hypothesis and it may play some role in the protection function, but it also has a weakness. Here again it seems unlikely that there would be sufficient surface concentration of antiozonant available either immediately or by replenishment to prevent crack formation entirely by a chain-relinking process.

It has also been suggested that amine antiozonants can function by utilizing the well-known reaction between tertiary amines and peroxides to produce free radicals.[147] Thus tertiary amine antiozonants would react with peroxidic products of the ozonolysis reaction to produce free radicals. These radicals would then become involved in crosslinking and oxidation reactions in the rubber. The net effect of such reactions would be to produce a surface that is less susceptible to ozone attack. If such a mechanism is in fact important, it is only significant for certain tertiary amines. The greater percentage of antiozonants probably do not function in this way.

A number of workers have observed the formation of a transient intense blue color when certain dialkyl *p*-phenylenediamines are treated with ozone. Delman et al.[148] have suggested that this color is caused by the formation of a Würster salt and that, therefore, antiozonant efficiency may be related to ease of oxidation. A similar suggestion by Barnhart and Newby[149] is based on the work by Michaelis[150] and Weissberger et al.[151] on the factors that stabilize radical ions. According to this suggestion, effective antiozonants are those with stable semiquinone forms. This form would be produced by reaction of the antiozonant with ozone. In a subsequent reaction the semiquinone form would react with the elastomer ozonide or peroxide to produce the quinone form and also to prevent chain splitting.

More recently Furukawa et al.[152] have advanced an additional mechanism for antiozonant effectiveness, paralleling somewhat the stable-radical view just described. In this proposal the relative effectiveness of an antiozonant is related to its oxidation potential. Compounds with oxidation potentials lower than 0.44 V are effective as antiozonants. An optimum oxidation potential of 0.25 V was also found for antiozonants. These workers went further and calculated indices S^r and S^n for free-radical and nucleophilic reactions, respectively. A good correlation was found between the nucleophilic index, S^n, and the ability of an antiozonant to prevent crack formation. At the same time the authors found a good correlation between S^n and the direct reactivity between the antiozonant and ozone.

The final mechanism for antiozonant action to be considered is perhaps the most attractive at this time. This mechanism suggests that antiozonants are effective according to their ability to react with (*a*) the products of the ozonolysis reaction, including ozonides, diperoxides, and so on, (*b*) the intermediates in the ozonolysis reaction especially the zwitterion, (**3**) described above,[153] or (*c*) a combination of these. The mechanism stems from the view of the cracking process described earlier. Thus if cracking is caused by the physical change accompanying the chemical change of transforming unsaturation into ozonolysis products, particularly ozonides, then effective antiozonants are those that interfere with this process.

If we depict a generalized amine antiozonant as R_2NH, for example, then the following kinds of reactions are reasonable ones based on our knowledge of the ozonolysis reaction and the chemistry of amines.

In (5.2) is shown a portion of the elastomer chain being attacked by ozone with the amine antiozonant interfering with ozonide formation by intercepting the zwitterion. In (5.3) the amine antiozonant is shown cleaving an already formed ozonide linkage via a nucleophilic displacement on the peroxide bonds. Variations of (5.3) would include a nucleophilic attack of the antiozonant on a surface diperoxide or a diperoxide between the surface layer and the underlying layer, that is, a crosslinking diperoxide.

As such processes continued to take place, the surface would become relaxed and mobile instead of maintaining the applied strain as unsaturation is converted to ozonolysis products. Further reaction of ozone with the relaxed surface would produce some ozonides and diperoxides unable to contribute toward a crack-prone surface. Thus the mechanism does not require a stoichiometric amount of antiozonant and thereby avoids one of the major objections to a number of other mechanistic proposals.

$$\sim\sim CH{=}CH \sim\sim + O_3 \longrightarrow \sim\sim CH{=}O + {}^{+}O(O^{-}){=}CH\sim\sim$$
$$\xrightarrow{R_2NH} \sim\sim CH{=}O + \sim\sim CH(OOH)(NR_2) \qquad (5.2)$$

$$\sim\sim CH{=}CH \sim\sim + O_3 \longrightarrow \sim\sim CH{=}O + {}^{+}O(O^{-}){=}CH\sim\sim \longrightarrow \text{ozonide } (\sim\sim CH\text{–}O\text{–}O\text{–}CH\sim\sim, \text{bridged by } O) \qquad (5.3)$$

$$\text{ozonide} + R_2NH \longrightarrow \sim\sim CH{=}O + O{=}CH\sim\sim + [R_2N{-}OH]$$

This proposal is supported by several other experimental observations. That some antiozonant becomes attached to the elastomer network as reported by Lorenz and Parks[146] is certainly consistent with this mechanism. The observation that antiozonants can readily decompose ozonides, with nickel dibutyldithiocarbamate being most effective in this respect,[146] also supports the suggested mechanism. The reaction of amine antiozonants as well as simple amines with isolated ozonides of model alkenes has been found to proceed extremely rapidly.[154]

As far as the chemistry is concerned, the proposal seems to be on a sound basis. When measured against the challenge of explaining antiozonant effectiveness in the known wide range of materials used as antiozonants, the proposal also fares reasonably well. According to this mechanism effective antiozonants are those that are capable of undergoing reactions with zwitterions or of undergoing nucleophilic displacements on peroxide bonds. At this point we are leaving aside nonchemical requirements such as compatibility and nonvolatility, although it should be borne in mind that some and perhaps even a major portion of the chemical architecture of some antiozonants serves to satisfy these requirements.

On this basis amines and hydroxyl-containing compounds are easily accounted for. Although their reactivity would normally be expected to be less, ethers and sulfur-containing compounds should also be capable of the necessary chemistry. Within any one group—the amines, for example—

varying effectiveness would be expected to be related to varying nucleophilicity as found by Farukawa et al.[152] In making decisions concerning nucleophilicity, particularly regarding ability to decompose ozonolysis products, both electronic and steric factors must be considered. In general this mechanism is consistent with the known relative effectiveness of antiozonants. Thus increasing the number of substituents at nitrogen in the *p*-phenylenediamines should increase nucleophilicity at least up to the point where increased steric requirements would tend to reduce effectiveness. The *p*-phenylenediamines as a class may be so effective because one nitrogen is able to contribute to the nucleophilicity of the other. The *N,N'*-dialkyl-*p*-phenylenediamines may be the best in this class because they contain the optimum balance of electronic and steric contributions toward nucleophilicity. At the same time, of course, the presence of the aromatic nucleus provides a system that is easily converted to an undesirable chromophore and hence causes staining. In this connection it should be pointed out that when antiozonants are designed to approximate the general architecture of the *p*-phenylenediamines but to omit the potential chromophore, then reasonably good, nonstaining antiozonants can be obtained.[155] An examination of the variety of compounds in Table 5.1 indicates that, to varying degrees, all might be expected to undergo one or the other or both of the reactions depicted in (5.2) and (5.3).

The described mechanism has several other factors in its favor. The observation[156] that the use of moist ozone provides a protective film on vulcanizate surface, for example, would be explained by postulating a reaction between zwitterions and water molecules to produce a mobile, relaxed surface. In this sense the water molecule performs the same function as an added antiozonant. Likewise the low susceptibility to cracking of polychloroprene rubber would be the result of a combination of the reduced reactivity of the unsaturation, the reduced tendency for ozonide formation from zwitterion–carbonyl recombination reactions, and the great tendency for the acid chloride produced in the ozonolysis to react with ambient moisture.

If the above view of the mechanism is correct, then in fact the real antiozonant—that is, the protective agent—is not added to the elastomer in the formulation but instead is produced in situ. The protective agent is a combination of ozonized elastomer and antiozonant. This means that the antiozonant is not immediately available but requires some finite time for its production. This point is important when discussing the evaluation of potential antiozonants. If a test formulation is thrust into an accelerated test chamber containing high (e.g., 50 pphm) concentrations of ozone, then the potential antiozonant may be judged a failure because it simply has not had time to perform the chemistry described above. The same antiozonant may perform in an acceptable manner in an outdoor exposure test, where ozone concentrations should be much lower. This factor may explain, at least in part, why discrepancies are frequently observed between accelerated and outdoor exposure tests (Chapter 10).

On this same point it has been shown that formulations containing potential antiozonants that perform poorly when tested in 50-pphm ozone chambers showed greatly improved performance in the same test chamber if they were preaged in 25-pphm chambers.[157] Presumably the preaging period in the lower ozone concentrations allowed for development of the relaxed surface as described above.

The last-discussed mechanism has many attractive features, but it is not without weaknesses. An examination of the best available antiozonants indicates that difunctionality is a highly desirable feature. The mechanism would explain such a feature by suggesting that the presence of one functional group increases the nucleophilicity of the other, as in the *p*-phenylenediamines. When the functional groups are so located that interaction between them is not possible, the mechanism would have to fall back on the concentration or proximity effect; that is, if the functional group performs an important chemical function in its immediate vicinity, then the likelihood of performing the function is increased if the number of available functional groups is increased. It may be that a combination of the chain-relinking mechanism and the relaxed surface mechanism is usually operative and that this accounts for difunctionality. At any rate, the mechanism question cannot be considered closed. Progress has certainly been achieved, but such nagging questions as the difunctionality feature still remain without complete explanations.

ACKNOWLEDGMENT

I wish to acknowledge the contributions of Dr. P. R. Story, Dr. L. D. Loan, Mr. George H. Bebbington, and Mr. Harold R. Messler to some of the antiozonant mechanism studies described here. I also wish to thank the National Science Foundation for partial support of this work through grant number GP 10895.

REFERENCES

1. I. Landler and P. Lebel, "Polyplastic," French Patent 1,176,772 April 15, 1959; *Chem. Abstr.*, **54,** 25970a (1960).
2. P. V. Koslov, M. M. Iovleva, and N. A. Plate, *Vysokomolekul Soedin.*, **1,** 1100 (1959); *Chem. Abstr.* **55,** 4034 (1961).
3. "Polyplastic," French Patent 1,166,652, Nov. 13, 1958; *Chem. Abstr.*, **54,** 26004d (1960).

4. "Polyplastic," French Patent, 1,186,992, Sept. 4, 1959; *Chem. Abstr.*, **54,** 25974e (1960).
5. J. R. Durland and H. Adkins, *J. Amer. Chem. Soc.*, **61,** 429 (1939).
6. C. C. Schubert and R. N. Pease, *J. Amer. Chem. Soc.*, **78,** 2044 (1956).
7. L. Long and L. F. Fieser, *J. Amer. Chem. Soc.*, **62,** 2670 (1940).
8. M. C. Whiting, A. J. N. Bolt, and J. H. Parish, *Advan. Chem. Ser.* **77-III,** American Chemical Society, Washington, D.C., 1968, p. 4.
9. G. A. Hamilton, B. S. Ribner, and T. M. Hellman, *Advan. Chem. Ser.* **77-III,** American Chemical Society, Washington, D.C., 1968, p. 15.
10. R. W. Murray and M. L. Kaplan, *J. Amer. Chem. Soc.*, **90,** 537, 4161 (1968).
11. E. Wasserman, R. W. Murray, M. L. Kaplan, and W. A. Yager, *J. Amer. Chem. Soc.*, **90,** 4160 (1968).
12. R. W. Murray and M. L. Kaplan, *J. Amer. Chem. Soc.*, **91,** 5358 (1969).
13. R. W. Murray, M. L. Kaplan, and J. W.-P. Lin, *Ann., N.Y. Acad. Sci.*, in press.
14. R. W. Murray, W. C. Lumma, Jr., and J. W.-P. Lin, *J. Amer. Chem. Soc.*, submitted for publication.
15. C. S. Foote, *Acct. Chem. Res.*, **1,** 104 (1968).
16. K. Gollnick, *Advan. Photochem.*, **6,** 1 (1968).
17. A. M. Trozzolo and F. H. Winslow, *Macromolecules*, **1,** 98 (1968).
18. Homer, *Iliad*, Vol. 8, 135; Vol. 14, 415, in *The Complete Works of Homer*, Modern Library Ed., Random House, New York, 1950.
19. Homer, *Odyssey*, Vol. 12, 417; Vol. 14, 307, in *The Complete Works of Homer*, Modern Library Ed., Random House, New York, 1950.
20. C. F. Schönbein, *Compt. Rend.*, **10,** 706 (1840).
21. A. R. Meetham, *Quart. J. Roy. Meteorol. Soc.*, **70,** 20 (1944).
22. Air Pollution Foundation, 704 South Spring Street, Los Angeles, "An Aerometric Survey of the Los Angeles Basin," July 1955.
23. A. W. Bartel and J. W. Temple, *Rubber Age*, **69,** 326 (1951).
24. A. J. Haagen-Smit, C. E. Bradley, and M. M. Fox, *Ind. Eng. Chem.*, **45,** 2086 (1953).
25. A. J. Haagen-Smit, *Ind. Eng. Chem.*, **44,** 1342 (1952).
26. A. P. Altshuller and J. J. Bufalini, *Photochem. Photobiol.* **4,** 97 (1965).
27. E. A. Schuck and E. R. Stephens, "Oxides of Nitrogen," in R. A. Metcalf and J. N. Pitts, Jr., Eds., *Advances in Environmental Science and Technology*, Vol. 1, Interscience, New York, 1968.
28. J. N. Pitts, Jr., A. U. Khan, E. B. Smith, and R. P. Wayne, *Environ. Sci. Technol.*, **3,** 241 (1969).
29. W. Thompson, *J. Soc. Chem. Ind.*, **4,** 710 (1885).
30. R. G. Newton, *J. Rubber Res. Inst. Malaya*, **14,** 27, 41 (1945); *Chem. Abstr.*, **39,** 29012 (1945).
31. Z. T. Ossefort and W. J. Touhey, *Rubber World*, **132,** 62 (1955).
32. K. Asano, *India Rubber J.*, **70** 307, 347, 389 (1925); *Chem. Abstr.*, **19,** 3616 (1925).
33. I. Williams, *Ind Eng. Chem.*, **18,** 367 (1926).
34. Yu. S. Zuev and S. I. Pravednikova, *Dokl. Akad. Nauk SSSR*, **116,** 813 (1957); *Rubber Chem. Technol.*, **32,** 278 (1959).
35. A. Van Rossem and H. W. Talen, *Kautschuk Gummi*, **7,** 79, 115 (1931); *Rubber Chem. Technol.*, **4,** 490 (1931).
36. W. J. Potter, *India Rubber World*, **103,** 41 (1940).
37. C. F. Schönbein, *J. Prakt. Chem.*, **66,** 282 (1855).
38. H. Staudinger, *Ber*, **58,** 1088 (1925).
39. R. Criegee, *Rec. Chem. Progr.*, (Kresge–Hooker Sci. Lib.), **18,** 111 (1957).

40. J. P. Wibaut, F. L. J. Sixma, L. W. F. Kampschmidt, and H. Baer, *Rec. Trav. Chim. Pays-Bas*, **69,** 1355 (1950).
41. J. P. Wibaut, and F. L. J. Sixma, *Rec. Trav. Chim. Pays-Bas*, **71,** 761 (1952).
42. R. Huisgen, *Angew. Chem., Int. Ed.*, **2,** 565, 633 (1963).
43. R. Criegee and G. Schröder, *Chem. Ber.*, **93,** 689 (1960).
44. F. L. Greenwood, *J. Org. Chem.*, **29,** 1321 (1964).
45. F. L. Greenwood, *J. Org. Chem.*, **30,** 3108 (1965).
46. P. S. Bailey, J. A. Thompson, and B. A. Shoulders, *J. Amer. Chem., Soc.*, **88,** 4098 (1966).
47. L. J. Durham and F. L. Greenwood, *Chem. Commun.*, **843** (1967).
48. L. J. Durham and F. L. Greenwood, *Chem. Commun.*, **24** (1968).
49. R. W. Murray, R. D. Youssefyeh, and P. R. Story, *J. Amer. Chem. Soc.*, **89,** 2429 (1967).
50. P. S. Bailey and A. G. Lane, *J. Amer. Chem. Soc.*, **89,** 4473 (1967).
51. P. R. Story, R. W. Murray, and R. D. Youssefyeh, *J. Amer. Chem. Soc.*, **88,** 3144 (1966).
52. P. D. Bartlett and M. Stiles, *J. Amer. Chem. Soc.*, **77,** 2806 (1955).
53. R. Criegee, *Advan. Chem. Ser. No. 21*, 133 (1959). American Chemical Society, Washington, D.C., 1959, p. 133.
54. R. Criegee in J. O. Edwards, Ed. *Peroxide Reaction Mechanisms*, Interscience, New York, 1962, p. 29.
55. R. Huisgen, *Angew. Chem.*, **75,** 604 (1963).
56. R. W. Murray, P. R. Story, and L. D. Loan, *J. Amer. Chem. Soc.*, **87,** 3025 (1965).
57. R. Criegee, private communication, 1968.
58. R. Criegee, S. S. Bath, and B. V. Bornhaupt, *Chem. Ber.*, **93,** 2891 (1960).
59. G. Riezebos, J. C. Grimmelikhuysen, and D. A. Van Dorp, *Rec. Trav. Chim. Pays-Bas*, **82,** 1234 (1963).
60. O. S. Privett and E. C. Nickell, *J. Amer. Oil Chem. Soc.*, **40,** 22 (1963).
61. L. D. Loan, R. W. Murray, and P. R. Story, *J. Amer. Chem. Soc.*, **87,** 737 (1965).
62. E. Briner, E. Dallwigk, and M. Ricca, *Helv. Chim. Acta.*, **41,** 1390 (1958).
63. E. Briner and M. Ricca, *Helv. Chim. Acta*, **41,** 2178 (1958).
64. E. Briner and S. Fliszár, *Helv. Chim. Acta* **42,** 1310 (1959).
65. E. Briner and S. Fliszár, *Helv. Chim. Acta*, **42,** 2063 (1959).
66. O. Lorenz and C. R. Parks, *J. Org. Chem.*, **30,** 1976 (1965).
67. G. Schröder, *Chem. Ber.*, **95,** 733 (1962).
68. P. Kolsaker, *Acta Chem. Scand.*, **19,** 223 (1965).
69. F. L. Greenwood and B. J. Haske, *Tetrahedron Let.*, **631** (1965).
70. F. L. Greenwood, *J. Amer. Chem. Soc.*, **88,** 3146 (1966).
71. R. W. Murray, R. D. Youssefyeh, and P. R. Story, *J. Amer. Chem. Soc.*, **88,** 3143 (1966).
72. N. L. Bauld, J. A. Thompson, C. E. Hudson, and P. S. Bailey, *J. Amer. Chem. Soc.*, **90,** 1822 (1968).
73. R. W. Murray, R. D. Youssefyeh, G. J. Williams, and P. R. Story, *Tetrahedron*, **24,** 4347 (1968).
74. R. W. Murray, *Acct. Chem. Res.*, **1,** 313 (1968).
75. S. Fliszár, J. Carles, and J. Renard, *J. Amer. Chem. Soc.*, **90,** 1364 (1968).
76. S. Fliszár and J. Carles, *ACS Polym. Preprints*, p. A59, April 1969.
77. P. R. Story, C. E. Bishop, J. R. Burgess, R. W. Murray, and R. D. Youssefyeh, *J. Amer. Chem. Soc.*, **90,** 1907 (1968).
78. S. Fliszár and J. Carles, *J. Amer. Chem. Soc.*, **91,** 2637 (1969).

79. R. Hagen and R. W. Murray, *J. Org. Chem.*, **36,** 1103 (1971).
80. G. R. Cuthbertson and D. D. Dunnom, *Ind. Eng. Chem.*, **44,** 834 (1952).
81. D. M. Smith and V. E. Gough, *Trans. Inst. Rubber Ind.* (*London*), **29,** 219 (1953).
82. H. Tucker, *Amer. Soc. Testing Mater. Spec. Tech. Publ., No.* 229, 30 (1958).
83. E. P. W. Kearsley, *Rubber Age*, **27,** 649 (1930).
84. *Rubber Age*, **62,** 82 (1965), attributed to W. R. Abell.
85. C. Harries, *Ber.*, **38,** 1195 (1905).
86. F. H. Kendall and J. Mann, *J. Polym. Sci.*, **19,** 503 (1956).
87. J. Crabtree and A. R. Kemp, *Ind. Eng. Chem.*, **38,** 278 (1946).
88. R. G. Newton, *Trans. Inst. Rubber Ind.* (*London*), **21,** 113 (1945).
89. J. S. Rugg, *Anal. Chem.*, **24,** 818 (1952).
90. H. A. Vodden and M. A. A. Wilson, *Trans. Inst. Rubber Ind.* (*London*), **35,** 82 (1959).
91. R. W. Murray and P. R. Story, in E. M. Fettes, Ed., *Chemical Reactions of Polymers*, Interscience, New York, 1964.
92. A. E. Eagles, *Proc. Inst. Rubber Ind.* (*London*), **13,** 94 (1966).
93. A. Dibbo, *Trans. Inst. Rubber Ind.* (*London*), **40,** T202 (1964).
94. J. C. Ambelang, R. H. Kline, O. Lorenz, C. R. Parks, and C. Wadelin, *Rubber Chem. Technol.*, **36,** 1497 (1963).
95. M. Braden and A. N. Gent, *J. Appl. Polym. Sci.*, **3,** 90 (1960).
96. M. Braden and A. N. Gent, *J. Appl. Polym. Sci.*, **3,** 100 (1960).
97. M. Braden and A. N. Gent, *J. Appl. Polym. Sci.*, **6,** 449 (1962).
98. A. N. Gent and J. E. McGrath, *J. Polym. Sci., Part* **A3,** 1473 (1965).
99. A. N. Gent and H. Hirakawa, *J. Polym. Sci., Part* **A2, 5,** 157 (1967).
100. M. Braden and A. N. Gent, *Rubber Chem. Technol.*, **35,** 200 (1962).
101. Kreusler and Budde, German Patent 18,740, Aug. 26, 1881.
102. H. A. Winkelman, *Ind. and Eng. Chem.*, **44,** 841 (1952).
103. L. L. Best and R. C. W. Moakes, *Trans. Inst. Rubber Ind.* (*London*), **27,** 103 (1951).
104. Y. S. Zeuv and V. D. Zaitseva, *Sov. Rubber Tech.*, **22,** 15 (1963).
105. B. I. C. F. Van Pul, *Rubber Chem. Technol.*, **31,** 866 (1958).
106. H. Bennet, Ed., *Commercial Waxes*, 2nd ed., Chemical Publishing Co., New York, 1956.
107. I. A. Levitin, E. N. Poloskin, U. D. Petrova, and E. D. Marchenko, *Kauchuk i Rezina*, **22,** 14 (1963); *Chem. Abstr.*, **59,** 4148a (1963).
108. S. W. Ferris, S. S. Kurtz, Jr., and J. S. Sweely, *Amer. Soc. Testing Mater. Spec. Tech. Publ. No. 229*, 72 (1958).
109. J. Crabtree and A. R. Kemp, *Ind. Eng. Chem. Anal. Ed.*, **18,** 769 (1946).
110. P. D. Sharp, *Rubber Age*, **77,** 884 (1955).
111. M. Harvey, U.S. Patent 2,678,892 (to Harvel Research Corporation) May 8, 1954.
112. F. J. Norton, *Rubber Age*, **47,** 87 (1940); *Rubber Chem. Technol.*, **13,** 576 (1940).
113. R. G. Newton, *India Rubber J.*, **121,** 257 (1951).
114. J. M. Buist, *Rev. Gen. Caoutchouc*, **31,** 479 (1954); *Rubber Chem. Technol.*, **28,** 230 (1955).
115. S. D. Shinkle, U.S. Patent, 2,648,613 (to U.S. Rubber Company), Aug. 11, 1953.
116. F. M. McMillan, U.S. Patent 2,514,661 (to Shell Development Company), July 11, 1950.
117. W. W. Rinne and E. J. Kvet, paper presented at Army Conf. on Elastomers, Ft. Belvoir, Va, April 1957.
118. F. L. Haushalter, *India Rubber J.*, **70,** 897 (1925); *Chem. Abstr.*, **20,** 678 (1926).
119. W. A. Wilson, *Rubber Age*, **90,** 85 (1961).
120. D. W. Young, D. J. Buckley, R. G. Newberg, and L. B. Turner, *Ind. Eng. Chem.*, **41,** 401 (1949).

121. R. A. Emmet, *Ind. Eng. Chem.*, **36,** 730 (1944).
122. H. A. Winkelmann, *India Rubber World*, **113,** 801 (1947).
123. W. F. Tuley, *Ind. Eng. Chem.*, **31,** 714 (1939).
124. J. E. Gaughan, *Rubber World*, **133,** 803 (1956).
125. D. C. Thompson, R. H. Baker, and R. W. Brownlow, *Ind. Eng. Chem.*, **44,** 850 (1952).
126. B. I. C. F. Van Pul, *Trans. Inst. Rubber Ind.*, **34,** 28 (1958); *Rubber Chem. Technol.*, **31,** 866 (1958).
127. B. I. C. F. Van Pul, *Trans. Inst. Rubber Ind.*, **34,** 37 (1958); *Rubber Chem. Technol.*, **31,** 874 (1958).
128. B. I. C. F. Van Pul, *Trans. Inst. Rubber Ind.*, **34,** 86 (1958); *Rubber Chem. Technol.*, **31,** 882 (1958).
129. B. C. Barton, U.S. Patent 2,324,056 (to U.S. Rubber Company), July 13, 1943; *Chem. Abstr.*, **38,** 2803 (1944).
130. C. P. Pinazzi and M. Billuart, *Rev. Gen. Caoutchouc*, **31,** 123 (1954); *Rubber Chem. Technol.*, **28,** 438 (1955).
131. K. E. Creed, R. B. Hill, and J. W. Breed, *Anal. Chem.*, **25,** 241 (1953).
132. R. F. Shaw, Z. T. Ossefort, and W. J. Touhey, *Rubber World*, **130,** 636 (1954).
133. E. W. Bergstrom, Rock Island Arsenal Laboratory Rept., No. 58-105 (1958).
134. W. L. Cox, *Symp. Effect of Ozone on Rubber, Amer. Soc. Testing Mater., Spec. Publ. No. 229*, 57 (1958).
135. F. C. Bruce, R. E. Isley, and B. Hunt, Ordinance Corporation Project No. TT1-718, Rep. No. 57, Burk Research Company, Oct. 1, 1957.
136. B. S. Biggs, *Rubber Chem. Technol.*, **31,** 1015 (1958).
137. F. A. V. Sullivan and A. R. Davis, *Rubber World*, **141,** 240 (1959).
138. W. D. England, J. A. Krimian, and R. H. Heinrich, *Rubber Chem. Technol.*, **32,** 899 (1960).
139. M. Braden, *J. Appl. Polym. Sci.*, **6,** S6 (1962).
140. M. Braden and D. Barnard, Summaries of Papers, National Rubber Producers' Assoc. Symp., July 1960; British Rubber Div., Publ. 13, p. 13 (1960).
141. G. H. Bebbington, Bell Telephone Laboratories, Inc., unpublished results, 1966.
142. G. T. Hodgkinson and C. E. Kendall, in *Proc. 5th Rubber Technol. Conf.*, T. H. Messenger, Ed., Institution of the Rubber Industry (London), 1962.
143. E. R. Erickson, R. A. Berntsen, E. L. Hill, and P. Kusy, *Amer. Soc. Testing Mater. Spec. Publ. No. 229*, 11 (1959).
144. R. M. Murray, *Rubber Chem. Technol.*, **32,** 1117 (1959).
145. Z. T. Ossefort, Ref. 143, p. 39.
146. O. Lorenz and C. R. Parks, *Rubber Chem. Technol.*, **36,** 201, (1963).
147. R. W. Murray and P. R. Story, presented at Div. Polymer Chemistry, American Chemical Society Meeting, St. Louis, 1961.
148. A. D. Delman, A. E. Ruff, B. B. Simms, and A. R. Allison, *Advan. Chem. Ser.*, **21,** American Chemical Society, Washington, D.C., 1959, p. 176.
149. R. R. Barnhart and T. H. Newby, in M. Morton, Ed., *Introduction to Rubber Technology*, Reinhold, New York, 1959, p. 140.
150. L. Michaelis, *Trans. 3rd Conf. Biol. Antioxidants*, **11** (1948); *Chem. Abstr.*, **44,** 4053d (1950).
151. R. L. Bent, J. C. Dessloch, F. C. Duennebier, D. W. Fassett, D. B. Glass, T. H. James, D. B. Julian, W. R. Ruby, J. M. Snell, J. H. Sterner, J. R. Thirtle, P. W. Vittum, and A. Weissberger, *J. Amer. Chem. Soc.*, **73,** 3100 (1951).
152. J. Furukawa, S. Yamashita, and T. Kotani, in R. F. Gould, Ed., *Stabilization of Polymers and Stabilizer Processes*, *Advan. Chem. Ser.*, **85,** American Chemical Society,

Washington, D.C., 1968, p. 110.

153. L. D. Loan, R. W. Murray, and P. R. Story, *J. Inst. Rubber Ind.* (*London*), **2**, 73 (1968).

154. L. D. Loan, P. R. Story, and R. W. Murray, unpublished results; R. W. Murray and H. Messler unpublished results.

155. R. W. Murray and P. R. Story, French Patent 1,494,578, Sept. 8, 1967.

156. J. H. Gilbert, in *Proc. 4th Rubber Technol. Conf.*, T. H. Messenger, Ed., 1962, p. 696.

157. G. H. Bebbington and R. W. Murray, unpublished results, 1967.

6

PROTECTION AGAINST IONIZING RADIATION

B. J. LYONS AND V. L. LANZA

Raychem Corporation, Menlo Park, California

A. INTRODUCTION

1. Definition of Terms and Units Used in Radiation Chemistry

Radiation chemistry is the study of the chemical effects of ionizing radiation. The action of radiation on materials is termed radiolysis, and the effects resulting, radiolytic effects. The ionizing radiation flux in a radiation field may be expressed in terms of the ionization produced (*i.e.*, the roentgen). However it is more important for the radiation chemist to know the amount of energy deposited in the irradiated material. This may be expressed in rads, the most universally accepted standard of absorbed dose, or in electron volts per gram. The relation between the various units used in radiation chemistry are given in Table 6.1.

Table 6.1 Definitions and Units

1 roentgen (r) is that quantity of x-ray or γ-radiation that produces in 1 cc of air at STP ions carrying 1 esu of electric charge. This corresponds to 2.08×10^9 ion pairs.

1 electron volt (eV) is the energy acquired by an electron in falling through a potential difference of 1 V.

1 rad corresponds to an energy deposition of 100 ergs/gram of absorber.

1 roentgen equivalent physical (rep) is the energy deposited by 1 r of x-rays or γ-rays in tissue and corresponds approximately to 93 ergs/gram or 0.93 rad.

Thermal energy, when applied to atomic particles, implies that their energy lies in the range resulting from normal random thermal motion of atoms. In general, energies less than 0.3 eV may be assumed.

1 r of x-rays or γ-rays delivers approximately 0.84 rad to air, 0.93 rad to water, and 0.96 rad to polyethylene. It therefore corresponds to approximately 0.96 rad when dealing with energy absorption in most polymers.

1 megarad (Mrad) $= 10^6$ rads $= 10^8$ ergs/gram $= 10$ J/gram
$= 6.25 \times 10^{19}$ eV/gram
$= 2.4$ cal/gram
$= 4.3$ Btu/lb
$= 1.26 \times 10^{-3}$ kWhr/lb

Much of the earlier experimental work on radiation protection was concerned with the evaluation of the effects of pile irradiation on polymers. A calculation of the energy deposited in samples irradiated in a nuclear reactor requires a knowledge of the number of neutrons and gammas of different energies and also of the cross section of the elements in the sample to radiations of each of these energies. The energy distribution in reactors varies from one position to another and can also be affected by the proximity of other absorbers. Moreover it rarely has been found possible to measure the gamma or fast-neutron flux inside reactors with any accuracy, and reactor doses are usually quoted in terms of the slow-neutron flux, which can be converted to the absolute absorbed dose as the conversion factors become known. The pile unit employed in many American (Oak Ridge) publications is a neutron flux of 10^{18} thermal neutrons/cm^2. The unit used in early work at Harwell, England, was 10^{17} thermal neutrons/cm^2 believed to correspond to an ionizing radiation dose of approximately 45 Mrad. This unit is only *approximately* one-tenth of the American unit, since the spectra of neutron and gamma radiation in the American and English reactors differed.

Radiolytic yields are expressed in terms of the chemical G value, which is defined as the number of molecules produced or changed by the deposition of 100 eV of energy. For aliphatic compounds the total yield of chemical events per 100 eV usually lies between 10 and 20. Aromatic hydrocarbons are usually found to be more than ten times less sensitive, and other aromatic derivatives occupy positions of intermediate sensitivity.

2. The Radiation Environment

Radiation environments of interest range from that inside a nuclear reactor to that of interplanetary space. The quality of radiation may range from electromagnetic radiation to energetic stripped nuclei up to and including iron.

a. Electromagnetic Radiation. X-rays are generated when an energetic electron hits a solid target, usually a metal of high atomic number. The electromagnetic spectrum produced is very broad; thus, because the absorption coefficient varies with wavelength, it is exceedingly difficult to estimate the absorbed dose given to a substance without direct absorption measurements. The electromagnetic radiations emitted by the nuclei of radioactive isotopes are called γ-rays. These isotopes occur naturally or they may be produced in nuclear reactors. γ-Rays are produced with discrete energy levels, for example, radioactive cobalt (Co-60) emits γ-rays with energies of 1.17 and 1.33 MeV. X-rays and γ radiations are highly penetrating, for example, the radiation (1.33 MeV) from Co-60 falls to half its initial intensity after passing through 11.5 cm of water.

b. Electrons. β-Rays (energetic electrons) are emitted by a number of radioactive isotopes, such as strontium-90 or tritium. In contrast to γ-rays, β-rays are emitted with a very wide range of energies from essentially thermal up to the maximum energy characteristic of the radio nuclide.

Energetic electrons may also be produced by electron accelerators using resonance and other forms of modified transformer or electrostatic generators as the source of high voltage, or by the intense electric field produced inside a waveguide by microwave radiation. Such sources are described in considerable detail in the literature[1–3] and will not be discussed further here. Electrons are less penetrating than γ- or x-radiation of equal energy, for example, the maximum penetration of a 1.3-MeV electron in water is about 0.7 to 0.8 cm.

c. Protons and α-Particles. Although α-particles and very energetic protons can cause displacement of nuclei by collision, their main effect is still to produce ionization and excitation. The relatively low velocity of protons and α-particles of quite high energies results in a high ionization rate and a correspondingly short range (for example, the range of a 2-MeV proton is only about 1% of the range of an electron of the same energy, and that of a 2-MeV α-particle is about seven times smaller than that of the proton).

d. Neutrons. Fast neutrons do not interact with orbital electrons but transfer energy through elastic collisions with atomic nuclei. Since this transfer is at a maximum for hydrogen atoms in polymers containing a preponderance of hydrogen atoms, the main effect is the production of energetic protons within the polymer. Since with most polymers of interest fast and slow neutrons do not induce significant amounts of radioactivity, the contributions from this source are not important. Howcvcr, significant induced radioactivity occurs in polymers containing chlorine.

e. Interaction of Radiation with Matter. The great majority of chemical changes result from electron–electron interactions, regardless of the nature of the incident radiation, provided atomic transmutation is not involved. This is because every primary interaction between the bombarding radiation and the absorbing material results in the ejection of energetic secondary electrons, which are themselves capable of ionizing many molecules. Observed differences in chemical effects of radiation are more properly ascribed to differences in ionization intensity than to differences in the nature of the incident radiation. Complete descriptions of the interaction of radiation with matter may be found,[1–3] as noted above.

f. Sequence of Chemical Events. We have seen that the first event of interest is the production by the primary particle of highly energetic secondary electrons, which themselves initiate many further ionizations until the ejected electrons have become "thermalized." Because the rate of deposition of energy by an electron increases as its velocity decreases, it follows that the majority of the free electrons are produced with energies close to thermal. As a result, most of the displaced electrons do not escape the coulombic field of the parent positive ion and geminate recombination occurs. The ionization may be written:

$$AB \rightsquigarrow AB^+ + e^-$$

where the symbol $\rightsquigarrow$ designates a radiation chemical event. It is known from mass spectrometry work that positive ions produced in high vacuum are often quite unstable and readily undergo molecular fragmentation or other ion–molecule reactions. However these reactions have only been observed to occur in the gas or liquid phase, and it is not at all certain that this type of process can occur as readily in highly viscous polymer media.

The energy transferred to an orbital electron by the bombarding particle may not suffice to produce an ionization; in such a case an excited molecule is produced:

$$AB \rightsquigarrow AB^*$$

The energy of excitation may fairly rapidly become localized in some particular bond of a molecule (and this is often, but not always, found to be one of the weaker bonds) resulting in homolytic bond scission:

$$AB^* \rightarrow A\cdot + B\cdot$$

Alternatively, the energy of excitation may be degraded to heat, or result in the emission of a quantum of light without chemical reaction of the excited molecule.

Highly excited molecules will arise as a result of charge neutralization:

$$AB^+ + e^- \rightarrow [AB^*]$$

This excited species carries excitation energy amounting to the lowest ionization potential of the molecule; for organic molecules this value may be greater than 10 eV, an energy much higher than any bond energy in the molecule; this means that homolytic bond scission with the production of free radicals is very likely.

Thus the observed chemical changes occurring in materials exposed to ionizing radiation may result from reactions involving positive ions, electrons, free radicals, and excited species.

Experimentally it is found that radiolytic reactions can be divided into two general classes according to their sensitivity to changes in the environment or to interference by added substances. Reactions that show a fairly marked dependence on radiation temperature or sensitivity to competition by added materials are usually described as secondary radiolytic events. Reactions that show very little dependence on temperature or other experimental parameters may be described as primary radiolytic events. An example of the last is the production of *trans*-vinylene unsaturation in polyethylene on irradiation. Since lifetimes of the radiolytic intermediates depend on the environment, this classification may well vary from one polymer to another; for example, ionic species that apparently decay extremely rapidly in polyethylene may survive for a considerably longer period in a more polar polymer. It should be borne in mind that this distinction between primary and secondary radiolytic events is empirical, and a radiolytic event is only assumed to be primary so long as it is found impossible to modify or inhibit it in some way.

3. Generalized Description of Radiolytic Changes in Polymers

There now seems general agreement that many, if not most, of the reactions that polymers undergo when exposed to ionizing radiation occur as a result of homolytic bond scission with the production of free radicals. In a polymer molecule of the general form

$$\left(\begin{array}{cc} a & \mathrm{H} \\ | & | \\ -\mathrm{C}- & \mathrm{C}- \\ | & | \\ b & d \end{array} \right)_n$$

side-chain scission (*i.e.*, C–*a*, C–*b*, C–*d*, or C–H bond scission) leads to the production of small molecular fragments (*a*, *b*, *d*, or H) that, once formed, can diffuse away from the polymer radical rapidly. Main-chain scission involving the carbon–carbon bond leads to the formation of free radicals that cannot casily diffuse away from one another because they are both polymeric; in other words, they are held in a cage formed by the surrounding polymer molecules. Thus it is hypothesized that there is a high probability that such free radicals will recombine. This perhaps is why in polyethylene, even though the carbon–carbon bonds in the main chain are considerably weaker than the carbon–hydrogen bonds, the major chemical products of radiolysis appear to arise as a result of carbon–hydrogen bond scission rather than carbon–carbon bond scission.

Ionic processes induced by ionizing radiation may include ion–molecule reactions or cationic and anionic polymerizations. Typical ion-molecule reactions that have been observed to occur in mass spectrometers are proton transfer reactions such as, in the case of methane

$$CH_4^+ + CH_4 \rightarrow CH_5^+ + CH_3$$

or condensation reactions typified by the reaction in methane:[4]

$$CH_3^+ + CH_4 \rightarrow C_2H_5^+ + H_2$$

Many attempts have been made to explain radiation-induced processes in polymers in terms of ion–molecule reactions such as the above.

Cationic dimerization or polymerization of olefins and monomers may be initiated by radiation. For example, this has been observed to occur with 1-hexadecene by Dainton and co-workers.[5] Anionic polymerization has only been observed when the positive charge is stabilized in some way, for example, in systems containing amines or amides.

Carbon–carbon unsaturation may result from two processes in polymers. *Trans*-vinylene or main-chain unsaturation (for example, in polyethylene) is believed to occur as a result of a primary chemical process involving the detachment of molecular hydrogen in a single step as originally suggested by Black in 1958:[6]

$$—CH_2—CH_2— \rightsquigarrow —CH{=}CH— + H_2$$

This formation reaction is very insensitive to temperature changes.[7–11] Unsaturation may also arise as a result of main-chain degradation. For example, the following scission reaction has been proposed[12] for polypropylene:

$$R—\overset{\overset{\large CH_3}{|}}{C}H—CH_2—R' \longrightarrow R—\overset{\overset{\large CH_3}{|}}{C}{=}CH_2 + R'H\cdot$$

Although the above form of degradation could result from a disproportionation reaction involving two polymeric radicals formed by homolytic main-chain scission, there is some evidence to suggest that main-chain scission frequently arises as a result of fragmentation reactions occurring subsequent to a side-chain scission. Thus it is possible that main-chain scission in poly(methyl methacrylate) occurs via the initial formation of a carboxyl radical that decarboxylates and rearranges to form two main-chain fragments.[13] The evidence for this process includes the close correspondence between gas evolution and chain-scission yields in this polymer.

It is observed that carbon–carbon unsaturation initially present in polymers decays more or less rapidly on exposure to ionizing radiation. This decay process, which is very rapid for vinyl and vinylidene groups but much less rapid for *trans*-vinylene groups, has been attributed to an energy-transfer process involving triplet formation,[8] or to an ion–molecule[5] or free-radical reaction.[14] Lyons[15] in 1963 pointed out that the kinetic behavior of the fast (vinyl) and slow (*trans*-vinylene) decay processes in polyethylene differs and suggested that *trans*-vinylene groups scavenged polymeric radicals to form allyl radicals, whereas vinyl, vinylidene, and other types of unsaturation decayed via a chain reaction with polymeric free radicals. The correlation between allyl radical formation and *trans*-vinylene decay was insufficiently precise at that time to support this hypothesis. However, recent work by Dole and co-workers[16] has shown a good correlation between the two processes. Recently, vinyl decay has been attributed to an ionic dimerization process,[17] but this explanation cannot account for the observed effect of temperature, the thermal history of the polymer, and orientation by drawing on the decay reaction.[10]

Side-chain scission in polyethylene may be written:

$$—CH_2—CH_2— \rightsquigarrow —CH_2—\dot{C}H— + H\cdot$$

$$H\cdot + —CH_2—CH_2— \rightarrow H_2 + —CH_2—\dot{C}H—$$

where it is believed that the hydrogen atom produced in the first reaction is very energetic. Crosslinking results from the combination of two polymer radicals

$$\begin{matrix} —\dot{C}H— \\ + \\ —\dot{C}H— \end{matrix} \longrightarrow \begin{matrix} —CH— \\ | \\ —CH— \end{matrix}$$

In more complex polymer molecules, side-chain scission may lead to the formation of polymer radicals; which, however, undergo scission rather than combining to form crosslinks as indicated above. It has been suggested in the case of polypropylene that chain-scission fragments containing terminal unsaturation react with the precursor of main-chain scission (which may be a polymer radical) to form a branched molecule[12] as in the following equation.

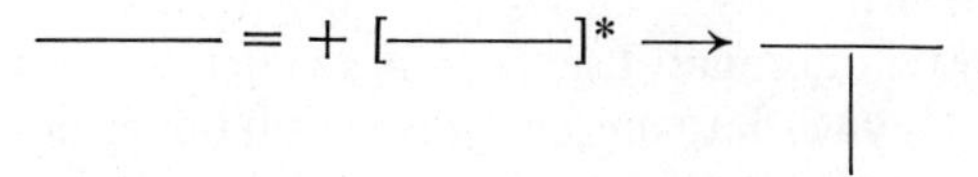

This mechanism has some relevance to one method of radiation protection described below.

B. RADIATION PROTECTION

The additivity rule first suggested by Manion and Burton[19] postulates that in the radiolysis of mixtures, the fraction of the total absorbed energy absorbed by each component of the mixture is proportional to the number of electrons the component of the mixture contributes to the total number of electrons present in the mixture. This may be expressed as follows:

$$G(\text{P}) = E_a G_a(\text{P}) + E_b G_b(\text{P})$$

where $G(\text{P})$ is the observed yield formed from a mixture whose components have yields of $G_a(\text{P})$ and $G_b(\text{P})$ and E_a, E_b are the electron fractions of components a and b in the system.

In a number of systems large deviations from the additivity rule are observed (e.g., in the classic case of benzene cyclohexane mixtures). The effect is illustrated schematically in Figure 6.1. Since the observed deviations are usually negative—that is, yields are lower than predicted—such deviations

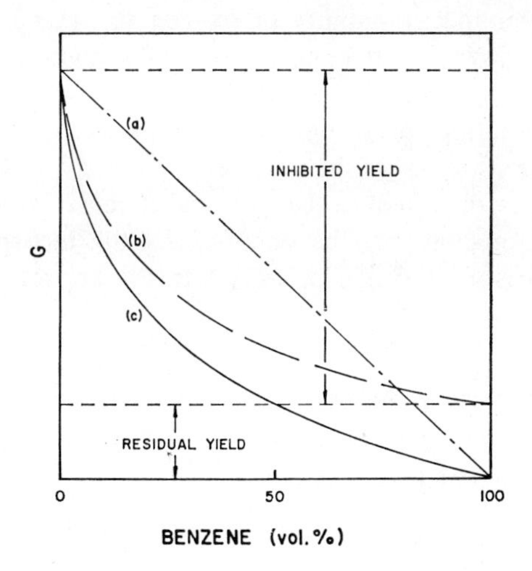

Fig. 6.1 Typical (schematic) inhibition curves: (a) yield expected from additivity rule; (b) observed yield corrected for dilution by benzene, (c) observed yield.

are usually described as radiation protection effects. Strictly speaking, only processes that restore the irradiated polymer to its original condition are radiation protective processes. However, many agents act by converting a chemical reaction that normally leads to marked deterioration in the physical properties of the polymer into another process that does not result in such a marked change in physical properties, and in this broader sense three types of protection may be envisaged. The inhibition may be sacrificial; that is, the additive or structure responsible for the protection itself reacts preferentially with an active species, which would otherwise lead to degradation of the polymer, to form products that can no longer protect the polymer. Inhibition may also be of the "sponge" type, in which a transfer of energy from the polymer to the protective substance occurs, the energy being then degraded by the protective agent without chemical reaction. A third type of protection is compensatory, that is, a polymer that crosslinks on irradiation may be simultaneously caused to degrade, or one that degrades may be caused to crosslink by suitable additions.

Sacrificial protection has been divided by Charlesby into two types, a true repair process:

$$P\cdot + HA \rightarrow PH + A\cdot$$

or, perhaps,

$$PH^{+} + A\cdot \rightarrow PH + \cdot A^{+}$$

and what Charlesby[37] terms stabilization repair:

$$P\cdot + X \rightarrow PX\cdot \text{ (unreactive)}$$

The extent of radiation-induced degradation often depends on the environment—for example, the availability of oxygen to induce radiolytic oxidation—and certain known additives protect against radiation-induced oxidation rather than the direct effects of the radiation itself.

Methods of radiation protection include internal protection, that is, modification of the structure of the polymer itself in a way serving to increase its radiation stability or to reduce the chemical effects of any radiolytic event. Radiation protection may also be obtained by the incorporation of small amounts of additives; following Charlesby,[20] this will be referred to as external protection.

C. KINETICS OF SCAVENGING

1. Scavenging of Free Radicals

a. General Kinetics. The kinetics of free radical scavenging has been examined by Charlesby and co-workers[20–27] and the following kinetic scheme is taken from Charlesby and Lloyd.[21,22]

Let I be the radiation intensity in Mrad/sec; [P·] the concentration of radicals on the polymer molecules, in moles/liter; and [A] the concentration of additive, also in moles/liter. The mechanism is postulated to be:

$$\text{polymer} \overset{k_1, I}{\rightsquigarrow} \text{P}\cdot$$

$$\text{scheme I} \quad \text{P}\cdot + \text{P}\cdot \xrightarrow{k_2} \text{P–P (second order radical decay process)}$$

$$\text{scheme II} \quad \text{P}\cdot \xrightarrow{k_2'} \text{inactive product (first order radical decay process)}$$

$$\text{P}\cdot + \text{A} \xrightarrow{k_3} \text{product}$$

The loss of additive is:

$$-\frac{d[A]}{dt} = k_3[\text{P}\cdot][\text{A}]. \tag{6.1}$$

Assuming steady-state conditions:

$$\text{scheme I} \quad k_1 I = k_2[\text{P}\cdot]^2 + k_3[\text{P}\cdot][\text{A}] \tag{6.2}$$

$$\text{scheme II} \quad k_1 I = k_2'[\text{P}\cdot] + k_3[\text{P}\cdot][\text{A}] \tag{6.3}$$

Scheme I. Let

$$x = \frac{k_3[\text{A}]}{4(k_1 I k_2)^{1/2}} \tag{6.4}$$

Then from (6.1), (6.2), and (6.4), and integrating,

$$(x^2 + 1)^{1/2} - \ln [1 + (x^2 + 1)^{1/2}] + \ln x + x = -k_3(k_2 I)^{-1/2} k_1^{1/2} r + c \tag{6.5}$$

where $r(= It)$ is the dose and c is a constant of integration.

At high concentrations of additive, when most active centers combine with the additive, (6.5) approximates to

$$2x = k_3\left(\frac{k_1 I}{k_2}\right)^{1/2}(t_0 - t)$$

and the loss of additive is given by

$$[\text{A}]_0 - [\text{A}] = k_1 I(t - t_0)(= k_1 r) \tag{6.6}$$

that is, the disappearance of A is 0 order in rate and $d\text{A}/dt$ gives G (active centers) directly.

At very low concentrations of additive most active centers disappear by recombination, x becomes very small, and (6.5) approximates to

$$1 + \tfrac{1}{2}x^2 - \ln (2 + \tfrac{1}{2}x^2) + \ln x + x = -k_3\left(\frac{k_1 I}{k_2}\right)^{1/2} t + c$$

that is,

$$\ln x + 0.307 = k_3\left(\frac{k_1 I}{k_2}\right)^{1/2}(t_0 - t) \tag{6.7}$$

or

$$\ln x = -k_3\left(\frac{k_1}{k_2 I}\right)^{1/2} r - 0.307 \tag{6.8}$$

Scheme II. The rate of loss of additive may be obtained from (6.1) and (6.3), whence

$$-\frac{d[\mathrm{A}]}{dt} = \frac{k_3[\mathrm{A}]k_1 I}{k_2' + k_3[\mathrm{A}]} \tag{6.9}$$

or

$$\frac{k_2'}{k_3}\ln\frac{[\mathrm{A}]_0}{[\mathrm{A}]} = k_1 r - ([\mathrm{A}]_0 - [\mathrm{A}]) \tag{6.10}$$

at high concentrations of additive the disappearance of A becomes independent of concentration and an equation similar to (6.6) pertains:

$$[\mathrm{A}]_0 - [\mathrm{A}] = k_1 r \tag{6.11}$$

whereas at low concentrations of additive the loss becomes exponentially dependent on the dose:

$$\ln\frac{[\mathrm{A}]_0}{[\mathrm{A}]} \approx \frac{k_1 k_3 r}{k_2'} \tag{6.12}$$

The rate of loss of additive is dose rate dependent in scheme I, but independent of dose rate in scheme II. Charlesby and Lloyd[21] found that although scheme I appeared to be followed in the cyclohexane–anthracene system, scheme II could be fitted best to the results obtained with anthracene–silicone mixtures[22] in that the loss of anthracene at low concentrations was found to be independent of dose rate in the range 6.4 to 211 rad/sec.

Probably the most plausible explanation of this independence of dose rate is that the radicals are produced in clusters, or at least in pairs, so that the chance of recombination within the cluster is much greater than that of random diffusion and recombination, and as a result the radical decay is first order.

Some results obtained by Charlesby and Lloyd for silicone mixtures are shown in Table 6.2.

Table 6.2 Decay of Anthracene in Polydimethylsiloxane

Source	$k_1 \times 10^3$ [moles/(liter)(sec)$^{-1}$]	$k_2 k_3^{-1} \times 10^4$ (moles/liter)
Co^{60}	3.0	1.56
Van de Graaf	1.8	1.56
Linear accelerator	1.3	1.56

G (crosslinks) in silicones may be calculated directly from k_1 (which is the concentration of active species produced by unit dose); values of 1.3 to 3.1 crosslinked units per 100 eV are obtained in good agreement with values quoted in the literature.

In a semicrystalline polymer such as polyethylene, the kinetic situation could be much more complicated. Any added material will segregate into the amorphous phase, whereas radicals are formed at random throughout the polymer during radiolysis. If the radicals can diffuse readily throughout the polymer, no change in kinetics will result. However, Lyons *et al.*[15] have pointed out that certain deviations from a simple kinetic scheme, such as II above which occur when larger concentrations of additives such as unsaturated compounds or scavengers are used in polyethylene, can be accounted for by postulating a change in the distribution of active species between the crystalline and amorphous phase.

This could possibly result from the reaction of hydrogen atoms (which can diffuse readily throughout the polymer) with the additive to form species (such as alkyl radicals) that can diffuse readily through the amorphous phase only. The main effect is to raise the range in which an exponential dependence of additive loss on dose is observed to higher concentrations of additive than would be expected on the basis of the theory outlined above. Using this modified kinetic scheme and following the isomerization of cyanurate to isocyanurate that occurs when dilute solutions of triallyl cyanurate in polyethylene are irradiated, G (free radicals) in a low-density polyethylene was calculated to be about 3.5 to 4, which is in good agreement with values from electron paramagnetic resonance measurements.

b. Variation of Protective Effect with Scavenger Concentration. Charlesby's treatment has been extended[40] to show the theoretical dependence of initial protective effect on scavenger concentration. It is assumed that the reaction of interest is the only one affected by the scavenger and that each scavenger molecule decayed corresponds to the inhibition of one radiolytic event. The radiolytic yield, G_0, is assumed to have two components, G which is insensitive, and G_0' which is sensitive to scavengers.

$$G_0 = G + G_0'$$

In the presence of a scavenger of initial concentration [A] the yield G_{A} is given by

$$G_{\mathrm{A}} = G + G_{\mathrm{A}}'$$

where

$$G_{\mathrm{A}}' = G_0' - G(-\mathrm{A})$$

Provided that the measurements are made over small dose increments, and from (6.9) ($k_1 \equiv G_0'$)

$$G_A' \approx G_0' - \frac{G_0' k_3[A]}{k_2' + k_3[A]}$$

or

$$\frac{G_A'}{G_0'} = 1 - \frac{k_3[A]}{k_2' + k_3[A]}$$

$$= \frac{k_2'}{k_2' + k_3[A]}.$$

Hence

$$\frac{G_0 - G}{G_A - G} - 1 = \frac{k_3[A]}{k_2'}$$

or

$$\frac{G_0 - G_A}{G_A - G} = \frac{k_3[A]}{k_2'} \tag{6.13}$$

Thus a plot of G_A against $(G_0 - G_A)[A]^{-1}$ will have an intercept of G and a slope of k_2'/k_3. However, because of the form of the abscissa, the measured slope can be unduly influenced by minor errors in G_A at low concentrations of scavenger. In these circumstances a better estimate of k_2'/k_3 may be obtained, knowing G, by a plot of $(G_0 - G_A)(G_A - G)^{-1}$ against [A].

Even if the assumptions made above are incorrect, then provided the scavenger has a similar effect on all the sensitive processes, the above relation will still hold. However, the slope of the relation will measure some average of the k_2'/k_3 values for these processes.

2. Scavenging of Ions

Kh. S. Bagdasar'yan and co-workers[28–32] have for some years been studying the mode of formation of cation radicals from aromatic amines and other compounds in frozen organic glasses (such as methylcyclohexane–isopentane[29] or methylene chloride[31]) and in poly(methyl methacrylate)[31,32] at liquid nitrogen temperature.

Radiolysis of poly(methyl methacrylate) is considered to involve the following steps:[32]

$$P \rightsquigarrow P^+ + e^- \tag{6.14}$$

$$P^+ + e^- \rightarrow [P^*] \rightarrow F_1^{\cdot} + F_2^{\cdot} \tag{6.15}$$

$$P^+ + A^{\cdot} \rightarrow P + {}^{\cdot}A^+ \tag{6.16}$$

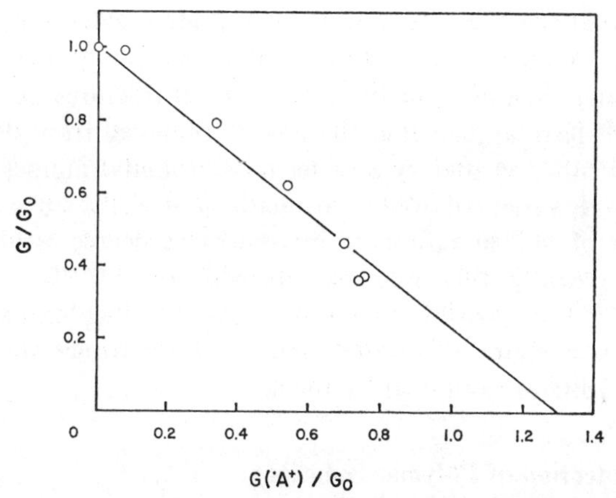

Fig. 6.2 Correlation between yield of ion radicals from *N,N,N',N'*-tetramethyl-*p*-phenylenediamine and yield of main-chain fractures in poly(methyl methacrylate).[32]

where P is a polymer molecule, P* the excited molecule produced by charge neutralization, $A^{\cdot}$ the additive free radical, and $^{\cdot}A^{+}$ the additive radical cation.

The total yield, G_0, of scissions in the absence of additive is given by

$$G_0 = G^* + \alpha G^+ \tag{6.17}$$

where G^* is the scission yield resulting from reaction of excited molecules (not formed by charge neutralization), G^+ is the yield of polymer cations, and α is the fraction of neutralizations that lead to scission. In the presence of additive the yield, G, of scissions is given by

$$G = G^* + \alpha[G^+ - G(^{\cdot}A^+)] \tag{6.18}$$

whence

$$\frac{G}{G_0} = 1 - \frac{\alpha G(^{\cdot}A^+)}{G_0} \tag{6.19}$$

Borovkova[32] and Bagdasar'yan measured the yield of radical cations not at room temperature (neutralization occurs too rapidly), but at 77°K. However G (scissions) G and G_0, were measured at room temperature. They conclude that, since (6.15) and (6.16) do not vary, or vary in the same manner with temperature, a comparison of these measurements is valid. It would seem to us that (6.16) would differ significantly in its temperature dependence from (6.15) because (*a*) the activation energy for diffusion of the additive is very different from that of an electron, (*b*) mobilities of positive and negative ions can and do differ considerably, and (*c*) electrostatic interaction comes into play in (6.15), which has essentially 0 activation energy.

Figure 6.2 shows the results obtained with *N,N,N',N'*-tetramethyl-*p*-phenylenediamine (TMPD). Borokova and Bagdasar'yan determine α to be

0.8, but note that their results would not exclude a value much closer to 1. They conclude from their data that the yield of unscavengeable excited species cannot be greater than 20% of the total yield of fractures and may be 0.

Kemp et al.[33] have argued that the spectra obtained from dilute mixtures of TMPD and other secondary and tertiary aromatic amines at low temperatures in a wide variety of organic aromatic or aliphatic solvents, and which resemble those of radical cations to a remarkable degree result from triplet states. (Very recently this was reconfirmed for TMPD by Allen and Capellos[35]). It is thus possible that Borovkova and Bagdasar'yan may have demonstrated scavenging of excited states to form triplet states of one or more of the additives examined by them.

3. Protection of Polymer Solutions

The kinetics of protection of polymer solutions have been studied by Charlesby and others,[24–27, 36–38] especially P. M. Kopp. They determined that two types of radiation action may operate in polymer solutions. In solutions of poly(ethylene oxide), for example, kinetic behavior and labeling experiments[23,24] clearly indicate that the action of the incident radiation is direct. In poly(vinyl pyrrolidone) solutions, after some initial uncertainty[23,24] Charlesby and Kopp confirmed by labeling experiments[36] that the incident radiation acts predominantly through an indirect effect. In both of these systems irradiation of aqueous solutions above 0.5% concentration leads to gelation of the polymer.

a. Kinetics of Protection against the Direct Effect in Solution.[26] The reaction scheme I and (6.1) and (6.2) apply if for $k_1 I$ one substitutes $k_1 Ic$. For gelation one crosslinking radical per weight average molecule is needed. The corresponding gelation time t_g is given by

$$\int_0^t k_2[\mathrm{R}\cdot]^2\, dt = \frac{10^3 c}{\mathrm{M_w}} = k_1 Ict_g \tag{6.20}$$

where c is the concentration of polymer and $10^3 c/\mathrm{M_w}$ is the number of weight average molecules per liter of solution. Of the polymer radicals only a certain fraction, $c/(c + c_0)$ will combine to give the external (intermolecular) crosslinks needed for gelation, the remaining radicals yielding internal (intramolecular) crosslinks, where c_0 is defined as the internal concentration of polymer.[38] Then

$$\int k_2[\mathrm{R}\cdot]^2 c(c + c_0)^{-1}\, dt = \frac{10^3 c}{\mathrm{M_w}} \tag{6.21}$$

If x is now given by

$$x = k_3[\mathrm{A}](4k_1\, Ick_2)^{-1/2} \tag{6.22}$$

and a new term τ is given by

$$\tau = -(k_3 k_2^{-1/2})(k_1 Ic)^{1/2} t \tag{6.23}$$

then (6.5) becomes

$$x + (x^2 + 1)^{1/2} + \ln x - \ln [1 + (x^2 + 1)^{1/2}] = \tau + k \tag{6.24}$$

which differs from the previous derivation in the substitution of $k_1 Ic$ (in x and τ) for $k_1 I$.

At high additive concentrations

$$[\mathrm{A}]_0 - [\mathrm{A}] \approx k_1 cr \tag{6.25}$$

which differs from (6.6) in the substitution of cr for r.

The radical concentration is given by

$$[\mathrm{R}\cdot] \approx \frac{k_1 Ic}{k_3([\mathrm{A}]_0 - k_1 Ict)} \tag{6.26}$$

and (6.20) may be rewritten

$$(k_1 cr_a)^{-1} = [\mathrm{A}]_0^{-1} + [\mathrm{A}]_0^{-2}(k_3^2 k_2^{-1})^{-1} t_g^{-1} \tag{6.27}$$

where r_a is the dose for gelation in the presence of additive. This equation predicts an approximately quadratic relation between r_a and $[\mathrm{A}]_0$ at very high concentrations of additive and this has been confirmed using thiourea as protective agent.[26] The relation can also be used to derive an approximate value for $k_3 k_2^{-1/2}$. Thus Charlesby and Kopp estimate a value of 0.25 ± 0.06 (liter/mole)$^{1/2}$ for thiourea in poly(ethylene oxide) solutions.

At low and intermediate values of additive concentration, a value is assumed for $k_3 k_2^{-1/2}$ based initially on the approximate estimate derived above. Using a theoretically derived relation between additive concentration and dose (Figure 6.3), the initial additive concentration $[\mathrm{A}]_0$ is converted to the equivalent x_0 and the corresponding time factor τ_0 $[= -k_3(k_1 Ick_2^{-1})^{-1/2} t_0]$ to read from the figure. The final time factor τ_a is then calculated from

$$\tau_a = -k_2 \left(\frac{k_1 Ic}{k_2}\right)^{1/2} t_0 + t_a$$

where t_a is the exposure time. Figure 6.3 then gives the corresponding value x_a.

The ratio

$$\frac{2(x_0 - x_a)}{\tau_0 - \tau_a} = \frac{\Delta[\mathrm{A}]}{k_1 Ict_a} = \frac{\Delta[\mathrm{A}]}{k_1 cr_a} \tag{6.28}$$

represents the fraction of the gelation dose used to produce polymer radicals that react with the scavenger. The remaining fraction $1 - \Delta[\mathrm{A}]/(k_1 cr_a)$

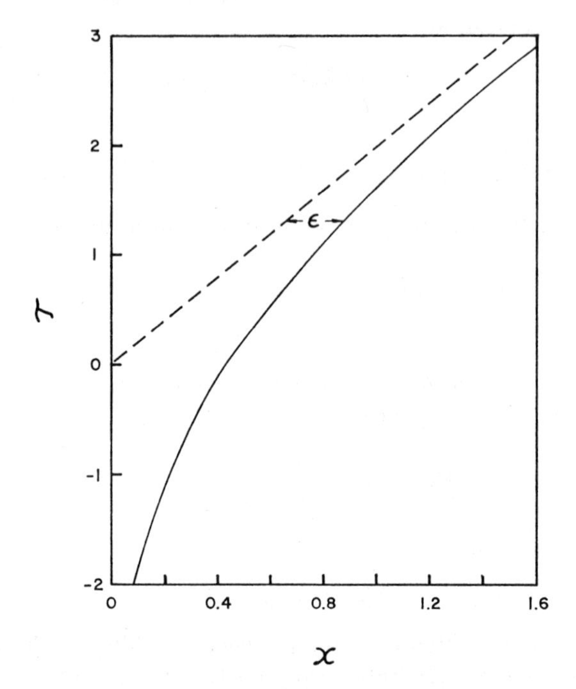

Fig. 6.3 Relation between τ and x.[26] Permission from Royal Society of London.

is used to produce radicals that crosslink to form the gel. Thus

$$\frac{1 - 2\Delta x}{\Delta \tau} r_a = r_g$$

By comparing the calculated and observed values of the gelation dose and altering the value of $k_3 k_2^{-1/2}$ to improve the agreement, an optimum value for this ratio may be obtained. Table 6.3 shows some results obtained by Charlesby and Kopp, assuming $k_3 k_2^{-1/2}$ for thiourea in aqueous poly(ethylene oxide) solution to be 0.2, 0.1, and 0.4, respectively.

b. Kinetics of Protection against the Indirect Effect in Solution.[36] When the principal mode of action is via the indirect effect, the kinetics becomes much simpler. Following Charlesby[36] we assume competition for H· or OH· (represented by W·) between additive, A, and polymer, P:

$$\mathrm{W\cdot + A \xrightarrow{k_a} products}$$

$$\mathrm{W\cdot + P \xrightarrow{k_p} P\cdot + WH}$$

Table 6.3 Calculated and Observed Gelling Doses for Poly(ethylene oxide) Solutions[26]

Poly(ethylene oxide) Concentration (%)	Assumed Value for $k_3 k_2^{-1/2}$	Thiourea Concentration (mg/liter)	Dose Rate (Mrad/hr)	Observed Gelling Dose r_a (Mrad)	Calculated Gelling Dose without Additive (Mrad)	Observed Gelling Dose without Additive (Mrad)
10	0.2	100	1	1.5	0.76	0.8
	0.1				1.14	
	0.4				0.49	
10	0.2	400	1	5.5	0.85	0.8
	0.1				1.75	
	0.4				0.28	
5	0.2	50	1	2.1	1.4	0.84
	0.4				1.03	
2	0.2			3.5	1.7	0.92
	0.4				0.96	
10	0.2	50	1	1.15	0.83	0.8
	—	200	1	2.4	0.73	0.8

The rate of production of polymer radicals is given by

$$\frac{d[\mathrm{P}\cdot]}{dt} = \frac{k_1 I k_p [\mathrm{W}\cdot][\mathrm{P}]}{k_p[\mathrm{W}\cdot][\mathrm{P}] + k_a[\mathrm{W}\cdot][\mathrm{A}]} \tag{6.29}$$

where $k_1 I$ is the rate of polymer radical production at incident radiation intensity I in the absence of scavenger.

In the presence of scavenger r_g is increased:

$$r_{g,a} = r_{g,0}\left(\frac{1 + k_a[\mathrm{A}]}{k_p[\mathrm{P}]}\right) \tag{6.30}$$

Thus the ratio $k_a k_p^{-1}$ may be readily estimated by measuring the concentration of scavenger required to double the gelation dose. Results obtained by Kopp

Table 6.4 Effect of Thiourea on Gelation Dose of Poly(vinyl pyrrolidone)[36]

Polymer Concentration (%)	Monomer Unit Concentration (moles/liters)	Thiourea Concentration (A) for Doubling Gelation Dose (mg/liter)	(mM/liter)	k_a/k_p = (P)/(A) (moles/liter)
16	1.13	37	0.49	2.9 × 10
5	0.45	10	0.132	3.4 × 10
2	0.178	10	0.132	1.35 × 10
1	0.089	1	0.013	6.85 × 10

and Charlesby[36] for thiourea in aqueous poly(vinyl pyrrolidone) solution are given in Table 6.4.

The agreement in the ratio is poorer at concentrations of poly(vinyl pyrrolidone) much below 2%. However, this is approaching the range (below 1%) in which no gel is obtained even in the absence of additive and a decrease in precision might be expected.

D. METHODS OF ESTIMATING SCAVENGING

1. Solubility Measurements

For substances that predominantly crosslink when exposed to ionizing radiation and are not already crosslinked prior to exposure, a very convenient method of estimating protection is afforded by solubility measurements. In systems such as the silicone fluids and aqueous solutions of poly(vinyl pyrrolidone) or poly(ethylene oxide) as studied principally by Charlesby and co-workers,[24–27, 36–38] the gel point and thus the gelling dose can be very clearly recognized; and such measurements, as shown previously, afford a simple and direct way of measuring scavenging efficiencies at low doses. Above the gel point the scavenging efficiency may be estimated using the relation:

$$s + s^{1/2} = \frac{p_0}{q_0} + \frac{1}{q_0 u_1 r} \tag{6.31}$$

where s is the soluble fraction and p_0, q_0 the chain scission and crosslinking probabilities per monomer unit, respectively, for unit radiation dose in a polymer of number average degree of polymerization u_1. This relation is only valid for a polymer of initially random molecular-weight distribution where p_0 and q_0 do not depend on the dose.[39] However, even if the observed relation is not linear, it may still be used to estimate scavenger efficiencies as

$$s + s^{1/2} = \frac{2}{\delta} \tag{6.32}$$

where δ is the crosslinking coefficient (the number of units crosslinked per weight average molecule) when only crosslinking occurs. Thus a comparison of the dose necessary to obtain a given gel fraction affords a direct comparison of scavenger efficiency (note that a comparison of differing gel fractions or $s + s^{1/2}$ values at the same dose will not necessarily yield a true measure of scavenger efficiency). When chain scission as well as crosslinking occurs, (6.32) becomes

$$s + s^{1/2} = \frac{2}{\delta'} \tag{6.33}$$

where δ' is the crosslinking coefficient and is related to δ above by the relation:[41]

$$\delta' = \frac{\delta}{1 + 2p_0\delta q_0^{-1}} \tag{6.34}$$

and if both chain scission and crosslinking are equally inhibited by scavengers, the situation is as outlined previously.

However, if crosslinking and chain scission are effected to differing degrees, a different technique of comparison should be used. The chain-scission and crosslinking probabilities in the absence of scavenger (p, q) become p' and q' in the presence of scavenger. If r and r' are the corresponding doses at which a given amount of gel is obtained, then p_0' and q_0' are given by

$$p' = p_0'r' \qquad \text{and} \qquad q' = q_0'r'$$

and if the polymer has an initially random distribution:

$$\frac{p_0}{q_0} + \frac{1}{q_0u_1r} = \frac{p_0'}{q_0'} + \frac{1}{q_0'u_1r'} \tag{6.35}$$

Thus a comparison of the tangents to a plot of $s + s^{1/2}$ against reciprocal dose at similar $s + s^{1/2}$ values will afford a measure of the protective effect on crosslinking, from which by also comparing the intercepts of the tangents at infinite dose without and with scavenger, the protective effect on chain scission may be calculated.[40] Provided that measurements at the same sol fraction are compared, this method will yield meaningful estimates even when the distribution is not random. However, a comparison of differing sol fractions at the same dose may lead to considerable error even if the polymer distribution is quite closely random.

The technique of Charlesby and Pinner[39] has been found to be very suitable for measurement of soluble fractions. We would, however, recommend that high-temperature extractions (e.g., of high-density polyethylenes with boiling xylene) include at least 0.5% of an efficient antioxidant, and that four 12-hr extractions, rather than two 24-hr extractions be performed, followed by a 30-min extraction with boiling antioxidant-free solvent.

2. Elastic Modulus

The estimation of crosslinking through elastic modulus measurements on samples swollen with solvent or, more frequently, by the more convenient method of stress–strain measurements on the polymer above its T_g or T_m has not been used to any large extent to measure protective effects. This is probably because some doubt exists about the possibility of obtaining equilibrium values, and there is uncertainty about the magnitude of corrections to be made for end effects, entanglements, and other factors.[42,43] In low-density polyethylene and in some rubbers, including natural rubber, however, the elastic modulus is found to be linearly related to the dose and hence can be used to estimate scavenger efficiencies. In high-density polyethylenes the elastic modulus at (say) 160°C is not necessarily linearly related to the dose[46] and furthermore depends markedly on the thermal history of the polymer before[44] or after[45] irradiation; as a result, modulus measurements do not provide a suitable estimate of protection efficiencies.

In studying scavenging, it is more convenient to write the theoretical expression for elastic modulus in the form[47]

$$f = \rho RTw^{-1}(\alpha - \alpha^{-2})q_0\left(r - \frac{2}{q_0u_1}\right) \tag{6.36}$$

where f is the elastic stress, R is the gas constant, T the absolute temperature, w the monomer molecular weight, and α the extension ratio. Where chain scission also occurs, the effect is to change the Flory correction factor for chain ends so that the equation becomes:

$$f = \rho RTw^{-1}(\alpha - \alpha^{-2})q_0\left[r - \frac{2(1 + p_0r)}{q_0u_1}\right] \tag{6.37}$$

whence

$$\text{modulus} = \frac{f}{\alpha - \alpha^{-2}} = \rho RTw^{-1}\left[q_0r\left(1 - \frac{2p_0}{q_0u_1}\right) - \frac{2}{u_1}\right] \tag{6.38}$$

Thus the occurrence of chain scission alters the slope of the modulus dose relation by the factor $(1 - 2p_0/q_0u_1)$, compared with the case where no chain scission occurs.[46] When scavengers differ in their effect on crosslinking and chain scission, it is possible to estimate the relative protective effects by obtaining the slope of the modulus–dose relation at a number of dose levels in the presence of additive (provided the scavenger concentration [A] is known and is low at these dose levels), then solving a set of simultaneous equations, each having the general form

$$\text{slope} \approx q_0[(1 - x[\text{A}] - 2p_0u_1^{-1}(1 - y[\text{A}])]K \tag{6.39}$$

and may be solved for x, y and, if they are not known already, q_0 and p_0. This procedure is obviously much more tedious than sol analysis.

3. Equilibrium Swelling

Equilibrium swelling has been used quite extensively for measuring the effects of scavengers on polymers, especially natural rubber, for which the interaction constant, μ in hydrocarbons has been measured accurately. The crosslinking density may be calculated from the Flory–Huggins equation

$$-\ln(1 - V_r) - V_r - \mu V_r^2 = \rho V_0 \text{M}_c^{-1} \text{V}_r^{1/3} \tag{6.40}$$

which is more familiar in the approximate form

$$\rho V_0 V_r^{0.6} = M_c(0.5 - \mu) \tag{6.41}$$

(valid for molecular weights between crosslinks, M_c, greater than 10^4) where V_0 is the molar volume of the swelling liquid, V_r is the volume fraction of polymer in the swollen material, and ρ is the density of the polymer.[43]

This derivation is only valid if no chain scission accompanies crosslinking. Since equilibrium swelling measures an equilibrium modulus (the solvent content at which the force of retraction just balances the osmotic pressure of the solution), the correction for chain scission is of the same form as that applied previously to elastic modulus measurements—that is, the incorporation of the Flory correction factor $(1 - 2M_c M_n^{-1})$ and its modification by a term expressing the chain-scission probability:

$$\rho V_0 V_r = \frac{(0.5 - \mu)w}{q_0 r(1 - 2p_0/q_0 u_1) - 2/u_1} \tag{6.42}$$

As in the case of modulus measurements, if both chain scission and crosslinking are affected by the scavenger, the contribution of each can be assessed by solution of a series of simultaneous equations.

4. Ultimate Properties

Changes in ultimate properties, especially elongation at rupture, have been frequently used to elucidate the protective effects of additives, especially in rubbers that may be already crosslinked prior to radiation exposure and consequently may not be suitable for investigation using solubility changes. Frequently the exposure time or dose required to reduce a given property to 50% of its initial value is used.

Flory[48] has pointed out that the extensibility of a crosslinked elastomer should vary as the square root of the distance between crosslinks. This conclusion has been verified experimentally by DiGuillio et al.[49] for peroxide-crosslinked ethylene–propylene rubbers, and Landel and Fedore[50] have shown that the same relation is followed by a wide range of elastomers. One of us[45] has shown that the extensibility of irradiated low- and high-density polyethylenes and ethylene copolymers above their crystalline melting points is linearly related to the reciprocal square root of the radiation dose (Figure 6.4).

Thus if the variation of the reciprocal square of the extensibility of an elastomer with the radiation dose with and without added scavenger is compared, an estimate of the number of crosslinks scavenged may be obtained:

$$\Delta c = (\alpha_{a,r}^{-2} - \alpha_{0,r}^{-2})K \tag{6.43}$$

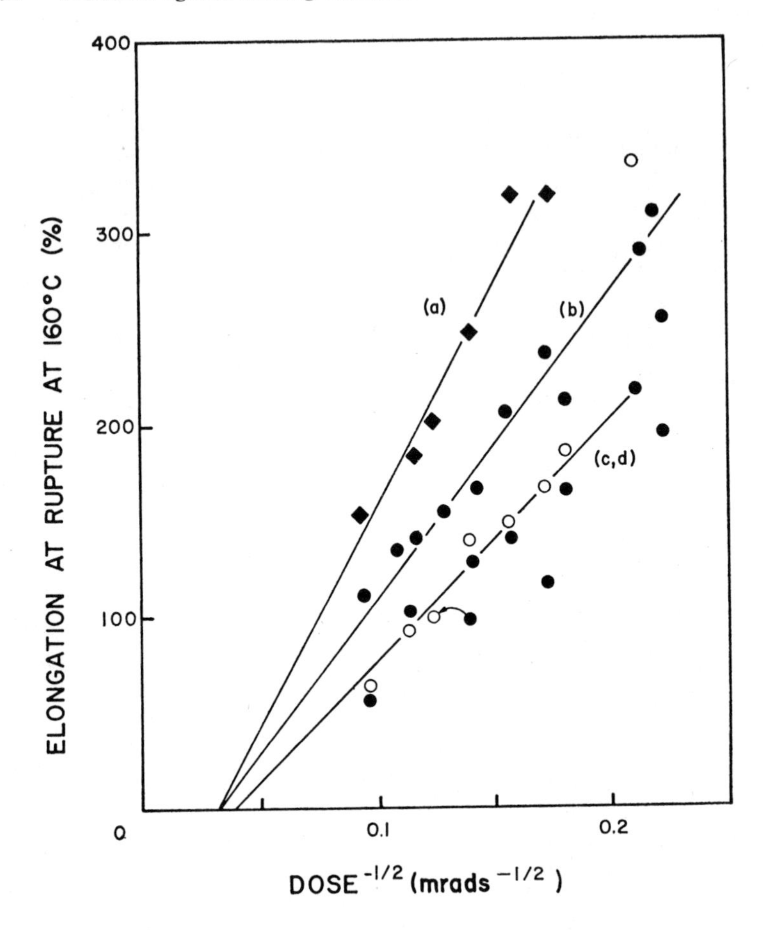

Fig. 6.4 Relation between elongation at rupture and radiation dose: (*a*) high-density polyethylene; (*b*) low-density polyethylene; (*c*) ethylene–methacrylic acid copolymer (closed circles); (*d*) copolymer, approximately 50% neutralized with sodium ion (open circles). Permission from American Chemical Society.

where Δc is the number of crosslinks inhibited by a concentration [A] of scavenger, $\alpha_{a,r}$, $\alpha_{0,r}$ are the extensibilities of the elastomer with and without the scavenger at radiation dose r, and K is a proportionality constant. The fraction of the incident energy diverted by the scavenger is given by

$$F = \frac{\alpha_{a,r}^{-2} - \alpha_{0,r}^{-2}}{\alpha_{a,r}^{-2} - \alpha_{a,0}^{-2}} \tag{6.44}$$

Provided that the initial concentration of scavenger is relatively small, Δc (or Δr, if equal extensibilities are compared) should vary approximately exponentially with the dose

$$\Delta c = \Delta c_{\max}(1 - e^{-k_1 k_3 r/k_2}) \tag{6.45}$$

from which the rate of reaction of additive and the maximum scavenging power of the additive may be obtained. Equation 6.45, of course, is simply another form of (6.12). It will be noted that no correction has been applied for inhibition of chain scission. This is because if a network has been formed in an initially uncrosslinked polymer, the extensibility is governed by the minimum distance between crosslinks, not the average. Furthermore the probability of finding a chain scission between the closest crosslinks (M_c min) must be 0 because if a chain scission occurs in the shortest active chain it must also have occurred with higher probability in all the other active chains and no network would have formed. The derivation should continue to be valid in an already crosslinked system so long as a network exists, even if the chain-scission probability exceeds that of crosslinking. Chain scission is mathematically the equivalent of reducing the initial molecular weight of the polymer, and elongation at rupture, which depends on the minimum distance between crosslinks, should not be grossly affected by molecular weight changes, provided a closed network exists. However, as one approaches the point at which the closed network ceases to exist, a marked change in elongation will occur. This treatment neglects the effect of branching (through the opening of closed loops) on elongation.

Equation 6.45 is also followed by the elastic modulus and sol + root sol inhibition, provided (*a*) that the modulus change with dose or sol + root sol change with reciprocal dose is linear in the absence of scavenger and (*b*) that chain scission is relatively unimportant.

Ultimate tensile strength measurements do not show such a simple dependence on crosslink density alone at a given strain rate and hence are less amenable to theoretical analysis.

5. Intrinsic or Inherent Viscosity

Estimating scavenging efficiencies by intrinsic or inherent viscosity has been used to considerable effect in studies on poly(methyl methacrylate) and polyisobutylene, which suffer extensive main-chain scission on irradiation. If the intrinsic viscosity can be related to the number average molecular weight of a polymer, changes in the main-chain scission rate caused by the additive may be readily calculated, provided that crosslinking reactions can be neglected. Once again, (6.45) should give a useful approximation to the observed scavenger effects from which the scavenging ability and rate may be estimated.

6. Stress Relaxation

Radiolytically induced chain scission in elastomers has been measured directly from stress-relaxation measurements. Although it has been experimentally verified that the rate of oxidative and thermal chain scission in elastomers is unaffected by strains of 100% or less,[51] no such verification appears to have been made in the case of radiolytically induced chain scission. It has been suggested that radiolytic chain scission in vulcanized natural rubber is strain dependent,[187] and this conclusion is discussed later. The stress relaxation shows a different dependence on initial crosslink density according to whether the scissions occur randomly along the chain or preferentially at crosslinks. In the first case the rate is inversely proportional to the crosslink density and in the second case it is independent of crosslink density.[51]

7. Miscellaneous Methods

a. Gas Evolution. Gas evolution has been used to estimate scavenging in some systems, especially where internal protection is involved. The measurement affords a simple and direct assessment of scavenger effects on reactions that result in the formation of gaseous products.

b. Electron Paramagnetic Resonance (EPR). The technique of electron paramagnetic resonance has been used to measure changes in polymer radical concentrations and in their rates of decay, which result from reaction with scavengers.

E. EXPERIMENTAL RESULTS

Radiation protection of polymers was reviewed by Charlesby in 1960.[52] We shall not duplicate any references reviewed by him unless they are particularly pertinent to the present discussion. More recently, reviews of narrower scope have also been made in the polymer field,[53–60] Bacq[18] has published a general account of progress, and there has been much activity in the medical and biological areas—only two very recent reviews of the many that have appeared are referenced.[61,62] This review is confined to synthetic polymers and biological polymers when studied in vitro.

1. Internal Protection

Polymers containing aromatic structural units [such as poly(ethylene terephthalate) or polystyrene] have considerable radiation stability. Apparently the benzene ring has the capacity to degrade the absorbed radiation without the formation of chemically active species.

Table 6.5 *G* Values for Main-Chain Scission in Isobutylene–Styrene Copolymers[63]

Styrene Content (%)	*G* (chain scissions) (events per 100 eV)
0	5.9
20	3.0
50	1.8
80	1.0

Charlesby and Alexander showed in 1955,[63] that the decrease in radiation susceptibility of isobutylene–styrene copolymers is more rapid with increased styrene content than can be explained on the basis of the reduction in isobutylene content. The *G*-values deduced by them for degradation of the isobutylene units after correction for the presence of styrene are shown in Table 6.5.

They conclude that the benzene ring protects not only its parent styrene unit, but also about one to two neighboring isobutylene units on either side. A slightly higher degree of protection has been claimed more recently.[64]

The protective effect of styrene in styrene–butadiene copolymers has been studied by a number of investigators.[65–69] Most agree that styrene reduces the crosslinking probability in excess of that expected on the basis of the aliphatic content of the polymer.[65–68] No protective effect was observed in physical mixtures of polystyrene and polybutadiene.[68] Some recent results have been interpreted as indicating that the aromatic group in three antioxidant-containing and peroxide-crosslinked styrene–butadiene copolymers enhances the crosslinking reaction and reduces chain scission.[69] However, the irradiations were apparently carried out at quite low dose rates (ca. 1.5 Mrad/hr) with free access of air, and it is possible that these results reflect the effect of the aromatic group on the sensitivity of the polymers to radiolytic oxidation. Terpolymerization of methyl methacrylate with styrene and butadiene results in some chain scission on irradiation, thus further lowering the net crosslinking.[68]

The effect of styrene content on the susceptibility of acrylonitrile–styrene copolymers has been examined.[70] The protective effect is found to be most efficient at a low (0.05 mole fraction) styrene content. Highly saturated ($> 80\%$) methyl mercaptan adduct polymers of emulsion-polymerized poly(1,3-butadiene) are extremely resistant to gamma radiation.[71] The protective effect of polystyrene grafted on polyethylene has been compared with that of physical mixtures of polystyrene, naphthalene, anthracene, and phenanthrene with polyethylene.[72] Both the mixture with polystyrene and the

Table 6.6 Hydrogen Evolved from Polyethylene Graft Copolymers with Styrene[72]

Electron Fraction of Styrene in Copolymer	G (hydrogen) (molecules per 100 eV)
0	3.0
0.014	2.7
0.075	2.1
0.180	1.7
0.238	1.6
0.375	1.3
1.0	0.02

graft reduce hydrogen evolution from polyethylene, but the graft is more effective. The variation of G (hydrogen) with the electron fraction of styrene grafted onto polyethylene is shown in Table 6.6.

The concentration of polymeric radicals[73] and the radioluminescence yield[74] in styrene–methyl methacrylate copolymers is found to decrease with increasing styrene content. The aliphatic hydrogen atoms in the styrene units are attacked by free radicals derived from the methacrylate moiety in the polymer.[42]

The reduced crosslinking yield from methyl phenyl silicones compared with dimethyl silicones has been frequently noted.[75–83] For equal concentrations of phenyl groups, the protective effect is claimed to be greater when the phenyl groups are grouped two to a Si–O segment than when they are distributed one to a segment.[79] More recently Schnabel[82] has contradicted this conclusion. Incorporation of aryl groups into the polymer backbone chain improves the resistance to radiation, compared with poly(dimethylsiloxane).[83]

The effects of ionizing radiation on ion-exchange resins based on styrene and the effect of changes in the ion-exchange groups have been reported.[84–91] Resins more highly crosslinked with divinyl benzene last longer in radiation fields.[84–86,89–91] Resins crosslinked with *m*-divinyl benzene are claimed to have crosslinks that are more radiolytically stable than those of resins crosslinked with the *o*- and *p*-isomers.[89] Cation-exchange resins containing pyridine nuclei are much more stable than other amine resins.[87] Resins prepared with aromatic structures in the polymer backbone are more stable than those containing aromatic nuclei as side chains.[90]

Cellulose and related materials have been stabilized against degradation by ionizing radiation by grafting vinyl or allyl monomers[92] or by esterification with aromatic acids.[93–96] Benzoate and cinnamate esters[93–96] show significant protective effects, but naphthoyl esters are less effectively protected.[96] There is disagreement over etherification; one group[93] finds significant protection with

Table 6.7 Radiation Parameters of Alkyl and Alkyl-Aromatic Polyamides[102]

Polymer Type	Water (%)	$1/q_0 u_1$	G (crosslinks) (events per 100 eV)	Chain Scission–Crosslink Ratio ($p_0 q_0$)
Poly(octamethylene isophthalamide)	Dried	103	0.60	0.60
	~5	141	0.42	0.62
Poly(*m*-phenylene sebacamide)	Dried	600	0.07	~0.5
	~6	650	0.06	~0.5
Poly(*m*-xylylene sebacamide)	Dried	20.5	0.5	1.14
	~6	9.3	1.05	1.08
Poly(octamethylene sebacamide)	Dried	12.2	1.2	0.65
	~4	12.3	1.2	0.62

the benzyl ether, whereas another group[96] reports no protection, observing that the benzyl group is cleaved from the chain by radiation.

Poly(vinyl alcohol) has been protected against radiation by benzalation and benzoylation.[97] The benzoate is found to offer more protection than the benzal derivative. Epoxy resins cured with phthalic anhydride are more stable to ionizing radiation than either anhydride or amine-cured resins.[98]

Aromatic polyamides are found to be much more stable to ionizing radiation than the aliphatic polyamides.[99–101] Work at this laboratory[102] on alkyl-aromatic polyamides has shown that the degree of protection is very markedly dependent on whether the amine or acid fragment is aromatic (Table 6.7). However, the major factor in determining the susceptibility of the molecule to radiolytic change appears to be the presence of hydrogen attached to the carbon atom α to the nitrogen atom, since lack of this accounts for an almost tenfold drop in radiation yield, whereas substitution of a phenylene group in the acid or a xylylene group in the amine moiety only halves the radiation yield.

2. External Protection against Crosslinking

a. Polyethylene. Cases of inhibition of crosslinking,[17,104–117] of radio oxidation,[118] and of oxidation by antioxidants that survive radiation[119–127] have been reported.

Polyethylene undergoes mainly crosslinking (G-value 2.0 ± 0.4[132]) *trans*-vinylene formation (G-value 1.9 for a high-density resin and 1.7 for a low-density resin[15]), and hydrogen evolution (G-value about 4[132]) on exposure to

Table 6.8 Early Experiments on Protection in Polyethylene[104]

Additive	Concentration (weight %)	Protection (%)
Di-β-naphthyl-	0.5	56
p-Phenylenediamine	0.4	43
2,6-Di-*t*-butyl-4-methylphenol	0.2	75
Ethylthiourea	0.2	84

ionizing radiation. The grounds for earlier conclusions[133] that appreciable main-chain scission occurs in this polymer have been questioned,[134] and more recent work[135,136] indicates that the chain-scission probability is very likely less than 3% of the crosslinking probability in a low-density polyethylene and may in fact be 0.

Lloyd[103] and Charlesby et al.[104] showed that the protective effect of antioxidants and other materials at low doses can be quite large. Some of their results are compared in Table 6.8. The protection factor is calculated from the increased dose necessary to obtain the same change (in melt index, elastic modulus, swelling, or solubility) when the additive is present. Okada et al.[105,106,109,113] have studied the effect of changes in the gaseous environment on crosslinking in polyethylene. The order of decreasing effectiveness as inhibitors is found to be nitrogen dioxide, oxygen, ammonia and sulfur dioxide, and chlorine. In these environments chemical reaction with the gas rather than crosslinking occurs.

Lyons and Dole[110] studied the reaction of nitrogen oxides with polyethylene films. They observed that the rate of consumption of nitric oxide during radiolysis is independent of the gas pressure, 1 molecule of nitrogen and 0.45 molecule of water being produced for every 3.4 molecules of nitric oxide consumed. The remaining nitrogen probably appears as nitrogen dioxide, which is found to react very rapidly with the polymer even in the absence of radiation. The proposed reaction mechanism is very similar to that suggested by Brown[111] for the heterolytic decomposition reaction between nitric oxide and isobutylene.

$$R\cdot + NO \rightarrow RNO$$

$$RNO + 2\,NO \rightarrow [R{-}N{=}N{-}O{-}\overset{\displaystyle O \atop \uparrow}{N}{=}O] \rightarrow N_2 + RONO_2$$

$$RONO_2 \rightarrow R{=} + HNO_3$$

$$\underline{HNO_3 + 0.5\,NO \rightarrow 0.5\,H_2O + 1.5\,NO_2}$$

$$R\cdot + 3.5\,NO \rightarrow R{=} + N_2 + 0.5\,H_2O + 1.5\,NO_2$$

The unsaturated groups react rapidly with the nitrogen dioxide produced.

The effects of paraffins, aromatic hydrocarbons, carbon tetrachloride, and sulfur on radical formation at low temperatures[107] have been investigated. No difference is noted with paraffins. With carbon tetrachloride, benzene, and toluene, lower radical yields are obtained.[107,108] Benzene inhibits crosslinking and vinyl decay but not *trans*-vinylene formation in a high-density polyethylene.[17]

At this laboratory the decay of scavenging and antioxidant ability of certain bisphenols and the decay of the initial antioxidant species with radiation dose has been studied and compared.[112] In all cases the decay is approximately exponential with dose; however the half-life for decay varies considerably (Table 6.9).

Table 6.9 Half-Dose for the Decay of the Initial Antioxidant Species, Antioxidant Activity, and Scavenging Ability in Polyethylene[112]

	Half Dose (Mrad)			Maximum Scavenging Yield (rads per molecule of scavenger)
Antioxidant[a]	Initial Species[b]	Antioxidant Activity[c]	Scavenging Ability[d]	
2,2′-Methylene-bis-(4-methyl-6-*t*-butyl-phenol)	5	6	44	≈11
4,4′-Methylene-bis-(2-methyl-6-*t*-butyl-phenol)	5	7	35	≈11
2,2′-Thiobis(4-methyl-6-*t*-butylphenol)	4	35	94	≈30
4,4′-Thiobis(2-methyl-6-*t*-butylphenol)	5	20	43	≈15

[a] All antioxidants were used at 0.5% concentration in a low-density polyethylene.
[b] Measured by decay of the characteristic absorbence band in the ultraviolet region.
[c] Measured by temperature programmed differential thermal analysis in a stream of oxygen.
[d] Measured by comparing the effect of the antioxidant on the variation of elastic modulus at 150°C with dose.

It will be noted that with the methylene-bridged bisphenols there is an excellent correlation between the decay of the initial antioxidant species, which can be followed using optical (ultraviolet) absorption measurements, and the decay of antioxidant activity. Since it is well established that the latter is associated with the phenolic function, it is reasonable to conclude that the ultraviolet measurements also follow decay of the phenolic function, which is found to decay at about the same rate on irradiation in all four compounds.

However, antioxidant activity in the thio-bridged bisphenols decays at a much slower rate on irradiation, suggesting that such activity in these cases is also associated with groups other than the phenolic function (*i.e.*, also with the thio group). In no case during irradiation does the residual antioxidant activity correlate with residual scavenging ability. With the methylene bisphenols, 50% or more of the scavenging ability survives at a dose sufficient to destroy substantially all antioxidant activity. This contrasts with Scanlan and Thomas's observation[130] that antioxidant activity and inhibition of peroxide crosslinking of natural rubber correlate well.

A number of interesting conclusions may be drawn from these results of Althouse. The average G-value over the first half-dose (5 Mrad) for the disappearance of the phenolic function is about 3. This yield is considerably greater than the yield of radicals formed in the amorphous phase of the polyethylene. Thus considerable scavenging must occur of radicals initially formed in the crystalline phase and subsequently migrating to the amorphous phase. Assuming a G(scavengeable radicals) of 4,[15] we can use a form of (6.45) to calculate, from the modulus data of Althouse, the maximum number of radicals scavenged by each antioxidant molecule. As Table 6.9 indicates, the maximum number for the methylene-bridged compounds roughly correlates with the maximum number of hydrogen atoms required to fully saturate the molecule. However, the thio-bridged bisphenols can inhibit radical recombination to a much greater extent. The unusual effectiveness of sulfur and sulfur-containing compounds has been noted elsewhere and will be discussed more fully below.

Black and Reynolds[114] have compared the protection afforded by a number of antioxidants at the levels of 0.5 and for 1% at a given dose, using the change in elastic modulus or other relative property as a measure of protection. The greatest protection is shown by tetramethylthiuram monosulfide and selenium dimethyldithiocarbamate.

The effect of small amounts of a number of antioxidants on the melt index of polyethylene at doses of about 1 Mrad have been compared.[115]

Charlesby and Kopp[116] found that addition of up to 5% of sulfur or 1% of selenium inhibits hydrogen formation in polyethylene some 10 to 15%. The inhibition of crosslinking (calculated from solubility data) is much greater, implying reactions other than the following equation:

$$S_x + H\cdot \rightarrow \cdot S_xH \tag{6.46}$$

They suggest a reaction similar to

$$\cdot S_xH + R\cdot \rightarrow RH + S_x \tag{6.47}$$

The yield of radicals (measured using EPR) at low temperatures is markedly depressed by the presence of sulfur. The sample also becomes highly colored. Charlesby and Kopp tentatively ascribe this to electron trapping.

Gardner and Harper[72] compared the effects of polystyrene, naphthalene, anthracene, and phenanthrene on hydrogen production in polyethylene, concluding that the compounds with the highest resonance energy protect the polymer to the greatest extent. They attribute the protection to energy transfer to the additive followed by dissipation without chemical reaction, but they observe that scavenging of hydrogen atoms cannot be excluded. In

Table 6.10 Protective Factors for Various Compounds in Polyethylene[131]

	Protective Factor[a] at Doses (Mrad)		
Additive[b]	10	15	20
Benzophenone	0.25	0.27	0.18
Hexachloroethane	0.36	0.40	0.30
$\alpha,\alpha,\alpha,\alpha',\alpha',\alpha'$-Hexachloroxylene	0.30	0.22	0.25
Hexachlorobenzene	~0	—	—
Hexabromobenzene	0.30	0.40	0.35
Tetrabromobenzene	0.25	0.40	0.40

[a] Defined as the fraction of the energy diverted by the additive and calculated from the change in elastic modulus at 160°C.
[b] All additives used at 1% concentration in a low-density polyethylene (DFD 6040).

view of Charlesby and Lloyd's observation of the decay of anthracene in polydimethylsiloxane[22] and hexane,[21] hydrogen atom scavenging may well occur.

Shibata et al.[117] conclude that the most effective protective agents in polyethylene are the phenolic antioxidants, especially 4,4'-thiobis-(6-*t*-butyl-3-methylphenol). Kostrov and Konkin[118] conclude that the main role of antioxidants in suppressing postirradiation grafting of vinyl monomers is as an inhibitor of oxidation during irradiation in air.

Photosensitizers such as carbon tetrachloride and chloroform inhibit radiolytic crosslinking; benzophenone, phosphorous trichloride, and ethylene tetrachloride are said to be almost without effect.[119] In this laboratory Lyons[131] has studied a number of similar compounds. Protective factors are shown in Table 6.10. The differences in activity between hexachloroethane, $\alpha,\alpha,\alpha,\alpha',\alpha',\alpha'$-hexachloroxylene, hexachlorobenzene, and hexabromobenzene are interesting. We would point out that although the chlorine atom and the substituted phenyl radical are active enough to abstract hydrogen atoms from the polymer, the pentachloroethyl (or trichloromethyl) and substituted benzylic radicals and bromine atoms are not capable of such abstraction reactions.

Charlesby and Black[120] examined the effect of a range of antioxidants on the radiooxidation that accompanies irradiations in air. They conclude that the antioxidants studied are only marginally effective, pointing out that it is uncertain whether any worthwhile reduction in radiooxidation can be achieved without markedly reducing the crosslinking yield in the polymer.

2,2′-Methylene-bis(4-ethyl-6-*t*-butylphenol)[125] is claimed to prevent odor formation in polyethylene films when exposed to ionizing radiation doses in the food sterilization range. The use of di-β-naphthyl-*p*-phenylenediamine as a radiation-resistant antioxidant for polyethylene seems to have been first suggested by Black and Brazier,[121] although a number of patent applications were filled about the same time.[121–124] These claims may seem surprising, since this compound has also been classed as an effective radiation-protective agent,[102] but although this diamine scavenges rapidly, the maximum number of radiolytic crosslinks it can inhibit is apparently about 1 per molecule.[46]

Gladkova et al.[126] have studied the antioxidant activity of a number of antioxidants at the 2 to 10% level in polyethylene after high doses of irradiation. The elongation at break of the aged material was measured after various periods of time in ovens at 150 and 200°C. After a 600-Mrad dose and on aging at 200°C, antioxidant activity is found to increase in the order *N*-isopropyl-*N*′-phenyl-*p*-phenylenediamine, 2,2′-methylene-bis(4-methyl-6-*t*-butylphenol), 2,2′-methyl-bis(4-ethyl-6-*t*-butylphenol), phosphite ester of a phenol–styrene condensation product, 4,4′-thiobis(6-*t*-butyl-*m*-cresol).

Irradiated polyethylene containing thiophosphates such as O,O,O-tri-*p*-tolylmonothiophosphate is claimed to have improved resistance to oxidation.[127] *N*-Stearoyl-*p*-aminophenol,[115] 2-mercaptobenzimidazole,[117] and nordihydroguaiaretic acid[117] are effective antioxidants in polyethylene after irradiation. Efficacy has also been claimed for zinc stearate and tin dioxide.[129] Organo-tin stabilizers, such as dibutyl tin maleate have a "stabilizing antioxidative" effect on irradiated polyethylene.[128]

b. Polypropylene. The protective effect of antioxidants and radical scavengers[137–143] on polypropylene and the effect of radiation on antioxidant effectiveness[143–145] in this polymer have been reported.

There has been a considerable divergence of opinion on the radiolytic behavior of polypropylene. For example, it has been concluded (*a*) that *G* (chain scissions) decreases and *G* (crosslinks) increases with increasing dose;[12,135] (*b*) that *G* (chain scissions) only decreases,[146] in both cases the initial chain-scission rate exceeding the crosslinking rate; (*c*) that both *G* values are independent of dose and the crosslinking rate predominates;[147,148] (*d*) that there is a large difference in the p_0/q_0 ratios measured for the atactic and isotactic polymer;[150] (*e*) that this ratio for the atactic polymer is similar to that for the isotactic polymer;[34,149] and (*f*) that slow-cooled and quenched isotactic polymers have markedly different p_0/q_0 ratios.[150] It seems fairly well established, however, that gel formation in isotactic polypropylene is remarkably sensitive to the thermal history of the polymer prior to and after irradiation and to the temperature of irradiation,[151,152] so the apparent inconsistencies may be caused by differences in crystallinity and, to a larger extent, crystallite size and amount of postirradiation radical reactions. Some of these inconsistencies have been attributed to postirradiation oxidation.[141]

Radiolytic crosslinking in isotactic polypropylene is inhibited by 2,6-di-*t*-butylmethylphenol[138,140,142] and bisphenols,[139] leading to better retention of physical properties on irradiation. Molecular iodine, methyl–mercaptan, and a number of low-molecular-weight compounds completely or almost completely inhibit crosslinking but reduce the chain-scission probability only by one-third.[141] Although 2,6-di-*t*-butyl-4-methylphenol, 3,5-di-*t*-butylpyrocatechol, and phenyl-β-naphthylamine all reduce alkyl free-radical concentrations in irradiated polypropylene to a marked degree at liquid nitrogen temperatures, only the first two are effective inhibitors of crosslinking.[145] In addition to the above, 4-methyl-, 4-*t*-butyl-, and 4-*t*-octyl-pyrocatechol are effective protective agents.[143] The fluorescense and absorption spectra of irradiated polypyropylene containing di-β-naphthyl-*p*-phenylenediamine have been studied.[142]

Anthracene, hydroquinone, and tin dichloride have little effect on gel formation.[137] The loss of antioxidant activity of bis(2-hydroxy-3-*t*-butyl-5-ethylphenyl)methane in this polymer has been studied.[144] Pyrocatechol α-naphthol cyclic phosphate, 4,4′-thiobis(2-*t*-butyl-4-methylphenol) and a mixture of the above phosphate with 2,6-di-*t*-butyl-4-methylphenyl cyclic phosphite are found to retain effective antioxidant activity in polypropylene after irradiation.[139]

c. Polyamides. It has been reported that the crosslinking[153–155] and chain-scission[154,155] rates in poly(hexamethylene adipamide) are dependent on the dose. More recently it has been concluded from solubility measurements that chain-scission and crosslinking *G* values are constant with radiation dose with a p_0/q_0 ratio of 0.5 to 0.7, the lower ratios being found with polyamides derived from longer chain acids and amines.[102]

Radiolytic crosslinking in polyamides is inhibited by small amounts of acrylic acid.[154] The inhibitory effect of various amines[156,157] and aminotriazines[156] on radiolysis and radiooxidation has been studied.

Di-β-naphthyl-*p*-phenylenediamine has little effect on the EPR spectra of irradiated poly(caprolactam)[157,159] or on the *G* values for hydrogen and carbon monoxide evolution.[160] This amine stabilizes the polyamide against radiooxidation.[158–160] The diformyl derivative of this amine is better than the amine itself for protection of polyamides against radiooxidation and oxidation subsequent to irradiation.[157]

d. Proteins and Similar Natural Products. Irradiated mixtures of various dry proteins with sulfur compounds have been studied using ESR techniques by Ormerod et al.[161,162,164,165] by Hsin et al.,[163] and by Gordy.[166] Ormerod and Singh have reported that metal ions markedly lower the free-radical concentration at 77°K and the alkyl–thio radical concentration at room temperature of dry γ-irradiated bovine serum albumin.[165] It has been suggested that when irradiation takes place at low temperatures, free electrons become trapped on added sulfur-containing compounds.[164] These anions react at higher temperatures to yield free radicals.

e. Poly(vinyl chloride). The effect of scavengers on the radiolytic response of poly(vinyl chloride)[167–183] and its copolymers[173] have been studied.

Poly(vinyl chloride), when irradiated with ample access of air, is found to darken slightly and degrade during irradiation;[184] however, crosslinking predominates if air is excluded from the sample[185] and the polymer rapidly darkens.

Wippler[167] studied the effect of additives on color formation in poly(vinyl chloride) during and after exposure to ionizing radiation. Metal stearates and diphenylthiourea stabilize against color formation without marked inhibition of crosslinking.[167,168] Organo-tin stabilizers inhibit crosslinking very markedly, but not color formation.[167] Incorporation of styrene and an alkyl ester of an α,β-unsaturated dicarboxylic acid into the resin greatly increases its radiation resistance.[167] Lower phthalate plasticizers stabilize more effectively than higher homologs or aliphatic esters.[170,174,178] Carbon (graphite) and lead fillers also diminish deterioration in physical properties on irradiation.[170] Other investigators have concluded that no filled polymer has radiation resistance as good as the polymer itself; the most radiation-resistant compound contained anatase, dioctyl sebacate, and an organo-tin–cadmium stearate–epoxidized-oil-stabilizer combination.[171]

The lower phthalate plasticizers have been variously reported as more[170,174,178] or less[167,171,176] effective radiation-protective agents than tritolyl phosphate.

Organo-tin stabilizers lower grafting rates of acrylonitrile and of styrene on poly(vinyl chloride), whereas lead stearate or tribasic lead sulfate are without effect.[181]

Electron acceptors, such as anthracene, *p*-terphenyl, *p*-benzoquinone, and chloranil are stated to reduce hydrogen and hydrogen chloride evolution in

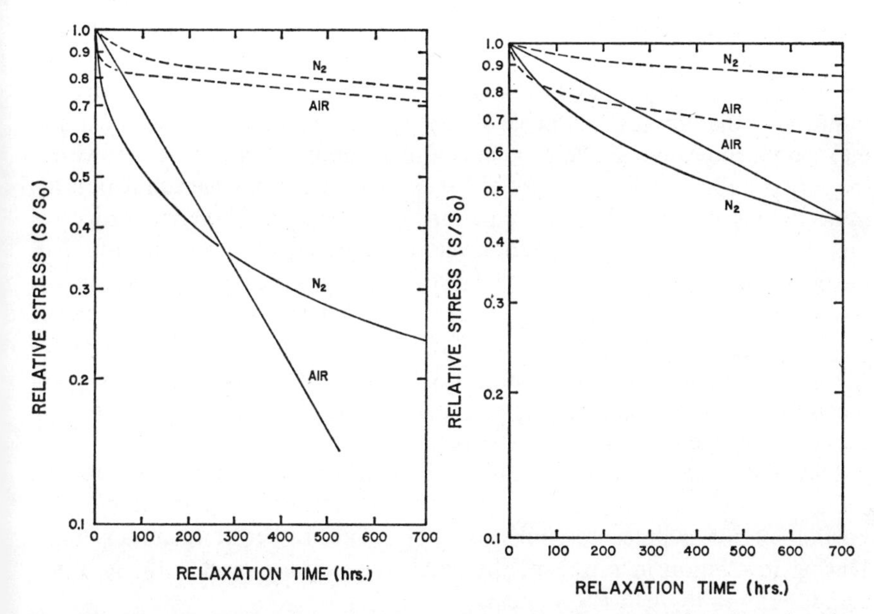

Fig. 6.5 The stress relaxation of carbon black-loaded natural rubber containing one part of phenyl-2-naphthylamine (control):[187] broken curves indicate phenomenon in the absence of ionizing radiation; solid curves indicate the presence of radiation.

Fig. 6.6 The stress relaxation of carbon black-loaded natural rubber containing five parts of *N*-cyclohexyl-*N'*-phenyl-*p*-phenylenediamine:[187] broken curves indicate phenomenon in the absence of ionizing radiation; solid curves indicate the presence of radiation.

this resin.[180,182] EPR measurements taken at 70°K suggest that anion radicals are formed.[180,182]

Dibutyl tin maleate and dichloride are the most effective and dibutyl tin dimyristate the least effective postirradiation thermal stabilizers of a number of organo-tin compounds examined, but none of these compounds is an effective stabilizer against γ-radiation.[177] Preferred thermal stabilizers for irradiated poly(vinyl chloride) compounds have been discussed.[183]

f. Natural Rubber. The effects of scavengers on the radiolytic changes that occur in natural rubber have been studied using already crosslinked[186–188,192,211,213,215,218] or uncrosslinked[189–191,193–200] gums. This polymer crosslinks [$G(X) \sim 0.9$] on irradiation. The chain-scission rate is 5% or less of the crosslinking rate.[198]

After an initial empirical survey,[186] Bauman and Born[187] published a detailed study of the effect of a number of scavengers on radiolytic stress relaxation in a sulfur vulcanizate. Typical results obtained by them are shown in Figures 6.5 and 6.6. To calculate the chain-scission rate they compared the slopes of the linear portions of the curves (i.e., at longer relaxation times), then corrected for the relaxation rate that would have been observed in the absence of radiation. They note that scavengers are somewhat less effective in nitrogen than in air. The linear decay of stress in air (see Figures 6.5 and 6.6) would indicate a single mechanism of scission operates here. However, the very pronounced nonlinearity of the experiments in nitrogen shown in the same figures indicates the participation of at least two different processes (apart from viscous effects). Moreover, the amount of stress relaxation in nitrogen during most ($> 60\%$) of the stress decay is greater than that in air. At very low irradiation times the relaxation in nitrogen can be one order of magnitude greater than that in air. Bauman and Born conclude that main-chain scission rates in irradiated, stressed, vulcanized natural rubbers are much higher than in unstressed, uncured rubbers. Their results may also be explained by the following hypothesis.[131]

The quite pronounced relaxation observed in the absence of ionizing radiation both in nitrogen and air probably indicates an appreciable proportion of the sulfide crosslinks are polysulfide rather than monosulfide. During irradiation in nitrogen, the alkyl free radicals react with the polysulfide bridges to yield RS· radicals.

$$RH \rightsquigarrow R\cdot + H\cdot$$

$$H\cdot + RH \rightarrow R\cdot + H_2$$

$$R\cdot + RSSR \rightarrow RSR + \cdot SR$$

Stress relaxation occurs as a result of the exchange reaction

$$RS\cdot + RSSR \rightleftharpoons RSSR + \cdot SR$$

Termination occurs by

$$R\cdot + \cdot SR \rightarrow RSR$$

After all the polysulfide links have been consumed in this way, stress relaxation occurs via the less favored reaction:

$$R\cdot + RSR \leftrightharpoons RSR + R\cdot$$

When an ample supply of oxygen is present, the alkyl radicals react to form peroxy radicals, which decompose to yield chain scissions (as was suggested by Bauman and Born[187]) in a chain reaction perhaps similar to the mechanism suggested by Chapiro[201] for poly(ethylene).

$$\sim\underset{O_2^{\cdot}}{CH}-CH_2\sim \longrightarrow \sim C(=O)H + \cdot OCH_2\sim$$

These Bauman and Born's values for G (chain scission) in sulfur-vulcanized natural rubber in the absence of air could be regarded as yields of sulfur crosslink scission (or exchange). This hypothesis removes the conflict between Bauman and Born's conclusions and those of other workers who find (G chain scissions) in unvulcanized natural rubber to be less than 10% of the crosslinking yield.[198,202] Some of the results obtained by Bauman and Born are given in Table 6.11 where, in line with the above hypothesis, the

Table 6.11 The Effect of Radical Scavengers on Scission in a Vulcanized Natural Rubber[187]

Additive (5 pph)	G (s-bond scissions) in Nitrogen (events per 100 eV)	G (main-chain scissions) in Air (events per 100 eV)
None (1 pph phenyl-2-naphthylamine)	2.7	13
N-Phenyl-*N′*-*o*-tolylethylenediamine	1.8	4.3
N-Cyclohexyl-*N′*-phenyl-*p*-phenylenediamine	1.2	1.4
6-Phenyl-2,2,4-trimethyl-1,2-dihydroquinoline	1.9	4.2
N,N′-Dioctyl-*p*-phenylenediamine	1.5	5.0
2-Naphthylamine	1.6	5.6
1,4-Naphthoquinone	2.0	5.6
Phenylhydroquinone	2.2	5.4
2-Naphthol	1.3	4.1
N,N′-Diphenyl-*p*-phenylenediamine (35%) + phenyl-1-naphthylamine (65%)	1.4	3.7
N,N′-Dicyclohexyl-*p*-phenylenediamine	1.5	3.0
p-Quinone	2.8	7.8

yield columns have been renamed. Recently stress-relaxation studies in nitrogen[290] have independently confirmed the hypothesis.

In a later paper Bauman[188] calculated G (crosslinks) from swelling measurements on these rubber samples irradiated in the relaxed state. Because no correction was made for concurrent chain scission, the values obtained in air may not be reliable. If the above hypothesis is correct, however, the values obtained in vacuum are not affected by such considerations and yield true estimates of crosslinking yields and inhibition. The results in vacuum are shown in Table 6.12.

Table 6.12 The Effect of Radial Scavengers on Crosslink Yields in a Vulcanized Natural Rubber[188]

Additive (5 pph)	*G* (crosslinks) in vacuum (events per 100 eV)
None (1 pph phenyl 1-2-naphthylamine)	1.9
N-Phenyl-*N'*-*o*-tolylethylenediamine	1.1
N-Cyclohexyl-*N'*-phenyl-*p*-phenylenediamine	1.3
6-Phenyl-2,2,4-trimethyl-1,2-dihydroquinoline	0.83
N,N'-Dioctyl-*p*-phenylenediamine	0.87
2-Naphthylamine	0.87
1,4-Naphthoquinone	1.1
Phenylhydroquinone	1.1
2-Naphthol	1.1
N,N'-Diphenyl-*p*-phenylenediamine (35%) + phenyl-1-naphthylamine (65%)	0.97

The rate of free-radical decay is greater in rubber having polysulfide crosslinks than in rubber with monosulfide crosslinks[191,217,221] but lower than that in rubber not containing sulfur compounds.[214] Turner[194] has carried out a detailed survey of the effect of scavengers on the radiation crosslinking of pale crepe. Crosslinking yields were estimated from equilibrium swelling measurements. A general survey was made of a variety of additives at one concentration and at one (in a few cases, two) dose levels. Some of the more effective inhibitors were studied at a number of concentrations. The dependence of the swelling ratio on the concentration of some of these additives is shown in Figures 6.7 and 6.8. He finds that even at quite high additive concentrations at least one-third of the crosslinks are still formed. Certain additives (e.g., hydroquinone) increase hydrogen evolution while inhibiting crosslinking. Turner suggests that this could arise through the contribution of reactions such as:

$$H\cdot + HOC_6H_4OH \rightarrow \cdot OC_6H_4OH + H_2$$

Although the inhibitory power of scavengers such as quinones, disulfides, and aromatic nitro compounds may be correlated with their reactivity toward free radicals, Turner[194] concludes that in some cases (notably 1,5-di-nitronaphthalene) the protective effect might involve energy acceptance.

In other work, radiolytic crosslinking in rubber is found to be inhibited by sulfur,[195,196,199,211,213] thiosulfinates and sulfoxides, tetramethylthiuram disulfide,[195,196,199,211,218] and nitromethane.[280] It has been suggested that the

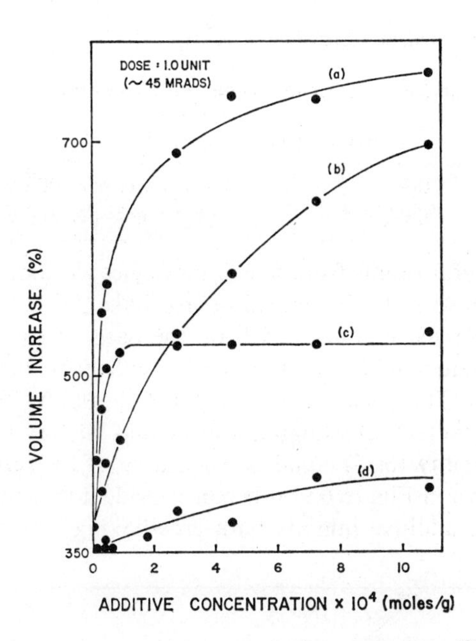

Fig. 6.7 The effect of concentration of additive on swelling:[194] (*a*) *m*-dinitrobenzene; (*b*) diphenylamine; (*c*) sulfur, (*d*) stearic acid.

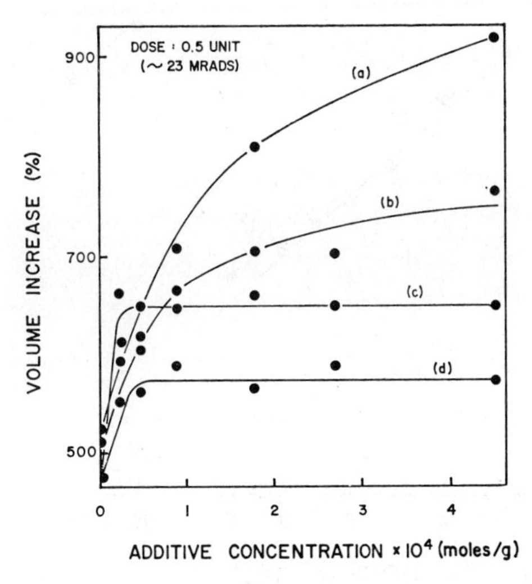

Fig. 6.8 The effect of concentration of additive on swelling:[194] (*a*) hydrazobenzene; (*b*) azobenzene; (*c*) dinitronaphthalene; (*d*) sulfur.

Table 6.13 Influence of Nitrobenzene on Crosslinking Fractures

	No Additive	Nitrobenzene (2.5×10^{-4} moles/gram)
G (crosslinks)	0.9	0.5_5 (events per 100 eV)
G (fractures)	0.04_5	0.08_5 (events per 100 eV)

protective effect of sulfur results from free S_8 molecules via energy acceptance and does not occur with sulfur combined in crosslinks.[193]

More recently it has been concluded[198] on the basis of solubility experiments that nitrobenzene inhibits crosslinking but enhances chain scission as shown in Table 6.13. The results were compared, however, at similar low dose levels when the actual crosslinking levels could be very low and deviations from the solubility (6.35) could be very large. If the results obtained are replotted as shown in Figure 6.9, it becomes evident that they are superposable—that is, the additive inhibits both crosslinking and chain scission

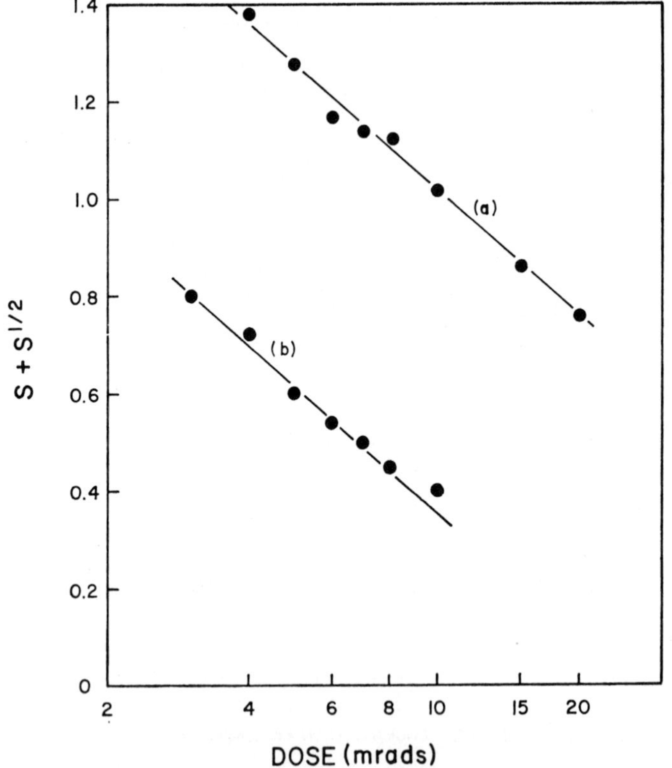

Fig. 6.9 Effect of nitrobenzene on gel formation in natural rubber: (*a*) with nitrobenzene; (*b*) without nitrobenzene. Results taken from Ref. 198.

to the same degree—and that the fraction of the incident energy diverted by the additive is approximately 0.8.

Zeplichal and Steiner[200] have studied the effect of aromatic hydrocarbons on the radiolytic crosslinking of natural rubber. They conclude that benzene does not have an important effect on the radiolysis, but that in its fused linear homologs the activity parallels the number of fused benzene nuclei. The angular homologs are the most active, phenanthrene and benzanthracene being among the best. In contrast, homologs with the most densely packed configuration, such as pyrene of benzopyrene, are not particularly effective. Of the polyphenyls, diphenyl is the most effective. Anthracene and benzanthracene divert up to 70% of the incident energy away from the polymer. These authors also examined[200] the effect of benzanthracene on vulcanizates and observed only a 20% diversion with the green stock and 50% diversion with the vulcanized stock. The reduced efficiency probably arises from the presence in the vulcanizate of appreciable amounts of antioxidant and cure accelerators or their residues, so that the total amount of scavengers is in the range where reduced efficiency has been demonstrated by Turner.[194] The protective effect thus should be, and is, less marked in the green stock which contains free sulfur and unreacted cure accelerators.

Dunn has suggested[189] the use of less soluble dithio compounds, such as zinc dimethyldithiocarbamate or zinc diisopropyldithiophosphate as antioxidants in irradiated rubbers, since these do not inhibit radiolytic crosslinking.

g. Polybutadiene and Poly(butadiene-co-styrene). It has been reported that G (crosslinks) in polybutadiene (sodium initiated) increases from an initial value of 2 to 10 at doses above 200 Mrad, whereas that in a butadiene–styrene (70/30) copolymer decreases from 2 to 0.6 over the same dose range.[203]

Sulfur,[204,213] dimethylthiuram disulfide,[204,218] and other scavengers[205] inhibit crosslinking in both sodium-initiated and stereoregular polybutadiene. *N*-phenyl-*N*-isopropylphenylamine is considered to be the most effective, diverting up to 70% of the incident energy.[205]

Sodium-initiated polybutadienes having more polysulfide crosslinks have larger stress-relaxation rates.[214] However, it has also been concluded that the sulfur crosslink has a greater stability to ionizing radiation than the carbon–carbon crosslink, and the stability increases with the number of sulfur atoms in the crosslink.[211] Protective efficiency in radiolysis does not correspond to antioxidant efficiency.[214]

The radiation resistance of butadiene–styrene copolymers can be increased by extension of the rubber with an aromatic oil and by incomplete curing with a sulfur-accelerated system[67] or by incorporation of disulfides,[206] thiophenols, and oxazoles.[67] The relation between the structure of radiation-protective agents and any difference in effect on radiolytic and subsequent oxidative changes has been examined in styrene–butadiene and butadiene rubbers.[210]

h. Nitrile and Other Rubbers. The effect of various proprietary ingredients on the changes in physical properties of poly(butadiene-co-acrylonitrile) during irradiation have been examined.[207,218] The effect of additions of sulfur on free-radical formation at low temperatures in poly(4-chlorobutadiene) and other rubbers has been studied.[208] A recent series of articles has concluded that radiation-protective compounds in general are only slightly effective with sulfur, oxime, or peroxide-cured nitrile rubbers.[209] Sulfur-containing compounds and amines have been used to inhibit crosslinking in ethylene-2-olefin copolymer rubbers.[216] Diphenylpicrylhydrazyl, contrary to expectations, enhances crosslinking in many rubbers; consequently an ion–molecule crosslinking mechanism is proposed.[219] Certain epoxy resins reduce crosslinking yields in poly(4-chlorobutadiene).[220] The order of (decreasing) favorable response to radiation-protective agents in cured rubber compounds is acrylic rubbers, acrylonitrile–butadiene, styrene–butadiene, natural, vinylidene fluoride–hexafluoropropylene, and polysulfide rubbers.[281] Surprisingly, sulfur has been found to enhance gel formation and tensile strength and to reduce the swelling in solvents of ethylene–propylene copolymers[227,228] and poly(vinyl ethyl ether)[228] on irradiation. Sulfur also reduced the ultimate elongation of the polyether, but increased that of the irradiated copolymer[228] after irradiation.

i. Polysiloxanes. Polysiloxanes have been fairly widely studied.[224] They crosslink on irradiation, the sensitivity decreasing as the aromatic content is increased, as we have seen. Chain scission has been reported to occur at less than 10% of the crosslinking rate during irradiation in vacuum;[225] there is one report of its occurrence during irradiation in air.[80]

Irradiation of polydimethylsiloxane and polydimethyldiphenylsiloxane in air, in nitrous and nitric oxide, and in helium leads to reduced crosslinking rates compared with irradiation under high vacuum.[223]

Although anthracene reacts readily with polymeric radicals in polydimethylsiloxanes,[22] it does not inhibit crosslinking.[25] Free-radical inhibitors such as di-*t*-butyl-*p*-cresol, quinone, and 4-*t*-butylanthraquinone at concentrations of 1% gave only 20% or less inhibition of crosslinking. Even at 22% concentration, di-*t*-butyl-*p*-cresol gave only just over 50% inhibition of crosslinking. Trimethylsilanol is similarly ineffective; but mercaptans, such as *n*-butylmercaptan (10%) inhibit up to 90% of the crosslinking. Benzene and tetralin are not very effective even at high ($\approx$50%) concentrations.[225] Iodine[25] and colloidal sulfur[25,34,222] give considerable protection against crosslinking and no effect of radiation intensity is observed. At least part of the effect of sulfur is attributable to its efficiency as a free-radical scavenger.[222] Iodine and sulfur have no effect on gas evolution; benzophenone, which also offers protection, reduces the gas yield.[25] The use of antioxidants in irradiated silicones has been discussed.[282]

j. Miscellaneous Polymers. Polystyrene and polyacrylic-based ion-exchange resins were found to be protected against radiation by aqueous solutions of *S*-2-aminothylisothiouronium bromide and 2-mercaptoethylamine.[226]

Inhibition of free-radical formation in polystyrene during radiolysis by pyrene has been observed.[286] However, no change in crosslinking or hydrogen production occurred. It was concluded that the latter products are formed through "hot hydrogen" or ion–molecule reactions. Pulse radiolysis of polystyrene containing various polynuclear aromatics leads to the production of solute triplet states. When triphenylamine was added, the solute positive ion was formed.[287,288]

3. External Protection against Degradation

a. Poly(methyl methacrylate). Early work of Charlesby and his collaborators demonstrated inhibitory effects of phenols, amines, thiourea, and other compounds[229] on radiolytic degradation in poly(methyl methacrylate). Use of radioactive scavengers showed that these became incorporated into the polymer.

Alexander and Toms[230] surveyed the effect of a wide variety of additives on radiolytic degradation. 8-Hydroxy-quinoline and naphthalene were examined in detail. Experiments with radioactive naphthalene showed that only a small amount became attached to the polymer during irradiation (one-fifth of the amount that would be bound if naphthalene were "repairing" breaks). They note that a direct competition between polymer and additive for energy put into the system (with no, or equal, energy-transfer probability) would lead to an exponential dependence of initial protection on additive concentration, and they demonstrate that this is not observed.

Gardner and Epstein[231] studied the effect of pyrene, *p*-terphenyl, xylene, benzene, and lead stearate on the susceptibility of this polymer to photolytic and radiolytic damage. They note that these compounds are much less effective at reducing radiolytic damage and that they are consumed during radiolysis, but apparently not during photolysis. They also conclude that most of the chain breaks in this polymer are caused by ions and only about 10% result from reactions of excitons (≡ excited species). Anthracene and terphenyl and other "electron acceptors" have also been studied by Kotov et al.[182] Ultraviolet absorption studies on diphenyl and naphthalene in poly(methyl methacrylate) subjected to pulse radiolysis indicate that both negative ions and triplet states are formed from these solutes.[228] Oxygen has been found to inhibit degradation in γ-irradiated poly(methyl methacrylate).[232] α-Naphthol and 4-methyl-2,6-di-*t*-butylphenol,[235] phenols, amides, benzene, and xylene[234] reduce the number of polymer radicals produced by radiation.

It is claimed that isotactic poly(methyl methacrylate) is more stable to ionizing radiation than the atactic form and that the isotactic compound protects the atactic form in mixtures.[236] However, the overall difference found between the two forms is only very slightly larger than the quoted margin of error. α-Methylstyrene copolymerized into the polymer backbone gave greater protection than an equal weight percent of cumene.[236]

The very interesting work of Bagdasar'yan and co-workers[28–32] was described in Section C.1.

Lyons[237] and also Pinner[238] have shown that addition of certain unsaturated compounds to poly(methyl methacrylate) suppresses radiolytic degradation in favor of a crosslinking reaction.

Some of the above authors have published details of the effect of additive concentration on the initial radiolytic degradation in this polymer. If their results[230,231,32,236] are compared with (6.13), it is found[42] that linear relations can be obtained with an intercept G/G_0 (the fraction of radiolytic events that cannot be inhibited) between 0.1 and 0.2. The results obtained with naphthalene[32,230] are shown in Figure 6.10, values for the fraction of scissions that cannot be scavenged and the ratio of the rate constants are given in Table 6.14, and typical relations obtained using these G/G_0 values are shown in Figure 6.11.

It thus appears possible that in poly(methyl methacrylate) more than one mechanism of radiolytic degradation operates and that approximately 20%

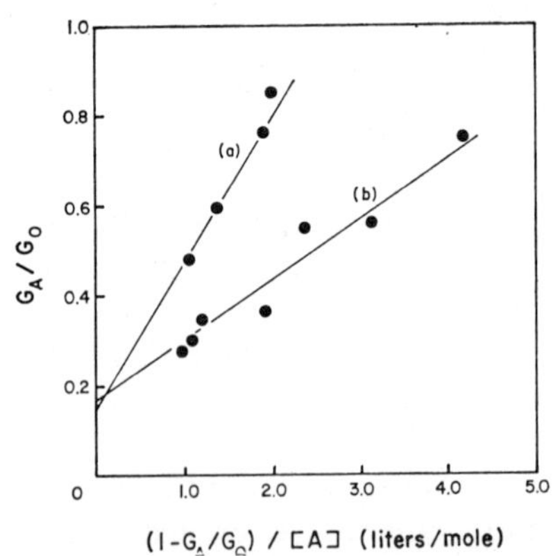

Fig. 6.10 The results of experiments with naphthalene in poly(methyl methacrylate) compared with (6.13). Results taken from (*a*) Ref. 230; (*b*) Ref. 32.

Table 6.14 **Radiation Parameters for Additives in Poly(methyl methacrylate)**

Additive	Unscavengeable Fraction of Chain Scissions (G/G_0)	$k_2'k_3^{-1} \times 10^2$ (liters/mole)	References
Naphthalene	0.17	12.2	32
	0.15	31.5	230
8-Hydroxyquinoline	0.25	7.7	230
Benzophenone	0.17	19.3	32
Benzoic acid	0.15	49.2	32
N,N,N′,N′-Tetramethyl-*p*-phenylenediamine	0.14	6.6	32
Aniline	0.21	5.7	32
Pyrene	0.24	9.5	231, 230
	0[a]	1.5[a]	231[a]
Cumene	0.25	30	236
α-Methylstyrene copolymer	(0.40)	20	236
Average	0.20 ± 0.05		

[a] In photolysis.

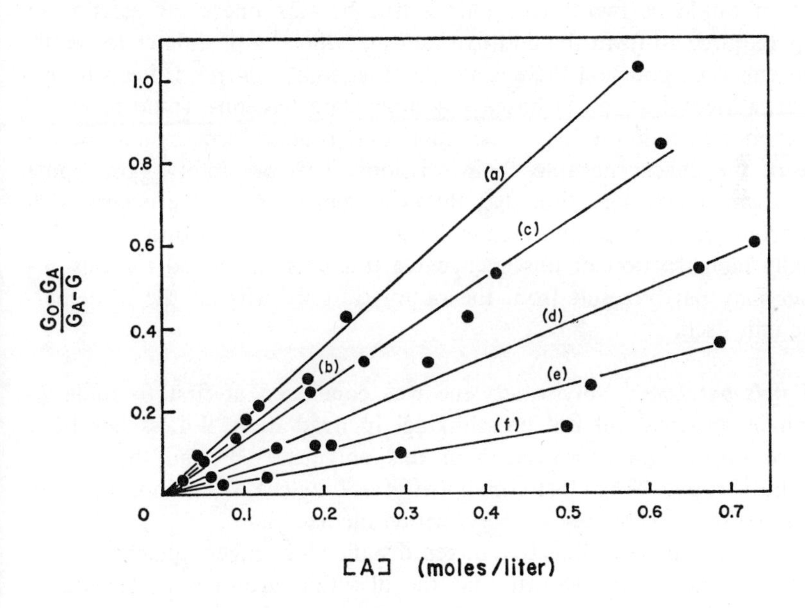

Fig. 6.11 The results of miscellaneous scavengers in poly(methyl methacrylate) compared with (6.13): (*a*) aniline; (*b*) *N,N,N′,N′*-tetramethyl-*p*-phenylenediamine; (*c*) 8-hydroxyquinoline; (*d, f*) naphthalene; (*e*) benzophenone. Results taken from Refs. 32 and 230.

of the chain scissions are not susceptible to the action of the additives listed in the table. The identity of these processes remains uncertain, however. Bagdasar'yan[32] suggests excited states might not be susceptible to scavenger effects, but Gardner[231] has shown that the excited states produced by photolysis are very efficiently scavenged. Moreover, as the table shows, the results for photolysis of pyrene-protected poly(methyl methacrylate) may be fitted to (6.19) if it is assumed that all species produced by photolysis that result in degradation of the polymer are scavengeable. Thus the excited species that may be produced during radiolysis are probably quite different from those produced during photolysis. It can be argued that this result shows that singlet and triplet states are not important during radiolysis because photochemical properties, such as the ability to transfer electronic energy (quenching), are determined only by the lowest excited-singlet and triplet states of molecules.[285] If more energetic excited states are formed, then the efficient internal conversion between electronic states will very quickly yield the lowest excited state. This is accompanied by a rapid removal of excess vibrational energy to give thermal equilibrium with the medium.

If the radical anion mechanism of Bagdasar'yan is correct and ion neutralization results in the formation of scission fragment radicals, then provided that there is no other source of free radicals, the maximum observed protection would be two-thirds. This is true because one chain scission in three is required to form the additive radical, which he postulates to be the ion scavenger. Of course if there is also another source of free radicals in this polymer, a lower fraction of unscavengeable chain scissions would result.

We cannot exclude the hypothesis that some primary process may be the source of the unscavengeable chain scissions. This possibility gains some support from the observation that the unscavengeable residue occurs with highly aromatic additives and also with the α-methylstyrene copolymer. The unusually high fraction of unscavengeable fractures calculated for this copolymer may partly result from the radiolytic sensitivity of the α-methylstyrene unit itself.

b. Polyisobutylene. Polyisobutylene was concluded at first to undergo main-chain scission, but not crosslinking, in irradiation.[247] Less has been published on radiation protection in this polymer[240–245] and the closely similar butyl rubber[218,239,245] (a copolymer of isobutylene with minor amounts of isoprene) than in the case of poly(methyl methacrylate).

Experiments with a radioactive mixed disulfide scavenger, phenyl-^{3}*H*-*sec*-butyl-^{35}*S*-disulfide, have shown that the alkylthio group is preferentially incorporated into the polymer molecule during radiolysis.[241] The chain scission yield for polyisobutylene is about 40% less in nitrous oxide than in vacuum. It was concluded that little or no crosslinking occurs during radiolysis.[242] Both tritiated thiophenol and nitrobenzene-^{14}C become bound to the polymer during radiolysis.[244] However, whereas thiophenol reduces, nitrobenzene increases G (chain scissions).[243,244] It was concluded that thiophenol reacts with polymeric radicals resulting from side-chain scission and nitrobenzene scavenges subsequent decomposition products as shown below.

$$\underset{\text{R}}{\overline{}} \rightsquigarrow \underset{\cdot}{\overline{}} + \text{R}\cdot$$

$$\underset{\cdot}{\overline{}} + \text{HS}\phi \rightarrow \underset{\text{H}}{\overline{}} + \cdot\text{S}\phi$$

$$\underset{\cdot}{\overline{}} + \cdot\text{S}\phi \rightarrow \underset{\text{S}\phi}{\underset{|}{\overline{}}}$$

$$\underset{\cdot}{\overline{}} \rightarrow \overline{} + = \overline{}$$

$$2\,\overline{}\cdot \rightarrow \overline{} \quad \text{(fracture repair)}$$

$$\overline{}\cdot + \phi\text{NO}_2 \rightarrow \text{products}$$

This mechanism could be tested by a study of competitive scavenging.

Sulfur becomes bound to butyl rubber during exposure to ionizing radiation.[213,218] Scavengers were found to be relatively ineffective in sulfur-vulcanized isoprene–isobutylene copolymers.[245] The degradation of polyisobutylene can be converted to a crosslinking reaction with certain additives.[246]

c. Cellulose and Derivatives. Cellulose degrades rapidly on exposure to ionizing radiation.[248] This damage is not prevented by the resins and other additives normally used with rayon fabrics.[249] Nitric oxide and *p*-benzoquinone prevent somewhat the deterioration of cellulose triacetate fibers.[250] Highly aromatic dyes are claimed to offer protection to cotton and polyester cotton blends against degradation by γ-radiation.[251] Large additions of triallyl citrate to cellulose acetate inhibit radiolytic degradation and result in an increase in tensile strength and a decrease in solubility after irradiation.[252]

d. Polytetrafluoroethylene. Exposure of polytetrafluoroethylene to various gases during radiolysis profoundly alters the EPR spectrum.[253–256] It has been suggested that radiation damage is altered by these gases.[253]

4. Protection in Polymer Solutions

a. Solutions of Biological Polymers in Water. The literature to 1959 of in vitro protection of macromolecules has been reviewed.[257] Acids, amino acids, and amines protect horse albumin solution against damage by ionizing radiation.[258] Cysteine (10^{-3} *M*) protects against sparsely ionizing (x-rays or Sr^{90} β-rays), but not against densely ionizing (Po^{210} α-rays) irradiation in human and bovine serum albumin.[259] In aqueous solutions of deoxyribonucleic acid (DNA), as in other polyelectrolyte solutions (see below), the protective effect of small additions of salts probably results from molecular contraction of the polyelectrolyte; when large amounts of salts are added, scavenging may also occur.[260] Dilute solutions of enzymes and DNA do not show an oxygen effect. However, when thiol-containing compounds are added, an oxygen effect (increased sensitivity) appears.[261] Solutions of the enzyme papain may be protected against radiolytic deactivation by blocking of the sulfhydryl group (e.g., by incubation with cysteamine monosulfoxide, sodium tetrathionate, or sodium *p*-chloromercuribenzoate).[262] The effect of an aminoethylisothiouronium salt, β-alanine, "sodium Nembutal®," and cysteamine on the radiolytic deactivation of pepsin solutions has been observed.[263] Dimethyl sulfoxide protects lyophilized beef liver catalase solution against radiolytic deactivation.[264] The use of gel formation in DNA solutions as a model for testing radiation-protective agents for biological systems has been discussed,[265] and free-radical processes in such model systems have been reviewed.[266] The effect of methionine on the radiolytic deactivation of ribonuclease has also been studied.[267] Sodium glutamate and benzoate protect albumin (source not given) solutions against radiolytic damage. The effect is at a maximum for about 0.01 to 0.1% of additive.[283]

b. Solutions of Synthetic Polymers. The kinetic studies of Charlesby and co-workers[36,37] have already been described. Charlesby and Kopp[36] studied the effect of a large number of additives of interest in biological protection on the gelling dose of 5% aqueous solutions of poly(vinyl pyrrolidone) and poly(ethylene oxide). With most additives, the dose to gel increases smoothly with increasing concentration of additive. With a few materials, the gelling dose increases to a certain value at low additive concentrations and remains constant at this value as the concentration is further increased.[36] A study with radioactive (^{35}S) thiourea showed that this compound becomes linked to the polymer both in the case of poly(vinyl pyrrolidone) (crosslinks via indirect effect)[36] and poly(ethylene oxide) (crosslinks via direct effect)[38] solutions. Radioactive colloidal sulfur becomes bound to poly(vinyl pyrrolidone) during radiolysis. The *G* value for sulfur incorporation (0.04) is much less than the reduction in the *G*-value for crosslinking:[37] each sulfur atom may be responsible for the inactivation of up to 17 polymer radicals.[32] The incorporation of sulfur into poly(ethylene oxide) could not be studied because of experimental difficulties. Tritium is not incorporated into either polymer in significant amounts from solutions during irradiation in the absence of sulfur. When sulfur is present, the amount of tritium incorporated into poly(vinyl pyrrolidone) corresponds to a G-value for hydrogen incorporation from water into the polymer of 4.6. No significant amounts of tritium become incorporated into poly(ethylene oxide) as a result of radiolysis under these conditions. The results are explained in terms of the following reaction mechanism:[37]

$$H_2O \rightsquigarrow H + OH \; (\equiv W\cdot)$$
$$W\cdot + PH \rightarrow P\cdot$$
$$PH \rightsquigarrow P\cdot + H\cdot$$
$$P\cdot + P\cdot \rightarrow PP$$
$$S_x + H\cdot \rightarrow S_xH\cdot$$
$$S_xH\cdot + H\cdot \rightarrow H_2 + S_x$$
$$S_xH\cdot + OH \rightarrow H_2O + S_x$$
$$S_xH\cdot + P\cdot \rightarrow PH + S_x \quad \text{True repair}$$
$$P\cdot + S_x \rightarrow PS_x\cdot \quad \text{Stabilization repair}$$

Charlesby's results indicate that in poly(vinyl pyrrolidone) solutions the protective process involves a true repair mechanism, as shown, but in poly-(ethylene oxide) solutions a stabilization repair process occurs.[37] A detailed discussion of the mechanism of protection of aqueous polyethylene oxide solutions by cysteamine was published[289] after this chapter was completed.

Addition of small amounts of sodium chloride to aqueous solutions of sodium carboxymethylcellulose more than halved the radiolytic chain-scission yield. Since this addition also profoundly changes the viscosity-concentration relation of the unirradiated material, it was concluded that the reduced yield resulted from changes in molecular configuration. A much larger protective effect is observed when methanol is added to the aqueous solution.[268] Addition of 5% methyl sulfate to a 5% aqueous solution of poly(vinyl alcohol) inhibits the increase in viscosity with increasing radiation dose.[269] The effect of various acids on the radiolysis of poly(vinyl alcohol) solutions has been briefly reported.[270]

The critical concentration (the concentration at which gel will just form and below which no gel forms whatever the dose) of aqueous solutions of poly(acrylic acid) increases with increasing degree of neutralization of the polymer. However, gelation occurs more readily with increasing sodium chloride concentration at low concentrations, although gel formation is suppressed by concentrations of salt above 0.01 N.[271]

In a study of the effects of ionizing radiation on polystyrene, poly(vinyl pyrrolidone) and poly(vinyl acetate) in various solvents, Henglein[272] has shown that the nature of the changes depends strongly on the type of solvent as well as on the solute concentration. He notes a strong inhibiting effect of methanol on the crosslinking of poly(vinyl pyrrolidone) in water. Tetranitromethane and other free-radical scavengers inhibit crosslinking of these polymers in solution. No relation was found to exist between the radiation sensitivity of the solvent and the rate of crosslinking of dissolved polymers, although crosslinking is favored in poor solvents.[272] A reexamination of poly(vinyl acetate) confirmed that crosslinking is favored in poor solvents at low concentrations, but retarded at high concentration. Similar results were obtained with poly(vinyl methyl ether) solutions and solutions of poly(vinyl alcohol) that had been partially acetalized with glyoxalic acid. Salt effects, as noted already, were also observed.[273] Inhibition of crosslinking of polystyrene solutions in chloroform by small amounts of ethanol has been noted.[274]

Iodine and hydroquinone are found to inhibit the radiooxidative degradation of poly(ethyl acrylate) solutions in benzene and carbon tetrachloride.[275] The radiolytic degradation of poly(methyl methacrylate) is less in mixtures of equal volumes of acetone and benzene than in either solvent alone.[276]

Irradiation of poly(vinyl chloride) as a solid solution in tetrahydrofuran in air results in crosslinking rather than the degradation, which occurs with the polymer alone.[277]

The radiolytic degradation of polyisobutylene in a number of solvents was not affected by addition of 2,2-diphenyl-1-picrylhydrazyl (DPPH) or by dissolved oxygen. However, added unsaturated hydrocarbons were found to inhibit chain scission.[278] Radiolytic crosslinking of natural rubber dissolved in various solvents was inhibited by DPPH or nitrobenzene.[279]

F. CONCLUSIONS

In many cases the criteria used in the selection of additives for evaluation as radiation-protective agents has been their known effectiveness as antioxidants. Another major consideration has been the established ability of a material to act as a photolytic quencher. However, as several investigators have shown, the species being scavenged during polymer radiolysis may not at all resemble that scavenged or decomposed in oxidation or photolysis. Thus sulfur compounds are effective inhibitors of both oxidative and radiolytic degradation, but apparently this effect exists in one case through a heterolytic process and in the other case through a homolytic process. Even when similar chemical processes are involved (e.g., in free-radical scavenging) the experimenter may obtain a very different ranking of protective activity against oxidation when compared with radiolytic degradation. This may be ascribed to differences in reactivity of the polymeric free radicals involved (peroxy and alkoxy as opposed to alkyl) and also to differing reactions involving the scavenger radical in an oxidizing and a radiolytic environment.

Usually oxidative experiments are carried out at higher temperatures than radiolytic experiments; in radiolysis solubility and diffusion may therefore play a much more important role. Although many phenols and amines inhibit radiolytic damage but do not survive as effective antioxidants after the system has been exposed to ionizing radiation, di-β-naphthyl-p-phenylenediamine, a very effective antioxidant, does not inhibit radiolysis to any considerable extent and retains most of its antioxidant activity in irradiated systems. Certain antioxidants such as the thiobisphenols inhibit radiolytic degradation and yet are able to exert appreciable antioxidant activity in irradiated systems.

There is now general agreement on the nature of the protective effect of benzene—a sponge-type protection involving the transfer of energy in some way (exciton transfer) and its degradation to heat without chemical reaction—but we cannot be equally confident about the role played by the various homologs of benzene. As we have seen, aromatic hydrocarbons such as the fused-ring homologs may be consumed when used to protect against radiolysis but not when used as photolytic quenchers.

In addition to the above groups, certain other materials have been found to be radiation-protective agents. Nitroaliphatic and aromatic compounds inhibit radiolytic damage in a number of polymers, possibly through "exciton" transfer, which is believed to occur with benzene, although radical scavenging reactions cannot be excluded. Aliphatic halogeno compounds and aromatic bromides and iodides are effective protective agents in some polymers, presumably through radical-scavenging processes. However, aliphatic and aromatic chloro compounds can sensitize certain rubbers to radiolytic crosslinking.

In many systems, even where the general nature of the protective action is known, the exact mechanism cannot be stated with any certainty. However, the theoretical work of the last decade, especially that of Charlesby and his collaborators, has established a satisfactory framework of testable chemical kinetics that can guide the experimentalist. Kinetics, of course, is not enough to demonstrate mechanisms, and work in this field would be considerably advanced by further detailed studies of the chemical fate of the protective agent. Again Charlesby, Turner, and others have demonstrated the enormous potentiality of radioactive tracer studies in elucidating the sometimes tortuous reaction sequences that may be involved.

The usefulness of studies employing simple scavengers to identify reactive intermediates in the radiolysis of gases, water, and the simpler organic liquids has been realized since the pioneering work of the 1940s. Particularly rapid progress has been seen in the last decade, for the recent elegant techniques of pulse radiolysis (in favorable circumstances) facilitate the identification of transient scavenger reaction products. It is disappointing to report that there has apparently been little interest in applying this methodology to polymer radiolysis. This may be largely the result of a lack of facilities: presumably very few workers in industry or academic life have access both to polymer-processing equipment and pulsed accelerators. However, there are recent signs[285] of progress in this direction. Whether by the use of such new techniques or by the more established methods outlined in this review, there does seem to us to be a considerable opportunity for progress in this field both in the development of more efficient protective agents and in the use of such materials to probe the chemical transformations that polymers undergo during exposure to ionizing radiation.

REFERENCES

1. A. Charlesby, *Atomic Radiation and Polymers*, Pergamon, London, 1960.
2. A. Chapiro, *Radiation Chemistry of Polymeric Systems*, Interscience, New York, 1962.
3. G. H. Hine and G. L. Brownell, *Radiation Dosimetry*, Academic Press, New York, 1956.
4. D. P. Stevenson, *J. Phys. Chem.*, **61,** 1453 (1957).
5. E. Collinson, F. S. Dainton, and D. C. Walker, *Trans. Faraday Soc.*, **57,** 1732 (1961).
6. R. M. Black, *J. Appl. Chem.*, **8,** 159 (1958).
7. A. Charlesby and W. H. T. Davison, *Chem. Ind.* (London), **232** (1957).
8. M. Dole, D. C. Milner, and T. F. Williams, *J. Amer. Chem. Soc.*, **80,** 1580 (1958).
9. E. J. Lawton, J. S. Balwit, and R. S. Powell, *J. Polym. Sci.*, **32,** 257 (1958).
10. B. J. Lyons, *ACS Polym. Preprints*, **8,** 672 (1967).
11. M. Dole, O. Saito, and H. Y. Kang, *J. Amer. Chem. Soc.*, **89,** 1980 (1967).
12. R. M. Black and B. J. Lyons, *Proc. Roy. Soc.* (*London*), **253A,** 322 (1959).
13. F. A. Bovey, *The Effects of Ionizing Radiation on Natural and Synthetic High Polymers*, Interscience, New York, 1958, p. 132.
14. R. Simha and R. A. Wall, *J. Phys. Chem.*, **61,** 425 (1957).
15. B. J. Lyons and M. A. Crook, *Trans. Faraday Soc.*, **59,** 2334 (1963).
16. M. A. Dole, private communication, 1968.
17. M. Dole, M. B. Fallgatter, and K. Katsuura, *J. Phys. Chem.*, **70,** 62 (1966).
18. Z. M. Bacq, *Chemical Protection Against Ionizing Radiation*, Thomas Press, Springfield, 1965.
19. J. P. Manion and M. Burton, *J. Phys. Chem.*, **56,** 560 (1952).
20. A. Charlesby, *Atomic Radiation and Polymers*, Pergamon, London, 1960, p. 493.
21. A. Charlesby and D. G. Lloyd, *Proc. Roy. Soc.* (*London*), **249A,** 51 (1958).
22. A. Charlesby and D. G. Lloyd, *Proc. Roy. Soc.* (*London*), **254A,** 343 (1958).
23. A. Charlesby, P. G. Garratt, and P. M. Kopp, *Int. J. Rad. Biol.*, **5,** 439 (1962).
24. A. Charlesby and P. M. Kopp, *Int. J. Rad. Biol.*, **5,** 521 (1962).
25. A. Charlesby and P. G. Garratt, *Proc. Roy. Soc.* (*London*), **273A,** 117 (1963).
26. A. Charlesby and P. M. Kopp, *Proc. Roy. Soc.* (*London*), **291A,** 129 (1966).
27. A. Charlesby, P. M. Kopp, and J. F. Read, *Proc. Roy. Soc.* (*London*), **291A,** 122 (1966).
28. Kh. S. Bagdasar'yan, V. A. Krongauz, and N. S. Kardash, *Dokl. Akad. Nauk SSSR*, **144,** 101 (1962); *Chem. Abstr.*, **57,** 8466f (1962).

29. Kh. S. Bagdasar'yan and Z. A. Sinitsyna, *Dokl. Akad. Nauk SSSR*, **147,** 1396 (1962); *Chem. Abstr.*, **58,** 9780h (1963).
30. A. I. Nepomnyashchii, V. I. Muromtsev, and Kh. S. Bagdasar'yan, *Dokl. Akad. Nauk SSSR*, **149,** 901 (1963); *Chem. Abstr.*, **59,** 2312f (1963).
31. R. I. Milyutinskaya and Kh. S. Bagdasar'yan, *Zh. Fiz. Khim.*, **38,** 776 (1964); *Chem. Abstr.*, **61,** 1420c (1964).
32. V. A. Borovkova and Kh. S. Bagdasar'yan, *Khim. Vys. Energ.*, **1,** 340 (1967).
33. T. J. Kemp, J. P. Roberts, G. A. Salmon, and G. F. Thompson, *J. Phys. Chem.*, **71,** 3052 (1967); **72,** 1464 (1968).
34. W. Schnabel and M. Dole, *J. Phys. Chem.*, **67,** 295 (1963).
35. C. Capellos and A. O. Allen, *J. Phys. Chem.*, **72,** 4265 (1968).
36. A. Charlesby and P. M. Kopp, *Int. J. Rad. Biol.*, **7,** 173 (1963).
37. A. Charlesby and P. M. Kopp, *Int. J. Rad. Biol.*, **9,** 383 (1965).
38. A. Charlesby, *High Polym. Jap.*, **14,** 209 (1965).
39. A. Charlesby and S. H. Pinner, *Proc. Royal Soc.* (*London*), **249A,** 367 (1959).
40. B. J. Lyons, unpublished work, 1968.
41. A. Charlesby, *Atomic Radiation and Polymers*, Pergamon, London, 1960, pp. 171 ff.
42. W. J. Burlant, *ACS Polym. Preprints*, **1** (2), 324 (1960).
43. P. J. Flory, *Principles of Polymer Chemistry*, Cornell University Press, Ithaca, N.Y., 1953, p. 579.
44. B. J. Lyons, *ACS Polym. Preprints*, **8,** 672 (1967).
45. B. J. Lyons and C. R. Vaughn, "Irradiation of Polymers," *Advan. Chem. Ser.*, **66,** American Chemical Society, Washington, D.C., 1967, p. 139.
46. B. J. Lyons, unpublished work, 1963.
47. A. Charlesby, *Atomic Radiation and Polymers*, Pergamon, London, 1960, p. 240.
48. P. J. Flory, *Principles of Polymer Chemistry*, Cornell University Press, Ithaca, N.Y., 1953, p. 487.
49. E. Di Guillio, G. Bellini, and G. V. Giandanoto, *Chim. Ind.*, **47,** 156 (1965); *Rubber Chem. Technol.*, **39,** 726 (1966) (English transl.).
50. R. F. Landel and R. F. Fedors, in *Proc. 4th Int. Congr. Rheology* (*1963*) *Part 2*, Wiley, New York, 1965, p. 543.
51. A. V. Tobolsky, *Properties and Structure of Polymers*, Wiley, New York, 1960.
52. A. Charlesby, *Atomic Radiation and Polymers*, Pergamon, London, 1960, p. 492.
53. D. A. Smith, *Trans. Inst. Rubber Ind.* (*London*), **33,** 11 (1957).
54. S. D. Gehman and T. C. Gregson, *Rubber Chem. Technol.*, **33,** 1375 (1960).
55. S. H. Pinner, *Brit. Plastics*, **35,** 518 (1962).
56. R. M. Black, *Rept. Progr. Appl. Chem.*, **50,** 144 (1965).
57. F. Zeplichal, *Rev. Gen. Caoutchouc Plastiques*, **43,** 971 (1966).
58. F. Zeplichal and N. Steiner, *Kautschuk Gummi Kunstst.*, **20,** 451 (1967).
59. Kh. S. Bagdasar'yan, *Elementarnye Protessy Khim. Vysokikh. Energ.*, *Akad. Nauk SSSR*, *Inst. Khim. Fiz. Tr.*, *Simp.*, *Mosk.*, *1963*, 160 (1965); *Chem. Abstr.* **50,** 10517e (1966).
60. *Effects of Radiation on Organic Materials*, U.S. Atomic Energy Commission, Oak Ridge National Laboratory Rept. ORNL 4229, 1968, pp. 183–187; *Nucl. Sci. Abstr.*, **22,** 25375 (1968).
61. H. Ringsdorff, Strahlentherapie, **132**, 627 (1967); *Chem. Abstr.*, **67,** 74209 (1967).
62. W. O. Foye, *Ann. Rept. Med. Chem.*, **1965,** 324 (1966); **1966,** 330 (1967).
63. A. Charlesby and P. Alexander, *J. Chem. Phys.*, **52,** 699 (1955).
64. Sheng-K'ang Ying, Yu Chi Hu, and Yu Fan Li, *Ko Fen Tsu T'ung Hsun*, **6,** 25 (1964); *Chem. Abstr.*, **64,** 830d (1966).

65. A. Charlesby and D. Groves, *Proc. 3rd Rubber Technol. Conf.* (*London*), **317** (1954).
66. R. Bauman and J. Glantz, *J. Polym. Sci.*, **26,** 397 (1957).
67. H. R. Anderson, *J. Appl. Polym. Sci.*, **3,** 316 (1960); *Rubber Chem. Technol.*, **34,** 228 (1961).
68. E. Witt, *J. Polym. Sci.*, **41,** 507 (1959).
69. R. K. Traeger and T. Castonguay, *ACS Polym. Preprints*, **6,** 992 (1965); *J. Appl. Polym. Sci.*, **10,** 491 (1966).
70. E. J. Weber and H. Hensinger, *Radiochim. Acta*, **4,** 92 (1965); *Chem. Abstr.*, **63,** 11723b (1965).
71. G. E. Meyer, F. J. Naples, and H. M. Rice, *Rubber World*, **140,** 435 (1959).
72. J. B. Gardner and B. G. Harper, *J. Appl. Polym. Sci.*, **9,** 1585 (1965).
73. M. Walicki, E. Turska, and J. Kroh, *Bull. Acad. Polon. Sci.*, *Ser. Sci. Chim.*, **12,** 805 (1964); *Chem. Abstr.*, **62,** 11919e (1965).
74. Z. Polacki and M. Grodel, *Nukleonika*, **10,** 469 (1965); *Chem. Abstr.*, **64,** 16855h (1966).
75. M. Prober, presented at 132nd American Chemical Society Meeting, New York, Sept. 1957.
76. K. H. Krause, *Kunststoffe—Plastics*, **48,** 564 (1958); *Chem. Abstr.*, **53,** 5720b (1959).
77. M. Koike and A. Danno, *Oyo Butsuri*, **30,** 97 (1961); *J. Phys. Soc. Jap.*, **15,** 1501 (1960); *Chem. Abstr.*, **60,** 9452b (1964).
78. R. K. Jenkins, *J. Polym. Sci.*, *Part B*, **2,** 999 (1964).
79. I. Ya. Poddubnyi and S. V. Aver'yanov, *Radiats. Khim. Polim.*, *Mater. Simp.*, *Mosk. 1964*, 306 (1966); *Vysokomolekul. Soedin* **8,** 1549 (1966); *Chem. Abstr.*, **66,** 116446 (1967).
80. R. K. Jenkins, *J. Polym. Sci.*, *Part A-2*, **4,** 41 (1966).
81. R. K. Jenkins, *J. Polym. Sci.*, *Part A-1*, **4,** 771 (1966).
82. W. Schnabel, *Makromol. Chem.*, **104,** 1 (1967).
83. S. B. Dolgoplosk, L. M. Chebysheva, A. L. Klebanskii, E. Yu. Shvarts, and L. P. Fomina, *Kauchuk i Rezina*, **22** (9), 1 (1963); *Chem. Abstr.*, **60,** 745h (1964).
84. E. D. Kiseleva, K. V. Chmutov, V. N. Krupnova, and N. V. Filatova, *Tr. 2-go*, [*Vtorogo*] *Vses. Soveshch. po Radiats. Khim.*, *Akad. Nauk SSSR*, *Otd. Khim. Nauk*, *Mosk.*, *1960*, 603 (1962); *Chem. Abstr.*, **58,** 4691b (1963).
85. E. D. Kiseleva, K. V. Chmutova, and V. N. Krupnova, *Zh. Fiz. Khim.*, **36,** 2707 (1962); *Chem. Abstr.*, **58,** 8432d (1963).
86. A. M. Semushin and I. A. Kuzin, *Zh. Prikl. Khim.*, **37,** 760 (1964); *Chem. Abstr.*, **61** 3255d (1964).
87. Yun-Sheng Lou, I. A. Kuzin, and A. M. Semushin, *Zh. Prikl. Khim.*, **37,** 893 (1964); *Chem. Abstr.*, **61,** 3262c (1964).
88. E. D. Kiseleva, K. V. Chmutov, and A. B. Pashkov, *Issled. Svoistv Ionoobmen. Materialov*, *Akad. Nauk SSSR*, *Inst. Fiz. Khim. 1964*, 163 (1964); *Chem. Abstr.*, **62,** 4161g (1965).
89. R. H. Wiley and G. Devenuto, *J. Appl. Polym. Sci.*, **9,** 2001 (1965).
90. K. V. Chmutov, E. D. Kiseleva, M. M. Klientovskaya, and V. N. Krupnova, *Radiats Khim. Polim.*, *Mater. Simp.*, *Mosk.*, *1964*, 389 (1966); *Chem. Abstr.*, **66,** 95740 (1967).
91. E. D. Kiseleva, K. V. Chmutov, M. M. Klientovskaya, and V. P. Li, *Zh. Fiz. Khim.*, **41,** 1590 (1967); *Chem. Abstr.*, **67,** 109240 (1967).
92. F. Muenzel (to Heberlein & Co. A.-G.), Swiss Patent 383,915 (1965 Appl. Aug. 23, 1961).
93. C. P. J. Glaudemans, E. Passaglia, and E. A. Wielicki, *Tappi*, **45,** 542 (1962); *Chem. Abstr.*, **57,** 8758f (1962).

94. G. O. Phillips, F. A. Blouin, and J. C. Arthur Jr., *Nature*, **202**, 1328 (1964); *Radiat. Res.*, **23**, 527 (1964).

95. J. C. Arthur Jr. and T. Mares, *J. Appl. Polym. Sci.*, **9**, 2581 (1965).

96. J. C. Arthur Jr., D. J. Stanonis, T. Mares, and O. Hinojosa, *J. Appl. Polym. Sci.*, **11**, 1129 (1967).

97. K. Kohdera, F. Kimura, and I. Sakurada, *Nippon Hoshasen Kobunshi Kenkyu Kyokai Nempo*, **8**, 43 (1966); *Chem. Abstr.*, **68**, 50424 (1968).

98. N. A. Birkin, A. N. Neverov, and V. K. Bocharnikov, *Mekh. Polim.*, **1967**, 476; *Nucl. Sci. Abstr.*, **22**, 30602 (1968).

99. A. S. Fomenko, E. P. Krasnov, T. M. Abramova, E. P. Dar'eva, E. G. Furman, and A. A. Galina, *Vysokomolekul. Soedin.*, **8**, 770 (1966); *Chem. Abstr.*, **65**, 2369g (1966).

100. A. S. Fomenko, E. P. Krasnov, T. M. Abramov, E. P. Das'eva, E. G. Furman, and A. A. Galina, *Khim. Prom. Ukr.* (Russ. ed.), **1968**, 31 (1968); *Chem. Abstr.*, **69**, 3425 (1968).

101. M. L. Kerber, O. Ya. Fedotova, and I. P. Losev, *Plasticheskie Massy*, **4**, 20 (1964); *Chem. Abstr.*, **61**, 738h (1964).

102. L. C. Glover and B. J. Lyons, presented at 155th American Chemical Society Meeting, 1968; *ACS Polym. Preprints*, **9**, 243 (1968).

103. D. G. Lloyd, reported in Ref. 1, p. 500.

104. A. Charlesby, W. H. T. Davison, and E. von Arnim, reported in Ref. 1, p. 502.

105. Y. Okada, T. Ito, and A. Amemiya, *Kogyo Kagaku Zasshi*, **64**, 355 (1961); *Chem. Abstr.*, **57**, 4834c (1962).

106. Y. Okada and A. Amemiya, *J. Polym. Sci.*, **50**, 522 (1961).

107. N. Ya. Buben, A. T. Koritskii, and V. N. Shamshev, *Tr. 2-go* (*Vtorogo*) *Vses. Soveshch. po Radiats. Khim. Akad. Nauk SSSR, Otd. Khim. Nauk Mosk., 1960*, 540 (1962); *Chem Abstr.*, **58**, 6937h (1963).

108. N. A. Slovokhotova, A. T. Koritskii, V. A. Kargin, N. Ya. Buben, and Z. F. Il'icheva, *Vysokomolekul Soedin.*, **5**, 575 (1963); *Chem. Abstr.*, **59**, 2967a (1964).

109. Y. Okada, *J. Appl. Polym. Sci.*, **7**, 695 (1963).

110. B. J. Lyons and M. Dole, *J. Phys. Chem.*, **68**, 526 (1964).

111. J. F. Brown Jr., *J. Amer. Chem. Soc.*, **79**, 2480 (1957).

112. V. E. Althouse, *ACS Polym. Preprints*, **4**, 256 (1963); presented at 144th National American Chemical Society Meeting (1963).

113. Y. Okada, *J. Appl. Polym. Chem.*, **7**, 1153 (1963).

114. R. M. Black and E. H. Reynolds, *Proc. Inst. Elec. Engrs.* (*London*), **112**, 1226 (1965).

115. J. Ferguson and B. Wright, *J. Appl. Polym. Sci.*, **9**, 2763 (1965).

116. A. Charlesby and P. M. Kopp, *Int. J. Appl. Radiat. Isotopes*, **17**, 352 (1966).

117. T. Shibata, Mj Hasebe, and T. Yoshida, *Furukawa Denko Jiho*, **1966**, No. 42, 70 (1966); *Chem. Abstr.*, **69**, 28248 (1968).

118. Ya. A. Kostrov and A. A. Konkin, *Khim. Volokna*, **1967** (2), 19; *Chem. Abstr.*, **67**, 12408 (1967).

119. A. A. Kachen, G. V. Chernyavskii, and V. A. Shrubovich, *Dopov. Akad. Nauk Ukr. SSR, Ser. B*, **29**, 626 (1967); *Chem. Abstr.*, **67**, 91281 (1967).

120. R. M. Black and A. Charlesby, *Int. J. Appl. Radiat. Isotopes*, **7**, 126 (1959).

121. R. M. Black and L. G. Brazier, British Patent 784,161, filed Jan. 27, 1955 (1957).

122. R. J. Prochaska, German Patent 1,028,775 (1958).

123. A. Charlesby, S. H. Pinner, and J. Burrows, British Patent 853,737 (1960).

124. General Electric Company, British Patent 830,899, Appl. date June 16, 1955 (1960).

125. H. N. Schlein and B. R. LaLiberte, U.S. Patent 3,194,668, Appl. date Apr. 8, 1963 (1965).

126. G. I. Gladkova, Z. S. Egorova, V. L. Karpov, S. S. Leshchenko, L. V. Mitrofanova, N. A. Slovokhotova, E. E. Finkel, and S. M. Cherntsov, *Plasticheskie Massy, 1965*, **9,** (8) (1965); *Nauchn.-Issled. Inst. Kabel'n. Prom.*, **9,** 131 (1963); *Chem. Abstr.*, **63,** 16543g (1965).

127. B. Graham, U.S. Patent 3,261,804, Canadian Patent Appl. Aug. 31, 1956 (1966).

128. V. L. Karpov, S. S. Leshchenko, L. V. Mitrofanova, and E. E. Finkel, *Tr. 2-go (Vtorogo) Vses. Soveshch. po Radiats Khim. Akad. Nauk SSSR, Otd. Khim. Nauk, Mosk., 1960*, 547 (1962); *Chem. Abstr.*, **58,** 6977g (1963).

129. E. E. Finkel, N. V. Vasil'eva, E. V. Markman, R. T. Malashkina, and Ya. Z. Mesenzhnik, *Dokl. Akad. Nauk Uz. SSR, 23(4)*, 24 (1966); *Chem. Abstr.*, **65,** 10735e (1966).

130. J. Scanlan and D. K. Thomas, *J. Polym. Sci., Part A*, **1,** 1015 (1963).

131. B. J. Lyons, unpublished work, reproduced by permission of Raychem Corporation, 1967.

132. Ref. 2, p. 433.

133. See Ref. 2, p. 418 ff, for a summary of the evidence.

134. Ref. 2, pp. 419–420.

135. B. J. Lyons, *J. Polym. Sci., Part A*, **3,** 777 (1965).

136. B. J. Lyons and A. S. Fox, *J. Polym. Sci., Part C*, **21,** 159 (1968).

137. R. A. Veselovskii, S. S. Leshchenko, and V. L. Karpov, *Dokl. Akad. Nauk SSSR*, **167,** 339 (1966).

138. N. A. Nechitailo, L. S. Polak, and P. I. Sanin, *Plasticheskie Massy* **1962** (7), 3; *Chem. Abstr.*, **57,** 16863c (1962).

139. P. A. Kirpichnikoff and L. D. Romanova, *Plasticheskie Massy*, **1965** (1), 8; *Chem. Abstr.*, **62,** 9298e (1965).

140. N. A. Nechitailo, P. I. Sanin, A. L. Gol'denberg, and P. S. Polack, *Plasticheskie Massy*, **1965** (7), 7; *Chem. Abstr.*, **63,** 10124f (1965).

141. D. O. Geymer, *Mackromol. Chem.*, **100,** 186 (1967).

142. R. N. Nurmukhametov, L. V. Bondareva, D. N. Shigorin, N. V. Mikhailov, and L. G. Tokarev, *Vysokomolekul. Soedin.*, **6,** 1411 (1964).

143. N A Nechitailo, J. Pospisil, P. I. Sanin, and L. S. Polak, *Plasticheskie Massy*, **1966** (1), 37; *Chem. Abstr.*, **64,** 11382e (1966).

144. I. Benesh and G. Kaplan, *Vysokomolekul. Soedin.*, **5,** 1580 (1963).

145. D. N. Aneli, N. A. Nechitailo, and P. I. Sanin, *Plasticheskie Massy*, **1967** (6), 14; *Chem. Abstr.*, **68,** 60031 (1968).

146. M. Dole, R. W. Keyser, and B. Clegg, *J. Phys. Chem.*, **67,** 300 (1963).

147. D. O. Geymer, *Mackromol. Chem.*, **99,** 152 (1966).

148. N. S. Marans and L. J. Zapas, *J. Appl. Polym. Sci.*, **11,** 705 (1967).

149. R. Salovey and F. R. Dammont, *J. Polym. Sci., Part A1*, **2155** (1963).

150. A. E. Woodward, *J. Polym. Sci., Part B*, **1,** 621 (1963).

151. E. J. Lawton, U.S. Patent 2,948,666, filed Nov. 21, 1956 (1960).

152. H. Sobue and Y. Tazima, *Nature*, **188,** 315 (1960).

153. C. W. Deeley, A. E. Woodward, and J. A. Sauer, *J. Appl. Phys.*, **28,** 1124 (1957).

154. J. Zimmerman, *J. Polym. Sci.*, **43,** 193 (1960).

155. J. Zimmerman, *J. Polym. Sci.*, **46,** 151 (1960).

156. A. I. Brodskii, A. S. Fomenko, T. M. Abramova, E. P. Dar'eva, E. G. Furman, and I. A. Spirina, *Dokl. Akad. Nauk SSSR*, **179,** 370 (1968); *Chem. Abstr.*, **69,** 3284 (1968).

157. L. A. Kotorlenko, A. P. Gardeniva, and V. G. Oleinik, *Ukr. Khim. Zh.*, **30,** 376 (1964); *Chem. Abstr.*, **61,** 16181d (1964).

158. A. I. Brodskii, A. S. Fomenko, T. M. Abramova, E. P. Dar'eva, A. A. Galina, E. G. Furman, L. A. Kotorlenko, and A. P. Gardeniva, *Vysokomolekul. Soedin.*, **7,** 116 (1965).

159. A. I. Brodskii, A. S. Fomenko, T. M. Abramova, E. P. Dar'eva, A. A. Galina, E. G. Furman, L. A. Kotorenko, and A. P. Gardeniva, *Radiats. Khim. Polim., Mater. Simp., Mosk., 1964*, 234 (1966).
160. A. S. Fomenko, T. M. Abramova, E. P. Dar'eva, A. A. Galina, and E. G. Furman, *Vysokomolekul. Soedin.*, **8,** 261 (1966).
161. D. Libby, M. G. Ormerod, A. Charlesby, and P. Alexander, *Nature*, **190,** 998 (1961).
162. P. Alexander and M. G. Ormerod, *Proc. Symp. Biol. Effects Ionizing Radiation Mol. Level, 1962*, 399; *Radiat. Res.*, **18,** 495 (1963).
163. Wen-Chuan Hsin, Ming-Jen Yao, Cheng-Lien Chang, and Ching-Hsin Hua, *K'o Hsueh T'ung Pao, 1964*, 1105; *Sci. Sinica.*, **15,** 211 (1966).
164. B. B. Singh and M. G. Ormerod, *Biochim. Biophys. Acta*, **109,** 204 (1965); **120,** 413 (1966).
165. B. B. Singh and M. G. Ormerod, *Int. J. Radiation Biol.*, **10,** 369 (1966).
166. W. Gordy and I. Miyagawa, *Radiat. Res.*, **12,** 211 (1960).
167. C. Wippler, *Rev. Gen. Caoutchouc*, **36,** 369 (1959).
168. C. Wippler, French Patent 1,187,214 (1959).
169. U.S. Rubber Company, British Patent 833,610 (1960).
170. L. P. Yanova, M. S. Monastyrskaya, S. A. Pavlov, and T. T. Gorbatova, *Izv. Vysshikh Uchebn. Zavedenii, Tekhnol. Legkoi Prom.*, **4,** 46 (1960); *Chem. Abstr.*, **55,** 11914e (1961).
171. I. D. Aitken, H. Wells, and I. Williamson, *At. Energ. Res. Estab.* (*Gt. Britain*), R-3381 (1961).
172. D. Kiessling, *Kolloid Z.*, **176,** 119 (1961); *Chem. Abstr.*, **55,** 2615c (1961).
173. A. A. Miller, *J. Appl. Polymer Sci.*, **5,** 388 (1961).
174. A. H. Selker, *Mod. Plastics*, **40** (1), 172 (1962).
175. T. Kimura, H. Murata, and K. Yoshida, *Kobunshi Kagaku*, **19,** 747 (1962); *Chem. Abstr.*, **61,** 7177f (1963).
176. V. M. Anikeenko, K. M. Kevroleva, R. M. Kessenikh, and V. G. Sotnikov, *Vestn. Elektroprom.*, **33** (6), 16 (1962); *Chem. Abstr.*, **57,** 10032e (1962).
177. M. N. Shteding and V. L. Karpov, *Vysokomolekul. Soedin.*, **4,** 1806 (1962).
178. K. Kimura, *Kobunshi Kagaku*, **20** (213), 65 (1963); **72**; *Chem. Abstr.*, **61,** 3263h (1964).
179. M. N. Shteding, *Zashchitn. Pokrytiya v At. Tekhn., Sb. Statei, 1963*, **35;** *Chem. Abstr.*, **61,** 2007b (1964).
180. B. V. Kotov and L. Zhalondkova, *Dokl. Akad. Nauk SSSR*, **159,** 640 (1964); *Chem. Abstr.*, **62,** 5354h (1965).
181. H. Langner, *Plaste Kautschuk*, **13,** 76 (1966); *Chem. Abstr.*, **64,** 14352a (1966).
182. B. V. Kotov and A. N. Pravednikor, *Radiats. Khim. Polim., Mater. Simp., Mosk., 1964*, 218 (1966); *Chem. Abstr.*, **67,** 11906 (1967).
183. M. Izumi, F. Suzucki, and T. Yuri, *Sumitomo Denki*, **90,** 44 (1965); *Chem. Abstr.*, **64,** 8389e (1966).
184. E. J. Lawton, A. M. Bueche, and J. S. Balwit, *Nature*, **172,** 76 (1953).
185. A. Charlesby, *Nature*, **171,** 107 (1953).
186. J. W. Born, *WADC Tech. Rept.*, *55–58*, Pt. 1 (1954); Pt. 11 (1955).
187. R. G. Bauman and J. W. Born, *J. Appl. Polym. Sci.*, **1,** 351 (1959).
188. R. G. Bauman, *J. Appl. Polymer Sci.*, **2,** 328 (1959).
189. J. R. Dunn, *Kautschuk Gummi*, **14,** WT 114 (1961); *Chem. Abstr.*, **55,** 14955h (1961).
190. A. Lamm and G. Lamm, *Kautschuk Gummi*, **14,** WT 334 (1961); *Chem. Abstr.*, **56,** 6137d (1962).
191. Z. N. Tarasova, V. T. Kozlov, and B. A. Dogadkin, *Radiat. Chem., Proc. Tihany Symp., Tihany, Hung.*, **1962,** 287 (1964).

192. Yu. Kvashenko, A. S. Kuz'minskii, and T. S. Fedoseeva, *Vysokomolekul. Soedin.*, **8,** 450 (1966).

193. T. S. Nikitina, A. S. Kuz'minskii, L. A. Oksent'evich, and V. L. Karpov, *Tr. 1-go, (Pervogo) Vses. Soveshch. po Radiats. Khim. Akad. Nauk SSSR, Otd. Khim. Nauk, 1957,* 292 (1958); *Chem. Abstr.*, **53,** 7647e (1959).

194. D. T. Turner, *J. Polym. Sci.*, **27,** 503 (1958); *Polymer*, **1,** 27 (1959).

195. A. S. Kuz'minskii and T. S. Nikitina, *Vulkanizatsiya Rezin. Izdelii, Yaroslavl, 1960,* **14**; *Chem. Abstr.*, **56,** 6137i (1962).

196. Z. N. Tarasova, M. S. Fogel'son, V. T. Kozlov, A. I. Kashlinskii, M. Ya. Kaplunov, and B. A. Dogadkin, *Vysokomolekul. Soedin.*, **4,** 1204 (1962).

197. At. Vasilev, Iv. Boyadzhov, Khr. Tsankov, and L. Khristanova, *Kozhi Obuvki Kauchuk Plastmasi (Sofia)*, **2,** 7 (1962); *Chem. Abstr.*, **57,** 7439f (1962).

198. D. T. Turner, *J. Polym. Sci.*, Part *B-1*, **97** (1963).

199. Z. N. Tarasova, V. G. Kozlov, and B. A. Dogadkin, *Magy. Kem. Lapja*, **19,** 354 (1964); *Chem. Abstr.*, **64,** 899h (1966).

200. F. Zeplichal and N. Steiner, *Kautschuk Gummi*, **20,** 508 (1967).

201. Ref. 2, Page 429.

202. A. Charlesby, *At. At. Technol.*, **5,** 12 (1954).

203. A. S. Kuz'minskii, T. S. Nikitina, E. V. Zhuravskaya, L. A. Oksentievich, L. L. Sumitsa, and N. I. Vitushkin, *Proc. 2nd Int. Conf. Peaceful Uses Atomic Energy, Geneva*, **29,** 258 (1958).

204. B. Jankowski and J. Kroh, *J. Appl. Polym. Sci.*, **9,** 1363 (1965).

205. A. S. Kuz'minskii, E. V. Zhuravskaya, E. Z. Novgorodova, L. V. Chepel, and B. A. Chapyzhnikov, *Radiats. Khim. Polim., Mater. Simp., Mosk., 1964,* 238 (1966); *Chem. Abstr.*, **67,** 116438 (1967).

206. H. R. Anderson, U.S. Patent 3,057,818, Appl. Oct. 2, 1959 (1962); U.S. Patent 3,112,291, Appl. Dec. 14, 1959 (1963); U.S. Patent 3,175,992, Appl. Aug. 21 1958 (1965); U.S. Patent 3,250,740, Appl. June 16, 1960 (1966); U.S. Patent 3,247,160, Appl. Dec. 19, 1960 (1966).

207. R. Harrington, *Rubber Age*, **90,** 265 (1961); *ibid.*, **97,** 88 (1965); Rept. No. HW-SA-3629, AEC Accession No. 30824.

208. V. T. Kozlov, Z. N. Tarasova, E. R. Klinshpont, V. K. Milinchuk, and B. A. Dogadkin, *Vysokomolekul. Soedin.*, **9,** 1541 (1967).

209. J. Tsurugi, K. Imamura, T. Baba, and M. Irabu, *Nippon Gomu Kyokaishi*, **40,** 382 (1967); 388 (1967); 394 (1967); *Chem. Abstr.*, **67,** 100796, 100797, 100798 (1967).

210. J. Tsurugi, K. Imamura, T. Baba, and M. Irabu, *Nippon Gomu Kyokaishi*, **40,** 636 (1967); *Chem. Abstr.*,**68,** 79388.

211. J. Tsurugi, K. Imamura, T. Baba, and M. Irabu, *Nippon Gomu Kyokaishi*, **40,** 1004 (1967); *Chem. Abstr.*, **68,** 79332.

212. A. S. Kuz'minskii, L. S. Fel'dshtein, E. V. Zhuravskaya, and L. I. Lyubchanskaya, *Tr. 2-go (Vtorogo) Vses. Soveshch. po Radiats. Khim., Akad. Nauk SSSR, Otd. Khim. Nauk, Mosk., 1960,* 576 (1962); *Chem. Abstr.*, **58,** 4710h (1963).

213. B. A. Dogadkin, Z. N. Tarasova, M. Ya. Kaplunov, V. T. Kozlov, I. A. Klauzen, and V. S. Matveev, *Tr. 2-go (Vtorogo) Vses. Soveshch. po Radiats. Khim., Akad. Nauk SSSR, Otd. Khim. Nauk, Mosk.*, 1960, 554 (1962); *Chem. Abstr.*, **58,** 2554h (1963).

214. A. S. Kuz'minskii, E. V. Zhuravskaya, L. I. Lyubchanskaya, and L. S. Fel'dshtein, *Radiat. Chem., Proc. Tihany Symp., Tihany, Hung., 1962,* 235 (1964).

215. Natural Rubber Research Association, British Patent 889,112, Appl. May 24, 1957 (1962).

216. Montecatini, Ital. Patent 644,767, Appl. Feb. 14, 1961 (1962).

217. V. T. Kozlov, Z. N. Tarasova, and B. A. Dogadkin, *Khim. Vys. Energ.*, **1,** 136 (1967); *Chem. Abstr.*, **67,** 74302 (1967).
218. A. S. Kuz'minskii and M. A. Zakirova, *Radiats. Khim. Polim., Mater. Simp., Mosk., 1964*, 388 (1966); *Chem. Abstr.*, **67,** 116439 (1967).
219. V. T. Kozlov, *Elementarnye Protsessy Khim. Vysokikh Energ., Akad. Nauk SSSR, Inst. Khim. Fiz., Tr. Simp., Mosk., 1963*, 220 (1965); *Chem. Abstr.*, **64,** 10774 (1966).
220. I. P. Cherenyuk, G. A. Blokh, and A. P. Meleshevich, *Radiats. Khim. Polim., Mater. Simp., Mosk., 1964*, 317 (1966); *Chem. Abstr.*, **67,** 116443 (1967).
221. B. A. Dogadkin, Z. N. Tarasova, M. S. Fogel'son, and A. I. Kashlinskii, *Dokl. Akad. Nauk SSSR*, **141,** 90 (1961); *Chem. Abstr.*, **56,** 10343i (1962).
222. P. G. Garratt and M. G. Ormerod, *Int. J. Radiat. Biol.*, **6,** 281 (1963).
223. R. K. Jenkins, *J. Polym. Sci., Part A-1*, **4,** 2161 (1966).
224. For a summary see Ref. 2, p. 473 ff.
225. A. A. Miller, *J. Amer. Chem. Soc.*, **83,** 31 (1961).
226. O. Costăchel, Gh. Furnică, and A. Drăgut, *Acad. Rep. Populare Romîne, Studii Cercetări Chim.*, **7,** 409 (1959); *Chem. Abstr.*, **54,** 16133f (1960).
227. P. E. Wei and J. Rehner, Jr., *Rubber Chem. Technol.*, **35,** 133 (1962).
228. J. Lal and J. E. McGrath, *Rubber Chem. Technol.*, **36,** 248 (1963).
229. Ref. 1, p. 498.
230. P. Alexander and D. J. Toms, *Radiat. Res.*, **9,** 509 (1958).
231. D. G. Gardner and L. M. Epstein, *J. Chem. Phys.*, **34,** 1653 (1961).
232. A. R. Schultz, P. I. Roth, and J. M. Berge, *J. Polym. Sci., Part A*, **1,** 1651 (1963).
233. F. A. Sliemers, J. F. Kircher, R. Lieberman, R. Markle, W. B. Gager, M. Luttinger, and R. I. Leininger, *U.S. At. Energy Comm.*, BMI-1610 (1963).
234. H. A. Atwater, *Amer. Soc. Testing Mater., Spec. Tech. Publ. 384*, 32 (1965); *J. Appl. Phys.*, **36,** 2220 (1965); reported in part by J. J. Banaszak, AD 611779 (1964).
235. I. I. Sapezhinskii, Yu. V. Silaev, and N. M. Emanuel, *Tr. Mosk. Obshch. Ispytatelei Prirody, Otd. Biol.*, **21,** 102 (1965); *Chem. Abstr.*, **63,** 5997g (1965).
236. G. A. Henry, *U.S. At. Energy Comm., COO-1224-15* (1967).
237. B. J. Lyons, *Nature*, **185,** 604 (1960).
238. S. H. Pinner and V. Wycherley, *J. Appl. Polym. Sci.*, **3,** 338 (1960).
239. T. Lemiszka and J. E. Shewmaker, U.S. Patent 2,982,706 (1961).
240. J. Zurakowska-Orszagh and Z. E. Sobolewska, *Polish Acad. Sci., Inst. Nucl. Res. Rept. No. 311/XVII* (1962); *Chem. Abstr.*, **58,** 12729d (1963).
241. G. Ayrey and D. T. Turner, *J. Polym. Sci., Part B*, **1,** 185 (1963).
242. Y. Okada, *J. Appl. Polym. Sci.*, **7,** 1791 (1963).
243. D. T. Turner, *J. Polym. Sci., Part A*, **2,** 1699 (1964).
244. G. Ayrey and D. T. Turner, *U.S. At. Energy Comm., CONF-415-1* (1963); *Polym.* **5,** 589 (1964).
245. J. Tsurugi, K. Imamura, T. Baba, and M. Irabu, *Nippon Gomu Kyokaishi*, **40,** 633 (1967); *Chem. Abstr.*, **68,** 79387 (1968).
246. G. C. Odian and B. S. Bernstein, *U.S. At. Energy Comm., TID-7643*, 245 (1962); *J. Polym. Sci., Part B*, **2,** 819 (1964).
247. See Ref. 1, pp. 325 ff. for a summary.
248. See Ref. 1, pp. 359 ff.
249. W. Albrecht and E. Völpel, *Melliand Textilber.*, **41,** 741 (1960); *Chem. Abstr.*, **54,** 18967e (1960).
250. G. A. Klein, M. P. Tikhomolova, and N. I. Kononenko, *Dokl. Akad. Nauk Uz. SSR* **20** (7), 13 (1963); *Chem. Abstr.*, **60,** 8177g (1964).
251. C. J. Westberry and W. L. Hyden, *Amer. Dyestuff Reptr.*, **56,** 13 (1967).

252. S. H. Pinner, T. T. Greenwood, and D. G. Lloyd, *Nature*, **184,** 1303 (1959).
253. H. N. Rexroad and W. Gordy, *J. Chem. Phys.*, **30,** 399 (1959).
254. T. Matsugashita and K. Shinohara, *J. Chem. Phys.*, **32,** 954 (1960)
255. R. J. Abraham and D. H. Whiffen, *Trans. Faraday Soc.*, **54,** 1291 (1958).
256. H. Ueda, Z. Kuri, and S. Shida, *J. Polym. Sci., Part A*, **1,** 3537 (1963).
257. P. Alexander, *Int. Ser. Monogr. Pure Appl. Biol. Mod. Trend Physiol. Sci. Div.*, **7,** 3 (1960).
258. K. Flemming, *Naturwissenschaften*, **48,** 555 (1961); *Chem. Abstr.*, **56,** 5079i (1962).
259. P. Alexander and D. Rosen, *Radiat. Res.*, **15,** 475 (1961).
260. S. Sugai and K. Kamashima, *Kobunshi Kagaku*, **18,** 191 (1961); *Chem. Abstr.*, **56,** 3056c (1962).
261. F. Hutchinson, *Radiat. Res.*, **14,** 721 (1961).
262. A. Pihl and T. Sanner, *Biochim. Biophys. Acta*, **78,** 537 (1963); *Chem. Abstr.*, **60,** 4431f (1964).
263. E. E. Ganassi, L. Kh. Eidus, and R. A. Arifulina, *Radiobiologiya*, **3,** 440 (1963); *Chem. Abstr.*, **59,** 11853b (1963).
264. W. Lohmann, A. J. Moss, Jr., and W. H. Perkins, *J. Nucl. Med.*, **6,** 519 (1965); *Chem. Abstr.*, **63,** 16740g (1965).
265. V. M. Merezhinskii, G. N. Morotseva, and V. P. Zhigalko, *Biokhim. Konf. Pribaltiisk. Resp. i Belorussk. 2nd, Sb., Posvyashch. 25-letiyu Vosstanovl. Sov. Vlasti. v Latv., Likovsk. i Est. SSR, Riga, 1965*, **54**; *Chem. Abstr.*, **65,** 18959e (1966).
266. N. M. Emanuel, E. B. Burlakova, K. E. Kruglyakova, and I. I. Sapezhinskii, *Izv. Akad. Nauk SSSR, Ser. Biol.*, **31,** 183 (1966); *Chem. Abstr.*, **65,** 7588f (1966).
267. H. Jung and H. Schuessler, *Z. Naturforsch.*, **21,** 224 (1966); *Chem. Abstr.*, **65,** 1009e (1966).
268. S. Sugai, *J. Phys. Soc. Japan*, **14,** 1573 (1959); *Chem. Abstr.*, **54,** 11642i (1960).
269. M. Matsumoto and A. Danno, *Large Radiation Sources Ind., Proc. Conf., Warsaw, 1959*, **1,** 331 (1960).
270. H. Narasaki and S. Fujiwara, *J. Polym. Sci., Part B*, **1,** 153 (1963).
271. I. Sakurada and Y. Ikada, *Bull. Inst. Chem. Res. Kyoto Univ.*, **41,** 103 (1963); *Chem. Abstr.*, **61,** 1964c (1964).
272. A. Henglein, *J. Phys. Chem.*, **63,** 1852 (1959).
273. I. Sakurada and Y. Ikada, *Bull. Inst. Chem. Res. Kyoto Univ.*, **40,** 1 (1962); **16**; **25**; *Chem. Abstr.*, **57,** 13954e (1962).
274. J. Durup, *J. Chim. Phys.*, **54,** 739 (1957).
275. H. Heyns and V. Desreux, *Large Radiation Sources Ind., Proc. Conf., Warsaw, 1959*, **1,** 257 (1960).
276. A. Polowinska, E. Turska, and J. Kroh, *Bull. Acad. Polon. Sci., Ser. Sci. Chim.*, **12,** 801 (1964); *Chem. Abstr.*, **63,** 3071e (1965).
277. C. Wippler, French Patent 1,171,117 (1959); *J. Polym. Sci.*, **29,** 585 (1958); C. Wippler and E. Gautron, *J. Polym. Sci., Part A*, **1,** 943 (1963).
278. A. Henglein and C. Schneider, *Z. Phys. Chem.*, **19,** 367 (1959).
279. K. Murayama and Y. Shinohara, *Kobunshi Kagaku*, **21,** 710 (1964); *Chem. Abstr.*, **62,** 10643e (1965).
280. I. M. Todorov, *Proc. 2nd Tihany. Symp. Radiat. Chem., Tihany, Hung., 1966*, **749** (pub. 1967).
281. E. E. Mooney and S. T. Semegen, *U.S. Dept. Comm., Office Tech. Serv., P.B. Rept.*, 158,552-2 (1959); *Chem. Abstr.*, **58,** 10379h.
282. T. W. Albrecht, *U.S. Dept. Comm. Office Tech. Ser, P.B. Rept.* 136,817 (1959); *Chem. Abstr.*, **54,** 11658 (1960).

283. M. Nisizawa, *J. Appl. Polym. Sci.*, **12,** 1781 (1968).

284. A. J. Swallow, "Radiation Chemistry," Vol. II, R. Gould, Ed., *Advanc. Chem. Ser. No. 82*, American Chemical Society, Washington, D.C., 499 (1968).

285. F. Wilkinson, in W. A. Noyes, G. S. Hammond, and J. N. Pitts, Eds., *Advances in Photochemistry*, Vol. 3, Wiley, New York, 1964, p. 241.

286. J. Wilske and H. Hensinger, *J. Polym. Sci., Part A-1*, **7,** 995 (1969).

287. S. K. Ho, S. Siegel, and H. A. Schwarz, *J. Phys. Chem.*, **71,** 4527 (1967).

288. S. K. Ho and S. Siegel, *J. Chem. Phys.*, **50,** 1142 (1969).

289. G. E. Adams, R. C. Armstrong, A. Charlesby, B. D. Michael, and R. L. Willson, *Trans. Faraday Soc.*, **65,** 732 (1969).

290. D. Evans, J. T. Morgan, R. Sheldon, and G. B. Stapleton, *J. Polymer Sci., Part A-2*, **7,** 725 (1969).

7

STABILIZATION AGAINST BURNING

P. C. WARREN

Bell Telephone Laboratories, Incorporated, Murray Hill, New Jersey

A. INTRODUCTION

Polymer combustion, representing the extreme of thermal oxidation, can be characterized by either a smoldering or vigorous and flaming degradation of the organic material. Although protection of the polymer from the ambient environment is solely of economic importance, stabilization against burning must ensure against human injury as well. In fact, the fire hazard of the common polymers has resulted in legislation to limit their application in several areas, particularly in the textile[1] and construction[2] fields. One can only foresee that such restrictions will become more and more stringent in the future. The inevitable safety requirement coupled with the ever-increasing need for plastics and textiles in new applications will surely put great demands on the methods of stabilizing polymers against burning.

The burning polymer is unique compared to the other forms of degradation treated in this book in that the rate of combustion is more dependent on physical phenomena. Gas turbulence and diffusion; conductive, convective, and radiative heat flow; polymer volatility, melting and dripping, and the surface/volume ratio are but a few of the nonchemical factors that enormously influence ignition and burning rates. The morphological characteristics of the solid state are of little consequence, however, because the interesting condensed-phase combustion reactions usually occur in the melt.

Historically, two general ways have been successfully used to render polymers less flammable. In the first of these, specific materials of high thermal and oxidative stability have been synthesized, not unlike the heat-resistant polymers treated in Chapter 3. Unfortunately, the high cost and lack of other desirable properties has generally restricted their use to speciality low-volume applications. In the second, additives have been combined either physically or chemically onto the surface or into the bulk of the normally flammable polymer. Contrary to the earlier case, the thermal stability of the fire-retarded polymer is not enhanced and can be significantly decreased.[3] Instead, these additives are specific inhibitors of the burning itself. It is this second approach that has proved to be the more practical and economically feasible, and therefore it receives the most attention in this chapter.

Relatively few fundamental ways exist to prevent polymer combustion, and virtually all were known long before synthetic materials became available. Phosphate salts, for instance, were reported as early as 1821 by Gay-Lussac[4] to be among the better fire retardants for cellulosics. In the past decade, much empirical research has gone into designing better fire retardants for specific polymers, but no basically new stabilizers have emerged from these studies. It is not the purpose of this chapter to list the many known inhibitors and their particular applications, since this has been accomplished elsewhere.[5–11] Nor is our purpose here to emphasize the development aspects of the problem, such as proper design of retardants for particular polymers, or to consider the effects of these inhibitors on other polymer properties. Rather, we shall concentrate on the fundamental physical and chemical mechanisms of both polymer combustion and its suppression, with the hope that this approach will eventually improve on the empirical efforts of the past.

It is evident that inhibition can take place in only two regions of the burning polymer, either in the flame or in the condensed phase, or possibly in both regions simultaneously. This review is organized around this natural division. After a discussion of the macroscopic details of polymer combustion, polymer flame chemistry and flame inhibition are considered. The degradation of the burning polymer itself is then treated, along with condensed-phase stabilization mechanisms. The process of polymer combustion is complex and little is known of the mechanism, particularly in the case of flame reactions. Fortunately, combustion research is not confined to polymers, and a large body of knowledge from other systems is at our disposal. We therefore draw heavily on such data where they are both relevant and justified.

B. MACROSCOPIC CONSIDERATIONS OF POLYMER FLAMMABILITY

In any fire, oxygen, heat, and fuel comprise the classic triangle necessary to sustain combustion. Partial removal of any of these three will usually slow the rate of combustion, and when the rate of chain termination exceeds that of chain branching, the flame will extinguish. Accordingly, it should be evident that tampering with any one or combination of these components has practical application in retarding polymer combustion.

The typical burning polymer is diagrammed in Figure 7.1. The material is initially heated in air from some outside source and eventually degrades to give volatile pyrolysis products. Depending on the flammability of these gases and their rate of evolution, they may ignite at some point when subjected to an external heat source, whereupon they will burn at a given rate. Some of the heat of combustion will be fed back to the degrading polymer, thus volatilizing more material. If the heat feedback is sufficient, the completed cycle will become self-sustaining after removal of the initial heat source. When the polymer burns downward in a candlelike manner, there will be some radiative heating but most will be via conduction.[12] In an actual fire, however, convective heating contributes to the feedback to make it extremely efficient. Coupled with the increased mixing rate owing to these

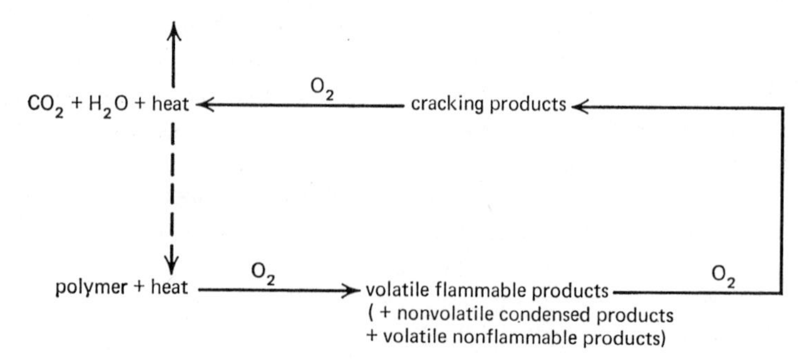

Fig. 7.1 Schematic representation of a burning polymer.

turbulent air currents, actual fire situations often become catastrophic in a very short time.

The equilibrium established by the burning polymer is therefore a complex sequence of processes, including the rate of evolution of pyrolytic gases, the rate of mixing with oxygen, the rate of reaction with oxygen, and finally, the rate of heat flow back to the polymer. Perturbation of this equilibrium can occur if the cycle is interrupted at any one or more of these stages. Specific means of producing such interruptions, thus constituting inhibition, are considered in detail in the appropriate section.

Some concise definitions and clarifications are offered at this point. No attempt is made that they be rigorous; rather, they are included for the sake of consistency throughout the chapter.

Premixed flame—a burning mixture wherein the fuel and oxidant gases are completely mixed before ignition. Most laboratory burners give fuel–air flames of this type.

Diffusion flame—a burning mixture wherein the fuel and oxidant gases mix (by diffusion) simultaneously with burning. An ordinary paraffin candle burning in air illustrates this type of flame.

Flammability limit or *limit*—the empirical fuel–oxidant composition that defines the borderline between burning and nonburning gaseous mixtures. Premixed fuel–air flames generally have two flammability limits, a fuel-rich limit and a fuel-lean limit. All compositions between the limits burn, and all those outside these extremes do not burn.

Additive—a fire-retardant stabilizer either physically blended with or chemically bound to a polymer. The former case is often referred to as an "additive" fire retardant and the latter as a "reactive" type, but these terms lose their relevance in a mechanistic discussion. For our purposes here, any material or functional group in the polymer that retards burning is considered an additive. For instance, the chlorine in poly(vinyl chloride) is an additive that is chemically bound to polyethylene.

1. Oxygen Index Measurements

Since flames are a complex interaction of radical reactions, heat feedback, and convective and diffusive flow, reproducible measurement of relative flammabilities of various fuels or of the effects of added inhibitors is difficult. Two general methods exist for both monomers and polymers, namely, "rate" and "limit" techniques. The rate measurements are more fundamental,[13] but the latter apply more directly to the practical case of determining the actual cessation of burning. Often the two methods do not correlate well, indicating that different quantities are in fact being measured. It is important to realize that both measurements are of overall processes and not of a single rate-determining step.

Burning rates in polymer systems are measured by burning the material under carefully defined and controlled conditions. Although these tests are usually cumbersome and not at all reproducible, they approximate actual fire situations and are very useful in that respect. They have been adequately reviewed elsewhere[9,14–17] and therefore are not repeated here.

Limit measurements, on the other hand, are particularly suited to measuring polymer flammability. Using this principle, Fenimore and Martin devised the oxygen index test method.[18,19] In this test the polymer flame is cooled and diluted by addition of inert gas to the oxygen–inert gas mixture forming the atmosphere until burning is extinguished; the fraction of oxygen just before extinction is defined as the oxygen index. It should be clear that this number is inversely related to the flammability of the polymer. The method is unique in that it is simple, can be applied to any kind of flame, is reproducible to about 1% at constant flow rate, and allows one to study the combustion of polymers in atmospheres other than air. Furthermore, it appears to be a linear method, as shown by the straight lines obtained when it is graphed against the vertical burning rate of a polymer inhibited with various concentrations of retardant.[17]

The oxygen index method was used initially to rate the flammability of various gaseous and liquid fuels,[20,21] and also to measure the effects of halogen inhibitors on gaseous diffusion flames.[22] It was shown in these early studies that the maximum flame temperature at this oxygen concentration correlated roughly with the limit temperature of the premixed flame, and also that the oxygen index of a diffusion flame was equal to that of a stoichiometric premixed flame.[21]

It should be emphasized that the oxygen index is not a flammability limit in the sense that this limit is applied to premixed flames, since there are concentration gradients throughout the initial unburned gases.[23] In fact, the diffusion flame has been characterized by a lean limit at the outer edge and a rich one at the inner edge, with the burning area located between these limits.[22] In any case, it is remarkable that the oxygen index "limit" gives the consistent values that it does. Since the method has been used extensively in studying mechanisms of polymer stabilization against combustion, it is analyzed in somewhat greater detail.

2. Parameters of the Oxygen Index Measurement

The actual oxygen index test for flammability consists of burning a small, vertical stick of polymer downward in a chimney containing a slowly rising oxygen–nitrogen atmosphere. Within certain limitations, the method is only mildly dependent on flow rate, sample size, or molecular weight. Several studies on polyethylene,[24] poly(methyl methacrylate),[24] polypropylene,[24] and polyester[26] resins show that the oxygen index varies unpredictably to the extent of 3 to 5% over a gas flow range between 3 and 12 cm/sec. Because of this slight dependence on the gas rise rate, the flow rate has been standardized at 4 ± 1 cm/sec.[24,27] The results obtained outside the extreme flow limits are much less reproducible, because the rate must be fast enough to clear burned gases from the immediate atmosphere (>3 cm/sec), but not so fast that the mixing rate at the leading edge of the flame is increased (>12 cm/sec).[21] Also, reported measurements of the oxygen indices of polycarbonate,[24–26] nylon, and polyester[26] all show consistently greater flammability in thin samples (<0.1 in.) and constant values in the intermediate range (0.1–0.5 in.). Thicker samples (>0.5 in.) give higher indices, and therefore appeared less flammable by this test.[18,19,28] Finally, an epoxy resin was shown to give the same oxygen index regardless of cure conditions, and above a certain minimum, polyethylene values were reported to be independent of molecular weight.[29]

The method is also dependent on pressure and temperature. For instance, polyethylene gave an oxygen index of 0.175 at room temperature and atmospheric pressure, but decreasing the pressure to 140 mm made it "less flammable," with a reading of 0.21.[18,19,28] In the pressure range of 1 to 0.2 atm, the percentage of increase in the oxygen index divided by the percentage of decrease in pressure was 0.30 for polyoxymethylene, 0.24 for polyethylene, 0.22 for chlorinated polyethylene ($C_2H_{3.8}Cl_{0.2}$), and 0.85 for a carbon rod.[30] Raising the temperature of the atmosphere around the polymer lowered the oxygen index of the sample, and interestingly, polymers of very different structure were affected similarly. For example, carbon, unfilled epoxy resins, polyethylene, and poly(methyl methacrylate) all gave straight lines with slopes of approximately 0.025 oxygen index units per 100°C when the temperature of the rising gas was plotted against the oxygen index.[30] An exception was polytetrafluoroethylene, which became abnormally more flammable with increasing temperature.[12,30] DiPietro and Stepniczka[26] showed the greater polymer flammability at higher temperatures to be caused by the initial temperature of the polymer itself, not the temperature of the atmosphere. They achieved a differential in the initial solid and gas temperatures by heating the apparatus by radiation, the solid being a better "black body" than the gas. This result is not unexpected, since the heating of the gas only slightly affects the relatively high flame temperatures (700–1000°C), but significantly reduces the heat feedback requirement from the flame because of the relatively low surface temperature required for volatilization (350–450°C). Miller[31] reported an analogous temperature function for textile flammability by plotting the maximum temperature against oxygen concentration. He interpreted the straight lines as a measure of the sensitivity of the burning process to oxygen content, suggesting that materials with steep slopes could be more easily extinguished than those with gradual ones.

The method is very dependent on sample orientation, sample melting and dripping, and the heat capacity and thermal conductivity of the inert gas. A strict requirement of the test is that the vertical sample burn downward, thus eliminating convective heating of the material or turbulent mixing in the flame. This accounts for the excellent reproducibility, but the tendency is also partially responsible for the poorer correlations with the various burning rate tests, which usually do include these variables. Accordingly, upward-burning vertical samples gave lower oxygen indices that were not at all reproducible.[18,19] Also, polymers that dripped easily gave false high readings, because heat was carried away without volatilizing the sample.[18,19] For instance, when a low-melting-point specimen of polypropylene giving an unexpectedly high oxygen index of 0.23 was burned on a ceramic wick, it gave the same value of 0.177 as a nondripping sample of higher molecular weight.[18,19] Similarly, nylon has an oxygen index of 0.26 and is therefore "self-extinguishing," but in fact burns in air when not allowed to drip.[31,32] Finally, the oxygen index was found to be dependent on both the heat capacity and thermal conductivity of the inert gas.[33] The oxygen–inert gas ratio just before extinguishment was found to vary linearly with the heat capacities of carbon dioxide, nitrogen, and argon.[33] Helium has a very low heat capacity, but the expected high flammability in oxygen–helium atmospheres is reduced somewhat because of the very high thermal conductivity of helium.[33,34]

The precision of the oxygen index test allows some common materials to be rated according to their relative flammabilities,[14,18,19,21] and several are listed in Table 7.1.

Table 7.1 Oxygen Indices of Various Materials

Materials	Oxygen Index	References
Gases		
Hydrogen	0.054	21
Carbon monoxide	0.076	21
Formaldehyde	0.071	21
Acetylene	0.085	21
Ethylene	0.105	21
Methane	0.139	21
Ethane	0.118	21
Propane	0.127	21
Liquids and Waxes		
Methyl alcohol	0.111	21
n-Octyl alcohol	0.132	21
Acetone	0.129	21
n-Pentane	0.133	21
n-Decane	0.135	21
Cyclohexane	0.134	21
Benzene	0.133	21
Paraffin (kitchen candle)	0.16	18, 19
Polymers		
Polyoxymethylene	0.150	18, 19
Poly(ethylene oxide)	0.150	18, 19
Poly(phenylene oxide)	0.28–0.29	18, 19
Polyethylene	0.175	18, 19
Polypropylene	0.175	18, 19
Polybutadiene	0.183	18, 19
Polystyrene	0.182	18, 19
Poly(methyl methacrylate)	0.173	18, 19
Nylon	0.24	25
Polycarbonate	0.26–0.28	18, 19
Polysulfone	0.30	25
Polytetrafluoroethylene	0.95	18, 19
Carbon (electrode)	0.635	18, 19
Poly(vinyl alcohol)	0.225	18, 19
Poly(vinyl fluoride)	0.226	14
Poly[3,3-bis(chloromethyl)oxetane]	0.232	18, 19
Poly(vinyl chloride)	0.45–0.49	18, 19
Poly(vinylidene fluoride)	0.437	14
Poly(vinylidene chloride)	0.60	18, 19
Cellulose Materials		
Cellulose (cotton)	0.186	25
Cellulose acetate	0.168	25
Cellulose butyrate	0.188	25
Wood (birch)	0.205	25
Wood (red oak)	0.227	25

The following comments are noted with reference to this list.

1. The oxygen concentration necessary for burning increases as the fuel is changed from gas to liquid to solid, paralleling the heat requirements for vaporization.

2. The polymers that contain little or no hydrogen are among the least flammable.

3. The chlorinated polymers are very difficult to burn, especially when compared to the fluoro- or hydroxy-substituted homologues.

4. Although 0.21 should be the border between flammable and nonflammable polymers in air, the more practical definition of 0.27 for self-extinguishing plastics[18,19] takes into account some convective heating, which is always present in an actual fire.

5. Some polymers burn entirely differently from others: poly(methyl methacrylate), polyethylene, polypropylene, and polytetrafluoroethylene all burn cleanly without smoke, char, or residue formation.[18,19] Poly(vinyl chloride), chlorinated polyethylene, or polycarbonate, however, burn with heavy smoking and char formation.[18,19,33] Although polystyrene does not leave a char when burned, it gives a very smoky flame.

6. A correlation has been reported; namely, that, with several exceptions, the more oxygen incorporated into the polymer, the lower the oxygen index.[29]

7. Finally, Martin[33] proposed a mathematical model of polymer combustion based on the leading edge of the diffusion flame in terms of the oxygen–nitrogen ratio at extinguishment.

3. Oxygen Index Mechanistic Experiments

Fenimore and co-workers[18,19,35] used the oxygen index method for mechanistic experiments to determine the *region* of inhibition in flame-retarded polymers, that is, whether they are inhibited in the condensed phase or in the flame. An understanding of these experiments is imperative for much of the mechanistic treatment in this chapter. Three of the most informative experiments are discussed: (*a*) additions of various gases to the atmosphere around the polymer, (*b*) the burning of polymers in nitrous oxide–nitrogen and oxygen–nitrogen atmospheres with comparison of the resulting oxygen indices, and (*c*) comparison of oxygen indices of the same additive in polymers of different structure.

Gases such as hydrogen chloride or chlorine were added to the atmosphere to determine whether the pyrolysis products from chlorinated polymers or additives were responsible for flame inhibition.[18,19] The amounts of chlorine in the gas phase were equated to the amounts in the polymer utilizing the fact that diffusion flames burn where the oxygen and fuel meet in stoichiometric concentrations. It was assumed that stoichiometric conversion of the polymer to carbon dioxide and water took place and that the diffusion rates

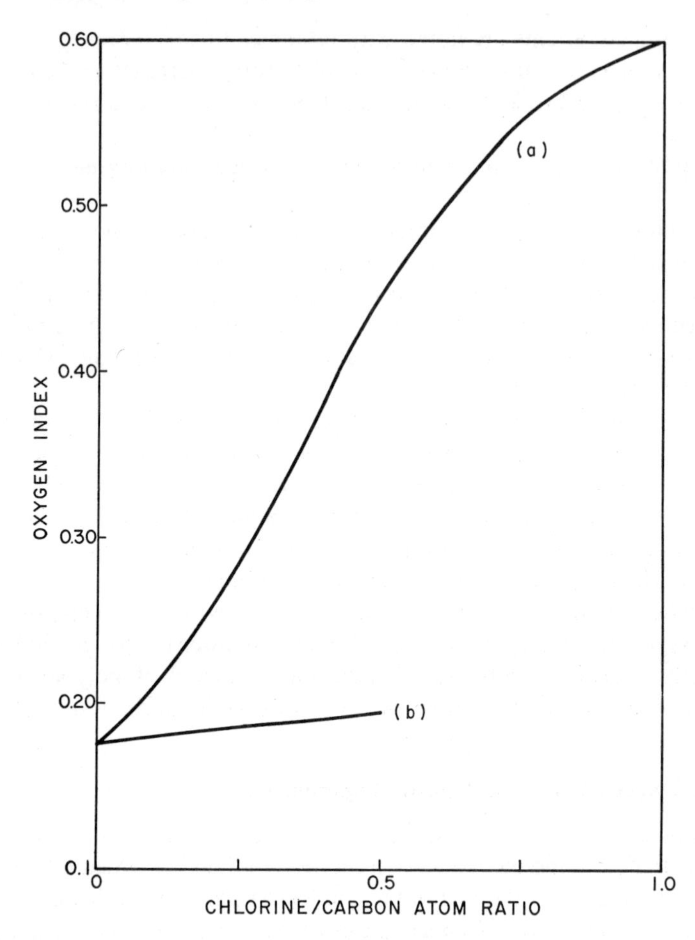

Fig. 7.2 Comparison of oxygen indices: (*a*) chlorinated polyethylene; (*b*) polyethylene burnt in a hydrogen chloride–oxygen–nitrogen atmosphere. (Reprinted from *Combustion and Flame*.)

of oxygen and hydrogen chloride or chlorine into the flame were similar. A comparison of the oxygen index versus chlorine concentration curves (Figure 7.2) showed the chlorine that was chemically bound to the polymer to be much more effective than chlorine added to the atmosphere around the burning polymer.

Determinations of polymer flammability in oxidant compositions other than oxygen–nitrogen were compared with the oxygen values.[35] For instance, if a fire-retarded polymer was inhibited similarly in both nitrous oxide–nitrogen and oxygen–nitrogen atmospheres, the inhibitor presumably functioned in the condensed phase, where it was insensitive to the change of oxidant around the burning polymer. An example of such a case is shown in Figure 7.3, where the oxygen indices of various concentrations of aluminum-oxide-trihydrate-filled epoxy resins were measured in these two oxidizing atmospheres.[29] A plot of the extinguishment values against concentration of filler showed similar curves in both cases. Such a finding is consistent with the expectation that a filler would operate only in the condensed phase.

When a flame inhibitor is incorporated into a polymer, however, one would expect the inhibitor to be very sensitive to the oxidant, since both are involved in the same reactions. Such is the case with antimony oxide-retarded polymers (Figure 7.4). The plot of antimony oxide concentration in chlorinated polyethylene versus the oxygen index shows it to be a totally ineffective inhibitor in the nitrous oxide atmosphere, as opposed to its excellent characteristics in oxygen.[35] Presumably antimony oxide works almost exclusively by poisoning flame reactions.

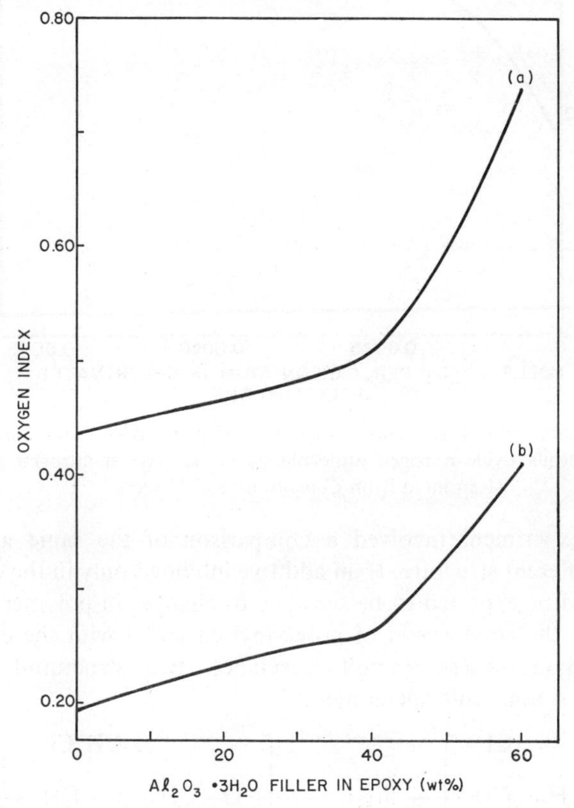

Fig. 7.3 Comparison of oxygen indices of an epoxy filled with hydrated alumina: (*a*) in nitrous oxide–nitrogen atmosphere; (*b*) in oxygen–nitrogen atmosphere. (Reprinted from *Combustion and Flame*.)

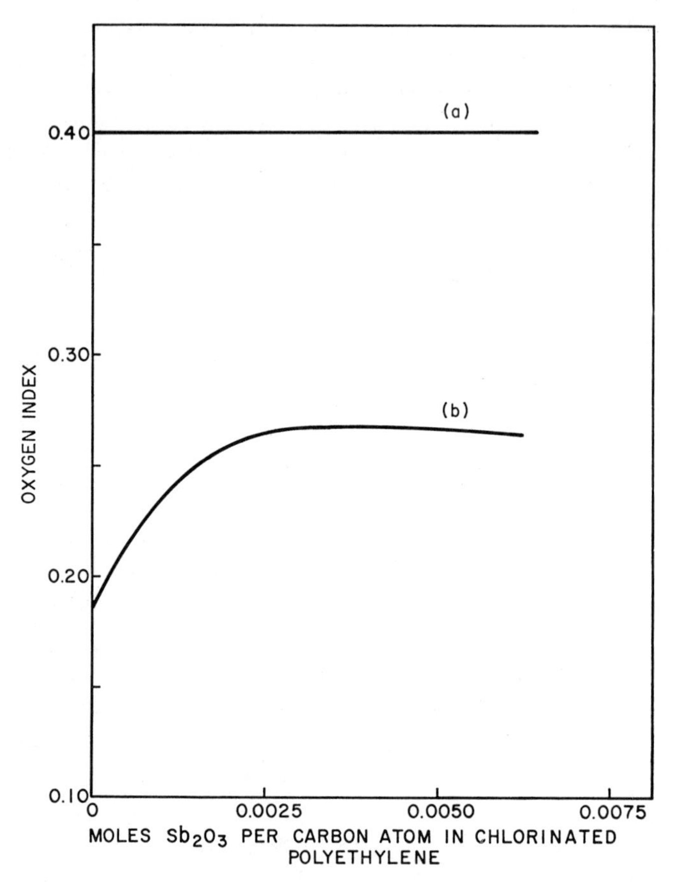

Fig. 7.4 Comparison of oxygen indices of chlorinated polyethylene containing antimony oxide: (*a*) in nitrous oxide–nitrogen atmosphere; (*b*) in oxygen–nitrogen atmosphere. (Reprinted from *Combustion and Flame*.)

The third experiment involved a comparison of the same additive in polymers of different structure. If an additive inhibited only in the condensed phase, it would be expected to be sensitive to changes in polymer structure. An example is the comparison of poly(vinyl chloride) with the chlorinated polyether, poly[3,3-bis(chloromethyl)oxetane], two structurally different polymers with similar chlorine content:

$$\left[-CH_2-\underset{}{\overset{\displaystyle Cl}{\overset{|}{C}H}}- \right]_n$$

Poly(vinyl chloride)

$$\left[-O-CH_2-\overset{\displaystyle CH_2Cl}{\overset{|}{\underset{\displaystyle CH_2Cl}{\underset{|}{C}}}}-CH_2- \right]_n$$

Poly[3,3-bis(chloromethyl)oxetane]

In general, unsubstituted polyethers and polyolefins extinguish under similar conditions, but the chlorinated derivatives of each are vastly different: poly(vinyl chloride) has an oxygen index of 0.45, whereas the chlorinated polyether has a value of only 0.23.[35] The large dependence of oxygen index on structure of chlorinated polymers again implies condensed-phase inhibition.

A flame-retardant material, however, would be insensitive to structural changes, since hydrocarbon polymers presumably burn with similar flame reactions. To illustrate this point, one can compare various concentrations of antimony oxide in poly(vinyl chloride) and poly[3,3-bis(chloromethyl) oxetane]. The increase in oxygen index in both cases was the same with an increased amount of antimony oxide.[35] The result is thus consistent with that derived from the nitrous oxide experiments; namely, that antimony oxide is primarily a flame retardant.

4. Experimental Techniques

A final note to this section should be made with regard to experimentation in polymer combustion research. High-temperature radical reactions are by themselves difficult to study, and in a flame many of them proceed simultaneously. Polymer combustion further complicates the situation, because it combines flame chemistry with a second difficult area, polymer thermal degradation. In general, two types of experiments can be performed. One can study the overall burning process, of which the oxygen index method is an example. Thermal gravimetric or differential thermal analyses are somewhat more specific in that these methods are confined to the solid phase, but they too allow study only of an overall process.

More difficult but also more informative are the probe experiments, giving the composition and thermal history of stable and transient species in a flame.[36,38] The flame is treated as a flow system in equilibrium and small volumes of the gas are extracted for analysis; the flame temperature is often simultaneously recorded with tiny thermocouples. Although this method can be applied only to the gas phase, probing just above the polymer surface gives information on the melt reactions also.

Convective flow, a third variable (in addition to composition and temperature) that is normally studied in flames, has received little attention in polymer systems to date. Eventually, the importance of such turbulence in real fires will necessitate the investigation of this quantity. In the following sections, data from both types of experiments are combined to draw as many consistent conclusions as possible about polymer combustion and its suppression.

C. POLYMER FLAME REACTIONS AND THEIR SUPPRESSION

One of the most practical and effective ways to retard burning in polymers is to incorporate additives that vaporize into the flame to function as inhibitors. On the reasonable assumption that organic polymers have flames similar to burning hydrocarbons, much of the mechanistic work on flames should be directly applicable to polymer flammability. Since excellent reviews on both hydrocarbon combustion[37–44] and flame inhibition[13,41,45–51] have been published, we confine ourselves here to information directly relevant to polymer flames.

1. Flame Chemistry

The various types of flames are shown in Figure 7.5. Flame processes are characterized as either nonsteady state or steady state in nature.[23] Ignition and quenching are examples of the former category, because the parameters of these phenomena change rapidly with time. The latter category of equilibrium burning can be separated into premixed and diffusion flames.[23,37] This is a fundamental division, for premixed flames are rate controlled and diffusion flame rates depend on how fast oxygen can migrate to the fuel.[23] Because of their more fundamental properties and greater stability, lean premixed flames constitute the bulk of gas-phase combustion research.[23]

Polymers appear to burn as homogeneous or heterogeneous diffusion flames, and therefore we should consider diffusion flame properties in some detail. When a stream of gaseous hydrocarbon burns in pure oxygen, the oxidation rate is diffusion controlled. Thus, it has been reported[37] that a given premixed ethylene–oxygen flame consumed oxygen at a rate of 4 mole/cc/sec, although the measurement of the same gas burnt as a diffusion flame gave an

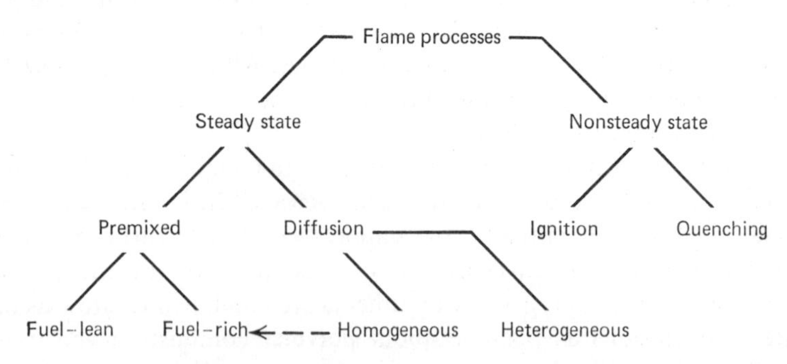

Fig. 7.5 Various types of flame processes.

oxygen consumption of only 6×10^{-5} mole/cc/sec. The reduction of oxygen pressure around the diffusion flame increased the rate of diffusion and decreased the rate of oxidation.[37] In the limiting case the diffusion rate becomes greater than the reaction rate, and indeed low-pressure diffusion flames behave as premixed flames.[38]

The addition of inert gas to the oxygen atmosphere achieves a similar effect by somewhat different means: the reaction rate is reduced by cooling and dilution without affecting greatly the diffusion rate of oxygen.[37] Accordingly, air flames have thinner reaction zones (owing to the narrowing of rich and lean flammability limits) and burn at much lower temperatures than those in pure oxygen.[37] In an actual fire we expect an even greater departure from the ideal diffusion case because of the increased mixing caused by turbulence, and therefore polymer combustion may conceivably resemble fuel-rich premixed flames more than diffusion flames.[41] With some confidence, then, we may apply appropriate premixed flame mechanistic arguments to polymer combustion systems.

a. Hydrocarbon Combustion Mechanism. The lean, premixed methane–oxygen flame can be analyzed as a model for the more complex polymer flames. Although this model suffers in that efficient burning causes maximum flame temperatures that are 1000°C higher than polymer flames, we must tolerate this difference, since no other combustion system is known in nearly so great mechanistic detail. Such a flame is conveniently divided into four regions:[39] (*a*) an initially cool region where the fuel–oxygen mixture is rapidly heated in a steep thermal gradient; (*b*) a thin reaction zone where methane is consumed, formaldehyde, water, and carbon monoxide appear, and hydrogen, oxygen, and hydroxy radicals are produced in greater-than-equilibrium concentrations; (*c*) a second oxidation zone where carbon monoxide is oxidized to carbon dioxide; and (*d*) a broad, diffuse region where the radicals combine by third-body collisions to reestablish equilibrium. By probing for the various components of the flame and mathematically separating out those contributions caused solely by diffusion, the compound and radical concentration profiles of the methane flame have been determined.[38] These quantitative descriptions of radical and compound concentrations in all regions of the flame form the experimental basis of the flame history, depicted in Table 7.2.

The initial attack on methane is primarily by a hydroxy radical,[38] followed by reaction of the methyl radical with an oxygen atom to give the intermediate formaldehyde.[50,52] A series of complex reactions then occurs to strip formaldehyde of its hydrogen, eventually producing the relatively unreactive carbon monoxide.[39,50] Further downstream the carbon monoxide is oxidized exclusively by attack of a hydroxyl radical.[38] The radical termination or

Table 7.2 The Mechanism of Methane Combustion

Propagation	
$CH_4 + OH\cdot \rightarrow CH_3^{\cdot} + H_2O$	(7.1)
$CH_4 + H\cdot \rightarrow CH_3^{\cdot} + H_2$	(7.2)
$CH_3^{\cdot} + O\cdot \rightarrow CH_2O + H\cdot$	(7.3)
$CH_2O + CH_3^{\cdot} \rightarrow CHO\cdot + CH_4$	(7.4)
$CH_2O + OH\cdot \rightarrow CHO\cdot + H_2O$	(7.5)
$CH_2O + H\cdot \rightarrow CHO\cdot + H_2$	(7.6)
$CH_2O + O\cdot \rightarrow CHO\cdot + OH\cdot$	(7.7)
$CHO\cdot \rightarrow CO + H\cdot$	(7.8)
$CO + OH\cdot \rightarrow CO_2 + H\cdot$	(7.9)
Chain Branching	
$H\cdot + O_2 \rightarrow OH\cdot + O\cdot$	(7.10)
Termination	
$H + H^{\cdot} + M \rightarrow H_2 + M^*, \ldots$	(7.11)

recombination steps must be accompanied by a third body to absorb the excess collision energy. It should be noted that the only important branching step in this entire flame process is the reaction of oxygen molecules with hydrogen atoms,[53] and therefore this step is very significant in flame inhibition. It is clear from this example of burning methane that hydrogen, not carbon, is the fuel species most responsible for flaming hydrocarbon degradation.

Fuel-rich flames differ from lean ones in several respects. For example, the initial attack on methane is by a hydrogen atom rather than a hydroxyl radical because of the low concentration of the latter species.[54] Second, carbon is often formed in fuel-rich hydrocarbon flames. Probe studies in various hydrocarbon flames have shown that soot and smoke formation are related to the amount of acetylene produced in the flame.[55] Radical acetylene condensations first give soot spheres 40 Å in diameter and of the empirical formula CH;[55] particles smaller than this do not survive oxidation by hydroxy radicals.[56,57] Further downstream the particles grow to an average diameter of 100 Å spheres of composition approximately C_2H. Ultimately the soot spheres coagulate to form smoke; the more of these spheres, the faster the aggregates are formed.[57] Materials that burn to give benzene apparently form soot by a somewhat different mechanism, because little acetylene is found in such flames.[55] In this case, benzene and acetylene have been postulated to copolymerize to form highly aromatic carbon particles.[55]

b. Polymer Flames. Hydrocarbon diffusion flames of methane,[55,58] propane,[59] paraffin,[60] *n*-eicosane,[34] linear[34,61] and branched polyethylene,[34] and polypropylene[62] have all been probed for simple products and the reported results are remarkably consistent. The work of Burge and Tipper[34] on low-density polyethylene summarizes these observations (Figure 7.6). Nitrogen was always detected in high concentrations in all parts of the flame, even close to the surface of the burning polymer. Its enormous coolant and dilution effect is consistent with the great dependence of oxygen index on the heat capacity of the inert gas. Oxygen concentrations were always very small near the center of the polymer surface, but they tended to increase rapidly toward the edge of the flame. The presence of small amounts of oxygen in most parts of the flame seems to warrant the earlier assumption that the fuel-rich premixed flame is a reasonable model of the flaming polymer. Carbon monoxide, carbon dioxide, and water were evident in all samples. Carbon dioxide and water increased, but carbon monoxide and hydrocarbon species decreased in concentration toward the edge and up the central axis of the

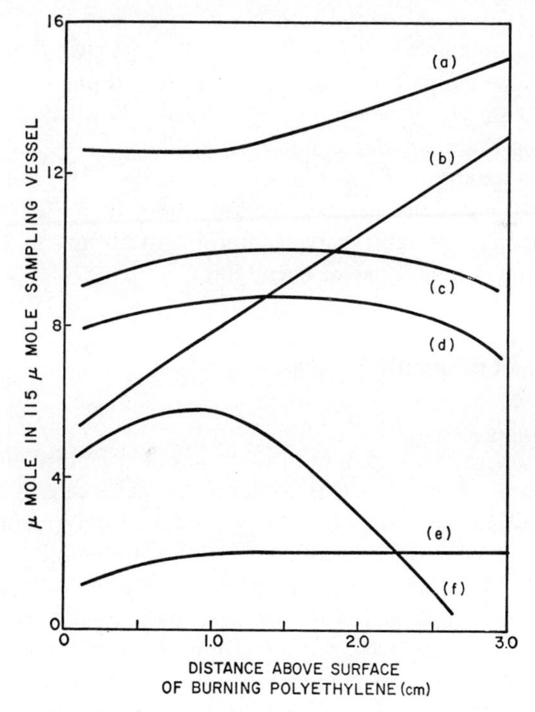

Fig. 7.6 Concentrations detected in a polyethylene flame: (*a*) carbon dioxide; (*b*) water; (*c*) oxygen consumed; (*d*) nitrogen × 0.1; (*e*) oxygen; (*f*) carbon monoxide. (Reprinted from *Combustion and Flame*.)

flame. The only organic products detected in all cases were the hydrocarbon pyrolysis products, alkanes and alkenes. Typically found directly above the surface of burning polyethylene were methane, acetylene, propene, and isobutylene, as well as smaller amounts of ethane, propane, *n*-butane, isobutane, and 2-butene. With the exception of traces of formaldehyde and peroxides near the edge of the flame, no oxygenated intermediates were observed.

Several other polymers have been probed for reaction products including poly(methyl methacrylate),[34,35] polyoxymethylene,[35] and polytetrafluoroethylene.[12,35] The first two polymers, although not studied in as great detail as the polyolefins, gave results similar to polyethylene, except that large amounts of monomer were found early in the flame. This is reasonable, since both materials are known to degrade almost exclusively via depolymerization reactions, as described in Chapter 3. Polytetrafluoroethylene also "unzips" during burning to give large amounts of monomer, along with carbonyl fluoride, tetrafluoromethane, carbon monoxide, carbon dioxide, and traces of hexafluoroethane and hexafluoropropene.[12]

Temperature as well as composition has been probed in polymer flames. Using small thermocouples, Volans[63] originally reported that polystyrene candles give surface temperatures of 230 to 540°C and maximum flame temperatures of 490 to 740°C when burned vertically downward in air. Polyethylene, burning under similar conditions, exhibited a surface temperature of ~400°C and a maximum flame temperature of 700°C.[14,61] Stuetz[62] recently recorded polyethylene flame temperatures of more than 1000°C and Fenimore[35] observed that the same polymer burning at 50 mm in 22% oxygen–argon gave a flame temperature of over 1500°C.

2. Flame Inhibition

a. Physical Inhibition. Flame inhibitors operate by either physical or chemical mechanisms, or in a few cases, by both. Chemical inhibitors are most practical, but materials such as water or inert gases should be mentioned as effective coolants and diluents. The oxygen index of poly(methyl methacrylate) is 0.175 in nitrogen, 0.135 in argon, 0.190 in helium, and 0.253 in carbon dioxide.[33] The dilution effect is constant among the various gases, but the amount of cooling is dependent on both heat capacity and thermal conductivity of the gas. With the exception of helium, the ratio of oxygen to inert gas was linearly dependent on the heat capacity of the above gases.[33] Helium is unique in that its very low heat capacity is counteracted somewhat by its great thermal conductivity; hence it cools more effectively than argon. The considerable dependence of the oxygen index on the inert gas is only true of atmospheric gases; any "blanketing" effect produced by nonflammable polymer pyrolysis products would presumably give only a proportionately slight inhibition.

b. Chemical Inhibition. Unlike the inert gas inhibitors, chemical flame retardants do not significantly change the heat output of the flame and thus do not affect the overall thermodynamics of the system.[59,64,65] Rather, they alter the rate of reaction by interfering with a crucial step in the burning process. There are apparently two types of deactivation: in one case, certain propagation and branching steps are slowed by "trapping" the essential free radicals, whereas in the second case, two radicals are catalytically recombined by a third body. Both mechanisms have practical application in retardation of polymer flames.

(*1*) *Radical Scavengers.* By far the most common flame retardant materials are the organic halide inhibitors. It should be mentioned, however, that the same compounds ordinarily are catalysts for low-temperature oxidations.[66–68] Indeed, Salooja[69] showed that various organic iodides, bromides, and chlorides promote the preflame oxidation, including the onset of ignition, of premixed mixtures of a variety of hydrocarbons, but inhibit flame processes. He[69] and others[66] have proposed mechanisms based on different propagating radicals in the two situations; low-temperature autoxidation mechanisms (Chapter 2) involve peroxy or alkoxy radicals as chain carriers, whereas in flames the higher energy hydrogen, oxygen, and hydroxy radicals predominate.

Wise and Rosser[50,70] originally proposed a flame-inhibition mechanism based on replacement of the radical chain carriers with less reactive halogen atoms:

$$H\cdot + HBr \rightarrow H_2 + Br\cdot \tag{7.12}$$

$$OH\cdot + HBr \rightarrow H_2O + Br\cdot \tag{7.13}$$

The highly reactive hydrogen or hydroxy carriers react preferentially with the hydrogen halide produced by pyrolysis of some organic halides to produce a free halogen atom. The small amount of hydrogen halide competes effectively with fuel species for the propagating radicals because of the relatively low activation energy of the reaction.[70–72] The halogen radical produced is much lower in reactivity than the propagating radicals and therefore helps postpone burning of the fuel. If and when the halogen does react with hydrocarbon, hydroxy, or hydrogen radicals, the hydrogen halide is regenerated for further inhibition:[50,70]

$$Br\cdot + OH\cdot \rightarrow HBr + O\cdot \tag{7.14}$$

$$Br\cdot + CH_4 \rightarrow HBr + CH_3\cdot \tag{7.15}$$

$$Br\cdot + H\cdot + M \rightarrow HBr + M^* \tag{7.16}$$

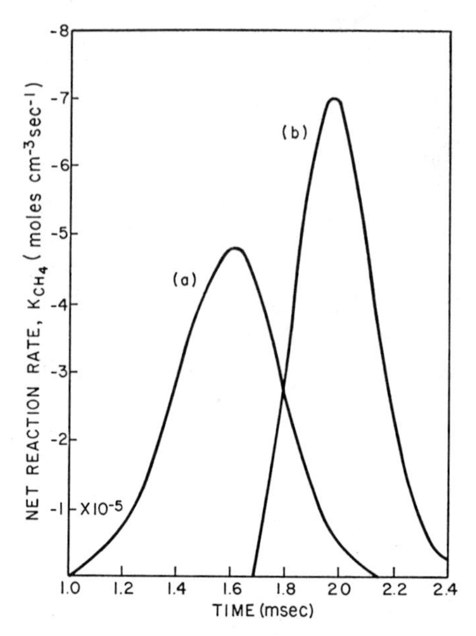

Fig. 7.7 Comparison of net reaction rates: (*a*) an uninhibited premixed methane flame, maximum temperature 1490°C; (*b*) a premixed methane flame inhibited with HBr, maximum temperature 1670°C. [Reprinted from *12th Symposium (International) on Combustion*.]

Most important, however, is that the hydrogen atom concentration is reduced and accordingly the branching reaction (7.10) is slowed. This mechanism is remarkably consistent with the following data.

1. The order of inhibition in virtually all reported flames is iodine > bromine > chlorine ≫ fluorine. This is in agreement with the reactivity order in both the replacement and regeneration steps outlined above.

2. Wilson[72] has shown that at 500°C, the inhibition reaction (7.12) occurs 1300 times faster than the branching reaction, (7.10). At higher temperatures, of course, this difference is less; the ratio is reduced to 240 at 700°C. It is clear that chemical inhibition should be most effective early in the leading edge of the flame.

3. The sharp contrast between the cooling effect of inert gases and the negligible effect of halogen gases on the average flame temperatures has already been emphasized.[64] The inhibition mechanism demands this be so; the summation of replacement reaction (7.13) and regeneration reaction (7.15) gives the same exotherm as (7.1), as if no inhibitor were present.[45] The reaction *rate*, however, is slowed for the aforementioned reasons. Interestingly, premixed methane flames inhibited with hydrogen bromide or methyl bromide burned narrower, higher off the burner and at a higher maximum temperature than the uninhibited flame (Figure 7.7)[71–73]

4. In a hydrogen bromide-inhibited premixed methane flame, the oxidation of methane did not begin until all the hydrogen bromide was consumed. In harmony with Wise and Rosser's mechanism, the concentration of hydrogen molecules increased and that of formaldehyde decreased in this early region relative to the uninhibited flame.[72]

Other mechanisms for organic halide inhibition have been proposed. One group of workers suggested a relation between the ion-capture efficiency of gaseous halides and the effectiveness of these same compounds as flame retardants.[23,74,75] This proposal has been recently criticized, since no such correlation could be found to exist in low-pressure methane–air or methane–oxygen diffusion flames.[76] Kaufman[77] showed in shock tube studies that organic halides are excellent catalysts for oxygen atom recombination, and others[73,59] have suggested that this is the reason for the effectiveness of organic halide extinguishants. Neither mechanism, however, appears to agree with the observations as well as that outlined earlier.

Diffusion flames of hydrocarbon fuels have been shown to give the same order of organic halide inhibition as in premixed flames.[78] These suppressants were reported to be an order of magnitude more effective when added to the air rather than the fuel side of the flame.[23,79] Using oxygen index measurements as his criteria, Creitz[22] reported methyl bromide to be 30 times better than nitrogen in extinguishing a butane flame on the air side, but only 1.4 times better on the fuel side. Since fire retardants in polymers that operate by adding halogen gases to the flame must add the inhibitor to the fuel side, the efficiency of such retardants would not be much better than that of inert gases.[59] The amount of inhibitors (added to the air side) required to extinguish a diffusion flame was said to be similar to that required for a premixed flame.[79]

Bromine and chlorine are the most important halogens utilized in the suppression of polymer combustion. Iodides are generally too unstable and fluoro-compounds are too inefficient as flame retardants to be practical. Empirical results show that aliphatic halogen is better than vinyl or aromatic halogen and that bromine is more effective than chlorine.[80,81] There is reasonable evidence that bromine inhibits primarily in the flame, whereas chlorine is more effective in slowing condensed-phase reactions.[35] For this reason the two halogens are treated separately.

When either tris(2,3-dibromopropyl) phosphate or tetrabromobenzene were incorporated into polyethylene, the oxygen index versus inhibitor concentration curves were markedly different when the polymers were burnt in nitrous oxide–nitrogen and in oxygen–nitrogen atmospheres.[35] The dependence of inhibition on the particular oxidant was taken as evidence for bromine inhibition primarily in the flame; this finding is consistent with the monomer results outlined earlier. Furthermore, Cullis and associates[82] reported that polyethylene and polypropylene rates of pyrolysis in nitrogen were only mildly affected either by hydrogen bromide in the atmosphere or by bromine substitution into the polymers.

Halogens are by no means the only radical catalysts that can inhibit flames. Hydrocarbons are themselves excellent inhibitors of rich premixed hydrogen–air flames.[83,84] Indeed, the rich fuel–air limit of hydrocarbon combustion is usually explained by hydrocarbon reaction with propagating radicals.[45] Presumably the radicals function like the halogen scavengers, reacting with free hydrogen atoms to form hydrogen molecules and less reactive alkyl radicals. Methyl and trifluoromethyl radicals are accordingly good inhibitors. Fenimore[73] concluded that the methyl bromide effectiveness was due to both bromine and methyl radicals, and Creitz[59] reported hexafluoroacetone to be an excellent source of trifluoromethyl radicals for flame inhibition. A recent observation by Hough[85] showed hexafluoropropanol to be several times better than hexafluoroacetone and similar to trifluoromethyl bromide in extinguishing a propane diffusion flame. None of the above mechanisms has had any reported application in retardation of polymer flammability.

(2) *Radical Recombination Catalysts.* For a long time it was realized that various metals, metal oxides, or metal salts act as flame inhibitors.[86] For instance, the effectiveness of sodium bicarbonate as a fire extinguisher was discovered to be far greater than could be explained by the liberation of carbon dioxide alone.[87] Also, glass surfaces are known to deactivate the hydrogen–oxygen combustion reactions at certain pressures.[88] These inhibitors can be classified as either homogeneous or heterogeneous,[50] and we shall see that the latter group has direct application to polymer systems.

Finely powdered sodium bicarbonate has been reported to be a homogeneous inhibitor in flames, the sodium atom being the active third-body intermediate:[50,89,90]

$$OH\cdot + Na\cdot \rightarrow NaOH \tag{7.17}$$

$$NaOH + H\cdot \rightarrow H_2O + Na\cdot \tag{7.18}$$

The requirements for such a recombination catalyst[50] are (*a*) that it be volatile and homogeneous in the gas phase, (*b*) that the intermediate have sufficient bond strength to survive until reaction with the second radical,but (*c*) that it not be so stable that it is inert to this reaction. Although such finely powdered salts have found wide application as fire extinguisher materials, their use in polymers is unprecedented, presumably because the first requirement is not fulfilled.

The introduction of heterogeneous, finely divided metal or metal-oxide particles in the flame, usually formed *in situ*, is one of the most effective ways of inhibiting flame reactions.[78] For instance, Lask and Wagner[91] showed the flammable iron pentacarbonyl and tetraethyl lead and nonflammable chromyl chloride to be at least an order of magnitude better than the best halogen inhibitors in decreasing the flame speed of a stoichiometric *n*-hexane–air flame by 30%. The inhibitory effect of these materials has been shown to drop off with decreasing pressure, but iron pentacarbonyl was still found to be 25 times more effective than carbon tetrachloride in extinguishing a 0.1-atm hydrocarbon diffusion flame.[92] Emission spectra of these flames inhibited by iron pentacarbonyl showed lines characteristic of iron and iron oxide, as well as a continuum attributed to hot particles well upstream of the maximum flame temperature in premixed flames.[92] Of relevance is the vigorous reaction reported by Kaufman[77] of iron pentacarbonyl with oxygen atoms in a shock tube. Similarly, one can cite the antiknock properties of tetraethyl lead; it is known that the lead or lead monoxide oxidation products are the effective agents in slowing internal combustion.[93] Indeed, solid lead monoxide is a highly effective inhibitor of *n*-butane oxidation over all temperature ranges.[94]

When an aqueous solution of a transition metal salt is sprayed into a hydrogen–air flame, it is well established that many nonvolatile metal-oxide particles are formed.[95] It has been proposed that radicals recombine on this oxide surface, giving up a fraction of the collision energy to the metal oxide:[50]

$$H\cdot + OH\cdot + \text{oxide surface} \rightarrow H_2O + \text{oxide surface}^* \tag{7.19}$$

By measuring the radiative temperature of a uranium oxide suspension by flame photometry, Bulewicz and co-workers[96] established that the temperature of the particles was up to 500°C in excess of the ambient flame temperature. Indeed, they found a linear relation between the greater-than-equilibrium concentrations of hydrogen and hydroxy radicals and this excess temperature. Similarly, Wise and Rosser[50] found that the heterogeneous recombination efficiency of such an oxide correlated well with the Debye characteristic temperature of the solid.

Heterogeneous flame inhibitors in polymer systems have a preliminary requirement to fulfill: such metallic additives must have sufficient volatility to be transported from the relatively low-temperature polymer surface up into the flame. Antimony oxide in the presence of chlorinated hydrocarbons satisfies this requirement. For instance, when antimony oxide was milled into polyethylene and the sample burned, no significant increase in oxygen index was noted.[35] When the same amount was added to a sufficiently chlorinated polyethylene, the oxygen index increased from 0.18 to 0.26.[35] Accordingly, insufficient chlorination gave lower oxygen indices. Chlorine is necessary for vaporization of the heavy antimony into the flame; the hydrogen chloride from a chlorinated hydrocarbon under burning conditions reacts with antimony oxide to give a volatile halide or oxyhalide. Upon reaching the hot flame, this intermediate decomposes into finely divided antimony or antimony oxide, which inhibits by the recombination mechanism detailed earlier. The excellent retardation of epoxy resin flammability with a triphenyl antimony additive showed that chlorine is necessary primarily for antimony mobility.[29] Presumably other metallic elements could function as inhibitors if they had sufficient volatility to reach the flame. It was demonstrated[19] that arsenic and bismuth oxides added to chlorinated polyethylene were somewhat less effective than antimony oxide. It should be mentioned that Wise and Rosser[50] proved arsenic and antimony oxides to be relatively ineffective in catalyzing oxygen atom recombinations, and therefore the volatility factor may be a principal requirement for heterogeneous inhibition in polymer systems.

In all reported cases of antimony–chlorine synergism, the effect leveled off at high concentrations of the oxide.[29,35] Fenimore[35] showed that the inhibition took place primarily in the flame rather than in the condensed phase by the following experiments.

1. Analysis of the burned residues of chlorinated polyethylene into which antimony oxide had been incorporated showed that when the burning was inhibited, the antimony had vaporized.
2. The gases sampled just above the burning surface of chlorinated polyethylene containing antimony oxide showed the identical distribution of hydrocarbons as the unmodified polymer, evidence that the condensed-phase reactions were independent of the added oxide.
3. The synergistic effect of added antimony oxide to two polymers of different structure, poly(vinyl chloride) and poly[3,3-bis(chloromethyl)-oxetane], was described as very similar.
4. The oxygen index versus antimony oxide concentration in chlorinated polyethylene showed different modes of inhibition in nitrous oxide–nitrogen and oxygen–nitrogen atmospheres.

Although phosphorus compounds are believed to inhibit polymer combustion primarily in the condensed phase, there is evidence that when the element is volatilized into the flame it mildly retards flame reactions. Fenimore[35] showed that the inhibition in polyethylene is independent of phosphorus structure, since equal concentrations of triphenyl phosphate, phosphine, and phosphite all gave similar increases in oxygen indices. Although comparison of the organo-phosphorus concentrations in polyethylene with the oxygen indices in nitrous oxide–nitrogen and oxygen–nitrogen atmospheres were inconclusive, the inhibition was apparently independent of condensed-phase structure. Tricresyl phosphate, in 0.02-mole per monomer unit concentration, gave increases in the oxygen index of 15, 14, and 8% in polyethylene, polyoxymethylene, and poly(methyl methacrylate), respectively.[35]

There are some scattered references to phosphorus as a flame inhibitor: Lask and Wagner[91] observed that trimethyl phosphate was five times better than carbon tetrachloride in reducing the flame speed of a hexane–air flame by 30%, and Jaques[97] showed that trimethoxyphosphine was twice as efficient as carbon tetrachloride in retarding the burning of a methane–air flame. This weak flame inhibition by phosphorus has unfortunately received little attention from a mechanistic viewpoint.

D. POLYMER CONDENSED-PHASE REACTIONS AND THEIR SUPPRESSION

Polymers differ substantially from burner flames in that they must be decomposed initially to give the gases that can subsequently ignite and burn. Since it is possible to extinguish a flame by retarding the flow of these flammable gases to the flame, inhibition in the condensed phase offers a second practical way to retard polymer flammability. For this reason the combustion of polymers at the solid-gas interface is discussed, followed by a review of condensed-phase inhibitors.

1. Polymer Combustion at the Surface

As would be expected, polymers of different structures burn with different condensed-phase oxidation mechanisms. We can conveniently divide all polymers into three groups: (*a*) those that do not interact significantly with oxygen at the surface, but merely pyrolyze and react with oxygen only later in the flame; (*b*) those that are catalytically degraded by oxygen to give off flammable gases at a greater rate than by mere pyrolysis; and (*c*) those that react significantly with oxygen directly at the surface.

Examples of the first case are those hydrocarbon polymers that "unzip" easily to give monomer such as poly(methyl methacrylate) or polyoxymethylene, discussed in Chapter 3. The combustion of such materials could well be approximated by burning the appropriate monomer on a burner as a diffusion flame. Fenimore and Jones[35] measured the thermal gradients just above the surface of the vertically burning polymers and combined these data with their regression rates in order to compare heat flux from the flame with the known heat requirements for monomer formation. Although several approximations had to be made with respect to dripping, the thermal conductivity of the gas above the polymer, and the resulting heat flux, they concluded that the conductive and radiative feedback was sufficient in both cases to fully pyrolyze the polymer at the measured rate of degradation. The implication that little heat was liberated at the polymer surface is consistent with the proposed lack of significant thermal oxidation in the condensed phase.

The majority of polymers, however, apparently belong to the second category, where thermal oxidation is an important factor in catalyzing surface degradation.[34] Heat-flow studies analogous to those described above were attempted with burning polyethylene[35,62] and polypropylene,[62] but no conclusions relevant to the role of surface oxidation were possible, since polyolefins pyrolyze by random chain-scission reactions (Chapter 3). In the probe experiments of polyolefins or hydrocarbon diffusion flames, little oxygen was found just above the surface of the burning material.[34,35,55,58–62] Burge and Tipper[34] proposed that the oxygen was consumed at the edge of the polymer by thermal oxidation, leaving little to diffuse into the middle of the burning surface. Fenimore and his associates,[35] however, suggested that atmospheric oxygen was all consumed in the flame before diffusing near the melt. Significant to the controversy were the large quantities of carbon monoxide, carbon dioxide, and water detected within one millimeter of the

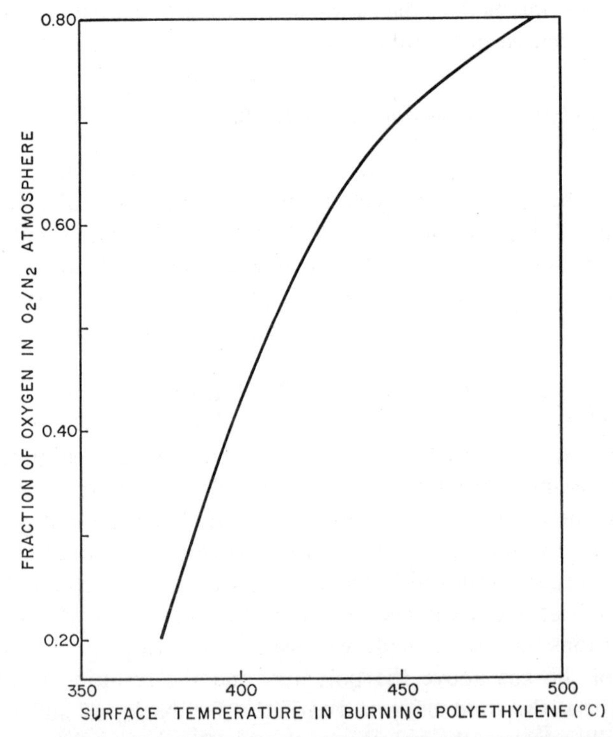

Fig. 7.8 Dependence of the surface temperature of burning polyethylene on oxygen concentration in the surrounding atmosphere. (Reprinted from *Combustion and Flame.*)

center of burning polymer by both investigators. Tipper argued that such final products could not have been formed so rapidly over so short a distance in the gas phase. Although he demonstrated that the formation of these oxidation products was much faster than that observed in ethane–oxygen mixtures at 620°C,[95] he and others[62] appear not to have considered their possible upstream diffusion from the flame edge.

Other data support the relevance of surface oxidation of burning polyethylene and related hydrocarbon polymers. Infrared analysis of *n*-eicosane

Table 7.3 Comparison of Rates of Pyrolysis of Vinyl Polymers with their Oxygen Indices

	Pyrolysis (in Nitrogen)	Combustion (Oxygen Index)
Increasing stability of vinyl polymer, X–[CH_2—CH]$_n$ ↑	X = H	X = Cl
	CH_3, F	OH, F
	C_6H_5	C_6H_5
	Cl, OH	CH_3, H

sampled from the burning surface showed the presence of carbonyl absorption bands, and similarly, diethyl mercury mixed with the burning hydrocarbon was oxidized in the melt to mercuric oxide.[34] An increase in the oxygen concentration of the atmosphere surrounding burning polyethylene was reported not to increase the maximum flame temperature or the oxygen concentration above the polymer surface, but it did significantly raise the burning surface temperature[34] (Figure 7.8). Furthermore, oxidation of uninhibited polymers at temperatures ~200° below those of burning polymers is rapid and autocatalytic (Chapter 2).

Finally, the oxygen index of vinyl polymers correlates badly with their pyrolysis rates,[19] as shown in Table 7.3. For instance, polyethylene is among the most stable of this series, yet it is the least stable toward combustion. Also, polyethylene and poly(methyl methacrylate) burn at the same oxygen index, yet the latter polymer is far less thermally stable than polyethylene. Similarly, polypropylene pyrolyzes more easily than polyethylene, and yet the two compounds burn at the same oxygen index.

At least two examples are known of the third and last case, where oxygen reacts almost exclusively at the polymer surface. Pure carbon can be considered a highly crosslinked, aromatic polymer (Chapter 3), and it obviously undergoes combustion directly at the surface. A less trivial example is the combustion of polytetrafluoroethylene, investigated by Fenimore and Jones.[12]

It had been previously demonstrated that this polymer ablated faster in oxygen than in nitrogen.[99] Also, polytetrafluoroethylene was abnormal relative to hydrocarbon polymers in that its oxygen index was very dependent on sample size, pressure, and temperature.[12] Heat flow experiments similar to those conducted on polyoxymethylene and poly(methyl methacrylate) showed that only about one-fourth of the heat flux required to depolymerize the polymer was supplied by conductive and radiative feedback from the flame.[11] Although elemental fluorine was not detected in burning polytetrafluoroethylene,[12,35] a mechanism based on its transient existence was founded on (*a*) qualitative tests such as iodine fluoride emission in polytetrafluoroethylene–iodine flames and molecular iodine formation when the combustion products were bubbled through hydriodic acid and (*b*) equilibrium calculations that implied that enough fluorine (functioning as an oxidizing agent) would be available to supply the extra heat.[11]

2. Stabilization via Condensed-Phase Inhibition

Condensed-phase inhibitors are believed to act only by retarding the flow of fuel to the flame. This is accomplished by a reduction of the heat of pyrolysis or catalysis of crosslinking reactions (at the expense of chain-cleavage reactions) to enhance residue or char formation. Although far more is understood about the former category, the latter pathway represents the approach many groups are taking to find novel retardants other than the antimony-halogen-phosphorus varieties. It is interesting that there are no reported cases of condensed-phase fire retardants that operate as radical-chain terminators analogous to low-temperature autoxidation stabilizers or flame inhibitors.

a. Physical Inhibition. Inert materials are often blended into a polymer and they can be moderately effective as fire retardants. Their mode of inhibiting combustion is exactly analogous to that of inert gases in the flame. Fillers help to cool the heated polymer by conducting heat away from the hot region more rapidly than such conduction occurs in the unfilled polymer. This cooling effect may only be effective in slowing polymer ignition, since Stuetz[62] recently demonstrated that, neglecting catalytic effects, metal wires embedded inside burning polyethylene increased the regression rate relative to the unfilled polymer. Also, Issacs[24,25] reported that 1 to 2.5% inert filler in polycarbonate actually decreased the oxygen index. When used in larger concentrations, fillers act as diluents of the flammable material. In almost all reported cases the oxygen index increased to a maximum at 20 to 40 weight

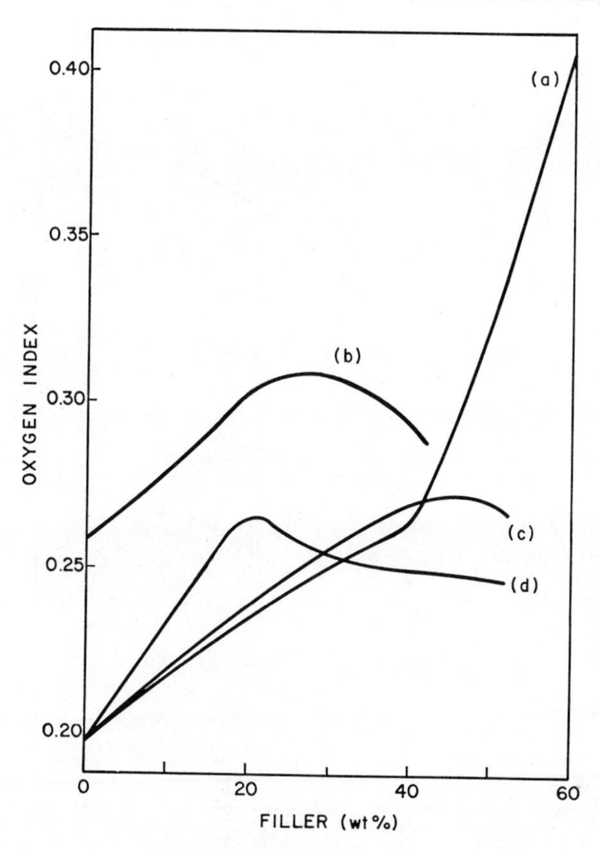

Fig. 7.9 Comparison of effects on the oxygen index: (*a*) hydrated alumina in epoxy; (*b*) glass fiber in polycarbonate; (*c*) quartz filler in epoxy; (*d*) anhydrous alumina in epoxy. (Reprinted from *Journal of Applied Polymer Science* and *Journal of Fire and Flammability*.)

percent additive,[24,29] after which the oxygen index decreased (Figure 7.9). Presumably the wicking action of filler residues becomes a dominating factor at these very high concentrations.

Materials that drip easily tend to burn slowly. This is probably because heat is removed from the flaming area, although ignition and burning of the drippings can amplify the burning hazard. Often the melting material does not ignite, however, and the aforementioned examples of nylon and low-melting-point polypropylene burn only with great difficulty. In at least one case, it was shown that a novel flame retardant previously thought to be a chemical inhibitor merely enhanced the dripping rate. Eichorn[100] reported that organic peroxide was synergistic with halogen inhibitors in retarding

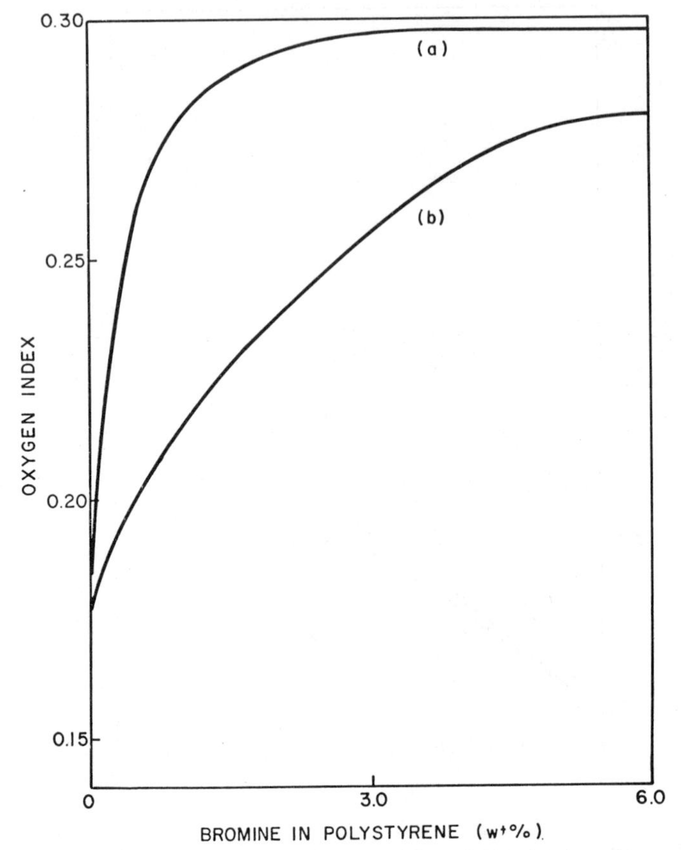

Fig. 7.10 Comparison on the oxygen index of effect of bromine in polystyrene: (*a*) containing 0.5% dicumyl peroxide; (*b*) without added peroxide. (Reprinted from *Combustion and Flame*.)

polystyrene ignition and combustion. Hindersinn,[101] however, observed no such increase in halogen effectiveness when the polymer was not allowed to drip. Fenimore[102] confirmed this inhibition of polystyrene ignition and combustion by a combination of tris(2,3-dibromopropyl) phosphate and dicumyl peroxide (Figure 7.10), and also concluded that facile melting and dripping were responsible for the effect. Significantly, peroxides were found to give no synergistic improvement to halogenated fire-extinguishing compounds.[103]

b. Endothermic Chemical Reactions. Cooling of the heated surface of a polymer may also be accomplished practically by concurrent endothermic reactions that occur below the catastrophic combustion temperature of the polymer, but not so low that the material is rendered thermally unstable at its normal service temperature. The important criteria for this type of stabilization are the temperature and magnitude of the endothermic reactions. It has been suggested from differential thermal analysis studies that these changes in no way correlate with the effectiveness of a flame retardant,[17,102] but we shall see that this cannot be true in many instances.

Such endothermic reactions also presumably lower the initial temperature of the flammable gases, but this would be expected to have a proportionately slight inhibitory effect, since the flame is mostly atmospheric nitrogen.

(*1*) *Dehydration Reactions.* Elimination of water from either organic or inorganic additives is a useful technique for absorbing heat within a burning polymer. For instance, aluminum oxide trihydrate is a filler for epoxy resins, being inert up to about 150°C. Above this temperature, however, it undergoes dehydration reactions with a large absorption of heat. Martin[29] has reported that this compound is an excellent fire retardant; an epoxy resin containing 60% showed a doubled oxygen index of 0.41 with no evidence of char or soot formation. Anhydrous aluminum oxide acted only as an inert filler being far less effective than the trihydrate[29] (Figure 7.9). Similar dependence of the oxygen index on the hydrated alumina concentration in both nitrous oxide-nitrogen and oxygen–nitrogen atmospheres was taken as evidence for primarily condensed-phase retardation[29] (Figure 7.3).

Water can also be eliminated from hydroxylated polymers. Poly(vinyl alcohol), for instance, has an oxygen index of 0.23 relative to the isomeric poly(ethylene oxide), which has a value of only 0.15. The latter polymer degrades primarily by depolymerization (Chapter 3) and, unlike polyvinyl alcohol, allows little chance for water formation in the condensed phase. Often a catalyst can effectively enhance this endothermic elimination reaction under conditions that would not ordinarily allow it to occur at a significant rate. Natural cellulose pyrolyzes primarily to levoglucosan,[17] which is easily further degraded to flammable products:

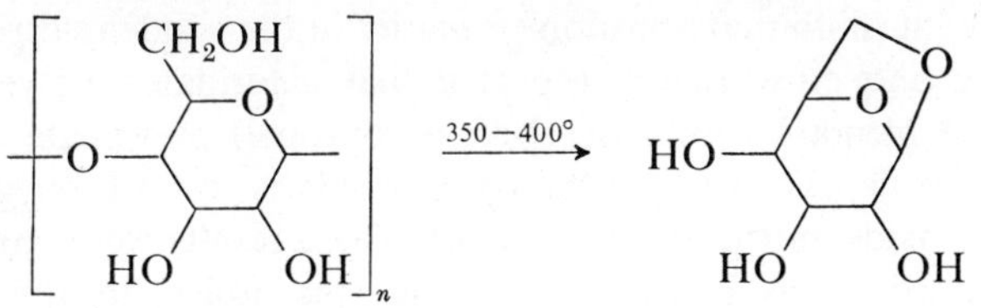

CH_2OH O HO OH $]_n$ —350–400°→ HO O O HO OH

Cellulose Levoglucosan

Treatment of cellulose with phosphoric acid lowers the decomposition temperature from between 350 and 400 to less than 300°C,[17] presumably because of catalyzed dehydration reactions occurring at a lower temperature than that required for levoglucosan formation:

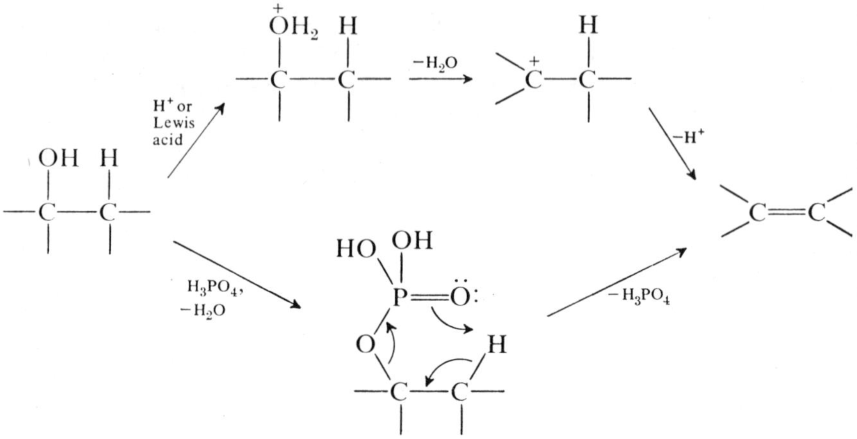

The unsaturated product must also pyrolyze at a higher temperature than the dehydration reaction, or the "heat sink" of water elimination will not prevent flaming decomposition of the products. Usually the carbon-rich residues crosslink to form thermally stable chars. Organic phosphates apparently promote the reaction as well as phosphoric acid. Also, phosphorus can apparently inhibit further fuel production by increasing the physical and thermal stability of the char residue; this phenomenon is discussed briefly later.

(*2*) *Dehydrochlorination Reactions.* Poly(vinyl chloride) and the various chlorinated hydrocarbon polymers appear to retard burning in a manner dissimilar to that of brominated fire retardants, which are primarily flame inhibitors. These chlorinated materials pyrolyze by virtually complete elimination of hydrogen chloride before catastrophic carbon–carbon bond scission occurs (Chapter 3). There is increasing evidence that the chlorinated organics might inhibit combustion primarily in the condensed phase.

Fenimore and co-workers[35] reported that chlorinated polyethylene and polyethylene blended with tris(2,3-dichloropropyl) phosphate gave similar concentration dependences on the oxygen indices in both oxygen–nitrogen and nitrous oxide–nitrogen atmospheres. These results were consistent with those obtained when hydrogen chloride was added to the atmosphere surrounding burning polyethylene;[18,19] the increase in oxygen index with increasing gas concentration was far less than that encountered when the halogen was incorporated directly into the polymer (Figure 7.2). Similarly, chloroethylene–ethylene or ethylene–hydrogen chloride diffusion flames burnt in the oxygen index apparatus gave only minor increases in the index with increasing chlorine concentration.[18,19]

All the above results are consistent with the well-documented weak flame inhibition of organic chlorides in hydrogen–air[45,105,106] or hydrocarbon–air[69,72] combustion. Condensed-phase reactions appear to account for the large differences in polymer and gas-phase inhibition when chlorinated additives are involved.[18,19,35]

Although dilution of hydrogen "fuel" by chlorine substitution might play a role in retardation, it must be a negligible effect as shown by the low oxygen index of poly(vinyl fluoride). Similarly, char or network formation via Diels–Alder reactions of the resulting polyacetylene cannot be responsible for several reasons: (*a*) poly[3,3-bis(chloromethyl)oxetane] burns with heavy char formation, yet has an oxygen index of only 0.23.[35] (*b*) poly(vinyl fluoride) and poly(vinyl alcohol) similarly char during combustion but again have low oxygen indices. (*c*) unpublished work in this laboratory has shown that when all the hydrogen chloride has been carefully pyrolyzed from poly(vinyl chloride), the resulting char burns readily in air.

Pyrolysis of poly(vinyl chloride) in either nitrogen[107,10] or air[108] atmospheres gives stoichiometric formation of hydrogen chloride at ~250 to 300°C. Further heating of the carbon–hydrogen residue at temperatures greater than 300°C in nitrogen gives primarily benzene and other aromatics, along with lesser amounts of various alkanes and alkenes.[107] This high percentage of benzenoid compounds is reasonable because of the facile reactions of the polyacetylene to form fused-ring systems, which make up the char (Chapter 3); recent reports have shown that the amounts of aromatic compounds increase with increasing chlorine substitution.[109,110] Burning chlorinated polyethylenes similarly gave more acetylene and benzene than the unchlorinated polymer[35] and in none of the above cases were chlorinated compounds other than hydrogen chloride ever observed.

The large differences in oxygen indices between poly(vinyl chloride) and other vinyl polymers have not been adequately explained. It is tempting to try to relate the low flammability of chlorinated polymers to the endothermic dehydrochlorination reactions which occur during pyrolysis. Differential thermal analysis of poly(vinyl chloride) in a nitrogen atmosphere does indeed show a sharp endotherm[111] at 250–300°C, but in air[108] a large exothermic reaction occurs in the same region. Other chlorinated, fluorinated, and hydroxylated polymers show the same tendencies in the two atmospheres. Comparison of differential thermal analysis data of poly(vinyl chloride), poly(vinyl fluoride), poly(vinyl alcohol), poly[3,3-bis(chloromethyl)oxetane], and poly(vinylidene fluoride) in this laboratory have showed no correlation between oxygen index and endothermic elimination reactions in either air or nitrogen.

c. Char Formation. The *in situ* formation of a thermally stable "barrier" during combustion at the polymer surface is an effective way to reduce heat and mass flow between the condensed and gas phases. The lower efficiency in energy and material transport lessens the amount of fuel fed to the flame, resulting in a reduction of flammability. Various additives, such as the organic and inorganic phosphorus or boron compounds, have been shown to help develop or accentuate char formation. Unfortunately, since little of the published work deals with mechanistic studies of these reactions, this discussion is brief.

The formation of char during combustion is very dependent on the particular polymer. Since the carbon-rich residue has a much higher molecular weight than the original material, some pathway for extensive crosslinking during pyrolysis is necessary.[112,113] For instance, the substituted vinyl polymers undergo elimination reactions to give polyacetylenes, which subsequently condense by Diels–Alder reactions[113] (Chapter 3). The unsubstituted polyolefins, on the other hand, do not char because only volatile fragments result from exclusive chain-scission reactions. Similarly, the polymers that depolymerize easily do not leave carbonaceous residues during combustion. It should be mentioned that char formation *per se* is not necessarily sufficient to increase the fire-resistance of the polymer. For example, the heavy char formed when poly(vinyl chloride) was pyrolyzed at 300°C burned easily in air. Clearly the structure as well as amount of char is important in creating a thermally stable barrier of low porosity.

Two different methods for promoting high molecular weight material under burning conditions have been reported. One way has been to slowly heat the polymer in air to give an oxidized, highly crosslinked network. For instance, Winslow and co-workers[113] found that preoxidized divinyl benzene (48%)–ethylvinyl benzene copolymer gave eight times more char than the untreated polymer when pyrolyzed to 600°C. Similarly, Hirsch[114,115] reported that when poly[*m*-phenylene bis(*m*-benzamido)terephthalimide] was heated in air for 150 minutes at 425°C, the originally flammable polymer did not burn in air. Although the hydrogen content was slightly decreased and the oxygen proportion was raised, he attributed the increased fire resistance to extensive crosslinking in the degraded material.

The incorporation of additives that catalyze condensation reactions or help stabilize the char is a second, more practical method to impart fire resistance in polymers via a char mechanism. Phosphorus compounds—and to a minor extent, boron compounds—significantly increase char formation in heavily functionalized polymers such as cellulose or the polyurethanes. Two inhibition mechanisms for phosphorus additives, flame poisoning and catalyzed dehydration, were mentioned earlier; since dehydration can be preliminary to the third (char stabilization) route, these two paths are obviously related.

Of the three possibilities, the dominant pathway is apparently dependent on the particular polymer. Fenimore[35] showed that various phosphorus additives in polyethylene inhibit primarily in the flame, and Martin proved that epoxies are retarded in the condensed phase.[29] The analysis of residues of cellulose inhibited with tris-(2,3-dibromopropyl) phosphate showed that virtually all the bromine was vaporized, and more than 90% of the phosphorus was incorporated into the char.[116] Phosphate ester formation with the primary hydroxyl group in cellulose was suggested by Lyons[117] to block the depolymerization to the volatile levoglucosan. Unfortunately, further reports relating to the phosphorylation mechanism are lacking. Much of the encyclopaedic accumulation of empirical results on boron and phosphorus retardants, however, can be found in the appropriate reviews.[97,118]

E. CONCLUSIONS

Some excellent mechanistic experiments on burning polymers, as well as data from other combustion systems, have permitted us not only to categorize the various fire retardants but to discuss their different modes of inhibition as well. We have generally neglected consideration of fire retardant *systems*; however, it must be emphasized that additive and synergistic combinations of the various inhibitors represent a large segment of practical research on fire-retardant materials. The antimony oxide–halogen mechanism was mentioned in the text in some detail. Another example of a combination is that used in intumescent coatings: a phosphorus dehydration catalyst, a hydroxylated material, and a blowing agent are combined in a surface coating.[119] The normally inert formulation readily forms an insulating foam under burning conditions. Recent reports have shown that a third combination, phosphorus and nitrogen, is synergistic as a cellulose retardant.[17,116,120,121] Finally, the combination of halogen and phosphorus forms the basis of many commercial fire retradants.[101] With the exception of antimony–halogen synergism, knowledge about the interrelated mechanisms of these systems have been strictly speculative.

In recent years the physiological effects of polymer flammability have been emphasized, in particular those caused by smoke and toxic gas formation.[123] A general conclusion is that, with few exceptions, fire-retarded polymers result in concentrations of smoke and toxic gas much greater than those occurring when untreated materials burn.[123,124] Inasmuch as these products are often as hazardous as the flame itself, the development of practical retardants is thus further complicated. The recent trend toward nonvolatile inhibitors that work by increasing the amount and stability of surface char appears to be a reasonable attempt to reduce the evolution of poisonous gas.

REFERENCES

1. Federal Flammable Fabrics Act of 1953; amended in 1967.
2. National Fire Codes, Vol. 4, National Fire Protection Association, Boston, 1970.
3. J. K. Smith, H. R. Rawls, M. S. Felder, and E. Klein, *Text. Res. J.*, **40,** 211 (1970).
4. J. L. Gay-Lussac, *Ann. Chim.*, **18,** 211 (1821).
5. M. W. Ranney, *Flame Retardant Polymers*, Noyes Data Corporation, Park Ridge, N.J., 1970.
6. N. E. Boyer and A. E. Vajda, *SPE* (*Soc. Plastics Engrs.*) *Trans.*, **4,** 45 (1964).
7. W. G. Schmidt, *Trans. J. Plastics Inst.* (*London*), **33,** 247 (1965).
8. J. A. Rhys, *Chem. Ind.* (*London*), **1969,** 187.
9. R. C. Nametz, *Ind. Eng. Chem.*, **62,** 41 (1970).
10. A. D. Delman, Naval Applied Science Laboratory, SF-020-03-06, AD838,689L, Aug. 1968.
11. H. Vogel, "Flammfestmachen von Kunststoffen," in *Technologie der Makromolekularen Chemie*, Kithig, Heidelberg, 1966.
12. C. P. Fenimore and G. W. Jones, *J. Appl. Polym. Sci.*, **13,** 285 (1969).
13. E. T. McHale, *Fire Res. Abstr. Rev.*, **11,** 90 (1969).
14. C. J. Hilado, *Flammability Handbook for Plastics*, Technomic, Stamford, Conn., 1969, pp. 45–75.
15. C. J. Hilado, *Ind. Eng. Chem. Prod. Res. Develop.*, **6,** 154 (1967).
16. A. A. Briber, *Polymer Conf. Ser.*, University of Detroit, June 1969.
17. J. E. Hendrix, T. K. Anderson, T. J. Clayton, E. S. Olson, and R. H. Barker, *J. Fire Flammability*, **1,** 107 (1970).
18. C. P. Fenimore and F. J. Martin, *Mod. Plastics*, **43,** 141 (1966).
19. C. P. Fenimore and F. J. Martin, *Combust. Flame*, **10,** 135 (1966).
20. A. R. Hall, J. C. McCoubrey, and H. G. Wolfhard, *Combust. Flame*, **1,** 53 (1957).
21. R. F. Simmons and H. G. Wolfhard, *Combust. Flame*, **1,** 155 (1957).
22. E. C. Creitz, *J. Res. Natl. Bur. Std., A*, **65,** 389 (1961).
23. F. J. Weinberg, *Optics of Flames*, Butterworths, Washington, 1963.
24. J. L. Issacs, General Electric Rept. No. TIS 69-MAL-13, Aug. 1969.
25. J. L. Issacs, *J. Fire Flammability*, **1,** 36 (1970).
26. J. DiPietro and H. Stepniczka, 28th Ann. Tech. Conf., Society of Plastic Engineers, New York, May 1970.
27. ASTM Standard No. 2863, adopted 1970.
28. General Electric Catalog No. 50-295010AAA, Instructions for Flammability Index Tester.
29. F. J. Martin and K. R. Price, *J. Appl. Polym. Sci.*, **12,** 143 (1968).
30. C. P. Fenimore, *Polym. Conf. Ser.*, Wayne State University, Detroit, June 1966.
31. B. Miller, *Symp. Polym. Flammability*, Hoboken, N.J., April 1970; reported in part in *Chem. Eng. News*, **48,** 34 (1970).
32. R. A. Blease, *Trans. J. Plastics Inst.* (*London*), *Conf. Suppl. No. 2*, 79 (1967).
33. F. J. Martin, *Combust. Flame*, **12,** 125 (1968).
34. S. J. Burge and C. F. H. Tipper, *Combust. Flame*, **13,** 495 (1969).
35. C. P. Fenimore and G. W. Jones, *Combust. Flame*, **10,** 295 (1966).
36. R. M. Fristom, C. Grunfelder, and S. Favin, *J. Phys. Chem.*, **64,** 1386 (1960).
37. A. G. Gaydon and H. G. Wolfhard, *Flames, their Structure, Radiation, and Temperature*, Chapman and Hall, London, 1953.

38. R. M. Fristom and A. A. Westenberg, *Flame Structure*, McGraw-Hill, New York, 1965.
39. R. M. Fristom, *Chem. Eng. News*, **41,** 150 (1963).
40. R. M. Fristom, "Flame Chemistry," in A. F. Scott, Ed., *Survey of Progress in Chemistry*, Vol. 3, Academic Press, New York, 1966.
41. C. P. Fenimore, *Chemistry in Premixed Flames*, Macmillan, New York, 1964.
42. R. C. Anderson, *J. Chem. Ed.*, **44,** 248 (1967).
43. J. L. Franklin, *Ann. Rev. Phys. Chem.*, **18,** 261 (1967).
44. A. Williams and D. B. Smith, *Chem. Rev.*, **70,** 267 (1970).
45. G. B. Skinner, ASD Tech. Rept. 61-408. AD 272122, Dec. 1961; AD 294475, Dec. 1962; AD 435541, Feb. 1964.
46. R. Friedman and J. B. Levy, WADC Tech. Rept. 56-568, AD 110685, Jan. 1957; AD 208317, Sept. 1958; AD 216086, April 1959.
47. R. Friedman, *Fire Res. Abstr. Rev.*, **3,** 128 (1961).
48. R. M. Fristom, *Fire Res. Abstr. Rev.*, **9,** 125 (1967).
49. C. T. Pumpelly, Z. E. Jolles, Ed., "Fire-Extinguishing and Fire-Proofing," in *Bromine and its Compounds*, Academic Press, New York, 1966.
50. H. Wise and W. A. Rosser, *9th Int. Symp. Combustion*, 733 (1963).
51. E. C. Creitz, *J. Res. Natl. Bur. Std., A*, **74,** 521 (1970).
52. C. P. Fenimore and G. W. Jones, *J. Phys. Chem.*, **65,** 1532 (1961).
53. C. P. Fenimore and G. W. Jones, *J. Phys. Chem.*, **63,** 1834 (1959).
54. C. P. Fenimore and G. W. Jones, *J. Phys. Chem.*, **65,** 2200 (1961).
55. A. S. Gordon, S. R. Smith, and J. R. McNesby, *7th Int. Symp. Combustion*, 317 (1959).
56. C. P. Fenimore and G. W. Jones, *J. Phys. Chem.*, **71,** 593 (1967).
57. C. P. Fenimore and G. W. Jones, *Combust. Flame*, **13,** 303 (1969).
58. S. R. Smith and A. S. Gordon, *J. Phys. Chem.*, **60,** 759 (1956).
59. E. C. Creitz, 28th Ann. Tech. Conf., Society of Plastic Engineers, New York, May 1970.
60. S. R. Smith and A. S. Gordon, *J. Chem. Phys.*, **22,** 1150 (1954).
61. S. J. Burge and C. F. H. Tipper, *Chem. Ind.* (*London*), **1967,** 362.
62. D. E. Stuetz, 28th Ann. Tech. Conf., Society of Plastic Engineers, New York, May 1970.
63. P. Volans, *Trans. J. Plastics Inst.* (*London*), *Conf. Suppl. No. 2*, 47 (1967).
64. R. F. Simmons and H. G. Wolfhard, *Trans. Faraday Soc.*, **51,** 1211 (1955).
65. W. A. Rosser, Jr., S. H. Inami, and H. Wise, *Combust. Flame*, **10,** 287 (1966).
66. C. F. Cullis, A. Fish, and R. B. Ward, *Proc. Roy. Soc.* (*London*), **276A,** 527 (1963).
67. M. Seakins, *Proc. Roy. Soc.* (*London*), **274A,** 413 (1963).
68. M. Seakins, *Proc. Roy. Soc.* (*London*), **277A,** 279 (1964).
69. K. C. Salooja, "The Influence of Halogen Compounds on Combustion Processes," in R. F. Gould, Ed., *Oxidation of Organic Compounds*, Vol. II, American Chemical Society, Washington, D.C., 1968.
70. W. A. Rosser, H. Wise, and J. Miller, *7th Int. Symp. Combustion*, 175 (1959).
71. W. E. Wilson, Jr., *10th Int. Symp. Combustion*, 47 (1965).
72. W. E. Wilson, Jr., J. T. O'Donovan, and R. M. Fristom, *12th Int. Symp. Combustion*, 929 (1969).
73. C. P. Fenimore and G. W. Jones, *Combust. Flame*, **7,** 323 (1963).
74. R. M. Mills, *Combust. Flame*, **12,** 513 (1968).
75. T. G. Lee, *J. Phys. Chem.*, **67,** 360 (1963).
76. W. J. Miller, *Fire Res. Abstr. Rev.*, **10,** 191 (1968).

77. F. Kaufman, *Prog. Reaction Kinetics*, **1,** 3 (1961).
78. R. Friedman and J. B. Levy, *Combust. Flame*, **7,** 195 (1963).
79. R. F. Simmons and H. G. Wolfhard, *Trans. Faraday Soc.*, **52,** 53 (1956).
80. I. N. Einhorn, *Polym. Conf. Ser.*, University of Detroit, June 1969.
81. J. A. Schneider, R. G. Pews, and J. D. Herring, *Abstracts*, 158th Meeting, American Chemical Society, New York, 1969.
82. M. D. Carabine, C. F. Cullis, and I. J. Groome, *Proc. Roy. Soc. (London)*, **306A,** 41 (1968).
83. D. R. Miller, R. L. Evers, and G. B. Skinner, *Combust. Flame*, **7,** 137 (1963).
84. R. R. Baldwin, N. S. Corney, and R. W. Walker, *Trans. Faraday Soc.*, **56,** 802 (1960).
85. R. L. Hough, Tech. Rept. AFAPL-TR-69-42, AD 692102, March 1969.
86. J. E. Dolan, *6th Int. Symp. Combustion*, 787 (1957).
87. C. S. McComy, H. Schoub, and T. G. Lee, *6th Int. Symp. Combustion*, 795 (1957).
88. S. W. Benson, *The Foundations of Chemical Kinetics*, McGraw-Hill, New York, 1960, pp. 452–459.
89. J. D. Birchall, *Combust. Flame*, **14,** 85 (1970).
90. W. A. Rosser, Jr., S. H. Inami, and H. Wise, *Combust. Flame*, **7,** 107 (1963).
91. G. Lask and G. Wagner, *8th Int. Symp. Combustion*, 43 (1961).
92. P. H. Vree and W. J. Miller, *Fire Res. Abstr. Rev.*, **10,** 121 (1968).
93. R. K. Sharma and J. D. Bardwell, *Combust. Flame*, **9,** 106 (1965).
94. J. Bardwell, *Combust. Flame*, **5,** 71 (1961).
95. M. L. Nielson, P. M. Hamilton, and R. J. Walsh, in Kuhn, Lamprey, and Sheer, Eds., *Ultrafine Particles*, Wiley, New York, 1963, p. 181.
96. E. M. Bulewicz, G. Jones, and P. J. Padley, *Combust. Flame*, **13,** 409 (1969).
97. J. K. Jacques, *Trans. J. Plastics Inst., Conf. Suppl. No. 2*, 33 (1967).
98. R. J. Sampson, *J. Chem. Soc.*, **1963,** 5095.
99. K. W. Graves, *AIAA J.*, **4,** 853 (1966).
100. J. Eichorn, *J. Appl. Polym. Sci.*, **8,** 2497 (1964).
101. R. R. Hindersinn, *Polym. Conf. Ser.*, Wayne State University, Detroit, June 1966.
102. C. P. Fenimore, *Combust. Flame*, **12,** 155 (1968).
103. H. Landesman, J. E. Basinski, and E. B. Klusmann, Tech. Rept. AFAPLTR-65-10, March 1965.
104. R. W. Little, *Flameproofing Textile Fabrics*, Reinhold, New York, 1947.
105. R. N. Butlin and R. F. Simmons, *Combust. Flame*, **12,** 447 (1968).
106. D. R. Blackmore, G. O'Donnell, and R. F. Simmons, *10th Int. Symp. Combustion*, 303 (1965).
107. M. M. O'Mara, *J. Polym. Sci., Part A-1*, **8,** 1887 (1970).
108. E. A. Boettner, G. Ball, and B. Weiss, *J. Appl. Polym. Sci.*, **13,** 377 (1969).
109. S. Tsuge, T. Okumoto, and T. Takeuchi, *Macromolecules*, **2,** 200 (1969).
110. S. Tsuge, T. Okumoto, and T. Takeuchi, *Macromolecules*, **2,** 277 (1969).
111. P. Dunn and B. C. Ennis, *J. Appl. Polym. Sci.*, **14,** 355 (1970).
112. F. H. Winslow, W. O. Baker, and W. A. Yager, *Proc. 2nd Conf. Carbon*, 93 (1956).
113. F. H. Winslow, W. O. Baker, N. R. Pape, and W. Matreyek, *J. Polym. Sci.*, **16,** 101 (1955).
114. S. S. Hirsch, *Polym. Conf. Ser.*, University of Detroit, June 1969.
115. S. S. Hirsch, *Symp. Polym. Flammability*, Hoboken, N.J., April 1970; reported in part in *Chem. Eng. News*, **48,** 34 (1970).
116. W. A. Reeves, R. M. Perkins, B. Piccolo, and G. L. Drake, Jr., *Text. Res. J.*, **40,** 223 (1970).
117. J. W. Lyons, *J. Fire Flammability*, **1,** 107 (1970).

118. W. G. Woods, *Polym. Conf. Ser.*, Wayne State University, Detroit, June 1966.
119. H. L. Vandersall, *Polym. Conf. Ser.*, Wayne State University, Detroit, June 1966.
120. G. C. Tesoro, S. B. Sello, and J. J. Willard, *Text. Res. J.*, **39,** 180 (1969).
121. J. J. Willard and R. E. Wondra, *Text. Res. J.*, **40,** 203 (1970).
122. H. H. Cornish, *Polym. Conf. Ser.*, University of Detroit, June 1969.
123. I. N. Einhorn, R. W. Mickelson, B. Shah, and R. Craig, *J. Cellular Plastics*, **4,** 188 (1968).
124. J. R. Gaskill and C. R. Veith, *Fire Technol.*, **4,** 185 (1968).

8

STABILIZATION AGAINST CHEMICAL AGENTS

B. D. GESNER

Bell Telephone Laboratories, Incorporated, Murray Hill, New Jersey

A. INTRODUCTION

The preceding chapters are concerned with the stabilization of polymers either against reaction with oxygen or ozone, or in the absence of a chemical reactant (Chapter 3). In Chapter 1, however, it was pointed out that a wide variety of chemical reactants can contribute to polymer deterioration. Certain of these reactants (e.g., water, oxides of nitrogen or sulfur, and other atmospheric contaminants) are often present in the normal exposure environment. In specific applications, polymers may be exposed to many other types of potential reactants, including bases, acids, solvents, adhesives, dyes, and so on. Deterioration attributable to these reactants is not a general phenomenon, however, either because they are not components of the normal environment or because their reactions are limited to specific polymers or classes of polymers. A discussion of the many specific reactions that can occur and how to prevent them would probably fill several volumes. This consideration of stabilization against chemical agents is limited to the reactions that occur more generally.

The approach to stabilization depends upon the degree of reactivity, availability, and selectivity of the degradant, and on the chemical and physical structure of the polymer. It is apparent from Chapters 2 and 4 that oxygen poses the more serious problem to polymers. This is undoubtedly a result of oxygen's high reactivity and availability and its relatively nonselective mode of attack. It is not surprising, therefore, that the most common stabilizers are antioxidants. The highly reactive trimer of atomic oxygen, ozone, is probably the second most damaging reactant to polymers. Ozone's high degree of selectivity and its low concentration in normal environments, however, reduce significantly the number of polymers that it degrades. The stabilization of polymers against ozone attack has been covered (Chapter 5).

More specialized reactions take place with other reactants, but these are of less importance because of the relative decreases in either or both the degrees of reactivity and availability, or a relative increase in selectivity. Water, for instance, has such relatively low volatility that it is not as readily available for diffusion into polymers as are oxygen and ozone. Furthermore, the reactivity of water toward polymers is low, relative to that of oxygen. Nonetheless, after oxygen and ozone, stabilization against water degradation probably ranks next in importance.

Other degradants, such as the air pollutants sulfur dioxide and nitrogen tetroxide, industrial acids and alkalis—particularly sulfuric acid, potassium, sodium hydroxides—and noxious gases such as chlorine and hydrogen chloride, are either of such low concentration or so infrequently encountered in the normal environment that, even though they are highly reactive, stabilization of polymers against them is of secondary importance.

In this chapter we review briefly how chemical reactants enter into polymers and the resultant effect on properties of the material. We also attempt to show the direction that stabilization against these agents and their damaging reactions has taken in more recent years.

B. PERMEATION

Chemical reactions occur only as a result of the direct interaction of the reactants. Where one reactant is a solid, reaction cannot occur except at the surface until the other reactant permeates into it. It is necessary, therefore, to consider the process of permeation through polymeric materials as a prelude to a discussion of chemical degradation.

1. Fundamental Process

The fundamental permeation process consists of three steps: deposition of the permeant on the substrate surface, surface dissolution of the collected permeant, and migration of the permeant through the bulk of the polymer. Whether penetration proceeds from one or both sides, the total process results in a steady-state concentration of permeant at a given temperature. These steps are dependent upon a number of characteristics of both polymer and permeant.

Solid polymers contain a steady-state distribution of molecular-size voids (holes). These holes can travel through the material by a reorganization of polymer–polymer bonding interactions. These bonds, resulting from van der Waal forces, rearrange themselves according to the prevailing thermal conditions. This secondary bond reorganization and the subsequent hole migration allow permeant to pass through the solid polymer. Pressure gradients and "solvation" effects increase the rate of permeation. Larger permeants require larger holes, and since the distribution of hole sizes is approximately of the Boltzmann type, the number of large holes is relatively small. This limits the rate of diffusion of large molecules into the polymer.[1] The temperature of the material is quite important, because it determines the effectiveness with which secondary bonds can reorganize in order to allow passage of penetrant. Thus diffusion proceeds at a much higher rate above the glass-transition temperature. The polarity and cohesive-energy density of the polymer are also important. When the penetrant is chemically similar to the polymer, plasticization occurs and leads to an increase in the diffusion rate. When it is chemically quite different from the polymer, clustering of penetrant occurs, resulting in retardation of diffusion.

2. Limiting Factors

The characteristics of the permeation process are altered by chemical reaction between polymer and permeant during diffusion or by structural modifications in the polymer that occur during processing. When the penetrant reacts with the polymer, permeability usually changes, as does the time required to reach the steady-state concentration. In extreme instances, chemical reaction can completely prevent diffusion into the polymer. On the other hand, reaction can modify both permeant and polymer.

There are three possible consequences to any permeant–polymer reaction. For slow reactions, the time to steady-state concentration is not altered significantly. For fast reactions, the permeant is removed rapidly, and accumulation of the permeant in the polymer does not take place until all reactive polymer sites have been consumed. Finally, the change in polymer, as a result of chemical reaction, alters the permeability parameters, and the steady-state concentration applicable to a polymer before and after chemical reaction will be different. Permeation and chemical reactions in polymers are therefore quite dependent on each other.

Polymer modifications that most commonly affect the permeation process are crosslinking, crystallization, and microporosity. Dependence of the permeation process on these three variables is limited, however, because each of them occurs in a very select group of polymers. Paul and DiBenedetto[2] have shown that from five to nine backbone carbons are involved in polymer segmental motions. It would seem therefore that an extremely large number of crosslinks would be necessary to cause any substantial change in the diffusion rate. However, Rogers and co-workers[3] have shown that the introduction of only one radiation induced crosslink per 50 backbone carbons reduces the nitrogen permeability of polyethylene by a factor of 2. Thus the role of crosslinking in the permeation process appears more important than would be expected from simple constraint of segmental motion, although the order of change is rather small.

The role of crystallization is somewhat like that of crosslinking in limiting segmental motion. Permeation is usually restricted to amorphous regions in polymers, and an increase in the degree of crystallinity further retards diffusion. Crystallites also tie down or inhibit individual segmental motions in the amorphous phase. Some permeants can improve diffusion by solvating crystallites, whereas crosslinked polymers require breaking of primary bonds to increase the diffusion rate.

Micropores or cracks in polymers allow a convective process for permeation. For instance, Park[4] found that the diffusion rate for methylene chloride in polystyrene was initially high but slowly decreased to a constant value as the steady state was reached. He suggested that initially porosity and microcracks in the polymer allow rapid permeation by convective flow. However, subsequent sealing of cracks by swelling stops convective flow, and true diffusion then takes place.

3. Summary

In summary then, the permeation process requires the presence of permeant at the polymer surface, the dissolution of permeant at the surface, and its diffusion throughout the polymeric material until a steady-state concentration is attained. Permeation therefore depends on polymer, permeant, and temperature. Because the reorganization of chain segments is temperature dependent, temperature strongly affects the diffusion rate. The permeation process may also be affected by polymer modifications including the degree of cross-linking, crystallinity and microporosity.

C. TYPES OF CHEMICAL DETERIORATION AND STABILIZATION

Up to this point we have not emphasized that polymer failure can involve more than the disruption of the intramolecular polymer bonding forces alone. To be consistent with the definition of failure outlined in the first chapter, we must consider the disruption of intermolecular bonding forces. In fact, the permeation process itself, in the absence of any polymer–permeant reaction, is often sufficient to cause failure in a polymeric material.

For polymers that come into contact with chemical degradants, the disruptions of inter- or intramolecular forces are quite different paths to failure. To understand the difference between the individual degradative paths we will distinguish between two very different types of bonds (Figure 8.1). A polymer molecule is held together by a series of covalent bonds between individual atoms, like the bonds between carbon atoms in polyethylene, for instance. The energy necessary to break these intramolecular bonds is of the order of tens of kilocalories per mole. These bonds will be referred to as primary bonds. On the other hand, long-chain molecules in crystalline or amorphous arrays are held together by a number of much weaker bonds of either the van der Waal or acid–base type. The energy necessary to disrupt these bonds rarely exceeds one kilocalorie per mole. These bonds, which will be referred to as secondary bonds, break reversibly, whereas primary bonds usually break irreversibly. Thus secondary bonds can be thermally cycled beyond their breaking point without irreversible damage to the polymer structure, and primary bonds cannot. Therefore it is reasonable to categorize the modes in which chemical degradants break down polymers according to which of the two types of bonds is ruptured. The first general type of degradation involves the breaking of primary bonds and the second, breaking of secondary bonds. Since disruption of either bond type can result in polymer failure, stabilization techniques must be formulated for both processes.

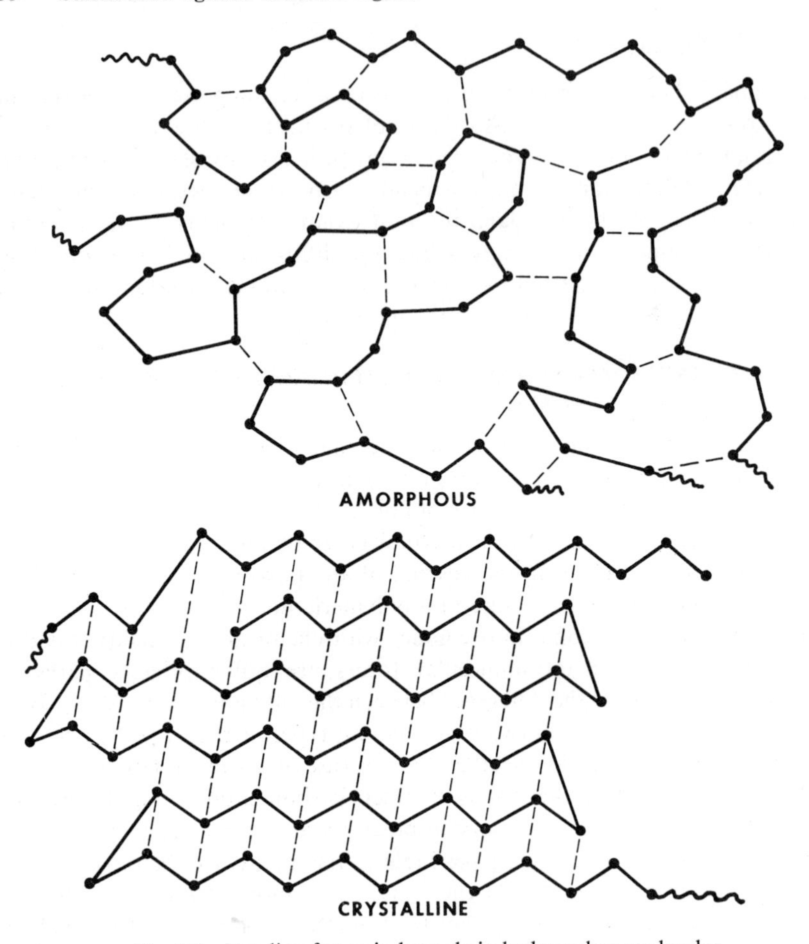

Fig. 8.1 Bonding forces in long-chain hydrocarbon molecules.

Stabilization against either type of degradation is accomplished by preventing penetration of the reactant into the polymer, by blocking polymer sites susceptible to the reactant, or by deactivating the reactant in an alternate, competitive reaction. The first method involves the addition of migratable oils or waxes into the polymer. For instance, rubber is protected against ozone by an exuded wax film. As an alternate method, protective films are formed over the surface. However, this method adds a potentially expensive processing step. The second method depends on the addition of blocking groups that protect active sites. Polymer modification is involved, however, so that these groups must be incorporated in such a way that important properties of the material are not significantly altered.

The final method involves incorporation of additives that react directly with the degradant. Whether an additive reacts with or simply blocks the reactant, the amount necessary to achieve stabilization usually has an adverse effect on polymer properties. Modification of the basic polymer structure can also result in important changes in physical properties. Surface treatment, although expensive, shows great promise because it does not alter the bulk of the polymer as do structural modification or incorporation of additives. Thus stabilization against chemical agents is no easy task. The choice of stabilization methods is narrow and each approach usually has inherent disadvantages.

1. Stabilization against Primary Bond Rupture

After oxygen and ozone, water is the most damaging reactant responsible for polymer failure. Water is an important degradant because of its presence in most environments, its reactivity at exposure temperatures, and its capability to permeate into almost all polymers. Furthermore, water is effective in the cleavage of both primary and secondary bonds, and it is often difficult to decide whether one or both bond-breaking processes has led to polymer failure. Polymer stabilization against primary bond rupture on contact with water will be the principal subject of this section. Protection against other specific primary bond degradants such as industrial acids and alkalies, and noxious gases will be discussed to a lesser degree.

In general terms, chemical agents reacting with primary bonds result in changes in the macromolecular identity of the polymer. Crosslinking, chain scission, addition of attacking agents, or any combination of these are the final results of primary bond rupture. Crosslinking, such as that which occurs with the acidolysis of poly(vinyl alcohol) yields intractable products of

$$2\ \text{---}CH_2\text{—}CH(OH)\text{---} \xrightarrow{H^{\oplus}} \begin{array}{c}\text{---}CH_2\text{—}CH\text{---}\\ |\\ O\\ |\\ \text{---}CH_2\text{—}CH\text{---}\end{array} + H_2O$$

increased brittleness. Tensile properties of the materials eventually decrease. On the other hand, chain scission, as with the hydrolysis of poly(ethylene terephthalate):

$$\text{---}\overset{\overset{\displaystyle O}{\|}}{C}\text{—}C_6H_4\text{—}\overset{\overset{\displaystyle O}{\|}}{C}\text{—}O\text{—}CH_2\text{—}CH_2\text{—}O\text{---} + H_2O \xrightarrow{H^{\oplus}}$$

$$\text{---}\overset{\overset{\displaystyle O}{\|}}{C}\text{—}C_6H_4\text{—}\overset{\overset{\displaystyle O}{\|}}{C}\text{—}OH + HO\text{—}CH_2\text{—}CH_2\text{—}O\text{---}$$

leads to softening, an increase in elongation, and a decrease in tensile strength. Addition of attacking agent to the polymer as, for example, the addition of chlorine to poly(1,4-butadiene):

$$---CH_2{-}CH{=}CH{-}CH_2--- + Cl_2 \rightarrow ---CH_2{-}CH(Cl){-}CH(Cl)CH_2---$$

generally results in an increase in modulus, particularly where the polymeric system is rubbery. As the modulus increases, elongation decreases and the impact strength of the material is reduced. Techniques for preventing or at least inhibiting these reactions depend almost entirely on the permeant and polymer characteristics. The stabilization method is usually predicated upon the concentration attainable by degradant in the system and its reactivity with the polymer.

a. Hydrolysis. Polymers with ester and amide links are very susceptible to water degradation, although it is often difficult to determine the extent to which primary bond cleavage is involved. Clarke and Miner[5] found that water immersion at 25°C had serious effects on the physical properties of cellulosic esters and nylons. Softening occurs on immersion of cellulose propionate, nylon 6,10 and nylon 6 (Table 8.1). In the amide systems, however,

Table 8.1 Effect of Water Immersion at 25°C for Nine Months on the Physical Properties of Some Plastics[a]

	Percentage Change in		
	Weight	Stiffness in Flexure	Hardness
Nylon 6	+9.0	−38	−35
Nylon 6,10	+3.0	−49	−10
Cellulose propionate	+2.0	−29	−34
Poly(methyl methacrylate)	+2.0	−8	−10
Polyacetal	+0.8	−14	+8
Polycarbonate	+0.4	+5	+3

[a] Courtesy of W. J. Clarke and R. J. Miner.[5]

weight gain decreases as the size of the alkyl portion is increased. Evidently an improvement in hydrophobicity is produced by the larger alkyl groups in the repeating unit. Clarke and Miner further observed that polyacetals, poly(methyl methacrylate), and polycarbonate are relatively unaffected in this environment. Two reasons have been proposed for the stability of polyacetals under these conditions. First, the water absorption is low, and second, acetal hydrolysis is specifically acid catalyzed, that is,

$$---OCH_2—OCH_2--- + H^{\oplus} \rightarrow ---OCH_2OH + {}^{\oplus}CH_2---$$
$${}^{\oplus}CH_2--- + H_2O \rightarrow ---CH_2OH + H^{\oplus}$$

Stabilization against hydrolysis in the presence of low concentrations of acidic impurities may be achieved by incorporation of basic inhibitors; but if the acid concentration becomes high, stabilization of polyacetals by basic inhibitors is difficult. For instance, the outdoor aging of these polymers has been shown by Kelleher[6] to proceed by chain scission during photodegradation. Rapid expulsion of formaldehyde leads to erosion of the material. Kern and co-workers[7] discovered that there is a buildup of acid degradation products during thermal aging that can stimulate hydrolysis. Evidently a complicated autocatalytic process is involved which depends on thermolysis for the accumulation of acid catalysts. These products in turn, after exhausting basic stabilizers, promote rapid catalytic hydrolysis of the polymer. The erosion observed by Kelleher might well be the result of a secondary process such as that indicated by Kern.

Poly(methyl methacrylate) absorbs a good deal more water than does polyacetal, and ester hydrolysis may be catalyzed by both acids and bases. Therefore, this polymer possesses hydrolytic stability for reasons that are necessarily different from, and indeed more complex than, those applying to polyacetals. The limited ester hydrolysis that occurs under weak acid and base catalysis and the fact that only pendant groups are being attacked can account for the difference. Limited pendant group attack does not substantially alter the properties that are dependent on the backbone structure. For instance, fibers of polyacrylonitrile, which is closely related to poly(methyl methacrylate) in terms of hydrolysis products, are surface hydrolyzed in dilute alkaline solutions for commercial dyeing. The physical properties of the fibers are unaffected, but their surfaces become more receptive to sulfonate dyes. That limited pendant group attack is the principal reason, however, for the relative inertness of poly(methyl methacrylate) to hydrolysis has been demonstrated by Baines and Bevington[8] who found that alkaline hydrolysis of methyl methacrylate copolymers is influenced by neighboring groups in the chain. Hydrolysis of poly(methyl methacrylate) ceases at 9% ester conversion, and further reaction is so sterically hindered that it does not occur. The center units of isotactic triads are hydrolyzed; other units are not affected, and the hydrolyzed product retains most of the desirable properties of poly(methyl methacrylate).

Impurities often catalyze unusual reactions in polymers. For example, Prince and Hornyak[9] found a marked difference in the reactivity of polyacrylonitriles prepared by two different polymerization techniques. Polyacrylonitrile made by gamma irradiation of the monomer (gamma-polymer) was rapidly hydrolyzed to polyacrylamide at 150°C and 5130 atm (Table 8.2). However, under the same conditions, polyacrylonitrile prepared by

Table 8.2 Effect of Exposure to Steam at 5130 Atm and 150°C for Three Hours on Polyacrylonitriles[a]

Polyacrylonitrile	Weight-Percent Added Peroxide	Hydrolysis (%)
Gamma-polymer	0	79.3
	0.1	98.7
Oxidation–reduction Polymer	0	0
	0.1	33.0

[a] Reprinted from Ref. 9, p. 165 by courtesy of *The Journal of Polymer Science*.

conventional peroxide catalysis (oxidation–reduction polymer) was not hydrolyzed. It appears that residual free radicals in the radiation-polymerized product initiated hydrolysis reactions. It is not unreasonable to suspect that, as in the case of the thermal aging of polyacetals,[7] accumulation of acids further catalyzed the hydrolysis. In any case, the sensitivity of the radiation polymerized system to hydrolysis requires further investigation.[10] This is particularly important to the electronics industry, since surface resistivity is very sensitive to polar contaminants when moisture is present.

The stability of polycarbonate as reported by Clarke and Miner[5] probably results from the low permeation and low reactivity of water under the conditions of their experiment. At higher temperatures, polycarbonate exhibits moderate penetration and reactivity with water. It is softened for instance on exposure at 100°C and 100% relative humidity[11] and can be hydrolyzed in dilute aqueous acid.[12] Goldberg[13] has shown that when chain flexibility is increased, as for example by copolymerizing different amounts of low-molecular-weight poly(ethylene glycol) and bisphenol *A* with phosgene, the rate of hydrolysis is increased. The added flexibility in the chain increases water absorption and presumably lowers the steric requirements for the attacking agent. It is probably for these two reasons that the related poly(ethylene terephthalate) is more susceptible to hydrolysis than is polycarbonate.[14]

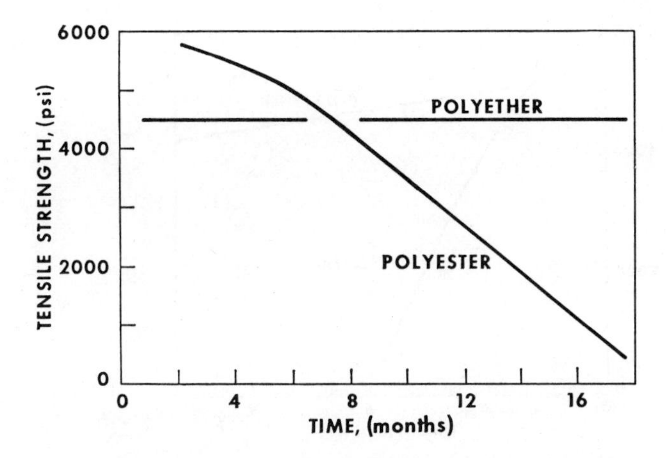

Fig. 8.2 Tensile strength versus exposure time at 25°C and 100% relative humidity for polyether- and polyester-based polyurethanes. Reprinted from Ref. 18, p. 711, by courtesy of *Rubber Age*.

Two techniques have been used to increase the hydrolytic stability of polycarbonates. Both rely on increasing the hydrophobicity of the polymer. Goldberg et al.[15] found that "polycarbonates" with increased aromatic ester and sulfonate linkages were more resistant to hydrolysis than unmodified polycarbonates. Kovarskaya and co-workers[16] produced a similar effect by replacing the isopropylidene portion of bisphenol *A* with aromatic groups. In both cases, the reactivity between polymer and the degradant was not changed significantly, but the wettability of both polymers was markedly reduced. Unfortunately, polymers with greater aromatic structure may become more sensitive to solar radiation.[17]

Ester links are known to be susceptible to acid- and base-catalyzed hydrolysis, whereas ether links are not. Athey[18] studied the effects of water on both ester- and ether-based polyurethanes. The polyether-based urethane shows no change in tensile strength after 18 months at 100% relative humidity and 25°C (Figure 8.2). The polyester-based material, on the other hand, loses 90% of its tensile strength. At 50°C in water the polyether material retains 70% of its tensile strength after six months, whereas the polyester system loses all tensile strength (Figure 8.3). Stabilization can be effected therefore by taking advantage of the much greater hydrolytic stability of polyethers, as compared to that of polyesters. Thus even though the initial properties of ester-based polyurethanes are better, the long-term property retention of the ether-based material makes it more desirable where greater stability is required.

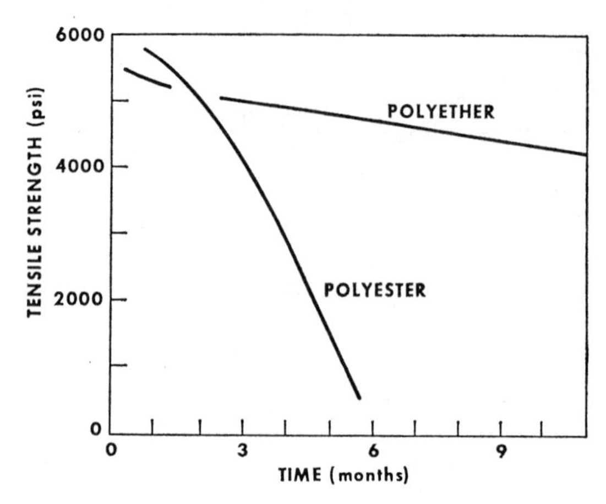

Fig. 8.3 Tensile strength versus immersion time in water at 50°C for polyether- and polyester-based polyurethanes. Reprinted from Ref. 18, p. 707, by courtesy of *Rubber Age*.

b. Other Reactions. Attack by other chemical agents is quite infrequent, because reagents of sufficient reactivity are not usually in contact with susceptible polymers. Furthermore, permeability to these reagents is in general very low, because their molecular dimensions are large relative to polymer void size. Thus degradation by these reagents occurs primarily at the surface.

One of the principal air pollutants, nitrogen dioxide, is very reactive toward a number of polymers. Ogihara and co-workers,[19] for instance, observed reaction with polyethylene, and Jellinek and Flajsman[20] reported nitration and crosslinking in polystyrene. Fortunately, the concentration of nitrogen dioxide in our atmosphere at present is quite small, and the normal antioxidants incorporated in polymers probably are sufficient to inhibit the resultant radical reactions.

A more serious problem exists for materials that are exposed constantly to a corrosive environment such as occurs during plating and printed wiring processes, in which strong acids and alkalies are commonly used. The floors of plating plants are constantly subjected to corrosive liquids, but normal concrete mortars are not able to withstand the variety of degradants encountered. Therefore polymeric mortars are used. This is a difficult application because industrial acids and alkalies are active degradants for the most resistant polymers. Sulfuric acid, for instance, sulfonates polymers with aromatic rings in their structure:

$$\text{---CH—CH}_2\text{---} + \text{H}_2\text{SO}_4 \longrightarrow \text{---CH—CH}_2\text{---} + \text{H}_2\text{O}$$

(In each structure a phenyl ring hangs from the CH; in the product the ring carries an SO_2OH group.)

Sulfation also may occur when ether or alcohol groups are present in polymers:

$$\text{---}CH_2\text{—}CH_2\text{—}O\text{—}CH_2\text{—}CH_2\text{—}O\text{---} + H_2SO_4 \rightarrow \text{---}CH_2\text{—}CH_2\text{—}O\text{—}SO_2OH + HO\text{—}CH_2\text{—}CH_2\text{---}$$

Nitric acid, which contains small concentrations of nitronium ion, that is,

$$2HNO_3 \gtrless NO_2^{\oplus} + NO_3^{\ominus} + H_2O$$

acts as a nitrating agent as, for instance, with polystyrene:

$$2\ \text{---}CH_2\text{—}CH(C_6H_5)\text{---} + NO_2^{\ominus} \longrightarrow \text{---}CH_2\text{—}CH(C_6H_4\text{—}NO_2)\text{---} + H^{\oplus}$$

and as an oxidant as with poly(butene-1):

$$2\ \text{---}CH_2\text{—}CH(CH_2\text{—}CH_3)\text{---} + 6\ NO_3^{\oplus} \longrightarrow 2\ \text{---}CH_2\text{—}CH(CH_2\text{—}C(=O)\text{—}OH)\text{---} + 3\ N_2O_4 + 2\ H_2O$$

Finally, alkalies catalyze ester hydrolysis:

$$\text{---}CH_2\text{—}CH_2\text{—}O\text{—}C(=O)\text{—}C_6H_4\text{---} + {}^{\ominus}OH \longrightarrow \text{---}CH_2\text{—}CH_2\text{—}OH + {}^{\ominus}O\text{—}C(=O)\text{—}C_6H_4\text{---}$$

Skiles[21] has summarized the effects of these degradants on a number of "acid-proof" plastic mortars (Table 8.3). The conditions of immersion are realistic for plating plant operations. Both phenolics are quite seriously affected by these reagents and would not be considered for general applications. The epoxy amine and furan mortars are more stable, but because of their sensitivity to nitric acid they would have to be used for limited application. Both polyesters are quite suitable for general application. The results of this study suggest that a judicious choice of mortar has more merit than attempting to stabilize a less desirable material. Fortunately, since these compounds are

Table 8.3 Effect of Immersion for 180 Days at 80°C on Silica-Filled Mortars[a]

Mortars	Percent Weight change in 78% H_2SO_4	15% HNO_3	10% NaOH
Epoxy phenolic	*D*[b]	*D*[b]	+2.0
Epoxy amine	+0.5	−18.7	+0.6
Phenolic, acid catalyzed	+14.1	*D*[b]	*D*[b]
Furan	+7.0	−54.7	+0.6
Polyester glycol	+0.7	+0.4	−9.6
Polyester bisphenol	+2.0	+1.0	−4.2

[a] Reprinted from Ref. 21, p. 322, by courtesy of the American Chemical Society, Division of Organic Coatings and Plastics.
[b] Decomposed within 30 days.

similarly priced, a more stable material can be selected without a significant increase in cost.

2. Stabilization against Secondary Bond Rupture

Catalytic amounts of degradant are responsible for primary bond breakage. Since concentration of active species is low, the concentration of stabilizer, too, is generally small. This is not so in secondary bond cleavage, in which much larger amounts of "reacting" species cause the observed change. Inhibiting these processes is more difficult because of the concentrations involved. Nonetheless, approaches to stabilization are quite similar.

Water is just as active in secondary bond breaking as it is in primary bond breaking. As a matter of fact, water plasticization is probably always a prelude to hydrolysis of esterlike polymer links. However, there are also many organic liquids that are quite capable of being absorbed by some polymer systems. Thus in the case of secondary bond cleavage, water is only one of many reactive agents.

Secondary bond rupture involves an intermolecular bonding force displacement. Chemical agents become solvated by the polymer:

$$
\begin{array}{c}
\text{---}\overset{\text{O}}{\overset{\|}{\text{C}}}\text{—NH---} \\
\vdots \\
\text{H} \\
| \\
\text{---}\underset{\underset{\text{O}}{\|}}{\text{C}}\text{—N---}
\end{array}
+ H_2O \longrightarrow
\begin{array}{c}
\text{---}\overset{\text{O}}{\overset{\|}{\text{C}}}\text{—NH---} \\
\vdots \\
\text{H} \\
| \\
\text{O—H} \\
\vdots \\
\text{O}\ \ \text{H} \\
\|\ \ \ | \\
\text{---C—N---}
\end{array}
$$

In this process neither the chemical agent nor the polymer is changed in a chemical sense. However, the polymeric material—that is, the total matrix—may be altered markedly. Among the phenomena associated with this process are environmental stress cracking, solvent crazing, plasticization, and delamination. Deplasticization and additive migration are associated with the reverse process, that is, the process in which a chemical agent is displaced by reorganization of polymer–polymer intermolecular bonding forces.

The failures accompanying these processes are just as critical as those observed in primary bond failure. For instance, in addition to the expected cracking, crazing, and softening phenomena arising from surface absorption processes, thermal treatment in the processing of contaminated polymers can lead to a rapid and uncontrolled volatilization of the absorbed species. A cellular structure results, and the desorbed materials may also stain the polymer surface.

Although the controlling parameters of these processes are not fully understood, polymers can be stabilized against this type of degradation. Because of the high concentration of degradant in secondary bond breakage, stabilization is usually achieved by modification of the polymer system. For instance, undesired plasticization can be controlled by altering the polarity of the polymer so that the degradant is repelled. Delamination can be minimized by introducing compatibility between the interfacial layers. Environmental stress cracking can be reduced by chain branching and crosslinking in the polymer. Unfortunately these modifications usually lead to alteration in polymer physical properties.

a. Plasticization. Polymer permeation depends on polymer void volume and the size of the penetrant. A permeant, however, is capable of plasticizing a polymer system only if the permeant–polymer interaction is favorable. That is to say, plasticization can occur if the free energy of interaction between permeant and polymer is at least as low as that between polymer molecules. Crystalline polymers, depending on the degree of crystallinity, contain varying combinations of crystalline and amorphous regions as depicted in Figure 8.1. Void volume and hence the diffusion rate are decreased by an increase in crystallinity. Assuming that the permeant can interact favorably with the polymer, plasticization of crystalline polymers will be slower than that of amorphous polymers. However, plasticization of crystalline polymers is more damaging because the degree of crystallinity of the system is lowered, and ultimately the polymer may become completely amorphous. In amorphous polymers, plasticization merely lowers the glass-transition point.

It is apparent from previous discussions that water plasticization precedes hydrolysis in polyesters, polyamides, and related polymers. For instance, in the work of Clarke and Miner[5] the short-term softening in cellulosics and polyamides after water immersion at 25°C has been attributed to plasticization. This is reasonable, particularly when one considers the rate of hydrolysis of simple esters in water under these conditions.[22] Clarke and Miner[5] also observed that, as the number of methylene units in a polyamide repeating unit are increased, the water-induced softening is less pronounced. Evidently hydrophobicity can be introduced into the chain by this means.

Polyimide film undergoes a 30% reduction in tensile strength after 400 hours in refluxing water (Figure 8.4). Several arguments support the contention that this change results from plasticization and not hydrolysis. First, the sample does not decrease in tensile strength after an equilibrium uptake of water occurs. Second, the original tensile strength is recoverable when the absorbed water is driven off. Third, Cannizzaro[24] found only an order of magnitude change in insulation resistance between a polyimide film exposed

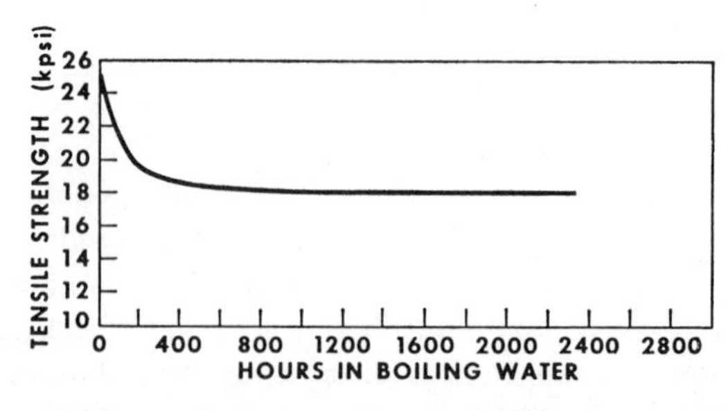

Fig. 8.4 Tensile strength versus exposure time in boiling water for a polyimide film. Reprinted from Ref. 23, p. 4, by courtesy of E. I. du Pont de Nemours and Company.

for 40 hours at 50% relative humidity and 73°F and a similar film exposed for 10 days at 91% relative humidity and 95°F. With polyimide–amide films, which are known to hydrolyze under these conditions, a change of 3 orders of magnitude was observed. The large change in polyimide–amide insulation resistance results from the accumulation of polar hydrolysis products. The small change in the surface resistance of polyimides must be caused by water absorption and not by hydrolysis. It has been found that water plasticization in polyimides can be reduced significantly by replacing some aromatic rings with alkyl groups.[25]

Plasticization by organic liquids is common only because of misapplication. For instance, it is not unusual to find vinyl tile in chemical laboratory flooring. Solvent spillage rapidly swells and disfigures the tile. Vinyl polymers cannot be stabilized economically against this environment and therefore should not be used. In another instance, electrical components and plastic shells have been handled in proximity. If degreasing agents that could contaminate the plastic packaging were used in processing the components, the plastic material might be damaged. In this instance, the plastic line has to be isolated from the degreasing agent, since again stabilization would not be economical.

b. Environmental Stress Cracking and Solvent Crazing. Environmental stress cracking and solvent crazing are probably the most general modes of failure resulting from secondary bond cleavage. Both types of degradation are a consequence of stresses built up in the plastic, and they proceed by similar mechanisms.

Environmental stress cracking, which appears to be limited to partially crystalline polymers, occurs when small amounts of certain nonsolvents are placed on biaxially stressed surfaces. Absorption of the agent into the substrate does not appear to play an important role. Since stress-crack initiation is a surface phenomenon, the viscosity, surface tension, and wettability of the active agent are important to the failure process. Solubility is not.

Since an impurity capable of increasing the stress at the surface is often responsible for initiation, the yield stress of the polymer is another important parameter. Stabilization can be attained in some instances by altering the degree and type of crystallinity in the polymer. Increases in molecular weight and in the degree of chain branching reduce crystallinity. For instance, it is known that polyethylenes with low melt indices and therefore high molecular weight are less prone to stress cracking (Figure 8.5). Chain branching also leads to a more stress-crack-resistant system.[26] Furthermore, the exclusion of low-molecular-weight fractions increases stress-crack resistance, so that narrow distribution and high molecular weight are desirable. The relation between the two is not rigorous, however, and interpretation of data from these parameters alone can be misleading.

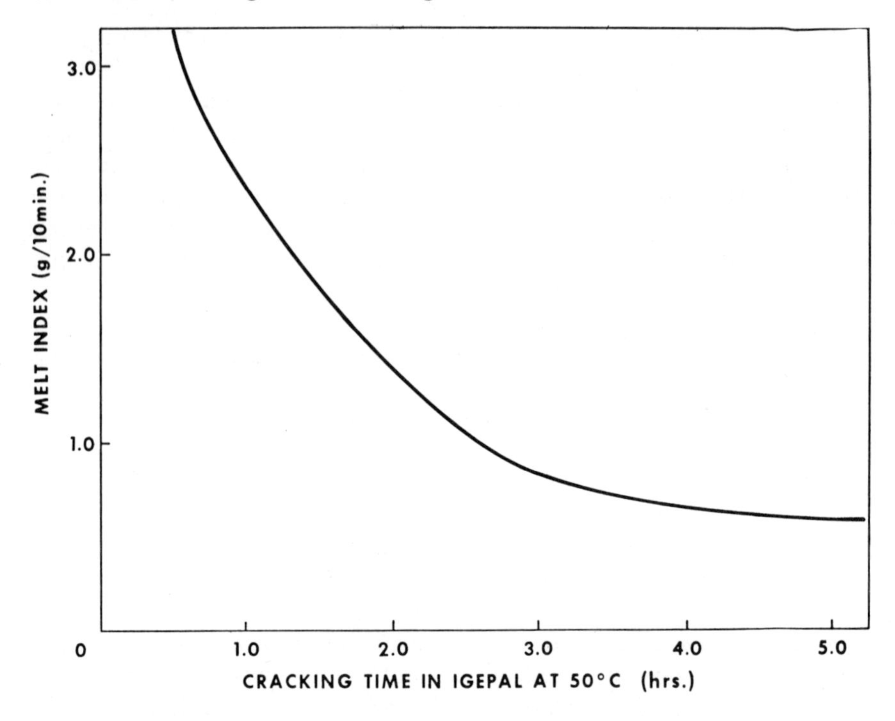

Fig. 8.5 Melt index versus time to visible stress cracking for polyethylene. Constructed from the data in Ref. 26, p. 400, by courtesy of the Society of Plastics Engineers.

Amorphous and therefore stress-crack-resistant materials can be produced by proper thermal treatment during processing. Howard and Gilroy[27] have shown that shock-cooled samples that contain very small crystallites because of this treatment have relatively good crack resistance. However, the reorganization of crystallites with time leads to vulnerability and, in addition, shorter cooling cycles such as shock cooling tend to yield internally stressed samples. Some compromise between these two conditions must be met for maximum stability.

Solvent crazing occurs in both amorphous and crystalline systems when small amounts of solvents are in contact with stressed polymers. However, the crazing process is reversible, since crazes disappear when the stress is removed.[28] The degree of stress and the rate of application of the solvent are the most important parameters in solvent crazing. For instance, Nielsen[29] has shown that poly(methyl methacrylate) plasticized with 8% dibutyl phthalate does not craze when exposed to benzene vapor at 27°C for several days, but it does craze immediately when plunged directly into benzene. The slow penetration of solvent relieves built up stress centers. If crazes do arise, they are healed by the penetrating solvent. On the other hand, the swelling action of the solvent in rapid penetration amplifies the stresses in the plastic and leads to spontaneous crazing. Stabilization is best attained by relieving internal stresses in polymeric materials. Long-term cooling cycles (annealing) directly after processing help to achieve this. However, if the material is stressed in use, annealing is not the practical solution.

c. Other Reactions. There remain but a few less well defined secondary bond breaking processes that contribute to polymer failure. These include additive migration, delamination, and chemical attack on additives. Although the latter process does not directly involve polymer degradation, it is worth mentioning because it could be responsible for failure of the polymer composition. Loss of plasticizer is accompanied by an increase in polymer rigidity, and failure occurs as a result of embrittlement. Migration of the plasticizer out of the polymer system usually occurs because of unusual volatility or pressure on the sample. The role of plasticizer volatility in the migration process has been demonstrated by Quackenbos,[30] who found a

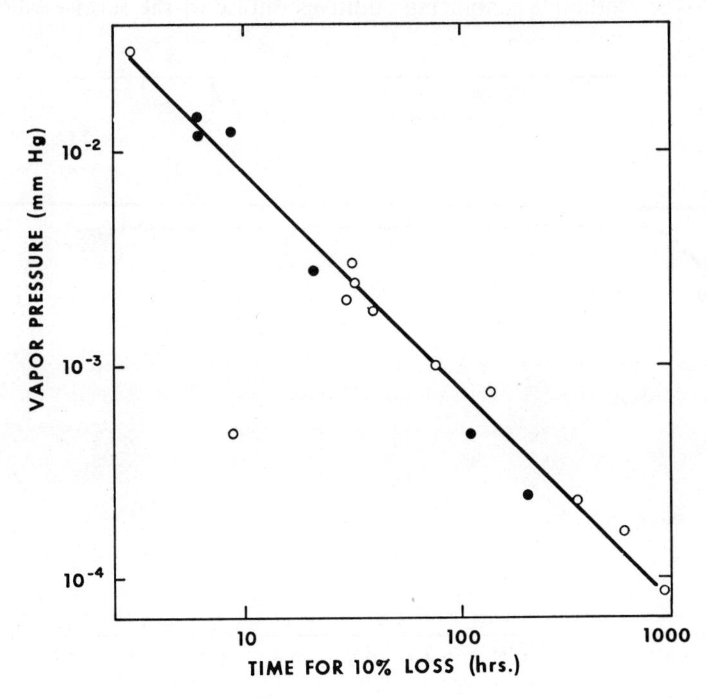

Fig. 8.6 Plasticizer vapor pressure versus time to 10% loss for poly(vinyl chloride) films at 98°C; solid circles represent mixed plasticizers. Reprinted from Ref. 30, p. 1341, by courtesy of *Industrial and Engineering Chemistry*.

linear relation between the vapor pressure of plasticizer and the time at which a 10% loss was observed in poly(vinyl chloride) (Figure 8.6). This implies that when the compatibility of different plasticizers is the same, poly(vinyl chloride) will retain the less volatile plasticizers longer.

Compression of plasticized polymers forces plasticizer out of the system. For instance, Frey[31] found that the migration of dioctylphthalate (DOP) out of poly(vinyl chloride) containing approximately 45% of DOP increased with an increase in applied pressure (Figure 8.7). Since plasticizer loss did not exceed 30%, it was concluded that the plasticizer retained was strongly bound to the polymer. Therefore, if the plasticizer can be maintained at or below this limiting concentration, migration by this process will be prevented.

The migration of other additives usually reduces the stability of polymers. For example, the migration of antioxidants, antiozonants, and light stabilizers may reduce the lifetime of the polymer. If, however, the volatility of the additive is relatively low, it accumulates at the surface where it could be quite effective. For instance, the migration of extender oils or waxes to the surface in tires provides some ozone resistance by forming an ozone diffusion barrier (Chapter 5). So, although stabilizing additives diffuse to the surface where

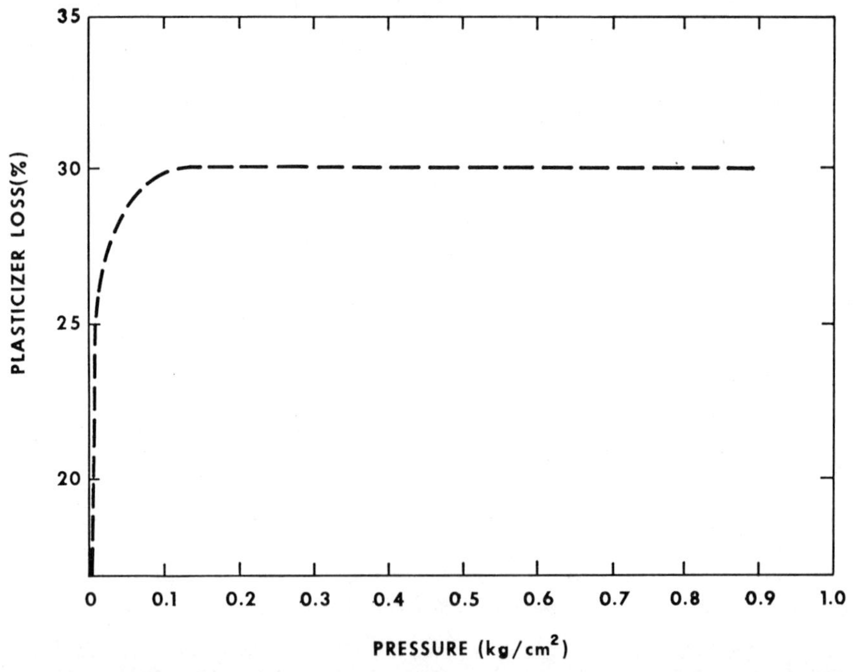

Fig. 8.7 Applied pressure versus plasticizer loss for supersaturated poly(vinyl chloride) sheet. Reprinted from Ref. 31, p. 84, by courtesy of *Kunststoffe*.

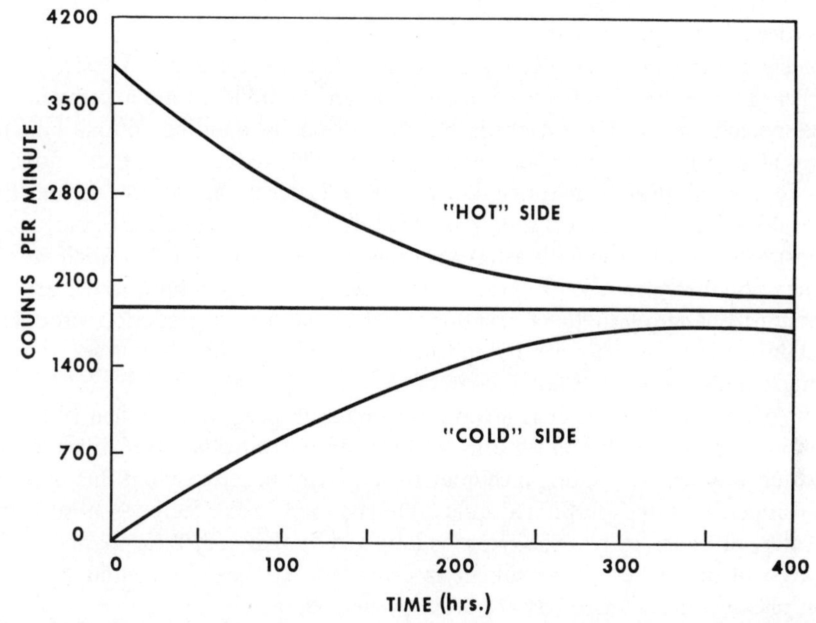

Fig. 8.8 Radioactive counts versus exposure time for *N,N'*-diphenyl-*p*-phenylenediamine migrating into styrene–butadiene rubber at 100°C. Reprinted from Ref. 32, p. 896, by courtesy of *Rubber Chemistry and Technology*.

they can be rubbed or washed away, they concentrate in the area of greatest sensitivity, and thus the migration process can be beneficial.

Lewis and co-workers[32] examined the migration of low-molecular-weight rubber additives. Using radiochemical techniques, they found an equilibration of additive throughout the system (Figure 8.8). Therefore the degree of stabilizer migration may increase as the polymer surface is exposed to extractants. This results because the system requires an equilibrium surface concentration that is maintained, as surface additive is washed away, by migration of additive from below the surface. Lewis and co-workers found that this process could be retarded by increasing the surface area of carbon black filler in the rubber. The absorbent character of the carbon black retards the attainment of an equilibrium situation. This filler may also retain the stabilizer in an inactive form.

Composites fail even when penetrated by chemically inactive contaminants. The accumulation of permeant at the interface in these heterogeneous systems reduces interfacial adhesion that results in delamination or cracking. Interfacial attack may occur through capillary flow. The chemistry involved at the interface of polymer and filler in reinforced plastics is quite complex.

The interface is a bonding region, but some protection of the filler surface is desirable. The penetration of water into the bonding region induces mechanical strain, and the resultant force separates the polymer from the filler. The mode of delamination depends on the bond strength between the connected layers. The system is best stabilized by building compatibilizing layers between the dissimilar materials, thereby reducing capillary flow.

In some materials, additive concentration is great enough to consider the results of direct chemical attack on the additives as it affects the total composition. This is true with ester-type plasticizers, hydrolysis of which would alter polymer-plasticizer interactions. Visual and electrical properties are the first to change with the formation of the resulting degradation products. Stabilization for polymer compositions containing these additives is very much the same as that for polyesters.

Volatilization of water is always a problem in processing. When rubber is cured at pressures of about 1 atm, a rapid and uncontrollable volatilization of water may occur, yielding a cellular structure. In order to avoid this, a paste of mineral oil and pulverized quicklime is added to the recipe to absorb the water, prior to curing. The recipe is adjusted so that very little change in the physical properties of the rubber is observed. To avoid a similar problem in plastics, molding pellets are often preheated.

D. CONCLUSION

We have seen that polymers are stabilized against chemical attack by somehow preventing the potential degradant from causing degradation. There are perhaps five means to accomplish this, and which one is chosen depends on the polymer and on the degradant.

A very effective way of stabilizing involves reducing the rate of diffusion of the degradant into the polymer. This may be accomplished by developing a permeation barrier at the polymer surface. For example, poly(vinyl fluoride) films laminated to plastics are excellent surface erosion protectors.[33] In addition, surface treatment by active gases such as that proposed by Schonhorn and Hansen[34] should play a very important role in stabilization against chemical attack. Neither process affects the bulk physical properties of the polymer, since the protective film is extremely thin relative to the thickness of the base material. However, both surface treatment techniques are expensive.

A second stabilization technique relies on removing degradant by alternate competitive reactions. As an example, polyacetals are stabilized against acidolysis with basic inhibitors.[35] Unfortunately, if the concentration of stabilizer is very high, the physical properties of the polymer are affected. In addition, the molecular weight of these basic stabilizers is low and they are so volatile that migration out of the system is a serious problem.

Third, where residual stresses in polymeric materials are a problem, control of the manufacturing process can alleviate the situation. Nearly stress-free samples can be produced by controlled cooling of the article in its mold or by subsequent annealing. Unfortunately both processes add to the time of manufacture and therefore the use of either would increase the price of the article.

The fourth and probably most familiar process for stabilizing against chemical attack requires structural modification of the polymer. Polycarbonates, for example, can be made more hydrophobic by increasing aromaticity in the system.[16] Unfortunately, any attempt to modify the structure of a polymer to decrease its permeability to some agent usually results in a significant change in the polymer physical properties.

The last and perhaps most sensible stabilizing technique is the judicious choice of polymer. This technique, which depends upon the availability of polymers for each application, can become quite expensive in many cases.

REFERENCES

1. H. L. Frisch, *J. Polym. Sci., Part C, Polym. Symp.*, **10,** 11 (1965).
2. D. R. Paul and A. T. DiBenedetto, *J. Polym. Sci., Part C, Polym. Symp.*, **10,** 17 (1965).
3. C. E. Rogers, J. A. Meyer, V. Stannett, and M. Szwarc, *TAPPI Monogr. Ser.*, **23,** 12 (1962).
4. G. S. Park, *Trans. Faraday Soc.*, **48,** 11 (1952).
5. W. J. Clarke and R. J. Miner, private communication, 1969.
6. P. G. Kelleher and B. D. Gesner, *Polym. Eng. Sci.*, **10** (1), 38 (1970).
7. W. Kern, H. Cherdron, V. Jaacks, H. Baader, H. Deibig, A. Giefer, L. Hohr, and A. W. Wildenau, *Angew. Chem.*, **73,** 177 (1961).
8. F. C. Baines and J. C. Bevington, *J. Polym. Sci., Part A-1*, **6,** 2433 (1968).
9. M. Prince and J. Hornyak, *J. Polym. Sci., Part A-1*, **5,** 161 (1967).
10. D. M. Kiefer, *Chem. Eng. News*, **47** (53), 46 (1969).
11. W. J. Jackson, Jr. and J. R. Caldwell, *Ind. Eng. Chem., Prod. Res. Develop.*, **2** (4), 246 (1963).
12. L. H. Lee, *J. Polym. Sci., Part A*, **2,** 2859 (1964).
13. E. P. Goldberg, *J. Polym. Sci., Part C, Polym. Symp.*, **4,** 707 (1963).
14. W. McMahon, H. A. Birdsall, G. R. Johnson, and C. T. Camilli, *J. Chem. Eng. Data*, **4,** 57 (1959).
15. E. P. Goldberg, F. Scardiglia, R. J. Schlott, and D. F. Hoeg, *3rd Bienn. Polym. Symp., ACS Div. Polym. Chem.*, Case Western Reserve University, Cleveland, June 22–24, 1966.
16. B. M. Kovarskaya, G. S. Kolesnikov, I. I. Levantovskaya, O. V. Smirnova, G. V. Dralyuk, L. S. Poletakhina, and E. V. Korovina, *Sov. Plastics* (English transl.), **1,** 41 (1967).
17. B. D. Gesner and P. G. Kelleher, *J. Appl. Polym. Sci.*, **13,** 2183 (1969).

18. R. J. Athey, *Rubber Age*, **96,** 705 (1965).
19. T. Ogihara, S. Tsuschiya, and K. Kuratani, *Bull. Chem. Soc. Japan*, **38,** 978 (1965).
20. H. H. G. Jellinek and F. Flajsman, *J. Polym. Sci., Part A-1*, **7,** 1153 (1969).
21. E. W. Skiles, *ACS Division of Organic Coatings and Plastics Preprints*, **23** (2), 321 (1963).
22. P. D. Bartlett and G. Small, *J. Amer. Chem. Soc.*, **72,** 4867 (1950).
23. Kapton (Du Pont registered trademark) Polyimide Film, Du Pont Tech. Inf. Bull. H-2, 1968.
24. J. R. Cannizzaro, *Solid State Technol.*, **31** (November 1969).
25. I. Halperin and V. A. Lanza, 8th Annual Wire and Cable Symposium, Atlantic City, Dec. 1969.
26. J. B. Howard, *SPE* (*Soc. Plastics Engrs.*) *J.* **15,** 397 (1959).
27. J. B. Howard and H. M. Gilroy, *SPE* (*Soc. Plastics Engrs.*) *J.*, **24** (1), 68 (1968).
28. R. P. Kambour, *Polymer*, **5,** 143 (1964).
29. L. E. Nielsen, *J. Appl. Polym. Sci.*, **1,** 24 (1959).
30. H. M. Quackenbos, Jr., *Ind. Eng. Chem.*, **46** (6), 1341 (1954).
31. H. E. Frey, *Kunststoffe*, **46,** 81 (1956).
32. J. E. Lewis, M. L. Deviney, Jr., and L. E. Whittington, *Rubber Chem. Technol.*, **42** (3), 892 (1969).
33. V. L. Simril and B. A. Curry, *Mod. Plastics*, **121** (July 1969).
34. H. S. Schonhorn and R. H. Hansen, *J. Appl. Polym. Sci.*, **12,** 1231 (1968).
35. M. Kubico, R. N. MacDonald, R. L. Stearns, and F. A. Wolff (to E. I. du Pont de Nemours and Co) U.S. Patent 2,810,708, October 22, 1957; *Chem. Abstr.*, **52,** 798d (1958).

9

BIOLOGICAL STABILITY OF POLYMERS

WALDIMERO COSCARELLI
Shulton, Incorporated, Clifton, New Jersey

A. BIOLOGICAL DEGRADATION OF POLYMERS

1. Natural Polymers

Polymers of natural origin are considered biodegradeable and are attacked by many forms of life. The principal agents in the soil environment are microorganisms, particularly bacteria and fungi, with some assistance from other forms of microscopic life. Higher forms of biological life, represented by insects and rodents, also play a dominant role in the degradative process. In the marine environment, microorganisms, especially bacteria, are significant, but from an economic standpoint it is the marine borers that are responsible for much of the destruction.

a. Cellulosics. Unwittingly, microorganisms play an essential role in the cyclic process necessary to maintain life. They participate in this accomplishment by breaking down complex plant and animal residues to carbon dioxide, water, and minerals. These breakdown products are utilized by plants during the photosynthetic process to produce sugars and starch, which serve as substrates for the synthesis of carbohydrates, fats, proteins, or other complex compounds. In the marine environment, microorganisms participate in the mineralization of organic materials and they in turn are a source of food for higher forms of animal life. Their contribution in ridding the environment of hydrocarbons is well documented.[1–3]

The natural destruction of wood by cellulose-decomposing microorganisms and by termites has received widespread attention because of the economic importance of wood as a building or structural material. The annual loss in this country due to decay is in excess of $400 million.[4] Fungi are the principal deteriorative agents in the standing tree, and they receive additional support from bacteria and termites once the tree is felled. Some of the fungi known to attack wood belong to the genera *Lenzites* and *Poria*, both basidiomycetes, and *Trichoderma*, which belong to the class known as Fungi Imperfecti. The sapwood of a standing tree is more resistant than the heartwood but becomes more susceptible once the tree is felled. The resistance of cellulose is associated with the presence of lignin, which is slowly degraded by microorganisms.

Lignin acts as a protective covering and once it is removed by chemical means, cellulose is rapidly degraded.

It should be mentioned that native cellulose is considered to be fairly resistant and only when it is modified by chemical or physical means does degradation proceed at an appreciable rate. Therefore, we can expect materials made from processed wood or cotton fibers to be more prone to biological attack, and accordingly they should be properly protected. In the marine environment, destruction of cordage and fishing nets is extensive. Marine fungi have been often overlooked, but they are quite capable of growing in submerged wood on the open sea.[5,6] However, marine borers are the biological agents that cause world-wide destruction of wooden pilings and wharves.

b. Rubber. Natural rubber, also called caoutchouc latex, guayule latex, or hevea latex, consists of polymerized olefinic hydrocarbons. It is produced by many plants and does not accumulate in nature. Since natural rubber, with its repeating isoprene units, is subject to oxidation, it is not certain whether it can be attacked by microorganisms without first undergoing some degree of oxidation.[7] Nevertheless, there are many microorganisms reported to attack untreated natural[1] and vulcanized rubber[8–10] whatever the environment–terrestrial, marine, or laboratory. Genera frequently mentioned for fungi are *Aspergillus* and *Penicillum* and for bacteria are *Actinomyces*, *Bacillus*, *Pseudomonas*, and *Thiobacillus*. The latter was implicated in the conversion of sulfur to sulfuric acid in vulcanized rubber.[11] Pitting and discoloration of rubber are the usual visible effects of microbial attack. In the terrestrial environment natural rubber is readily attacked by insects, particularly if the material is exposed in the topsoil area.

c. Lignin. Lignin is a plant residue that decomposes very slowly and therefore tends to accumulate in the soil. It contributes to the formation of humus in soil. It is extremely resistant under anaerobic conditions, but aerobically it is degraded by a number of basidiomycetes.

d. Others. Plant polymers such as starch, pectin, and related polysaccharides are rapidly decomposed by many types of microorganisms.[12]

Keratin and chitin are fairly resistant polymers but do not accumulate in the soil. Keratins are a class of insoluble proteins found in the vertebrate epidermis and its appendages: hair, feathers, claws, and horns. Keratin is more resistant in its natural state but is rapidly decomposed by fungi after it is denatured by heat or alkali treatment.[13–15] Chitin, a polymer of *N*-acetylglucosamine, is the basic cell wall material of most fungi and the structural polysaccharide of higher invertebrates. Enzymes capable of breaking down chitin are produced by some actinomycetes, a few fungi, and the snail, *Helix pomatia*.

2. Derived or Synthetic Polymers

The biological degradation of synthetic polymers is mediated by the same forms of life that affect natural polymers. Unlike natural polymers, whose deterioration is not only visibly apparent but readily measured either as a function of microbial growth or disappearance of substrate, deterioration of synthetic polymers (as discussed in Chapter 10) is usually measured by changes in their physical, chemical, or electrical properties. The appearance of fungal growth on the surface of plastics or elastomers is often misconstrued for insipient deterioration of the material in question when, in reality, growth is caused by the presence of extraneous susceptible matter. On the other hand, deterioration caused by bacteria is often overlooked because their growth characteristics on the surface of the materials do not lend themselves to obvious recognition. All these factors contribute to the confusion that exists in the literature on the questionable susceptibility of certain materials to microbial attack.

a. Derived Cellulose. The microbial degradation of chemically modified cellulose depends on at least three factors: the nature of the substituent, the degree of substitution, and the orientation of the fiber. The overriding factor appears to be the degree of substitution. Complete resistance is usually obtained regardless of the substituent, providing that there is at least one substituent on each anhydroglucose unit.[4] For example, cellulose acetate is reported to be resistant to microbial attack if fully acetylated. However, if the percentage of acetate is not high enough, resistance to attack is lowered. Cellulose triacetate, on the other hand, is unaffected by microorganisms, and its resistance is independent of the degree of polymerization. Other factors being equal, less oriented cellulose is more susceptible to microbial attack than more highly oriented cellulose. Cellulose acetate–butyrate and cellulose propionate are considered resistant but lose this property when plasticized with a biodegradeable plasticizer such as dibutyl sebacate. After exposure to soil burial, changes in physical properties are noted. There is a decrease in ultimate elongation and an increase in tensile strength. These changes are probably caused by plasticizer loss.

b. Laminates and Casting Resins. Laminated phenolics are resistant to attack if the layers are composed of an inorganic material such as fiberglass cloth. If the base material is a cellulose derivative, the resistance to microbial and termite attack is diminished to a degree that depends on the amount of fiber present. As a general rule, the incorporation of cellulose derivatives in plastic formulations tend to decrease the resistance to biological attack.

Casting resins are not affected by microorganisms, but some are attacked in the ocean environment by marine borers.[16]

c. Elastomers. The susceptibility to microbial degradation of compounds containing natural rubber or synthetic elastomers has received considerable attention.[1,8] The picture is rather confusing because susceptibility of the elastomer is sometimes mistaken for the susceptibility of the finished material. Frequently the processing method is a contributing factor to deterioration.[17] Under accelerated laboratory conditions, Blake and co-workers[10] found that raw Buna-*S* was attacked by fungi but did not support growth after acetone extraction, although fungi did grow on the evaporated extract. Butyl rubber and neoprene appeared not to support fungal growth. Nevertheless, electrical failures of coated wires, as evidenced by a drastic decrease in insulation resistance, did occur. It was discovered that failures were caused by fungal mycelia penetrating the elastomers and making contact with the conductors. It was concluded that the susceptible material was not the elastomer but some other substance in the formulation. In a simulated marine environment, using enrichment cultures of bacteria as inocula, neoprene, GR-*S*, butyl, and silicone rubber were attacked as judged by BOD (biological oxygen demand) studies.[2,16] Anaerobically only butyl appeared to be affected by sulfate-reducing bacteria. In ocean exposure, elastomers are subject to borer activity as well as microbiological attack.

Bennicelli[18] reported that raw polymers and copolymers of silicones and chlorosulfonated polyethylene were not affected by fungi. Polysulfide, which was employed as a coating for gasoline storage tanks, was found to be damaged by bacteria and fungi.[19] In contradiction, Zobell[1] and Heap[7] have reported that polysulfide and "Hypalon" are resistant to attack.

Polyurethanes are finding increasing use because their physical and chemical properties make them adaptable to products that have commercial application. The group is difficult to evaluate from a microbiological point of view because the formulations are complex, and unless each ingredient in the formulation is known, it is almost impossible to predict resistance to microorganisms. These problems are discussed in reviews by Kaplan[20] and Darby.[21] There is considerable evidence to suggest that ester-linked polyurethanes are susceptible to microorganisms, whereas ether-linked polyurethanes are quite resistant.[22]

Polyester failures are also attributable to the hydrolytic action of water or moisture on the ester linkages, resulting in a scission of the polymer. Darby and Kaplan[21] tested 22 ether-linked polyurethanes and three ester-linked polyurethanes. All polyesters were heavily attacked. They found that most polyethers were highly resistant to degradation, and those that were slightly or moderately attacked were low-molecular-weight unbranched alkane diols

(1,4-butanediol, 1,5-pentanediol, and 1,6-hexanediol) and the higher molecular weight propylene glycols (molecular weight 1020 and 1320). Of the diisocyanates used, polymers made with linear diisocyanates (1,6-hexamethylene-diisocyanate) proved to be more resistant than cyclic diisocyanates.

In the soil environment, elastomers are subject to vigorous insect activity. Usually the flexible formulations are more prone to attack than the more rigid urethanes. In the marine environment, they can be readily attacked by borers and, if susceptible, by marine microorganisms.

d. Polyolefins and Others. Considering how many plastics are utilized today, there are very few reports published on the biological stability of plastics. It is not, of course, that no one is concerned, rather, because of the inherent resistance of plastics to microbial degradation, there is usually very little to report, the one important exception being plasticized poly(vinyl chloride).

Polyethylenes have excellent resistance to soil and marine microorganisms and marine borers.[7,16,23] Surface growth usually stems from extraneous surface matter and does not affect the material.[24] In highly infested soils, insects attack the edges of high-density polyethylenes; low-density materials may be attacked in a nonspecific manner, often showing abortive attempts of biting.

Polyamides, poly(ethylene terephthalate), polypropylene, polytetrafluoroethylene, polymonochlorotrifluoroethylene, polystyrene, poly(methyl methacrylate), polyformaldehyde, and poly(phenylene oxide) are just a few of the plastics that are resistant to microbial attack.[24] In contrast, many of these are subject to borer activity; in particular, the polyamides, poly(methyl methacrylate), and the polyformaldehydes. Of these poly(methyl methacrylate) appears to be more susceptible.[25]

Most of the literature dealing with the deterioration of plastics is concerned with plasticized poly(vinyl chloride). There is common agreement that poly(vinyl chloride) resins are not degraded by microorganisms.[16,26] However, evidence based on some early work with fungi[27,28] has been presented to show that many plasticizers are degraded, and as a result plastic formulations containing susceptible plasticizers are presumed to be susceptible to biodegradation. Many studies have since supported this claim. Booth and Robb[29] subjected 13 plasticized poly(vinyl chloride) formulations to soil inoculated with bacteria. All were attacked to some degree. Formulations plasticized with sebacates, azelates, and adipates were more susceptible than those containing phthalates or phosphates. Others have reported that the incorporation of nitriles[25] or ricinoleates[30] renders the formulations susceptible. As a rule, all aliphatic dibasic acids and modified soybean oils are readily degraded. Although carbon plasticizers normally supply the carbon essential to growth, Klausmeier[31] found that many microorganisms did not utilize the plasticizer for growth but alternatively produced an enzyme that would attack the plasticizer if the organism was grown in the presence of another carbon source such as glucose.

Phthalates are the plasticizers of choice because of their properties and because it is generally accepted that they are reasonably resistant to biological degradation. However, by the use of the enrichment culture techniques, whereby the plasticizer is added to soil, for example, to encourage the proliferation of microorganisms capable of degrading the plasticizer, organisms can be isolated which in pure culture can degrade the enrichment plasticizer. As a consequence, many plasticizers, particularly many of the phthalates, which were originally reported to be resistant, are now considered to be susceptible. These laboratory findings are now being supported by long-term soil burial studies.[32] Internally plasticized poly(vinyl chloride) is reported to be resistant to microbial action.[63]

Other forms of biological life also affect plasticized poly(vinyl chloride). In the soil environment they are vulnerable to attack by termites and rodents.[33] When exposed to shallow ocean environment, poly(vinyl chloride) is subject to marine borer penetrations.[16] Little else is known about the biological resistance of other types of vinyl polymers. They are presumed to behave similarly to poly(vinyl chloride). The reported susceptibility of poly(vinyl acetate)[34] and the copolymer of vinyl chloride and vinyl acetate[35] to microorganisms supports this contention.

B. CAUSATIVE ORGANISMS

Biological agents responsible for the degradation of polymers under natural environmental conditions include microorganisms such as bacteria and fungi and other classes of higher life, embracing forms representative of insects, rodents, and marine borers.

1. Bacteria

Bacteria comprise the great majority of microorganisms, and of this class, heterotrophs are by far the largest and the most important group concerned with deterioration. Characteristically they require preformed organic carbon substrates for their energy sources. They vary in structure and physiology and are quite capable of adapting to adverse environmental conditions. Some grow aerobically, needing elemental oxygen; others require little or no oxygen. Some enhance their fitness to withstand unfavorable conditions by forming spores. All require nitrogen, and most depend on fixed forms of inorganic nitrogen. A few require organic nitrogen and even fewer are capable of fixing atmospheric nitrogen. Although bacteria are abundant in number, they are small in size and cannot be seen without the aid of a microscope. A typical cell measures about 1 or 2 μ wide and up to 5 μ long.

Taxonomically, the Gram stain is used to classify bacteria into two major categories. They are either Gram-positive or Gram-negative. This differential staining procedure was originally developed as an empirical method for demonstrating the presence of Gram-positive bacteria in animal tissues, which are Gram-negative. However, recent developments have shown that the Gram stain reflects basic structural and chemical differences between the cell walls of Gram-negative and Gram-positive bacteria. Thus the stain has taken on added significance in the identification of bacteria.

Since bacteria are difficult to classify according to morphological characteristics because of their size, other means, in addition to differential staining, must be employed. Through the judicious use of selective nutritional media, bacteria can be grown and observed for specific biochemical reactions. The results can then be compared with those reported in the literature in order to provide meaningful identification.

Bacteria are readily accepted as deteriorative agents of natural polymers such as cellulose or collagen mainly because their action results in a complete loss of integrity and dissolution of the original substrate. Growth on most synthetic polymers is less obvious, and degradative effects must be measured by methods more critical than visual observations. In contrast to the cottonlike growth of fungi, bacterial growth is difficult to detect without the aid of a microscope and often goes unnoticed or is perfunctorily dismissed as slime. Therefore, only a few material engineers accept bacteria as deteriorative agents of synthetic polymers; the idea is of doubtful significance to some and is ignored by most.

2. Fungi

Fungi, especially filamentous fungi, have received more attention with respect to deterioration problems than any other single group of organisms. Their ability to attack recalcitrant organic substrates was highlighted during the tropical phases of World War II when serious damage to equipment—particularly items made of wood, fabric, or leather—resulted in significant economic losses.

Fungi are easily recognized because of their morphological appearance. The cottonlike mass, known as the mycelium, is visible to the naked eye. It consists of a branching system of walled tubes called hyphae, which continually extend by growing at the tips and by lateral branching. The diameter of fungus hyphae varies from one fungus to another, but most hyphae fall within the range of 2 to 20 μ in diameter and can grow several inches in length. Taxonomically they can be broadly classified by the spores they produce and the structures that bear these spores.

Filamentous fungi grow best in an aerobic environment where there is an ample supply of oxygen. Their basic nutritional requirements are similar to those of bacteria; but unlike bacteria, which must be bathed in an aqueous environment for growth, fungi are better equipped to grow in adverse environments because of their ability to translocate food and water through their hyphae.

Interpreting the appearance of fungal growth on polymers, especially high-molecular-weight synthetic polymers, should be approached with caution. Surface growth may be substantial but superficial, and it frequently occurs at the expense of unassociated detritus. On the other hand, sparse surface growth often belies the penetrating action of the fungal hyphae, as Blake et al.[8,9] demonstrated in their work with electrical wire insulated with various elastomers and plastics.

Some of the criteria now used in evaluating degradation are oxygen consumption, carbon dioxide production, electrical and physical tests, and multiplication of microorganisms.

3. Macroorganisms

Several groups of macroorganisms are responsible for the degradation of high-molecular-weight organic materials. As we have previously noted, among these groups and worthy of special attention are the rodents, the insects, and the marine borers, which cause tremendous economic losses by their attacks on cellulosic and plastic structures.

a. Rodents. Rodents, particularly rats, mice, pocket gophers, and squirrels, are responsible for most of the damage to material, equipment, and installations. They vary in size and physical appearance, but all possess chisel-like incisor teeth that are especially suited for gnawing in quest of food or shelter. Evidence of their presence usually consists of shavings, shredded material, or the rodents' characteristic tooth marks. Gnawing is vital for their existence because it maintains their ever-growing incisors to lengths necessary for everyday existence. Rodents are the largest order of mammals, comprising more than 2000 species; with very few exceptions—such as house rats and muskrats—they are exclusively herbivorous, subsisting on grains, nuts, barks, and so on.

Structures encased in plastic are vulnerable to rodent attack. Tree squirrels are the culprits associated with damage to overhead cables and pocket gophers are believed responsible for damage to underground cables. Such damage invariably paves the way for secondary deteriorative effects by exposing wires and metallic shields, resulting in short circuits or corrosion. Despite popular belief, there is no evidence that rodents derive any nutritional benefits from plastics. A more plausible explanation for these attacks is that rodents, especially the burrowing type, express their curiosity about unknown structures by gnawing at these potential food sources (or obstructions to their normal habits). Cables newly buried in soils infested with burrowing rodents are more prone to attack than those that have been buried for a number of years. The incidence of attacks is related to soil compactness and to the number of burrowing tunnels that are bisected during cable-laying operations. The greater the soil compaction, the lower the incidence of attack.[36]

b. Insects. Insects represent the largest group of living organisms known to man. More than 200,000 kinds have been described, and entomologists believe that the total number of species will be in excess of six million when all are discovered. Nutritional requirements vary considerably within such a large class, but in general insects eat almost all kinds of vegetables and animal matter, dead or alive. They are equally voracious in attacking materials of vegetable or animal origin. Beetles and termites are some of the more injurious members of this group, and their activities portend serious economic problems.

The powder-post beetles derive their name because they attack hard, dry wood, tunneling through timber in successive generations until the interior is completely reduced to fine powder. One species, *Scobicia declivis*, was given the name lead cable borer because of its habit of eating holes through the lead sheathing of aerial telephone cables. In southern California this type of insect was reported to have made 125 holes in a 100-ft span of cable.

Of the soil-inhabiting termites, the subterranean termites, represented by the species *Reticulitermes*, is considered the most deleterious type. They attack wood that is in or on the soil and characteristically excavate channels as they follow the grain. The net effect is insidious, because the extent of damage often goes undetected until too late. Termites are capable of attacking plastics and are particularly attracted to plastics that contain cellulosic fillers. Laminated wood-flour phenolics belong to this class; these materials are extremely susceptible and are more vulnerable as the percentage content of the filler increases. Since it appears that food is the motivating factor, it is not clear why poly(vinyl chloride) and polyethylene are attacked. Plastics are not metabolized by insects but are probably damaged during their search for food. Visual observations of termite damage to plastics differs markedly from the type of damage observed in wood. Poly(vinyl chloride) and polyethylene appear abraded, and under the microscope tiny tufts of plastic can be observed. On soft, malleable plastic materials, damage is apt to occur anywhere regardless of the geometry of the structure. On rigid plastic materials, however, only the edges appear to be vulnerable.

c. Marine Borers. In the ocean environment, the marine borers are the principal biological agents that degrade structures made of wood. Marine borers are primarily mollusks or crustaceans and have the capacity to excavate hard materials—particularly rock, wood, or shell—in which they find shelter and protection.

The molluscan borers *Teredo* and *Bankie*, popularly known as shipworms, have long wormlike bodies with only the head in a bivalve shell, which is used as a cutting tool. They enter wood during their embryonic stage leaving only a pinhole in the surface as a clue, but once established they burrow quickly, enlarging the tunnel as they increase in size. Burrowing clams and pholads belong to another important family of boring mollusks. These animals are capable of boring into rock and shale in search for shelter. Teredos[37–40] and pholads[41,42] are reported to have attacked polyethylene and lead-sheathed submarine cables.

Limnoria, the gribble, is a crustacean borer with a reputation for causing widespread destruction to wood. Activity against submarine cable has also been reported. Chilton[43] reports the presence of *Limnoria* in the splice of a submarine cable at about 60 fathoms off the New Zealand coast. *Limnoria* was responsible for the failure of the Holyhead–Dublin cable in 1875, and Jona[44] frequently found *Limnoria* in cables recovered from the Adriatic Sea.

C. EFFECT OF ENVIRONMENTAL FACTORS ON BIOLOGICAL DEGRADATION OF POLYMERS

Environmental factors affect the degradation of polymers, and some of these factors have a profound influence on the survival and growth of biological life. In many instances it is difficult to determine whether biological activity or the environment is responsible for the deterioration of materials exposed outside a laboratory-type environment. For example, ester-linked polyurethanes are considered to be highly susceptible to fungal attack.[20,21] However, it is also known that water is able to hydrolyze the ester linkages. In a similar vein, Dickenson[45] believes the deterioration of rubber proceeds primarily through a nonbiological oxidation step and that microorganisms attack the oxidation products.

For the purpose of discussion it is convenient to segregate the various environmental factors that affect biological life. Not to be overlooked, however, are the interrelations among them. The chemical and physical factors that manifest themselves in the laboratory are present in the natural environment, and it is the net result of these interrelations that has a profound effect on biological activity.

1. Water

The association of water and microorganisms in relation to the degradation of natural and synthetic polymers is well known. Bacteria and some lower forms of fungi are dependent upon an aqueous environment for growth, whereas higher fungi can develop in the presence of high moisture levels, usually 95% or more. Fungi tolerant of low moisture levels (85–90%) are also tolerant of high osmotic pressures.

The hydroscopic nature of an insoluble substrate often determines its susceptibility to microbial degradation. Wood is considered a hydroscopic material and subject to fungal attack, but if it contains less than 20% moisture it is rarely attacked by fungi. On the other hand, the hydroscopic nature of nylon does not render it susceptible to microbial attack. Quite the contrary, nylon is considered highly resistant to microbial attack. In arid environments where evaporation exceeds the amount of precipitation, microbiological degradation is almost nonexistent.

High moisture levels in a soil environment are associated with low oxygen tension, and these conditions favor the development of anaerobic bacteria. The importance of anaerobes to the deterioration of materials has received very little attention. We know that under anaerobic conditions plant residues accumulate, indicating that microbial decomposition proceeds at a slower rate, possibly because of the limited kinds of bacteria present.

2. Temperature

Microorganisms have narrow optimum temperature requirements for growth and deviations from these optima result in a reduction in their metabolic rate and, consequently, in their growth. Generally speaking, fungi are considered to have temperature optima between 20 and 28°C[46] whereas bacteria grow best between 28 and 37°C. Some microorganisms grow best at extreme temperature ranges. Microorganisms whose optima lie below 20°C are known as psychrophiles, microorganisms whose optima lie above 45°C are known as thermophiles, and the rest are considered mesophiles.

Since microorganisms consist largely of water, the minimum temperature at which they can grow is set by the temperature at which water freezes. In this connection metabolism ceases, but death does not necessarily ensue; many microorganisms can survive in a frozen state for long periods of time. The maximum temperature an organism can survive is determined by the thermal stability of its essential constituents, proteins, and nucleic acids. There is evidence to suggest that thermophiles can grow at elevated temperatures because their enzymes are more heat stable than the enzymes of mesophiles.

Deterioration of susceptible materials proceeds best at so called "moderate" temperatures (15–40°C). However, microbiological deterioration can occur at extreme temperatures. At ocean-bottom temperatures (5°C) elastomers[16,47] are slowly degraded.

3. pH Value

Most microorganisms can grow within a broad pH span, but they develop best at optimized reaction levels. Fungi are considered to be more tolerant of an acid pH than are bacteria. Fungi grow well at pH values between 4 and 7. Some, however, such as the dermatophytes, grow abundantly at pH 9. Bacteria in general grow best at slightly alkaline reaction levels (7.4–8.5), but a few, such as the sulfur-oxidizing group, *Thiobacillus*, prefer to grow at pH values close to 2. One of the theories offered to explain the effect of pH on the microbial cell concerns the cell wall, which influences what goes in or out of the cell because it is amphoteric and because it is altered by pH. The ability of *Thiobacillus* to tolerate acid environments is ascribed to this phenomenon. Some fungi are tolerant of heavy metals at low pH values but succumb to metal poisoning as the reaction level approaches neutrality.[48] It is well known that many biologically active substances have almost no biological effect in the nonionized state.[49] However, by shifting the ratio of neutral molecules to the ionized state, which is a function of the ionization constant and the pH value of the environment, cidal or at least static activity may be obtained. The action of Crystal Violet against certain forms of bacteria is typical of this type of action. The explanation having the widest acceptance is that only the ionized form is able to enter the cell under these circumstances.

4. Oxygen

The availability of molecular oxygen to microorganisms is of special significance because most reports on the degradation of high-molecular-weight organic materials deal with aerobic forms. Filamentous fungi, with few questionable exceptions, are considered strict aerobes and do well in environments where there is free movement of air. Under submerged cultural conditions in the laboratory, air must be supplied to the liquid medium at a rate sufficient to maintain the appropriate level of oxygen needed by the fungus.

The three classifications that bacteria fit into, with respect to oxygen are aerobic, microaerophilic, and anaerobic. Obviously the amount of oxygen present serves only to limit the kinds of bacteria that can develop. In the aerobic process molecular oxygen serves as a hydrogen and electron acceptor and in the process is reduced to water.

In a general sense, anaerobes cannot utilize molecular oxygen as an electron acceptor. In its place other oxidized compounds serve as electron acceptors. Sulfate serves this purpose for sulfate-reducing bacteria, as does nitrate with nitrate reducers. Sulfate-reducing bacteria are an economically important group because of their ability to grow on complex organic substrates and because they serve as depolarizing agents during the corrosion of steel. Foul odors are associated with their activities, since hydrogen sulfide is produced from the sulfate. Bacteria considered to be microaerophilic require only minimal amounts of oxygen, but they require more than normal levels of carbon dioxide, too, which is of even greater significance.

D. MECHANISMS OF DEGRADATION

The microbial degradation of susceptible materials is accomplished through the action of biosynthetically produced proteins called enzymes. They may be located on the cell wall or within the protoplasmic structure of the microbial cell. Some enzymes are secreted into the surrounding environment; others are retained within the cell and are released only when the cell is lysed or mechanically disrupted. They are highly specific in their ability to catalyze biochemical reactions, which proceed rapidly under mild physiological conditions.

Organisms capable of utilizing substances of high molecular weight do so by two probable means. One involves the breakdown of the substrate into assimilable molecules by liberating enzymes into the surrounding medium. It is also believed that there must be close contact between the organism and the high-molecular-weight substance to enable enzymes on the cell surface to function.

Organisms require a minimum number of enzymes in order to grow or maintain themselves. These are always present, irrespective of the surrounding environment. These enzyme systems are known as constitutive enzymes. Many organisms have the potential to synthesize enzymes in response to specific compounds. These enzymes are known as inducible enzymes, and the substrate is called the inducer. In the absence of the substrate, the enzymes are usually absent or present in immeasurable amounts. It is highly probably that susceptible high-molecular-weight materials are broken down by inducible enzymes.

1. Cellulose

Native cotton cellulose is composed of very long, unbranched chains, DP > 8000, of *D*-gluco-pyranose linked 1–4β with a molecular weight in excess of one million. It is a highly crystalline structure with the bundles of chains held together by hydrogen bonds and van der Waals forces to form microfibrils about 50 Å in diameter.

Many microorganisms, both aerobic and anaerobic, fungi and bacteria, are capable of breaking down cellulose. It was originally thought that the breakdown of cellulose was mediated by the action of two enzymes. The first would be an extracellular enzyme that would cleave cellulose as far as the 1–4β-linked disaccharide, cellobiose. Then a second enzyme would hydrolyze the cellobiose to glucose. Mandels and Reese[50] isolated these components from the cellulase of *Trichoderma viride* and called the former C_1 and the latter C_x. They found that C_1 could solubilize cotton fibers and make it accessible to C_x for further enzyme action. C_x was found to be inactive against cotton fibers, but was active against carboxymethylcellulose. Fortunately, improvements in separation techniques, ushered in by the development of gel filtration, paved the way for the work by Selby.[51,52] He reported that component C_x was actually composed to two enzymes, one a carboxymethyl cellulase and the other a cellobiase. When these two enzymes were recombined with the C_1 fraction, complete activity was restored toward cotton. In summary, he found that (*a*) C_1 is almost inactive in the absence of C_x, (*b*) there is a lack of synergism between components of C_x, and (*c*) all three enzymes must be present to account for activity toward cotton.

As mentioned earlier, the degree of substitution and the location of the substituents have a direct bearing on the resistance of cellulose. Reese[53] suggested that enzymic hydrolysis occurs where two adjacent unsubstituted glucosyl residues are present. Wirick[54] and Eriksson[55] suggested that three or more adjacent unsubstituted glucosyl residues are necessary for enzyme attack. Klop and Kooiman[56] found that the glycosidic bond between two glucosyl residues could be hydrolyzed if the aglycone group was unsubstituted or 6-substituted, but not if it was substituted at the *C*-2 or *C*-3 position.

2. Rubbers and Elastomers

Although many communications appear concerning the microbiological breakdown of natural rubber, there are no readily available reports on the mechanisms involved in its breakdown. There is no "rubberase," as one might expect for a natural product that has had extensive commercial usage. One reason may be that natural rubber is really quite resistant in comparison to say, cellulose, and relatively long-term experiments are necessary to elucidate the mechanism of attack. Another reason might be that shortages of natural rubber during World War II made it imperative to find substitutes, which have replaced it in a number of commercial items, thereby minimizing its importance.

Recently Nickerson[57] isolated a black yeastlike fungus and pink bacterium from percolator studies using vulcanized natural rubber or synthetic polyisoprene rubber as organic substrates. He observed pitting and penetration of the rubber surface, which becomes finely dispersed into coacervate globules. His opinion is that the exocellular enzymes are bound on these globules. Cell-free preparations caused extensive pitting on vulcanized rubber after a 24-hour incubation period. The limiting factor appeared to be the particle size of rubber.

There are some general observations that may be applicable and pertinent in the field of rubber degradation. Natural rubber is acknowledged to be susceptible to oxidation. It is believed that microorganisms attack the oxidation products at the end of the molecular chain (Chapter 2).

Paraffinic hydrocarbons are attacked more readily and by more species than cyclic compounds.[1] Hydrocarbons having double bonds seem to be more susceptible to microbial oxidation than saturated hydrocarbons. Normal or straight-chain compounds are more resistant than iso- or branched-chain hydrocarbons. In a recent review, van der Linden and Thijsse[58] refuted the contention that double-bonded hydrocarbons are more susceptible than their saturated counterparts. The opposite may be true. With bacteria, it was found that olefins are attacked at the saturated end of the molecule. In addition, the terminal double bonds can be oxidized. In general, since enzymes oxidizing the aliphatic hydrocarbons are not very specific with respect to molecular configuration, byproducts and alternate pathways are possible.

The many facets of processing the polyurethanes make it difficult to assess them from a microbiological point of view. Ester-linked urethanes are seemingly vulnerable to esterases, enzymes that are common among microorganisms, but there are no reports yet of studies with cell-free systems to support such contentions. Ether-linked urethanes were found[21] to be significantly more resistant. The structure of ether-linked urethanes, found to be moderately susceptible, suggests that enzymatic attack occurs only if there is a sufficiently long unbranched carbon chain length between the urethane linkages of the polymer. It is also believed that no appreciable attack will occur without three adjacent methylene groups; the presence of fewer than three groups results in a more resistant material. Shorter alkane diols and branched diols yielded resistant materials, possibly because they prevent enzymes from gaining access to susceptible groups.

3. Vinyls

As a result of early studies[24,27] on the microbial deterioration of plasticizers and resins in a mineral medium, a microbiological viewpoint has evolved which is used as a guide by the plastics industry in selecting plasticizers for vinyl formulations. It is assumed that resins are microbiologically inert, and plastic formulations containing susceptible plasticizers are presumed to be susceptible to biodegradation, whereas nondegradable plasticizers contribute toward the resistance of vinyl formulations.

Loss of plasticizer by physical, chemical, or biological means, results in undesirable changes in physical properties of the material, and in time these values will approach those attributable to semirigid materials.

Most plasticizers are susceptible to microbial degradations, especially if a portion of the plasticizer is made from a natural product. Lack of agreement in the literature on what is a susceptible or resistant plasticizer can probably be attributed to the improper use of test organisms, which can be easily obtained by enrichment culture techniques. Many plasticizers that were originally considered to be inert have been found subsequently to be biodegradeable. The commonly used plasticizers are tricresyl phosphate, the phthalates, nitrile rubber, and polyesters. Four-year soil-burial studies reveal that nonmigratory plasticizers performed best, and these include tricresyl phosphate, polyester, nitrile rubber and dipentaerythritol ester.[32] Phthalate esters were generally attacked, with branched-chain types performing better than plasticizers prepared from normal alcohols. Flexibility is believed to be lost through the enzymatic degradation of the plasticizer at the soil–plastic interface. Diffusion of the plasticizer to the surface appears to be the rate-limiting reaction. Yeager[26] believes that formulations containing antimicrobial agents are rendered ineffective because of improvements in the resin systems that do not allow for sufficient migration of the plasticizer and the antimicrobial systems to the surface. However, it appears that there would be less need for the use of antimicrobial agents if the migration of the plasticizers were kept to a minimum.

In laboratory studies, Williams et al.[59] found that acetone powders of fungi could break down the plasticizer dibutyl sebacate to sebacic acid. He reported that vinyl films plasticized with dibutyl sebacate were attacked by acetone preparations of enzymes in a similar manner. He attributed this activity to the presence of esterases that could be inactivated by copper sulfate, sodium fluoride, and heat. Esterases from horse liver were equally effective. In another communication Klausmeier[60] reported that a *Fusarium sp*. utilized the monobutyl ester of dibutyl phthalate but could not utilize the monobutyl phthalate. In studies not published, the author found a pseudomonad-type bacterium that was capable of utilizing *n*-octyl-*n*-decyl-*o*-phthalate completely as the sole source of carbon. Phthalic acid, which remains after the alcohol moieties have been split off, is readily metabolized via the protocatechuic acid pathway.[61,62]

E. PREVENTION OF DEGRADATION

1. Theoretical Considerations

In the preceding sections degradation of polymers by biological life was discussed. In the case of microorganisms the degradation process resulted in nutrients for the organisms, although this is not always a prerequisite for attack. Forms of higher life other than termites receive little or no nutritional benefit from their attacks on synthetic polymers. Nevertheless, by examining all the factors that enter into the resistance or susceptibility of a polymer or material, it may be possible to shift the balance in favor of a more resistant form.

a. Microorganisms. The resistance of insoluble polymers, or high-molecular-weight organic materials to biological degradation may be caused by one or more factors. For microorganisms it may mean that they lack the appropriate enzymes required to solubilize long molecular chains into assimilable sizes. Good performance of materials exposed to soil burial for many years is a positive sign that microorganisms lack these enzymes, providing of course that the soil is reasonably fertile and contains a normal complement of microorganisms. It also means that they are resistant to enzymes or byproducts that are produced as a result of growth on some other nutritive source. There are those who feel that as part of the evolutionary process of natural selection and mutation it is but a matter of time before microorganisms have the capabilities to break down materials synthesized by man. They point to cellulose, one of nature's oldest polymers, which is attacked by many bacteria, both aerobic and anaerobic, and by a great many fungi. Lignin, on the other hand, perhaps nature's newest polymer, is attacked only aerobically by higher fungi and accumulates under anaerobic conditions.

Microorganisms depend on the biocatalytic action of enzymes to maintain their living processes; enzymes are the biological entities involved in the breakdown of polymers. Antimicrobial agents may function as enzyme inhibitors, neutralizers, or poisons, and they may disrupt the integrity of microbial cells.

Enzymes may be classified according to their location in the cell or conditions governing their synthesis. Enzymes found in the interior of the cells are known as intracellular or endocellular enzymes. Ectocellular enzymes are found on the surface of the cell and should be distinguished from extracellular enzymes, which are released by the cell and accumulate in the surrounding environment. Constitutive enzymes are always present in the cell and do not depend upon the presence of their intended substrate for synthesis. Induced enzymes, however, are only synthesized in response to a specific substrate. Enzymes may also have a nonprotein portion that is indispensable for activity. If the nonprotein portion is tightly bound, it is called a coenzyme. Many enzymes also require a metallic ion for activity. In such cases they are known as cofactors.

The conformation of an enzyme is a function of its amino acid sequence, the manner in which the polypeptide chains are covalently crosslinked (usually because of the involvement of the S–S bonds of cystine), and the folding of the peptide chain into a three-dimensional configuration. The latter, which is essential for biological function, is the feature usually lost during denaturation. There is a good deal of evidence to suggest that many enzymes are multisited and are therefore capable of accepting more than one molecule of substrate. Each active site may be composed of one polypeptide chain or of more than one. Hemoglobin, for example, consists of four polypeptide chains; myoglobin contains a single polypeptide chain.

Prevention of microbial deterioration of susceptible materials is best accomplished by having a basic understanding of the metabolic processes of microorganisms and the physical and chemical factors that affect these processes.

b. Higher Forms of Life. No such generalizations as those described for microorganisms can be made for higher forms of biological life. The reasons for the attack by animals on organic polymers are obscure. Some feel that these materials represent obstacles during the animals' search for food. It is known that some rodents gnaw as a biological necessity for self-preservation. Whatever the reasons, they are of little consolation to the materials engineer, who must be concerned with preventive measures or face the dire consequences of material failure. He must, if at all possible, prevent animals from coming into contact with his material or structure. This may be accomplished by the use of repellents either incorporated into the material or placed in the surrounding environment. Should this fail, he can engineer into his structure a barrier, preferably of metal, that would discourage even the most persistent species. A third possibility would be to design the geometry of his structure so that the physical ability of animals to attack it would be seriously limited.

2. Chemical Modifications and Inhibitors

a. Chemical Modificaitons. In the previous sections, the importance of chemical groupings was discussed with respect to microbial resistance. Then the specificity of enzyme conformation with respect to activity was noted. The basic objectives of chemical modifications should be directed toward fabricating a polymer or formulation or substituents within a formulation embodying one or both factors. As a consequence, microbial resistance would be imparted as a result of chemical or stereochemical considerations rather than through the use of antimicrobial agents. Inherently resistant chemically modified materials will remain so until microorganisms undergo evolutionary changes and are able to synthesize new enzymes.

There is much to be learned from the wealth of information that has been published on the chemical modifications of cellulose. As mentioned earlier, the nature of the substituent, the degree of substitution, and the orientation of the fiber all play vital roles in the resistance of modified cellulose. Substituents that increase hydrophobicity of the polymer or encourage cross-linking are desirable. The crystallinity of a polymer is more than casually related to resistance. As a general rule, the more crystalline the polymer, the more difficult it is to break down. Polymers or formulations that remain susceptible despite chemical modification must be protected by inhibitors.

b. Inhibitors. There are many desirable properties that all good inhibitors should have. Some of these may be sacrificed if the intended use of the material and the expected service life are known, for no one inhibitor is expected to possess all these characteristics.

Inhibitors should be effective against a wide variety of microorganisms at low concentrations, but they must be nontoxic to man. They ought to be readily available and economical, but not affected by the elements. They must also be compatible with materials they come into contact with and should not contribute any characteristic odor or color or change any of the physical properties of the material they are protecting. Finally, they must be slightly soluble in an aqueous environment in order for the toxic inhibitor to act.

The chemical reactivity of the inhibitor lies in its capacity to interfere with the vital reactions of the cell. Apart from surface-active agents, which disrupt the semipermeable nature of the cell membrane, most inhibitors are effective as enzyme inhibitors. They can be classified as (*a*) competitive inhibitors, which are structurally related to the enzyme substrate and therefore compete for its active site and (*b*) noncompetitive inhibitors, which do not necessarily compete with the substrate for the active site but prevent catalytic action by the enzyme.

Inhibitors are used indiscriminately, without consideration for their mode of action. It can be assumed, however, from data obtained from classical enzyme studies, that heavy metals will poison sulfhydryl groups of enzymes. Likewise, sulfur compounds and many other organic compounds with sequestering action may deactivate metal-containing enzymes.

Toxic inhibitors are more than adequately reviewed by Sui[4] and Greathouse and Wessel[64] for cellulosics and by Baseman[65] for plastics. It is beyond scope of this presentation to discuss these compounds in any detail. Some of the larger classes of inhibitors frequently associated with materials include the metallic ions such as mercury, copper, and arsenic either alone or attached to some organic molecule, the phenolics, the quaternary ammonium compounds, the sulfur-containing compounds, and the organo-tins.

The arduous task of experimental testing with a selected number of inhibitors is the best method for choosing the appropriate inhibitor. Through the judicious use of antimicrobial agents, the service life of a susceptible material may be extended beyond what was reasonably thought possible.

REFERENCES

1. C. E. Zobell, *Enzymol*, **10,** 443 (1964).
2. C. E. Zobell, C. W. Grant, and H. F. Haas, *Bull. Amer. Assoc. Petrol. Geol.*, **27,** 1175 (1943).
3. L. D. Bushnell and H. F. Haas, *J. Bacteriol.*, **41,** 653 (1941).
4. R. G. H. Sui, *Microbial Decomposition of Cellulose*, Reinhold, New York (1951).
5. E. B. G. Jones, "The Distribution of Marine Fungi on Wood Submerged in the Sea," in A. H. Walters and J. J. Elphick, Eds., *Biodeterioration of Materials*, Elsevier, New York, 1968, pp. 460–485.
6. S. P. Meyers, "Degradative Activities of Filamentous Marine Fungi," in A. H. Walters and J. J. Elphick, Eds., *Biodeterioration of Materials*, Elsevier, New York, 1968, pp. 594–609.
7. W. M. Heap and S. H. Morrell, *J. Appl. Chem.*, **18,** 189 (1968).
8. J. T. Blake and D. W. Kitchin, *Ind. Eng. Chem.*, **41,** 1633 (1949).
9. J. T. Blake, D. W. Kitchin, and O. S. Pratt, *AIEE Trans.*, **69,** 748 (1950).
10. J. T. Blake, D. W. Kitchin, and O. S. Pratt, *Appl. Microbiol.*, **3,** 35 (1955).
11. A. C. Thaysen, H. J. Bunker, and M. E. Adams, *Nature*, **155,** 322 (1945).
12. T. Curren, *Can. J. Microbiol.*, **15,** 1241 (1969).
13. J. J. Noval and W. J. Nickerson, *J. Bacteriol.*, **77,** 251 (1959).
14. A. J. E. Barlow and F. W. Chattaway, *J. Invest. Dermatol.*, **24,** 65 (1955).
15. C. G. C. Chesters and G. E. Mathison, "The Decomposition of Wool Keratin by *Keratinomyces ajelloi*," *Sabouraudia*, **2,** 225–237 (1963).
16. W. Coscarelli, "Deterioration of Organic Materials by Marine Organisms," in H. Heukelekian and N. C. Dondero, Eds., *Principles and Applications in Aquatic Microbiology*, Wiley, New York, 1964, pp. 113–145.
17. J. T. Blake, D. W. Kitchin, and S. O. Pratt, "The Microbiological Deterioration of Rubber Insulation," *AIEE* Tech. paper, 53-TP-59 (1952).
18. C. Bennicelli, "Report of Investigation of Fungus Resistance of Some Raw and Compounded Vulcanized, Natural and Synthetic Elastomers," U.S. Naval Shipyard, Brooklyn, Materials Laboratory, Project 5129-4, Prog. Rept. 1 (1957).
19. F. H. Allen, *Ind. Eng. Chem.*, **45,** 374 (1953).
20. A. M. Kaplan, R. T. Darby, M. Greenberger and M. K. Rogers, *Develop. Ind. Microbiol.*, **9,** 201 (1968).
21. R. T. Darby and A. M. Kaplan, *Appl. Microbiol.*, **16,** 900 (1968).

22. H. G. Hedrick, *Develop. Ind. Microbiol.*, **10,** 222 (1969).
23. F. E. Kulman, *Corrosion*, **14,** 23 (1958).
24. United Kingdom Ministry of Supply, *Rept. Plastics in Tropics*, **2,** "Polythene," 18 pp. (1953).
25. L. R. Snoke, *Bell Syst. Tech. J.*, **36,** 1095 (1957).
26. C. C. Yeager, *Develop. Ind. Microbiol.*, **9,** 222 (1968).
27. S. Berk, H. Ebert, and L. Teitell, *Ind. Eng. Chem.*, **49,** 1115 (1951).
28. W. H. Stahl and H. Pessen, *Appl. Microbiol.*, **1,** 30 (1953).
29. G. H. Booth and J. A. Robb, *J. Appl. Chem.*, **18,** 194 (1968).
30. J. V. Harvey and F. A. Meloro, "Studies on the Degradation of Plastic Films by Fungi and Bacteria," *U.S. Quartermaster Depot, Phil., Microbiol. Ser.*, Rept. 16 (1949).
31. R. E. Klausmeier, "The Effect of Extraneous Nutrients on the Biodeterioration of Plastics," *S.C.I. Monogr. No. 23, Microbiological Deterioration in the Tropics*, pp. 232–243 (1966).
32. J. B. DeCoste, "Soil Burial Resistance of Vinyl Chloride Plastics," presented at American Chemical Society 155th National Meeting, San Francisco (1968).
33. H. J. Hueck, *Plastica*, **12,** 24 (1959).
34. W. Pöge, *Plaste Kautschuk*, **8,** 74 (1961).
35. S. H. Ross, "Biocides for a Strippable Vinyl Plastic Barrier Material," Rept., R-1396, U.S. Ordnance Corps Proj. TB4-006E, 23 pp. (1957).
36. A. M. Popon, *Vestn. Svyazi*, **19,** 37 (1959).
37. W. F. Clapp and R. Kenk, "Marine Borers: A Preliminary Bibliography, Parts I and II," Library of Congress, Techn. Inf. Div., Washington, D.C. (1956).
38. W. T. Henley, "Notes on Everyday Cable Problems: Distribution of Electricity," *W. T. Henley's Telegraphic Works Co. Ltd.* (*London*), **8,** 1896–1898 (1935).
39. C. S. Lawton, *AIEE Trans.*, **78,** 5 (1959).
40. B. C. Heezen, M. Ewing, and G. L. Johnson, "Cable Failures in the Gulf of Corinth: A Case History," Lamont Geol. Observ., Palisades, N.Y. 1960.
41. P. Bartsch and H. A. Rehder, "The West Atlantic Boring Molluscs of the Genus *Matersia*," *Misc. Coll. 104 Smithsonian Inst.*, **2,** Washington, D.C. (1945).
42. L. R. Snoke and A. P. Richards, *Science*, **124** (1956).
43. C. Chilton, *Ann. Mag. Nat. Hist.*, **18** (1916).
44. E. Jona, *Atti Soc. Ital. Progr. Sci.*, **6,** 263 (1913).
45. P. B. Dickenson, *Rubber J.*, **147,** 54 (1965).
46. D. L. Strider and N. N. Winstead, *Phytopathology*, **50,** 583 (1960).
47. P. L. Steinberg, *Bell Syst. Tech. J.*, **40,** 1369 (1961).
48. R. L. Starkey and S. A. Waksman, *J. Bacteriol.*, **45,** 509 (1943).
49. A. Albert, *Selective Toxicity*, Wiley, New York, 1960.
50. M. Mandels and E. T. Reese, *Develop. Ind. Microbiol.*, **5,** 5 (1964).
51. K. Selby and C. C. Maitland, *Arch. Biochem. Biophys.*, **118,** 254 (1967).
52. K. Selby, "Mechanism of Biodegradation of Cellulose," in A. H. Walters and J. J. Elphick, Eds, *Biodeterioration of Materials*, Elsevier, New York, 1968, pp. 62–78.
53. E. T. Reese, *Ind. Eng. Chem.*, **49,** 89 (1957).
54. M. G. Wirick, *J. Polymer Sci.*, **6,** 1965 (1968).
55. K. E. Eriksson and B. H. Hollmark, *Arch. Biochem. Biophys.*, **133,** 233 (1969).
56. W. Klop and P. Kooiman, *Biochim. Biophys. Acta*, **99,** 102 (1965).
57. W. J. Nickerson, "Microbiol Transformations of Natural Occurring Polymers," in *Fermentation Advances*, Academic Press, New York, 1969, pp. 631–634.
58. A. C. van der Linden and G. J. E. Thijsse, *Advan. Enzymol.*, **27,** 469 (1965).
59. P. L. Williams, J. L. Kanzig, and R. E. Klausmeier, *Develop. Ind. Microbiol.*, **10,** 177 (1969).

60. R. E. Klausmeier and W. A. Jones, *Develop. Ind. Microbiol.*, **2,** 47 (1961).
61. D. T. Gibson, *Science*, **161,** 1093 (1968).
62. D. W. Ribbons and W. C. Evans, *Biochem. J.*, **26,** 310 (1960).
63. W. H. Stahl and H. Pessen, *Mod. Plastics*, **31,** 111 (1954).
64. G. A. Greathouse and C. Wessel, *Deterioration of Materials—Causes and Preventive Techniques*, Reinhold, New York, 1954.
65. A. L. Baseman, *PLASTECH* (Plastics Tech. Evaluation Center), **12,** 33 (1965).

10

METHODS FOR MEASURING STABILIZER EFFECTIVENESS

W. LINCOLN HAWKINS

Bell Telephone Laboratories, Incorporated, Murray Hill, New Jersey

A. INTRODUCTION

Despite the large number and variety of tests developed to determine the effectiveness of stabilizers in the protection of polymers, unequivocal results are obtained only by field trial. That is, we cannot be certain that a stabilized polymer will perform adequately as part of a component or system until after the complete design, or a representative part of it, has been fully tested in service. Nonetheless, many useful tests have been developed to measure the effectiveness of polymer compositions. Correctly interpreted, data derived from these tests correlate reasonably well with results obtained by field trials. Within the limitations discussed later, some of these tests can be used to derive an approximate prediction of the time to failure.

For obvious reasons, it is seldom practical to evaluate stabilized polymer compositions by field trial. In addition to the problems often inherent in fabricating the complete article or device, the time required to develop data for even moderately stable compositions can be prohibitive. Therefore, considerable effort has been directed to the development of accelerated tests and to the difficult problem of predicting failure from the data obtained. Field trials are as diverse as the applications in which polymers are used, taking into account every factor in the design that affects stability of the material. Individual consideration of these factors is beyond the scope of this chapter. The following sections deal with accelerated tests and are oriented primarily to considerations of material rather than design failure.

Accelerated tests usually reduce the time to failure by intensifying the exposure environment. Failure is then determined by following the change in a selected property of the material. As a general principle, reliability of accelerated tests decreases as the deviation from the service environment increases and as the property measured becomes less related to failure.

Polymers are exposed to complex environments, consisting of chemical reactants capable of producing an irreversible change in the polymer and energy sources that promote these reactions. An environment is seldom constant, since both the concentration of individual reactants and the intensity of energy sources are subject to considerable variation. In many applications, stabilized polymers are exposed out of doors, and thus are subject to the effects of the weather with its many vagaries. In other applications, materials are designed for indoor use, which may result in exposure to unusual energy sources or to chemical reactants not encountered under outdoor exposure conditions. Even initial degradation products become part of the exposure environment to the extent that they inhibit or promote further reaction. Detailed descriptions of specific environments are included in the following sections of this chapter as they relate to the corresponding accelerated test procedures.

The criterion for failure must be established for each individual application, since different properties do not always change at the same rate.[1,2] Because deterioration is usually greater near the surface, the properties related to the surface are more sensitive to change than the properties associated with the bulk of the material. Ideally, the property measured should be the one responsible for failure. In practice, however, other properties are often measured because more sensitive analytical procedures are available. It then becomes necessary to establish the relation between the property measured and the property actually responsible for failure. For example, measurement of the rate of chemical reactions leading to deterioration in polymers is adaptable to established procedures. But to interpret these reaction rates as a direct prediction of the failure resulting from changes in mechanical, dielectric, and other properties could lead to erroneous conclusions.

It is not the purpose of this chapter to discuss in detail the many tests employed to determine stabilizer effectiveness, but rather to present a critique of the various appoaches to accelerated testing, developing the limitations of each while at the same time emphasizing how various tests can be applied to generate useful data. The environmental factors intensified to achieve acceleration are used as the basis for classification and discussion of test methods. Details of individual tests can be obtained through the appropriate references.

B. WEATHERABILITY TESTS

Outdoor weathering tests are used extensively to measure the stability of polymers, but these tests are invariably slow and the data obtained are subject to the inconsistency of the weather. Although some degree of acceleration is possible through selection of the exposure site, most accelerated tests for stabilized polymers are made in artificial weathering devices. However, the complex phenomenon of natural weather cannot be duplicated accurately nor truly accelerated in laboratory tests. Ultraviolet radiation, heat, rainfall, and atmospheric contamination are all contributing factors in outdoor weathering. Each of these variables can be controlled and intensified in the laboratory, but many indoor weatherability tests are accelerated simply by using an artificial source of ultraviolet radiation, maintained at constant intensity throughout the test. It is evident that unless all other factors that contribute significantly to failure of the material out of doors are incorporated into the artificial weathering test, accurate correlation with natural weathering results cannot be expected. Even the most carefully designed laboratory test cannot be expected to correlate precisely with outdoor weathering results that are themselves subject to uncontrollable variations.

1. The Outdoor Weathering Environment

Various chemical reactions take place in polymers during outdoor exposure, and many of these contribute to failure. In addition to the ubiquitous oxygen and water, ozone and various atmospheric contaminants are often present, particularly in industrial areas. Although oxidation and hydrolysis are the most common reactions contributing to failure, other reactions are important in the deterioration of specific polymers, as for example, the reaction of stressed rubber with ozone. In addition, atmospheric contaminants can have a catalytic effect on many chemical reactions. For example, the increased rate of hydrolysis of some polymers is attributed to the presence of trace amounts of acids or bases (Chapter 8). The weathering of other polymers is accelerated by mechanical stress,[3] and microorganisms contribute to the deterioration of certain polymers (Chapter 9). When any of these factors is important in determining the rate of deterioration, correlation among results obtained in their absence will be erratic. Many factors must therefore be considered in attempting to predict the service life of polymers from accelerated tests.

Sunlight is the primary energy source in the outdoor environment. Its ultraviolet component is the primary initiator of photochemical reactions, and its infrared component initiates thermally induced reactions. Often photochemical and thermal reactions take place simultaneously. The ultraviolet region between 3000 and 4000 Å, commonly referred to as the near-ultraviolet region, is responsible for the deterioration of many polymers. This region represents only about 5 to 7% of the total energy emitted by the sun,[4] and, as a result of scatter and absorption by the atmosphere, only a fraction of this radiation reaches the earth's surface. Ultraviolet energy is largely absorbed by ozone[5] and its intensity is considerably more sensitive to variable conditions in the atmosphere than is the total solar radiation. Figure 10.1 shows the daily fluctuation in ultraviolet energy in contrast to that of the total solar energy, as recorded at Stamford, Conn. The rapid drop in ultraviolet intensity during the afternoon is attributed to haze formation. Extraterrestrial applications of polymers must take into account the higher levels of radiation beyond the earth's atmosphere, such as are encountered in the Van Allen belt.

Specific polymers respond to different wavelength regions in the near ultraviolet, and absorption at these wavelengths is usually responsible for the initiation of deterioration. Hirt and Searle[7] have devised the term "activation

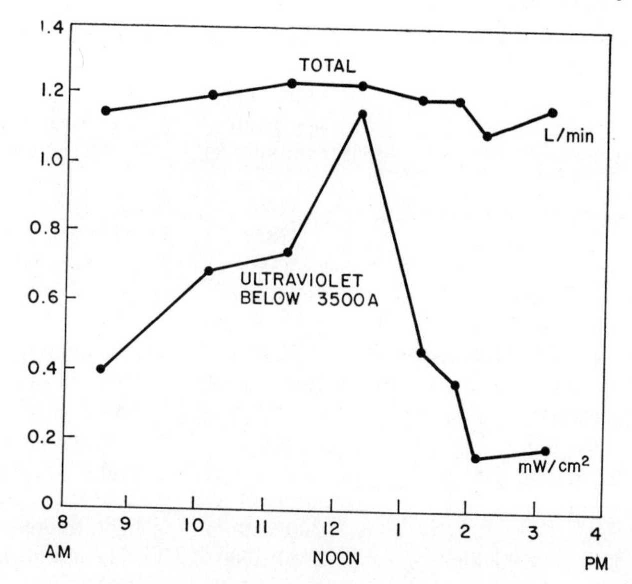

Fig. 10.1 Comparison of the daily variation in total and ultraviolet energy of the sun.[6] Reprinted with permission from Interscience.

spectrum" to describe the wavelength response of polymers in the near-ultraviolet region. These spectra are obtained by dispersing ultraviolet radiation from a xenon arc across the polymer sample by means of a fast quartz spectrograph. The activation spectrum maximum is then defined as the wavelength region causing the maximum deterioration as determined by the increase in absorption or fluorescence. Data for a variety of clear polymers (Table 10.1) shows that the peak sensitivity is maximized at approximately 3300 Å with only minor differences between polymers. The relation between the wavelength of absorbed radiation and the rate of deterioration is an important consideration in the choice of ultraviolet sources to be used in artificial weathering devices.

Under extreme conditions, absorption of infrared energy can develop temperatures up to 70°C[8] in black-pigmented compositions. Thus increases of several decades over ambient temperature can occur. Using the approximation that reaction rates double with each ten-degree increase in temperature, rates of photo oxidation, hydrolysis, and so on, could be increased considerably. The infrared intensity at the earth's surface also varies with weather conditions, although to a lesser extent than that observed in the ultraviolet region. Temperature control is therefore an important factor in developing accelerated weathering tests.

Table 10.1 Activation Spectra Maxima for Various Polymers[7]

Polymer	Activation Spectrum Maximum (Å)	Means of Measurement
Polyesters (various formulations)	3250	Visible light[a]
Polystyrene	3185	Visible light, fluorescence
Polyethylene	3000	Infrared (carbonyl)
Polypropylene (nonheat stabilized)	3700	Infrared (carbonyl)
Poly(vinyl chloride), homopolymer	3200	Visible light, fluorescence
Poly(vinyl chloride), copolymer with vinyl acetate	3270 and 3640	Visible and untraviolet[b] light
Poly(vinyl acetate) film	< 2800	Visible and ultraviolet[b] light
Polycarbonate film	2850–3050[c] and 3300–3600[d]	Visible and ultraviolet light, fluorescence
Cellulose acetate butyrate film	2950–2980	Visible and ultraviolet[b] light
Styrene acrylonitrile film	2900 and 3250	Visible and ultraviolet[b] light

[a] 4000 or 4300 Å.
[b] 2750–3000 Å.
[c] Includes several formulations and film thicknesses.
[d] Longer wavelength maxima were detected by the ultraviolet technique only. Fluorescence intensity decreased under the long-wavelength irradiation.

Tests designed to predict the weatherability of polymers are of three general types, based on exposure to the outdoor environment, simulation of sunlight with artificial light sources, or the early detection of weathering in its initial stages.

2. Outdoor Weathering Tests

The simplest approach to accelerating outdoor weathering is to select as the exposure site the location at which the polymer degrades most rapidly. To accomplish this, weathering sites are usually located in areas of intense sunlight, high humidity, or both. Exposure in selected sites has been used extensively as a materials test in which stabilizers are rated according to their effectiveness in the protection of individual polymers. This approach has also been employed as a field trial in which polymer compositions are exposed as an integral part of the article or device in which they are to be used. In either application it is desirable to know how much faster the material deteriorates at the selected exposure site than at other geographic locations. If this ratio, referred to as the acceleration factor, is known, reasonable estimates of the service life in various locations can be based on results obtained at the test site. Distortions will occur, however, when other factors contribute to deterioration of a polymer, to the extent that these factors vary between locations.

The degree of acceleration that can be obtained by selection of the exposure site is always small. For example, polyethylene has been shown by brittleness tests to fail only about twice as rapidly in Arizona as in northern New Jersey.[9] Martinovich and Hill[10] have reported that the deterioration of polyethylene and copolymers of ethylene with butene-1 in Arizona after 1 year is equivalent to that occurring in two years in Oklahoma and 3.5 years in Ohio. An acceleration factor of 1.5 has been reported[11] for poly(vinly chloride) between Florida and New Jersey; and a factor of 2 has been given between Arizona and New Jersey.

Several important factors other than location influence the rate at which polymers weather. These factors also must be standardized in order to obtain reproducible results. One of the most important is the angle and direction in which the sample is mounted. Because many of the early exposure tests originated in locations having approximately a 45° latitude angle, the angle most generally used is 45° from the horizontal with samples facing south. It is logical to expect that the adverse effect of weathering would increase in proportion to the amount of solar energy reaching the surface of samples and this might be expected to correspond to the latitude angle. However, Newland and co-workers[12] have reported that in Kingston, Tenn., the maximum intensity in the near-ultraviolet region occurs when samples are mounted at about 30° from the horizontal, although the latitude angle is 36.5°. Exposure of several different polymers in Florida and in Arizona showed no significant difference in color development for samples exposed at 45° and at the respective latitude angles.[13] Although Darby and Graham[14] have found that the most rapid rate of deterioration for plasticized poly(vinyl chloride) occurs when samples are mounted horizontally, this was explained by the more rapid leaching of plasticizers from samples mounted in this position. It has been the general practice in the testing of paints to mount panels vertically or at 5° from the horizontal, perhaps to relate more accurately to service conditions.

In an attempt to standardize the angle of mount, the American Society for Testing and Materials[15] has adopted a 45° angle with samples facing due south. Additional standards are also set for construction of exposure racks and mounts. Adherence to these standards is useful in correlating data obtained at different locations, but the data do not always represent the angle of maximum exposure or most rapid deterioration. A typical exposure site for outdoor weathering is shown in Figure 10.2. Test samples are usually mounted in a fixed position, but they may be rotated using a motor-driven equatorial mount[16] to increase the daily level of incident radiation. However, such devices are cumbersome and only slightly increase the deterioration rate.

The background beneath samples in an exposure site also contributes to the rate of weathering.[17] Different backgrounds (e.g., asphalt, cement, earth) vary in the extent to which they reflect ultraviolet irradiation and absorb infrared energy. The distance at which samples are mounted from the background is thus important, and standards for positioning samples are also

Fig. 10.2 A typical outdoor exposure site.

included in ASTM D-1435.[15] Samples may or may not be shielded from direct rainfall, depending on the purpose of the test or on the service conditions expected. Duplication of weathering results by different investigators as well as the determination of reliable acceleration factors requires the maximum standardization of test conditions.

Caryl and Helmick[18] have developed a device that uses mirrors to intensify the solar radiation reaching test samples. This machine, acronymously known as EMMA (Figure 10.3), combines an Equatorial Mount with Mirrors for Acceleration. It consists of a follow-the-sun rack equipped with ten flat mirrors so positioned that the incident angle of the sun's rays is approximately 90° throughout the day. These mirrors, which are 6 × 72 in. sheets of electropolished aluminum treated to minimize corrosion, reflect the solar radiation to samples in the target area. They reflect between 70 and 80% of the ultraviolet energy, so that samples receive a total of about eight times as much radiation as they would have absorbed in the same interval using only an equatorial mount. As shown in Figure 10.4, the ultraviolet energy distribution with the EMMA machine is quite similar to that of natural sunlight at 45°S.[19] Forced air, directed onto and under the test samples to prevent

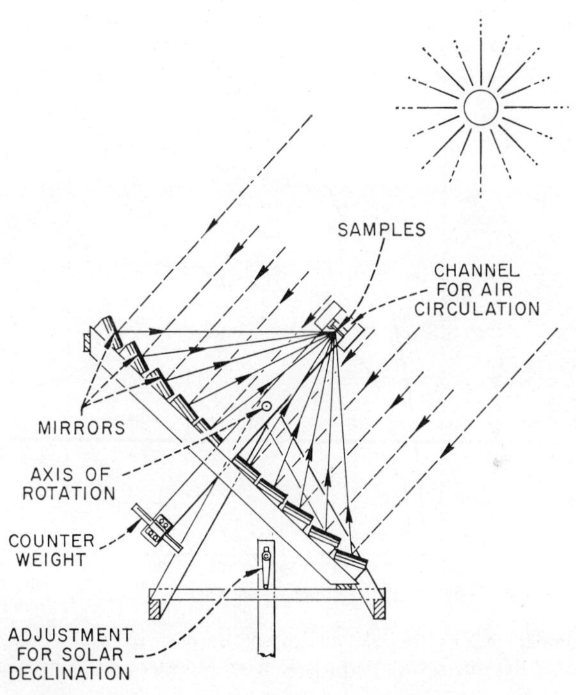

Fig. 10.3 Schematic representation of the EMMA device.

overheating, which could reach 300°F, maintains sample temperatures in approximately the same range as developed with a fixed mount at 45°S.[20]

The EMMAQUA is a modification of EMMA in which samples are sprayed with triple-distilled water for 8 minutes out of each hour of operation. In addition to simulating rainfall or humidity, this spray is effective in reducing the temperature of test samples. The rate at which additives, for example, plasticizers and stabilizers, are lost[21] through leaching by the water spray must be taken into account, however, if the loss is not comparable to that encountered during normal outdoor exposure.

Both EMMA and EMMAQUA have been used extensively to study the weathering of finishes and various other polymeric materials.[20,22,23] Excellent correlations between EMMAQUA data and normal exposure in Florida at 45°S have been reported by Kennedy and Beardsley[23] in their studies on 47% PVC paints based on vinyl acetate–ethylene, acrylic, and vinyl maleate emulsions. An accelerated factor of approximately 8 was obtained. Using the EMMA machine, Garner and Papillo[24] observed acceleration factors varying between 2 and 11 in comparison to 45°S for a series of polymer compositions, the degree of acceleration varying with the formulation. Ranking of the weatherability of formulations of a single polymer is usually good. However,

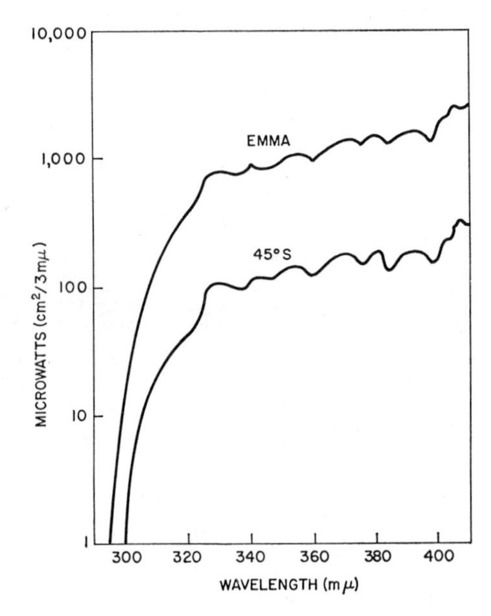

Fig. 10.4 Comparison of ultraviolet energy spectra of direct sunlight at 45°S and in the EMMA device.[19] Reprinted with permission from Industrial and Engineering Chemistry.

Martionovich and Hill[10] observed that increasing concentrations of an ultraviolet absorber in high-density polyethylene from 0.2 to 1% had almost no effect, as indicated by EMMA, although considerable improvement was observed with increasing concentrations when the same samples were exposed in Arizona at 45°S. Papillo[24] concluded that although EMMAQUA was not perfect in its correlations with normal outdoor weathering, of 63 samples tested in this accelerated device for 6.5 weeks, accurate prediction of the weathering effect observed after nine months in Florida was obtained for 85% of the samples. The prediction was borderline in 12% of these samples, and results failed to correlate with Florida data in only 3%.

3. Indoor Weathering Tests

Indoor laboratory tests have been developed in an attempt to accelerate the testing of weatherability and also because better control of exposure conditions is possible in the laboratory. Since absorption of ultraviolet energy is principally responsible for the outdoor weathering of most stabilized polymers, indoor tests are usually designed around an artificial source of ultraviolet radiation. This light source is operated at constant output within experimental limitations, and other variables including temperature can be controlled to different degrees. Although indoor tests do not really accelerate

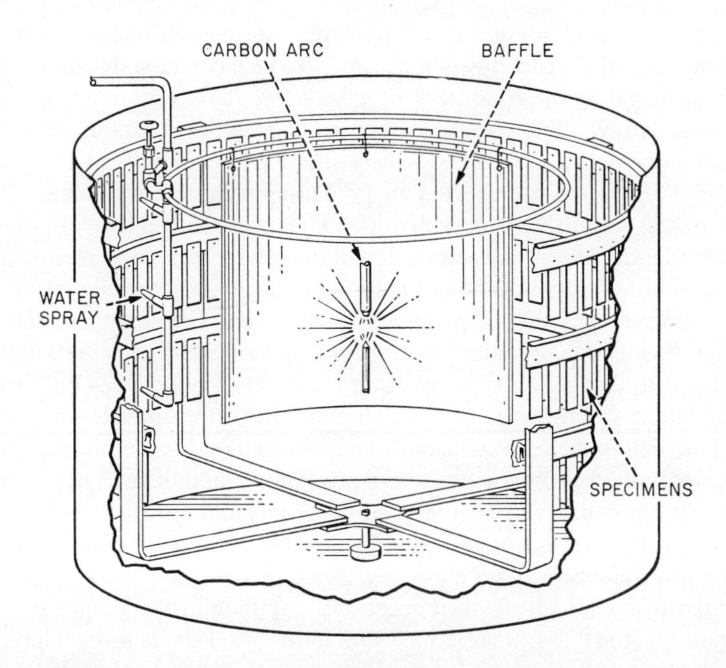

Fig. 10.5 Schematic of an artificial weathering device.[8] Reprinted with permission from Industrial and Engineering Chemistry.

nor do they reproduce natural weather, nonetheless these tests can provide useful data. Accurate predictions of weatherability are possible only to the extent that each important variable in the outdoor environment is duplicated.

The basic design for most artificial weathering devices consists of a radiation source about which test samples rotate (Figure 10.5). Additional features are sometimes included to simulate other important factors in the outdoor environment. The baffles shown in Figure 10.5 were used in this model to reduce sample temperature, but they also provide a day–night cycle. Forced-air jets or water sprays simulate rainfall and also lower sample temperature.

However, loss of stabilizers by leaching or evaporation[21] may occur at an abnormal rate when these are used. The distance between test samples and the light source is used to control both the temperature and the amount of incident radiation.

The principal variation among indoor tests is found in the types of ultraviolet sources used. It is generally agreed that reliability in the correlation between indoor and outdoor test data is dependent on how well the artificial light duplicates the solar spectrum. Radiation at the wavelengths responsible for degradation of individual polymers must be present in the source and should have intensities comparable to those found in sunlight. Wavelength peaks not present in the solar spectrum can distort results either by promoting reactions that would not occur in the polymer during outdoor exposure or by initiating abnormal reactions in additives, particularly in stabilizers.

The principal artificial sources of ultraviolet radiation are carbon arcs, fluorescent lamps, mercury arcs or lamps, and various types of xenon arcs. Each of these sources has been widely used, and each has its proponents. However, the successes reported in predicting weatherability have usually been based on results with individual polymers or limited classes of polymers. No single light source has been demonstrated to be reliable in predicting the weatherability of the multitude of materials classed as polymers, particularly when additives—stabilizers, pigments, and so on—are present in the composition. Absorption of radiant energy by the polymer and each of its additives contributes to failure. Thus, to evaluate the effect of an artificial light source, the activation spectrum (refer to Section B.1) of the complete composition should be determined. Maxima in the activation spectrum can result from absorption by the host polymer or any of its additives. Therefore, the presence of stabilizers and other additives may require duplication of the solar spectrum over a wider region of the near ultraviolet, sometimes extending into the visible region.

Comparison of the ultraviolet spectra of common artificial light sources is shown in Figure 10.6. Data obtained with unmodified single sources have been used in this comparison. In many instances, the energy distribution from individual sources can be modified by means of filters to give a closer approximation of the solar spectrum. Combinations of different sources have also been used effectively to supplement deficiencies in the output of the primary source.

a. Xenon Arcs. Xenon arcs, as evident from Figure 10.6, give the most accurate reproduction of the solar spectrum. Despite the good correlation, however, peaks that are not found in natural sunlight do occur, particularly at the longer wavelengths of the near-ultraviolet and visible regions. Higher capacity xenon arcs are usually water jacketed. Both the circulating water and

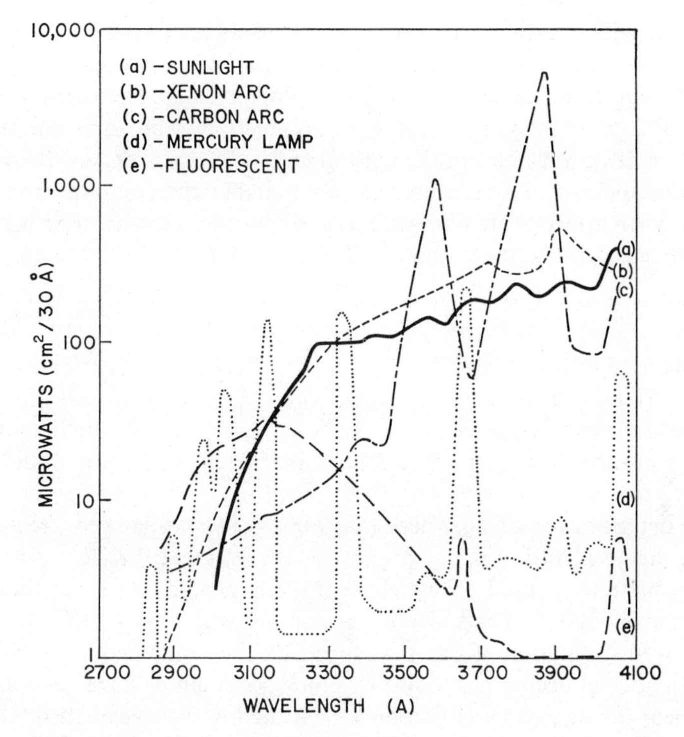

Fig. 10.6 Ultraviolet energy distribution of artificial ultraviolet sources.

the pyrex or quartz envelope that encloses a xenon arc serve as filters. Additional filters have also been used successfully to improve the correlation between the energy distribution of these arcs and sunlight.[6] Xenon arcs show a characteristic loss of intensity, primarily because of a decrease in transmission of the envelope and filters. The loss is rapid at first, decreasing in rate after the first several hours. Preaging of arcs and filters has been used to offset this initial decrease in intensity. Nonetheless, since output from xenon lamps continues to decrease, indoor accelerated tests using these arcs as the ultraviolet source give more reliable results when the rate of deterioration is expressed as a function of incident radiation rather than time of exposure. Even greater accuracy is obtained by expressing deterioration as a function of the incident ultraviolet radiation.[25]

The intensity of ultraviolet radiation varies with the size of the xenon arc, and therefore the distance of samples from the source should be greater for more intense arcs. In most commercial equipment, the distance is such that the intensity reaching samples approximates the maximum that would occur out of doors. Greater acceleration results if this distance is decreased, but correlation with outdoor weathering is adversely affected when the radiant energy exceeds a critical level that is characteristic of individual polymers.[26]

Many commercial devices have been developed using xenon arcs of various types to accelerate the indoor testing of polymers,[6,26] and some are modified to simulate other variables in the natural exposure environment. Because the energy distribution of xenon arcs closely matches the solar spectrum, particularly when appropriate filters are used, indoor accelerated tests with these light sources correlate reasonably well with outdoor tests.

b. Carbon Arcs. Carbon arcs, used extensively in studies on the weatherability of paints, have also found wide application as ultraviolet sources in the accelerated testing of stabilized polymers. The spectral distribution shown in Figure 10.6 is for an enclosed carbon arc without additional filters. This source is deficient in the 3000- to 3500-Å region within which most clear polymers absorb damaging radiation,[7] and energy peaks not found in the solar spectrum occur at 3600 and 3800 Å. Although this carbon arc accelerates the deterioration of polymers that absorb at wavelengths greater than 3500 Å, the acceleration is not proportionally as great for clear polymers.

Filters have been used effectively with carbon arcs to reduce the energy output above 3500 Å. The sunshine carbon arc with Corex "D" filters gives an improved simulation of the solar spectrum, but appreciable energy above the sunlight level occurs near 3800 Å. Fluorescent lamps have been used[27] to supplement the deficiency of carbon arcs at the lower wavelengths. Although reasonable correlation of outdoor weathering data with results obtained with modified carbon-arc devices has been reported,[27] distortion may occur when polymer compositions contain additives that absorb at the higher wavelengths.

c. Fluorescent Lamps. Fluorescent lamps have the lowest energy output of the usual sources of ultraviolet radiation (Figure 10.6). Their intensity is considerably lower than that of sunlight in the critical near-ultraviolet region. These lamps, however, have the advantages of being relatively inexpensive and easy to maintain. Their peak output is at 3130 Å, so their effect is somewhat greater for clear polymers than that obtained with either carbon or xenon arcs. The deficiency at longer wavelengths can be improved by addition of fluorescent blacklamps.[28,29] Blacklamps are available with peak emission at 3520 and 3650 Å,[6] and combinations of these with standard lamps give a closer, though not exact, duplication of the solar spectrum. Although even the best combinations of fluorescent sunlamps and blacklamps do not duplicate the solar spectrum to the same extent as a filtered xenon lamp, thermal effects are less prominent because fluorescent lamps are much cooler than arcs. When absorption at higher wavelengths is not an important factor, as in the weathering of clear polymers, good correlations have been obtained with fluorescent lamps and outdoor weathering.[30]

d. Mercury Arcs. Mercury arcs or vapor lamps, although often used as ultraviolet sources in indoor accelerated tests, do not give an accurate reproduction of the solar spectrum (Figure 10.6). The energy output is concentrated in the mercury emission lines with a pronounced background deficiency in the 3000- to 4000 Å region. High-pressure arcs with appropriate filters more closely approximate the solar spectrum. Good correlation between results using a high-pressure arc and outdoor data has been obtained with vinyl polymers.[31]

Important characteristics of common ultraviolet sources used in accelerated indoor testing, summarized by Clark and Harrison,[32] are shown in Table 10.2. Examination of this table supports the preceding discussion of accelerated weathering devices and the conclusion, expressed by Kamal and Saxon,[33] "acceleration is not equivalent to correlation with a particular natural weathering exposure"' Nonetheless, indoor weathering tests continue to develop useful data within the limitations of their individual capabilities, particularly in comparing the effectiveness of stabilizers.

4. Early Detection of Weathering

The major effort in the prediction of weatherability has been directed toward development of tests in which deterioration is accelerated by exposing polymers to a more severe environment. These methods give reasonable predictions of service life for limited classes of polymers, but they are not infallible, primarily because it is impossible to duplicate or accelerate natural weathering. A more reliable prediction of weathering may be obtained by exposure to the actual outdoor environment, and detection of the incipient effects of weathering by a sensitive analytical procedure. This approach requires sensitive test methods and established relations between incipient deterioration and ultimate failure. The first of these requirements has been met, in part, by application of techniques designed to measure rates of reactions responsible for deterioration. The second requirement, however, is more difficult to satisfy, since it is not always possible to relate the rate of chemical reactions to the changes in mechanical or other properties that cause polymer failure. These relations have been established in only a few instances.[34–36]

Because exposure times in the early detection method are short, perhaps only a few months, samples are exposed to weather conditions that are unlikely to correspond to the average conditions encountered over the useful lifespan of the material. For accuracy, therefore, changes occurring in the polymer during the short exposure times should be related to the particular variable in the environment causing failure. For example, changes in chemical composition must be related to the ultraviolet radiation reaching the sample

Table 10.2 Comparison of Artificial Ultraviolet Sources.[32]

Type		Solar Simulation	Intensity Uniformity		"Suns"
			Spatial	Temporal	
Carbon arc	Single enclosed	Poor in UV (high 360/380 mμ UV peaks)	±20%	Very variable	1/2
	Sunshine	"	±20%	Variable	1
Fluorescent sunlamp/ blacklamp (FS/BL)		Poor in visible/ IR (low energy in visible/IR above 350 mμ)	±20%	"Relatively constant"	1 (UV only)
Xenon arc	Atlas	Poor in IR (high 800-1000 mμ IR peaks)	±20%	Decreases about 30%/ 1000 hr.	1
	Xenotest	Good (excess IR filtered out)	±10%	Decreases about 15%/ 1000 hr.	2
	Spectrolab	Best available	±5%	Constant (±1%)	1-20

and not to the time of exposure. Weather bureau data could then be used to calculate the time interval during which the amount of radiation required to cause failure would reach the polymer. If only data on total solar radiation are available, the factor developed by Coblenz[37] can be used to convert the langleys recorded to appropriate units of ultraviolet radiation.

Transmission spectroscopy in the infrared region has been studied as a possible technique in the early detection of the weathering of polyethylene.[38] Carbonyl groups, which absorb strongly in the infrared at 5.81 μ, are formed during both the photoinduced and the thermally induced oxidation of polyethylene and many related polymers (see Chapters 4 and 2). Tamblyn and coworkers[38] have measured the rate at which ketonic carbonyl is formed in a series of polyethylenes of varying molecular weights during the early stages of outdoor exposure. They have attempted to correlate these data with loss in tensile elongation. No correlation was evident in the unprotected polymers. However, a rank order rating of a series of stabilizers in a single polymer, as determined by infrared specroscopy after four months' exposure, approximated the rating obtained by measurement of tensile elongation after 14.5 months.

Internal reflection spectroscopy (IRS) is a useful technique for studying chemical reactions occurring at the surface of polymers. It is generally applied as illustrated schematically in Figure 10.7. A beam of light is propagated through a crystal by multiple internal reflections from opposing flat surfaces, in this case, polymer strips. The beam penetrates slightly into the strips, losing energy at those wavelengths where the polymer absorbs radiation. The attenuated radiation that emerges is measured to give an absorption spectrum

Reprinted With Permission From Interscience.

Correlation	Exposure Area, In.^{2}	Maintenance, Hr.	"Down Time"	Lamp Replacement, Hr.
Poor to good	1000	High-ea. 20	Weekends + 1 hr./day	20
Fair to good	1500	High-ea. 20	"	20
Good for colorless or unplasticized plastics	1500	Low-ea. 100	Negligible	100
Good	1500	Low-ea. 100	1 hr./week(1%)	2000
Good	650	Low	Negligible	1500
Not available	200(at 1 sun)	Low	Negligible	1500

characteristic of the sample. This phenomenon is also referred to as attenuated total reflection (ATR). A light beam of any useful wavelength may be used in the internal reflection method. The depth of penetration of the beam into the sample is dependent on the crystal composition and the angle of incidence of the beam onto the sample. By controlling these variables, the reaction taking place at different depths can be studied.

Internal reflection spectroscopy has been used to study changes in the surface chemistry of polymers as an approach to the early prediction of weathering. T. J. Gedemer,[39] using artificially weathered polycarbonates, showed that degradation rates can be determined by this technique, and Hearst[40] used the method to detect chemical changes in photodegraded organic coatings. Chan and Hawkins[41] have applied IRS to the early detection of weathering in natural and black formulations of polyethylene and in ABS resins. An increase in surface carbonyl was observed before changes in low-temperature brittleness, a mechanical property sensitive to weathering, became evident. The method has also been used to rate stabilizers in polyethylene.[41] However, to predict the service life of stabilized polymers from

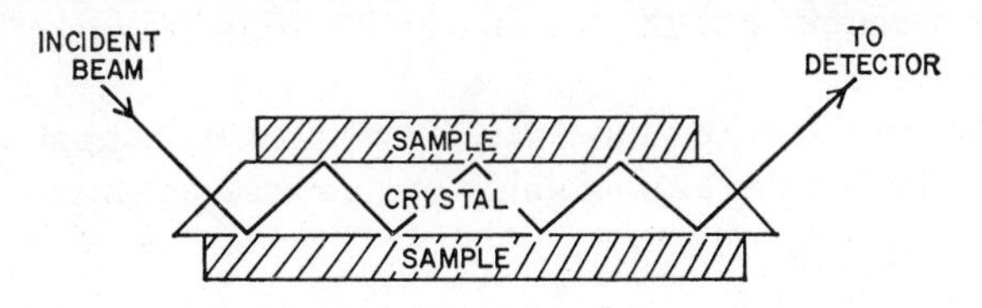

Fig. 10.7 Schematic representation of the internal reflection spectroscopy technique.[41] Reprinted with permission from the Wilks Scientific Corporation.

short-exposure internal reflection data, it is necessary to know the rate at which surface reactions progress into the polymer bulk and how the observed chemical changes relate to actual failure.

It is characteristic of the developing spectrum of a polymer exposed to near-ultraviolet irradiation that maxima observed in the ultraviolet region simply broaden until they extend into the visible region. This has led Kamal and Saxon[33] to suggest that changes in the near-ultraviolet spectrum, after short exposure intervals out of doors, could be used to predict the resistance of polymers to discoloration. This suggested method is supported by Bruck's[42] observation that, in the irradiation of polyamides, changes in the ultraviolet spectrum are detected before visible color becomes apparent.

Changes in mechanical and dielectric properties have also been considered for detecting the early stages of weathering. Kaufman[3] and Tamblyn et al.[38] have observed that failure of polyethylenes occurs more rapidly out of doors when the samples are under stress. This suggests that early detection of the weathering of this polymer could be accomplished by using stressed samples, but only if the relative rates of deterioration of the stressed and relaxed polymer were established.

Although the concept of early detection of weathering has not been extensively explored, it is highly probable that the development of new, sensitive tests for surface properties will result in wider application of this method.

5. Summary of Accelerated Weathering Tests

The intensive effort that has been directed toward the development of reliable tests for predicting the outdoor service life of polymer compositions is evidence of the importance of this problem to the polymer industry. As more effective stabilizers are developed, there will be a need for even greater degrees of acceleration. The ideal test would predict, in a few months at most, how long a stabilizer would protect a polymer against the outdoor environment. Although only polymer compositions with exceptional stability now have an outdoor service life in excess of 20 years, resistance of outdoor exposure of this order should be anticipated as the science of polymer stabilization develops. For this reason, acceleration factors of 20 or more would be desirable.

The preceding discussions have developed three basic approaches to the accelerated testing of the weatherability of stabilized polymers:

1. Intensification of outdoor exposure.
2. Laboratory simulation of weathering.
3. Early detection of outdoor weathering.

Each of these methods has been used with limited success to predict the weatherability of specific types of polymers, but each has its own limitations and none has general applicability.

Stabilizers are added to extend the useful life of the polymers that would fail prematurely during outdoor exposure. Thus the problem of predicting the service life of polymer compositions in specific designs resolves itself first into a material-oriented test in which the relative stability of the stabilized material is rated against that of the unmodified polymer. The second step is considerably more complex, since each design parameter that could contribute to failure must be taken into account. The available methods of testing are not as accurate as would be desired, but it is possible, within the limitations of present technology, to establish a program that gives reasonably reliable results in predicting the time to failure of stabilized polymers in design applications. Such a program consists of three separate steps.

The first step is to establish the service life of the unprotected polymer. Since many unstabilized polymers deteriorate rapidly in the outdoor environment, accurate data can often be obtained by using a standardized outdoor exposure test.[15] Exposure sites should be selected for the maximum effect, and several would be desirable in a comprehensive program to provide variations in important environmental factors. Reliability in this first step depends on control of the following variables: polymer selection, environmental factors, and criteria for failure.

The variability in polymers from different sources requires that composition of the base polymers be standardized to the fullest extent. Important properties must be determined so that this polymer can be reproduced in a continuing program and results obtained by different investigators can be compared. Data reported on polymers of ill-defined composition are of negligible value.

Since the outdoor environment is never constant, important variables must be monitored throughout the exposure period. Stability would then be expressed as a function of absorbed radiation, and computer technology might be applied to integrate other important variables and to give an average exposure condition. Thus failure would be measured primarily in units of absorbed energy, and this result would be translated into exposure time using long-term weather reports from the exposure site.

For a general program of stabilizer evaluation, failure of the polymer should be based on those visual, mechanical, and dielectric properties, important to the service application. Sample size and geometry must also be standardized for later comparison with the stabilized polymer compositions.

Only a limited amount of data are available on the weatherability of polymers under carefully controlled conditions, and when this type of datum is not available, stability of the base polymer must be included as the first step in the test procedure.

The second step involves use of an accelerated test to compare the extent to which various additives improve stability of the base polymer. Selection of the accelerated test is perhaps the most critical factor. Test procedures are continuously developing, and their relative values must be revised constantly. From the preceding discussion, it is evident that accuracy in this step is dependent both on duplication of every contributing factor in the service environment and on the determination of failure. Failure can be ascertained either by direct measurement of the property in which a change would cause failure or by measuring an alternate property that can be related accurately to actual failure. (Perhaps supplementary tests could be used to determine the importance of other factors, such as evaporation or leaching out of stabilizers.) The results are adjusted, where necessary, to account for the important secondary factors. The extent to which each stabilizer improves stability of the base polymer is then applied to estimate the service life of the stabilized polymer compositions. This extrapolation, however, is not infallible. Errors can occur when secondary factors, not common to steps 1 and 2, contribute to failure.

The third and final step requires input best supplied by the design engineer. Completion of steps 1 and 2 only rates the stabilized compositions on the basis of material failure. How the material functions in a specific design depends on the stress introduced, the presence of unusual environmental factors specific to the use envisioned, and contact with or proximity to other materials that affect its stability. Important in the last category are metals that catalyze deterioration and materials capable of extracting or modifying the stabilizer.

Although many imperfections persist in the individual steps, by careful control of experimental conditions, reasonably reliable data can be obtained with methods currently available. As techniques improve, this type of program could be used with increasing confidence to predict service life under all exposure conditions.

C. ACCELERATED TESTS FOR THERMAL STABILITY

Determination of the effectiveness of antioxidants in protecting polymers against thermal oxidation is considerably less difficult than solving the problem of predicting weatherability. Thermal energy is often a contributing factor in outdoor weathering, as previously noted, and many polymers are exposed in service to conditions in which only thermal reactions would be responsible for failure. Since fabrication is almost always carried out at elevated temperatures, usually in the absence of light, most polymers acquire a thermal history. During this period, significant deterioration can occur, or processes that will continue during long-term exposure can be initiated.

Accelerated tests to determine thermal stability usually involve simply raising the temperature above that of the service environment. In most applications, thermal degradation takes place in the presence of air, and the most common reactions are those of oxidation. However, in many fabrication processes oxygen is essentially excluded, and thermal degradation in the absence of oxygen must also be considered.

1. Tests in the Absence of Oxygen

Several methods have been developed to measure the thermal stability of polymers in the absence of oxygen. The usual apparatus consists of a heater that may be temperature programmed or operated isothermally and a system for measuring change occurring during heating. The measurement of weight changes by thermogravimetric analysis (TGA) is the most common technique. However, various other changes that occur during heating have also been used to measure stability. These include measurement of the pressure generated by volatile fragments, temperature-dependent transitions, and determinations of properties such as viscosity of the polymer residue at various stages of degradation.

Custom-designed apparatus was used in most of the earlier studies. Much of this work was concerned with the kinetics of deterioration and with reaction mechanisms through qualitative and/or quantitative analysis of degradation products. Highly efficient vacuum equipment was designed to exclude oxygen, and several methods were developed to measure small weight changes during heating. These custom designs, many of which have been described in detail by Madorsky,[43] have contributed significantly to the fundamental understanding of polymer pyrolysis.

More recently, commercial equipment has become available for the thermogravimetric analysis of polymers. The method, essentially the same as in custom-designed equipment, uses a highly sensitive thermobalance (e.g., a Cahn electrobalance) and programs the temperature cycle reproducibly at a variety of selected rates. Commercial equipment is seldom designed to operate at high vacuum, but oxygen is excluded by introduction of inert gases into the reaction chamber. Products of the reaction can be analyzed throughout the heat cycle by collecting the volatile decomposition products and analyzing them by gas chromatography.[44] Commercial TGA equipment is quite well adapted to routine determination of the thermal stability of polymers in the absence of oxygen.

Grassie and co-workers[45–47] have studied the thermal decomposition of a variety of polymers by measuring the pressure generated by volatile products under isothermal conditions. McNeill[48] has extended this method of thermal volatilization analysis (TVA) to record changes in pressure during a programmed temperature increase. The apparatus employed consists of a heating unit into which the glass sample tube is placed, a liquid nitrogen trap to collect evolved gases as the system is continually evacuated, a Pirani gauge to determine changes in pressure, and a strip recorder. Thermal volatilization analysis is sensitive to small changes in pressure. TVA thermograms show distinct peaks for different volatile fractions, in contrast to TGA thermograms, which show inflections for different stages of degradation. In general, the TVA technique is better adapted to the detection of trace amounts of volatile materials, as, for example, traces of residual solvent. Both TGA and TVA measurements are somewhat dependent on the heating rate, which tends to shift position of the peaks.

Thermoanalysis of polymer deterioration in the absence of oxygen can also be accomplished by differential scanning calorimetry (DSC)[49] or by differential thermal analysis (DTA).[50,51] As applied to polymer deterioration studies, both methods depend on an exothermic reaction that occurs when bonds in the polymer molecule are ruptured. They differ in that the DTA method detects the temperature difference between the degrading sample and an inert reference material (e.g., aluminum oxide), whereas the DSC technique records the electrical energy required to maintain reference and sample thermocouples at the same temperature. Although applicable to studies of the thermal stability of polymers, these methods have been more widely used in determining transition temperatures such as T_m and T_g.

Since additives have very little effect in stabilizing polymers against the high temperatures at which pyrolysis occurs, most studies on thermal stability in this temperature range have centered around the effects of changes in polymer structure. Hence test methods employed are essentially those used to study unmodified polymers.

2. Tests in the Presence of Oxygen

Stabilizers designed to protect polymers against thermal oxidation, commonly referred to as antioxidants, may be effective during processing, long-term aging, or both, and accelerated tests have been developed to rate antioxidants in each of these stages in the lifetime of the polymer. Although determination of oxidative stability is usually accelerated by increasing the temperature above that encountered in service, reliability increases as the test temperature approaches that of the service environment. As will be seen later in this section, the relative effectiveness of stabilizers may not be the same in a polymer above and below its T_m. To predict the service life of individual polymer compositions, the stability is determined at several elevated temperatures and the accelerated data are then extrapolated to determine the time to failure at service conditions.

a. Milling Tests. Several procedures have been developed to determine the stability of polymer compositions under compounding or fabricating conditions. A standard ASTM procedure describes a milling test[52] for measuring the stability of polymers. In a modification of this test, Baum and Perun[34] have measured changes in physical behavior and chemical composition occurring during milling as a test for stabilizer effectiveness. Polymer compositions were milled continuously at 170°C until sticking occurred on the rolls as the material oxidized. Rating of stabilizers by this simple procedure correlated well with the ranking as determined by carbonyl formation using infrared spectroscopy (Table 10.3). However, several anomalies were observed

Table 10.3 Relative Efficiency of Antioxidants Using the 170°C Milling Procedure[34] [a]

Antioxidant	Time to 0.1% Carbonyl (min)	Time to Sticking (min)
Hydroquinone	33	37
2-Ethylhexyl octyl phosphite	38	41
o-Aminophenol	40	41
p-tert-Amylphenol–formaldehyde resin	42	50
N,N'-di-*sec*-Butyl-*p*-phenylenediamine	42	60
Lauryl gallate	47	60
4,4'-Thiobis(3-methyl-6-*tert*-butylphenol)	82	83
N,N'-diphenyl-*p*-phenylenediamine	90	90

[a] Reprinted with permission from Plastics Technology.

in comparing the relative effectiveness of these compositions with oven-aging data obtained at 110°C. This is probably a result of differences in the rate of stabilizer loss at the two test conditions. The milling test is useful in rating antioxidants under fabricating conditions, but its reliability in predicting their effectiveness during long-term aging is questionable.

As polymers degrade during processing, changes in viscosity occur. Devices[53] have been developed to measure the change in torque that takes place as polymers are subjected to heat and shear stress. The material is sheared between counterrotating rotors driven by a dynamometer at constant rpm and the resistance to the applied stress is recorded. The more effective the antioxidant, the longer the torque remains at or near its original level. As the stabilizer fails, oxidation of the polymer occurs and there is an abrupt change in the torque. Since this determination can be made in a controlled atmosphere, it is quite useful in estimating stabilizer effectiveness as it would occur in fabricating equipment with limited access to air. Variations in the speed of rotation and in the applied stress combine to make another parameter for simulating fabrication processes.

b. Oven-Aging Tests. In one of the simplest accelerated tests for measuring antioxidant effectiveness, polymer compositions are ovenaged at selected temperatures and examined periodically.[34] Changes in visual appearance, mechanical properties, and chemical composition have been used to follow the course of degradation. If nondestructive measurements are made, samples can be returned to the oven for continued aging. However, since removal from the oven and a return to room temperature can produce a discontinuity in the observed rate, it is advisable, whenever possible, to use separate samples for each time interval with no interruption in aging.

Although simple in concept and in operation, oven-aging tests have several inherent sources of error. Forced-air ovens, which are often used, can result in an abnormal rate of antioxidant loss by evaporation. Even stabilizers of relatively high molecular weight are subject to loss by evaporation.[21] Transfer of stabilizers between different compositions is also a definite possibility, particularly in a static environment. Ovens may also become contaminated in time with antioxidants and their decomposition products, which could distort test results. The popularity of oven aging can perhaps be attributed to the wide variety of properties that can be measured in the evaluation of stability. It is often practical to measure the exact property responsible for failure.

c. Oxygen-Uptake Tests. Measurements of the rate at which polymers react with oxygen are basically adaptations of the Warburg respiration apparatus for bioassay.[54] Adaptations include both manometric and volumetric determinations of the oxygen uptake. As an example of the former, the Norma–Hoffman bomb[55] has been used extensively in studies of elastomer degradation. In this procedure, samples are heated in pure oxygen at an initial pressure of several atmospheres and the change in pressure in the bomb is recorded. Little change is observed with stabilized polymers until the antioxidant fails and then rapid oxidation of the polymer ensues. This time interval, termed the induction period, is evidenced by an abrupt drop in pressure. Using an oxygen pressure of 100 $lb/in.^2$ in the bomb, Kennerly and Patterson[56] have reported an acceleration factor of 35 over a test at the same temperature run in air at 1 atm. Reproducibility in this device is not good, however, and it is both cumbersome and difficult to operate. Furthermore, since the oxidation rate is diffusion dependent[57] for many polymers, particularly in thick samples, results obtained at elevated pressures must be correlated with service condition, which usually involves atmospheric pressure.

Oxygen uptake at approximately 1 atm has been used by many investigators[58–60] in the study of antioxidant effectiveness. Although many variations have been reported, the test generally consists of a temperature-controlled heating device—either an oven, an oil bath, or an aluminum block—in which tubes containing the test samples are placed and connected to an external volumetric or manometric measuring device as shown schematically in Figure 10.8. The atmosphere may be air or pure oxygen, and an absorbent is usually placed in the sample tube to remove products of the degradation. In the volumetric measurement, leveling bulbs are adjusted to maintain a pressure of approximately one atmosphere.

Unless sample thickness is selected properly, the measured rate of oxygen absorption may be diffusion controlled and thus would not reflect the true

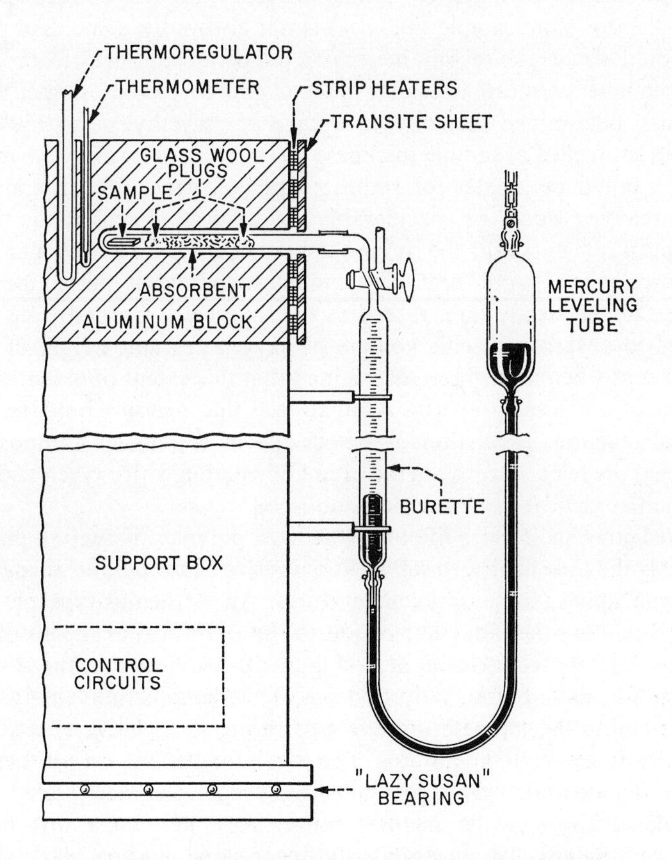

Fig. 10.8 Schematic of a typical oxygen-uptake apparatus.

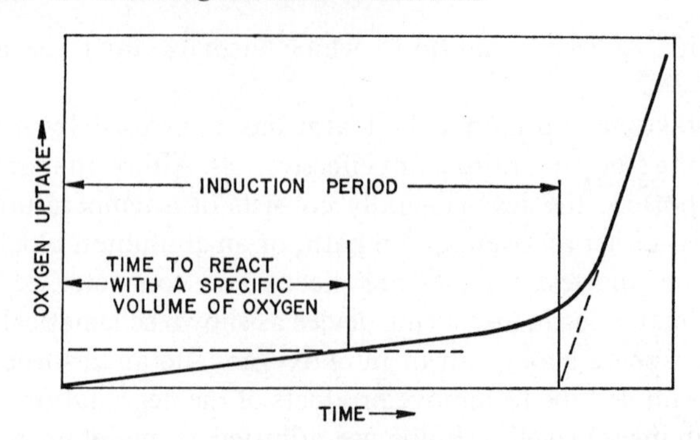

Fig. 10.9 Schematic representation of the oxidation rate curve for an inhibited polymer.

oxidation rate of the material. Diffusion rates vary with different polymers, and so the optimum sample thickness is not always the same. Sample thickness should also decrease with increasing temperature, since greater amounts of oxygen must permeate to maintain the maximum oxidation rate. Biggs and Hawkins[57] determined that oxygen uptake of polyethylene at 150°C is not diffusion controlled at sample thickness up to 5 mils. However, the maximum thickness could be greater for stabilized polyethylene, since the amount of oxygen reacting would be considerably less in an inhibited reaction.

A typical rate curve for the oxygen uptake of a stabilized polymer is shown schematically in Figure 10.9. The induction period is determined as the intercept of the steady-state rate with the time axis. Alternatively, the time required to absorb a specific volume of oxygen per unit weight of polymer may be used when it has been determined that this extent of oxidation results in failure of a critical property. As in all tests that measure only the rate of a chemical reaction, prediction of service life of a polymer composition requires that the relation between the extent of reaction with oxygen and changes in properties critical to failure be established.

In predicting the service life of a stabilized polymer, induction periods, or preferably the time to absorb sufficient oxygen to cause failure, are determined by oxygen uptake at several temperatures. An Arrhenius-type plot is then prepared relating the induction period to the reciprocal of the absolute temperature. Data by Hawkins et al.[61] (Figure 10.10) show that the slope is not the same for all stabilized polyethylenes. This demonstrates the importance of determining the temperature effect in rating the relative effectiveness of stabilizers at service temperatures. The curve for the 3% carbon black composition demonstrates the errors that can occur when accelerated data are extrapolated through the melting range. Although most low-molecular-weight compounds give an essentially linear slope, carbon black and other

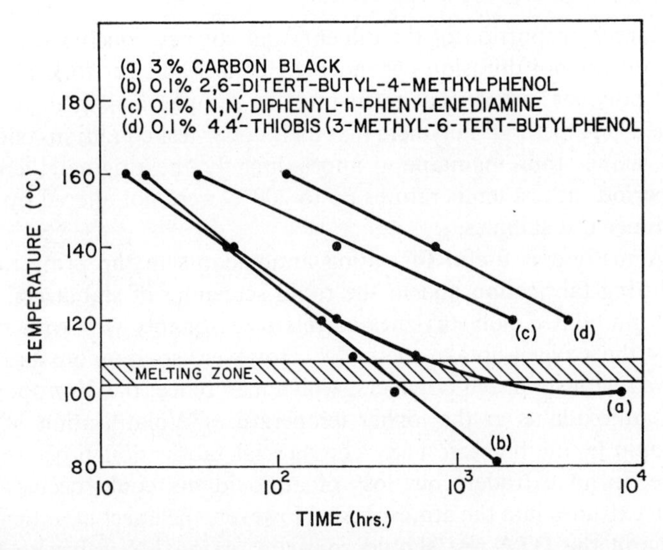

Fig. 10.10 The relation between stability and temperature for representative stabilizers.

stabilizers of comparable particle size[62] give plots that diverge from linearity in passing through the melting zone.

Because of the relative simplicity of the measurement and because a large number of samples can be run simultaneously with little possibility of interactions, oxygen-uptake tests have found wide application. The test is well suited to rating of the effectiveness of stabilizers, and when the proper relations are established, it can be used to give reliable predictions of the service life of polymer compositions. Hansen[63] has suggested the addition of catalytic amounts of copper to polyolefins to further accelerate the test and to permit the use of lower temperatures, but the general applicability of this modification has not been established.

d. Differential Thermal Analysis. Rapid evaluation of antioxidants in polymer composition has been reported[50,51] using differential thermal analysis. The method is the same as that described in Section C.1 except that the inert gas surrounding the sample is rapidly replaced with oxygen when the test temperature is reached. A temperature range from 150 to 210°C has been used in studying oxidative stability, and induction periods are usually measured in minutes. At these elevated temperatures, one must be assured that sufficient oxygen permeates into sample to prevent diffusion control of the oxidation. Rudin and co-workers[50] have overcome this difficulty by mixing ground polymer with 10 parts of glass powder. The induction period was not affected by using a larger proportion of the diluent, and so they concluded that the determination was not diffusion dependent. Howard and co-workers[64] have used 10-mil polymer samples cut to fit into the sample cups in their application of the DTA method. Polymers that melt under test conditions conform to the cup shape, thus maintaining approximately the original thickness. Induction periods at test temperatures up to 200°C were not altered by starting with thinner test samples.

The DTA method is useful for rating antioxidants in the protection of polymers during fabrication and in the rapid screening of stabilizers. Data taken with uninhibited polyethylenes correlate reasonably well with results obtained by the oxygen-uptake method.[64] However, some anomalies have been observed with stabilized polymers, which may reflect the disproportionate loss of antioxidants at the higher temperatures. Volatilization of antioxidants would be inhibited in many commercial fabrication processes, as, for example, in an extruder; but loss of antioxidants could occur as the composition extrudes into the atmosphere. However, the effect of antioxidant volatilization in the DTA test should compare reasonably well with losses that occur during compounding on a mill. Thus since loss of antioxidants depends on fabrication conditions, the effect of volatilization must be considered in relating DTA data to stability during processing.

The extrapolation of DTA data to temperatures encountered during long-term aging is less reliable than using accelerated data obtained at lower temperatures. Although DTA data for polyethylene stabilized with some stabilizers fit approximately the linear plot obtained by oxygen uptake at lower temperatures, DTA results obtained with black formulations cannot be extrapolated to temperatures below the polymer-melting range. For this composition, only data obtained below the T_m would give a correct estimation of how effectively carbon black protects polyethylene against oxidation during low-temperature aging. However, even with the limited number of anomalies encountered, the method has been applied with some success[64] in the rapid screening of antioxidants.

The rate at which oxygen combines with stabilized polymers can be measured over a wide temperature range by thermogravimetric analysis. However, weight changes are usually small and a high sensitivity is required in the measurement. This method is also more difficult to run than the DTA or oxygen-uptake tests and, in contrast to the latter, it is not adapted to multiple determinations. An unusually sensitive method for detecting weight changes during oxidation has been reported[65] in which the variation in frequency of a piezoelectric crystal is used to detect the weight changes in a polymer film deposited on the crystal. Although this method has a high degree of sensitivity, it would be extremely difficult to deposit films of stabilized polymers on the crystal surface.

e. Flammability Tests. The large number of tests developed to measure flammability reflects the complexity of the burning process and the variety of characteristics that are important in the combustion of polymers. Tests have been devised to measure ease of ignition, rate of flame propagation, smoke density and composition, resistance to flame passage, rate of pyrolysis, effect of the burning polymer on other materials, and other less important aspects of burning. Obviously, tests intended to evaluate different characteristics will not rate the flame resistance of polymers in the same order. Hilado[66] has suggested a system of tests, each to measure a different characteristic, as a practical solution to this problem. Unfortunately, results from methods intended to measure a single aspect of polymer combustion often fail to agree.[66,67]

Details of the most common flammability tests have been described by Hilado[68] and others[69,70] and classified according to the type of data generated. Several of these tests have been adopted as ASTM standards,[71] others have been approved by Underwriters' Laboratories,[72] and several are used as federal specifications by the U.S. Department of Commerce.[73] Recently considerable interest has evolved in the Fenimore–Martin oxygen index test,[74–76] which measures the percentage of oxygen necessary in a nitrogen–oxygen atmosphere to sustain combustion of samples previously ignited with a hydrogen flame (see Chapter 8). This test gives reproducible results that express flame resistance as the minimum ratio, $O_2/O_2 + N_2$, at which burning continues. This ratio can be determined with an accuracy up to three significant figures. Values greater than 0.27 indicate that the polymer is self-extinguishing.[76] However, despite its accuracy in reproducing data, this test, like all the others, predicts flammability only under its own specific conditions and does not give a reliable prediction of the resistance to burning in an actual fire.

D. MISCELLANEOUS ACCELERATED TESTS

When one considers the great number and variety of applications in which polymers are used, it is at once recognized that some polymers will be exposed to conditions in which failure would result from reactions other than oxidation. Dependent on polymer structure, and to a lesser extent upon exposure conditions, reaction with water, ozone, organic solvents or non-solvents, acids, bases, dyes, adhesives, fillers, and so on, can be the primary cause of failure. When any of these reactions is responsible for failure, tests can be developed to measure resistance to that particular reactant; but in most instances these tests are quite specialized, and often field trials are the more practical approach. Reactions with ozone or water are exceptions, however, since these reactants are almost always present to some extent in the normal environment, and each is responsible for the failure of many polymers. Thus tests for resistance to ozone attack and to hydrolysis are of sufficient generality for inclusion in this section. Although only a limited number of polymers are degraded by living organisms, these species are always present in the environment and so general procedures for evaluating resistance to biodegradation are also included.

1. Tests for Ozone Resistance

Because of its importance in the deterioration of elastomers, resistance to ozone has been investigated extensively, and many tests have been developed to measure the stability of polymers during exposure to ozone. In most of these, the concentration of ozone, the strain imposed on the samples (static or dynamic), and the temperature are controlled. However, even when the maximum control is exerted over these variables, correlation between accelerated and outdoor tests is not always good. Both ultraviolet radiation and moisture influence outdoor tests, and these variables are difficult to duplicate in the laboratory. Obviously, it is most important in accelerated tests that the ozone concentration be accurately measured and controlled within narrow limits.

The test method most commonly used to measure ozone resistance was developed by Crabtree and Kemp[77] and has been adopted by the American Society for Testing and Materials.[78] The apparatus consists of an aluminum cylinder through which flows a mixture of ozone and oxygen with an ozone concentration of 50 pphm. This concentration is much greater than found in the normal environment, but it compares with values actually measured in urban areas under intense smog conditions, which vary from 12[79] to 90[80] pphm. Because the ozone concentration is comparable to that encountered in some outdoor exposures, results from this test give reasonably good correlation with outdoor testing results. Ozone concentration is measured at frequent intervals by titration after absorption in a spray-jet device or a countercurrent absorption column. A selected temperature between 40 and 50°C is used and controlled to within ±1°C. Samples are held under strain as described in ASTM test method D518-61,[81] and failure is determined by visual observation of cracks. This test is generally used to evaluate the effectiveness of antiozonants in polymers under static strain. Testing under dynamic stress has not been standardized to the same extent, but testing under these conditions is used widely in the tire industry.

2. Tests for Resistance to Moisture

Exposure to moisture or high humidity causes polymer deterioration by secondary bond rupture (platicization), hydrolysis of primary bonds, or both, as described in Chapter 8, and both effects can be measured by established procedures. The methods employed to measure absorption vary from the simple determination of weight gain to elaborate techniques for measuring permeation.[82–84] Plasticization also results in changes in the degree of crystallinity in crystalline polymers and lowering of the glassy transition point in amorphous polymers. Changes in both of these properties can be determined by conventional methods.

Measurement of the rate of hydrolysis is usually adapted to the susceptibility of specific polymers. Neutral conditions are used for less reactive polymers, and the reaction is accelerated by increasing the temperature or by exposure to high humidity conditions. Either acidic or basic catalysts are used to accelerate hydrolysis of more resistant polymers. Catalysis is useful in accelerating the reaction, but tests performed in this way may not correlate with performance unless similar catalytic effects occur during normal exposure.

Various methods have been used to measure the hydrolysis of polymers.[85] Birdsall and McMahon[86] have followed the hydrolysis of poly(ethylene terephthalate) by intrinsic viscosity determinations. Where adaptable, end-group or product analyses are also useful methods for evaluating the resistance to hydrolysis.

3. Tests for Biodegradation

Many living organism attack polymers, including microorganisms that exist in the soil, in water, and in the atmosphere (Chapter 9). The most effective method for accelerating biological degradation is to expose test samples to an environment where these organisms are concentrated, or to subject them to a bioassay test using a laboratory environment in which pure or mixed cultures of bacteria are placed in contact with the polymer. Chemical, mechanical, or dielectric tests can be applied to determine the extent of deterioration. The rate of oxygen consumption by laboratory cultures has been used by Snoke[87] to measure the rate of microbial deterioration. As might be expected, experimental control is better under laboratory conditions, but results obtained under these conditions do not always correlate with outdoor test results.

The soil-burial test[88–90] has been widely used to determine the resistance of polymer compositions in underground applications. Some degree of acceleration is made possible in this test by selecting burial sites favorable to organism activity. Exposure to marine organism in sea water is also accelerated by selecting as the immersion site locations of high marine biological activity.

E. PROPERTIES MEASURED IN EVALUATING STABILITY

Changes that occur in polymers during exposure to the environment are usually determined by comparison of the original properties with those measured at intervals during deterioration. In many instances, test methods used to follow the deterioration of a polymer are the same as those used for the unexposed material. Since, in theory at least, any measurable property that changes during exposure could be used to evaluate stability, many and various properties are measured. The properties of polymers and the methods for their measurement have been described in several reviews.[91–93] These include determination of chemical composition, modulus, tensile strength, elongation, dielectric strength, and visual appearance. Choice of the property or at least the type of property to be used in each test should be determined by the end use to which the polymer will be subjected. In addition to the usual property measurements, several tests have been developed specifically to evaluate polymer stability. A brief summary of these tests follows, with emphasis on those most applicable to stabilized polymers.

The effects of exposure of a polymer to the environment are commonly evaluated by visual appearance. Discoloration and cracking or crazing occur at the surface of many polymers as they deteriorate. These changes in appearance, when not visible to the naked eye, can often be detected by microscopy[94] and reproduced in photomicrographs.[95] Evaluation of stability based on the rate of color formation, however, can lead to erroneous conclusions. Many polymers develop color that fades on continued exposure. The bleaching of cellulosic materials is a notable example, and the oxidative fading of the initial color developed in poly(vinyl chloride) has been discussed in Chapter 3. Many stabilizers discolor polymers, and this staining effect may increase during exposure as a result of the formation of chromophoric products from the stabilizer. Although color formation in some stabilized polymers does indicate changes, and perhaps a loss in effectiveness of the stabilizer, it does not necessarily mean that the polymer has deteriorated. Nonetheless, if a polymer composition fails in service because it develops color, it is not important which component of the composition was responsible, except, of course, as the result directs a change in the formulation.

Chemical changes that take place during deterioration afford a most effective measurement of stability. Changes in chemical properties or composition vary widely with different polymers, but several reaction products are common to many. These include peroxy, carbonyl, carboxyl, and olefinic groups. Measurement of the concentration of these reaction products can be made by a variety of analytical procedures. For example, colorimetric tests based on the reactions of dyes and amines with carboxyl[96,97] and carbonyl[98] groups have been used. Peroxy groups have been determined by iodometric analysis[99] as well as by infrared spectroscopy.[100–103] A more sensitive method for the determination of peroxy groups has been reported by Mitchell and Perkins.[104] In this test, sulfate groups formed by interaction of hydroperoxide groups with sulfur dioxide are measured in the infrared region at 1195 cm^{-1}.

A few tests for stability of polymer compositions make use of changes in mechanical properties that are not commonly measured in polymer characterization. Important among these are tests based on stress relaxation,[105] which have been widely used in studying the aging of elastomers. Samples are placed under stress in an oxidizing environment, and the decrease in stress required to maintain constant extension is measured. Stress relaxation occurs as molecular chains or crosslinks in the polymer are ruptured. The cantilever beam test[106,107] also employs measurement of changes in mechanical properties during exposure. In this test the angle of break of samples under load is used to determine heat and light stability.

The direct evaluation of stabilizer effectiveness has been approached in several ways, all unique to a particular class of stabilizers. The oxidation potential of phenols, an important class of thermal antioxidants, has been related to antioxidant effectiveness,[108] suggesting that these stabilizers might be performance rated on the basis of known or measurable oxidation potentials. Electron-spin resonance has also been suggested[109] as a technique for determining radical products formed from amine and phenol antioxidants. Whenever the mechanism by which a stabilizer functions is known, determination of the rate at which products are formed from the stabilizer, either alone or in a polymer composition, could be used to establish effectiveness.

REFERENCES

1. V. E. Gray and B. C. Cadoff, "Survey of Techniques for Evaluating Effects of Weathering on Plastics," in M. R. Kamal, Ed., *Appl. Polym. Symp., No. 4,* Interscience, New York, 1967, p. 85.
2. J. B. Titus, "The Weatherability of Polyolefins," *PLASTECH* (*Plastics Techn. Evaluation Center*) *Rept. 32,* Picatinny Arsenal, Dover, N.J., 1968, pp. 17, 19, 45, 46.
3. K. S. Kaufman, Jr., "A New Technique for Evaluating Outdoor Weathering Properties of High Density Polyethylene," in M. R. Kamal, Ed., *Appl. Polym. Symp., No. 4,* Interscience, New York, 1967, p. 131.
4. N. Z. Searle and R. C. Hirt, *J. Opt. Soc. Amer.*, **55,** 1413 (1965).
5. R. W. Singleton, R. K. Kunkel, and B. S. Sprague, *J. Text. Inst.*, **35,** 228 (1965).
6. R. C. Hirt and N. Z. Searle, "Energy Characteristics of Outdoor and Indoor Exposure Sources and their Relation to the Weatherability of Plastics," in M. R. Kamal, Ed., *Appl. Polym. Symp., No. 4.* Interscience, New York, 1967, p. 64.
7. R. C. Hirt, N. Z. Searle, and R. G. Schmitt, *SPE* (*Soc. Plastics Engrs.*) *Trans.*, **1,** 21 (1961).

8. J. B. Howard and H. M. Gilroy, *Polym. Eng. Sci.*, **9**, 286 (1969).

9. H. M. Quackenbos and H. Samuels, "Practical Problems in Predicting Weathering Performance," in M. R. Kamal, Ed., *Appl. Polym. Symp.*, *4*, Interscience, New York, 1967, p. 161.

10. R. J. Martinovich and G. R. Hill, "Practical Approach to the Study of Polyolefin Weatherability," in M. R. Kamal, Ed., *Appl. Polym. Symp.*, *No. 4*, Interscience, New York, 1967, p. 144.

11. J. B. DeCoste and V. T. Wallder, *Ind. Eng. Chem.*, **47**, 314 (1955).

12. G. C. Newland, R. M. Schulken, Jr., and J. W. Tamblyn, *ASTM Mater. Res. Std.*, **3**, 487 (1963).

13. R. C. Neuman, *ASTM Mater. Res. Std.*, **9** (6), 38 (1969).

14. J. H. Darby and P. R. Graham, *Mod. Plastics*, **39** (5), 148 (1962).

15. *ASTM Book of Standards*, *D 1435*, **27**, 509 (1969).

16. H. A. Gardner and G. G. Sward, *Paint Testing Manual*, 12th ed., 1962, p. 262.

17. G. R. Rugger, "Weathering," in D. V. Rosato and R. T. Schwartz, Eds. *Environmental Effect on Polymeric Materials*, Vol. I, Interscience, New York, 1968, p. 350.

18. C. R. Caryl and W. E. Helmick, U.S. Patent 2,945,417, July 19, 1960.

19. B. L. Garner and P. J. Papillo, *Ind. Eng. Chem., Prod. Res. Develop.*, **1** (4), 249 (1962).

20. C. R. Caryl, "Methods of Outdoor Exposure Testing," in W. E. Brown, Ed., *Testing of Polymers*, New York, 1969, p. 379.

21. W. L. Hawkins, Mrs. M. A. Worthington, and W. Matreyek, *J. Appl. Polym. Sci.*, **3**, 277 (1960).

22. C. R. Caryl and A. E. Rheineck, *J. Paint Technol.*, **34** (452), 1017 (1962); **37** (481), 129 (1965).

23. R. J. Kennedy and H. P. Beardsley, "Exterior Paints for Wood Based on Vinyl–Ethylene Emulsions," presented at the Annual Symposium of the Northwest Society for Paint Technology, May 6, 1967.

24. P. J. Papillo, *J. Paint Technol.*, **40** (524), 359 (1968).

25. G. C. Newland and J. W. Tamblyn, "Actinometry of Sunlight at Kingsport, Tennessee," in M. R. Kamal, Ed., *Appl. Polym. Symp.*, *No. 4*, Interscience, New York, 1967, p. 119.

26. V. Schafer, "Accelerated Light Exposure in the Xenotest," in M. R. Kamal, Ed., *Appl. Polym. Symp.*, *No. 4*, Interscience, New York, 1967, p. 111.

27. J. W. Tamblyn and G. M. Armstrong, *Anal. Chem.*, **25**, 460 (1953).

28. R. C. Hirt, R. G. Schmitt, N. Z. Searle, and A. P. Sullivan, *J. Opt. Soc. Amer.*, **50**, 706 (1960).

29. L. P. Cipriani, P. Giesecke, and R. Kinmonth, *Plastics Technol.*, **11**, 34 (1965).

30. G. A. Thacker, L. I. Nass, and L. B. Weisfield, *SPE* (*Soc. Plastics Engrs.*) *J.*, **21**, 460 (1953).

31. L. D. Maxim and C. H. Kuist, *Offic. Dig., Fed. Soc. Paint Technol.*, **36** (474), 723 (1964).

32. J. E. Clark and C. W. Harrison, "Accelerated Weathering of Polymers," in M. R. Kamal, Ed., *Appl. Polym. Symp.*, *No. 4*, Interscience, New York, 1967, p. 108.

33. M. R. Kamal and R. Saxon, "Recent Developments in the Analysis and Prediction of the Weatherability of Plastics," in M. R. Kamal, Ed., *Appl. Polym. Symp.*, *No. 4*, Interscience, New York, 1967, p. 9.

34. B. Baum and Mrs. A. L. Perun, *Plastics Technol.*, **7** (4), 29 (1961).

35. F. H. Winslow, M. Y. Hellman, W. Matreyek, and S. M. Stills, *Polym. Eng. Sci.*, **6** (3), 1 (1966).

36. C. S. Schollenberger and K. Dinbergs, *SPE* (*Soc. Plastics Engrs.*) *Trans.*, **2**, 31 (1961).

37. W. W. Coblentz and R. Stair, *J. Res. Natl. Bur. Std.*, **17**, 1 (1936).

38. J. W. Tamblyn, G. C. Newland, and M. T. Watson, *Plastics Technol.*, **4** (5), 427 (1958).
39. T. J. Gedemer, *Appl. Spectry.*, **19** (5), 141 (1964).
40. P. J. Hearst, *ACS Polym. Preprints*, **28** (1), 672 (1968).
41. M. G. Chan and W. L. Hawkins, *ACS Polym. Preprints*, **9** (2), 1638 (1968).
42. S. D. Bruck, *Polym.*, **7**, 321 (1966).
43. S. L. Madorsky, *Thermal Degradation of Organic Polymers*, Wiley, New York, 1964, pp. 10–25.
44. J. Chiu, *Anal. Chem.*, **40** (10), 1516 (1968).
45. N. Grassie and H. W. Melville, *Proc. Roy. Soc.* (*London*), **199A**, 1 (1949).
46. N. Grassie and I. C. McNeill, *J. Chem. Soc.*, **1956**, 3929.
47. D. H. Grant and N. Grassie, *Polym.*, **1**, 125, 445 (1960).
48. I. C. McNeill, *J. Polym. Sci.*, *Part A-1*, **4**, 2478 (1966).
49. S. M. Ellerstein, "The Use of Dynamic Differential Calorimetry for Ascertaining the Thermal Stability of Polymers," in *Analytical Calorimetry*, Eds., R. S. Porter and J. F. Johnson, Plenum, New York, 1968, p. 279.
50. A. Rudin, H. P. Schreiber, and M. H. Waldman, *Ind. Eng. Chem.*, **53** (2), 137 (1961).
51. B. B. Stafford, *J. Appl. Polym. Sci.*, **9**, 729 (1965).
52. *ASTM Book of Standards, D 1248*, **26**, 79 (1969).
53. J. B. DeCoste, *SPE* (*Soc. Plastics Engrs.*) *J.*, **21** (8) 1 (1965).
54. O. Warburg and F. Kubowitz, *Biochem. Z.*, **7**, 214 (1929).
55. *ASTM Book of Standards, D 525*, **17**, 198 (1969).
56. G. W. Kennerly and W. L. Patterson, Jr., *Ind. Eng. Chem.*, **48**, 1917 (1956).
57. B. S. Biggs and W. L. Hawkins, *Mod. Plastics*, **31**, 121 (1953).
58. J. R. Shelton and H. Winn, *Ind. Eng. Chem.*, **38**, 71 (1946).
59. J. R. Shelton, *Amer. Soc. Testing Mater. Spec. Tech. Publ.* **89**, 12 (1949).
60. J. E. Wilson, *Ind. Eng. Chem.*, **47**, 2201 (1955).
61. W. L. Hawkins, W. Matreyek, and F. H. Winslow, *J. Polym. Sci.*, **41**, 1 (1959).
62. W. L. Hawkins, W. Matreyek, and F. H. Winslow, *J. Appl. Polym. Sci.*, **5** (16), 515 (1961).
63. R. H. Hansen, W. M. Martin, and T. DeBenedictis, *Mod. Plastics*, **42**, 137 (1965).
64. J. B. Howard, private communication, 1970.
65. W. F. Fischer and W. H. King, Jr., *Anal. Chem.*, **39**, 1265 (1967).
66. C. J. Hilado, *Ind. Eng. Chem. Prod. Res. Develop.*, **6** (3), 154 (1967); **7** (2), 81 (1968).
67. A. J. Briber, *SPE* (*Soc. Plastics Engrs.*) *Tech. Papers*, **13**, 1041 (1967).
68. C. J. Hilado, *Flammability Handbook for Plastics*, Technomic, Stamford, 1969.
69. A. J. Hammerl, *Reinforced Plastics*, **2** (5), 22 (1963).
70. S. S. Feuer and A. F. Torres, *Chem. Eng.*, **69** (7), 138 (1962).
71. *ASTM Book of Standards*, refer to Vols. 26–28.
72. L. M. Kline, *UL Res. Bull. No. 55* (Feb. 1, 1964).
73. U.S. Gen. Services Admin., *Fed. Test Method Std. 406*, Method 2021 (Oct. 1961).
74. C. P. Fenimore and F. J. Martin, *Combust. Flame*, **10** (2), 141 (1966); *Mod. Plastics*, **43**, 141 (1966).
75. J. L. Isaacs, *J. Fire Flammability*, **1**, 36 (1970).
76. *ASTM Book of Standards, D 2863*, **27**, 719 (1970).
77. J. Crabtree and A. R. Kemp, *Ind. Eng. Chem.*, *Anal. Ed.*, **18**, 769 (1946).
78. *ASTM Book of Standards, D 1149*, **28**, 563 (1969).
79. Kettering Lab., Univ. of Cincinnati, "Concentrations of Oxidizing Substances in the Atmosphere of a Number of American Cities," Sept. 1954.
80. Air Pollution Foundation, 704 South Spring Street, Los Angeles, "An Aerometric Survey of the Los Angeles Basin," July 1955.

81. *ASTM Book of Standards, D 518*, **28**, 300 (1969).
82. C. E. Rogers, "Solubility and Diffusivity," in *The Physics and Chemistry of Organic Solids*, Eds. D. M. Fox, M. M. Labes, and A. Weissberger, Interscience, New York, 1963, p. 509.
83. V. Stannett and H. Yasuda, "Permeability," in R. A. V. Raff and K. W. Doak, Eds., *Crystalline Olefin Polymers, Pt. II*, Interscience, New York, 1964, p. 131.
84. R. M. Barrer, *Diffusion in and Through Solids*, Cambridge Press, London, 1941.
85. D. A. S. Ravens and Mrs. J. E. Sisley, "Cleavage Reactions A, Hydrolysis," in E. M. Fettes, Ed., *Chemical Reactions of Polymers*, Interscience, New York, 1964, p. 551.
86. W. McMahon, H. A. Birdsall, G. R. Johnson, and C. T. Camilli, *J. Chem. Eng. Data*, **4**, 57 (1959).
87. L. R. Snoke, *Bell Syst. Tech. J.*, **36**, 1095 (1957).
88. A. Burges, *Micro-organisms in Soils*, Hutchinson, London, 1958, p. 15.
89. R. A. Connolly, *Bell Labs. Rec.*, **40**, 124 (1962).
90. J. B. DeCoste, *Ind. Eng. Chem. Prod. Res. Develop.*, **7**, 238 (1968).
91. L. E. Nielsen, *Mechanical Properties of Polymers*, Reinhold, London, 1962.
92. A. E. Lever and J. A. Rhys, *The Properties and Testing of Plastic Materials*, Chemical Publishing Co., New York, 1962.
93. *ASTM Book of Standards*, refer to Vols. 27–29.
94. J. A. Lindquist, V. E. Widmer, and G. G. McKinley, *SPE (Soc. Plastics Engrs.) Trans.*, **2**, 152 (1962).
95. A. L. Smith and J. R. Lowry, *Mod. Plastics*, **35** (3), 134 (1958).
96. S. R. Palit and P. Ghosh, *J. Polym. Sci.*, **58**, 1225 (1962).
97. P. G. Campbell and J. R. Wright, *Ind. Eng. Chem. Prod. Res. Develop.*, **5**, 319 (1966).
98. V. E. Gray and J. R. Wright, *J. Appl. Polym. Sci.*, **7**, 2616 (1963).
99. E. L. Stanley, "Acrylic Plastics," in G. M. Kline, Ed., *Analytical Chemistry of Polymers*, Pt I, Interscience, New York, 1959, p. 8.
100. J. P. Luongo, *J. Polym. Sci.*, **42**, 139 (1960).
101. J. P. Luongo, *J. Appl. Polym. Sci.*, **3**, 302 (1960).
102. B. Baum, *J. Appl. Polym. Sci.*, **2**, 281 (1959).
103. H. C. Beachell and G. W. Tarbet, *J. Polym. Sci.*, **45**, 451 (1960).
104. J. Mitchell, Jr. and L. R. Perkins, "Determination of Hydroperoxide Groups in Oxidized Polyethylene," in M. R. Kamal, Ed., *Appl. Polym. Symp., No. 4*, Interscience, New York, 1967, p. 167.
105. A. V. Tobolsky, *Properties and Structure of Polymers*, Wiley, New York, 1960.
106. J. W. Tamblyn, G. C. Newland, and M. T. Watson, *Plastics Technol.*, **4**, 427 (1958).
107. P. G. Kelleher, R. J. Miner, and D. J. Boyle, *SPE (Soc. Plastics Engrs.) J.*, **25**, 53 (1969).
108. J. L. Bolland and P. Ten Have, *Disc. Faraday Soc.*, **2**, 252 (1947).
109. J. K. Becconsall, S. Clough, and G. Scott, *Trans. Faraday Soc.*, **56**, 459 (1960).

INDEX